Advances in NMR and MRI of Materials

Advances in NMR and MRI of Materials

Editor

Igor Serša

MDPI • Basel • Beijing • Wuhan • Barcelona • Belgrade • Manchester • Tokyo • Cluj • Tianjin

Editor
Igor Serša
Condensed Matter Physics
Jožef Stefan Institute
Ljubljana
Slovenia

Editorial Office
MDPI
St. Alban-Anlage 66
4052 Basel, Switzerland

This is a reprint of articles from the Special Issue published online in the open access journal *Molecules* (ISSN 1420-3049) (available at: www.mdpi.com/journal/molecules/special_issues/NMR_MRI_materials).

For citation purposes, cite each article independently as indicated on the article page online and as indicated below:

LastName, A.A.; LastName, B.B.; LastName, C.C. Article Title. *Journal Name* **Year**, *Volume Number*, Page Range.

ISBN 978-3-0365-7621-3 (Hbk)
ISBN 978-3-0365-7620-6 (PDF)

© 2023 by the authors. Articles in this book are Open Access and distributed under the Creative Commons Attribution (CC BY) license, which allows users to download, copy and build upon published articles, as long as the author and publisher are properly credited, which ensures maximum dissemination and a wider impact of our publications.
The book as a whole is distributed by MDPI under the terms and conditions of the Creative Commons license CC BY-NC-ND.

Contents

About the Editor vii

Preface to "Advances in NMR and MRI of Materials" ix

Marianna Porcino, Xue Li, Ruxandra Gref and Charlotte Martineau-Corcos
Solid-State NMR Spectroscopy: A Key Tool to Unravel the Supramolecular Structure of Drug Delivery Systems
Reprinted from: *Molecules* 2021, 26, 4142, doi:10.3390/molecules26144142 1

Urša Mikac and Julijana Kristl
Magnetic Resonance Methods as a Prognostic Tool for the Biorelevant Behavior of Xanthan Tablets
Reprinted from: *Molecules* 2020, 25, 5871, doi:10.3390/molecules25245871 19

Natalia E. Kuz'mina, Sergey V. Moiseev, Elena Y. Severinova, Evgenii A. Stepanov and Natalia D. Bunyatyan
Identification and Quantification by NMR Spectroscopy of the 22R and 22S Epimers in Budesonide Pharmaceutical Forms
Reprinted from: *Molecules* 2022, 27, 2262, doi:10.3390/molecules27072262 31

Rui Wang, Jiaxiang Xin, Zhengxiao Ji, Mengni Zhu, Yihua Yu and Min Xu
Spin-Space-Encoding Magnetic Resonance Imaging: A New Application for Rapid and Sensitive Monitoring of Dynamic Swelling of Confined Hydrogels
Reprinted from: *Molecules* 2023, 28, 3116, doi:10.3390/molecules28073116 43

Ioan Ardelean
The Effect of an Accelerator on Cement Paste Capillary Pores: NMR Relaxometry Investigations
Reprinted from: *Molecules* 2021, 26, 5328, doi:10.3390/molecules26175328 55

George Diamantopoulos, Marios Katsiotis, Michael Fardis, Ioannis Karatasios, Saeed Alhassan and Marina Karagianni et al.
The Role of Titanium Dioxide on the Hydration of Portland Cement: A Combined NMR and Ultrasonic Study
Reprinted from: *Molecules* 2020, 25, 5364, doi:10.3390/molecules25225364 65

Mihai M. Rusu, David Faux and Ioan Ardelean
Monitoring the Effect of Calcium Nitrate on the Induction Period of Cement Hydration via Low-Field NMR Relaxometry
Reprinted from: *Molecules* 2023, 28, 476, doi:10.3390/molecules28020476 85

Urša Mikac, Maks Merela, Primož Oven, Ana Sepe and Igor Serša
MR Study of Water Distribution in a Beech (*Fagus sylvatica*) Branch Using Relaxometry Methods
Reprinted from: *Molecules* 2021, 26, 4305, doi:10.3390/molecules26144305 97

Magdalena Broda and Daniel J. Yelle
Reactivity of Waterlogged Archeological Elm Wood with Organosilicon Compounds Applied as Wood Consolidants: 2D ^{1}H–^{13}C Solution-State NMR Studies
Reprinted from: *Molecules* 2022, 27, 3407, doi:10.3390/molecules27113407 107

Aleš Mohorič, Gojmir Lahajnar and Janez Stepišnik
Diffusion Spectrum of Polymer Melt Measured by Varying Magnetic Field Gradient Pulse Width in PGSE NMR
Reprinted from: *Molecules* 2020, 25, 5813, doi:10.3390/molecules25245813 125

Kevin Lindt, Bulat Gizatullin, Carlos Mattea and Siegfried Stapf
Non-Exponential ^{1}H and ^{2}H NMR Relaxation and Self-Diffusion in Asphaltene-Maltene Solutions
Reprinted from: *Molecules* **2021**, *26*, 5218, doi:10.3390/molecules26175218 . . . . **135**

Madison L. Nelson, Joelle E. Romo, Stephanie G. Wettstein and Joseph D. Seymour
Impact of Xylose on Dynamics of Water Diffusion in Mesoporous Zeolites Measured by NMR
Reprinted from: *Molecules* **2021**, *26*, 5518, doi:10.3390/molecules26185518 . . . . **171**

Marina G. Shelyapina, Rosario I. Yocupicio-Gaxiola, Iuliia V. Zhelezniak, Mikhail V. Chislov, Joel Antúnez-García and Fabian N. Murrieta-Rico et al.
Local Structures of Two-Dimensional Zeolites—Mordenite and ZSM-5—Probed by Multinuclear NMR
Reprinted from: *Molecules* **2020**, *25*, 4678, doi:10.3390/molecules25204678 . . . . **179**

Aiswarya Chalikunnath Venu, Rami Nasser Din, Thomas Rudszuck, Pierre Picchetti, Papri Chakraborty and Annie K. Powell et al.
NMR Relaxivities of Paramagnetic Lanthanide-Containing Polyoxometalates
Reprinted from: *Molecules* **2021**, *26*, 7481, doi:10.3390/molecules26247481 . . . . **195**

Rodrigo Henrique dos Santos Garcia, Jefferson Gonçalves Filgueiras, Luiz Alberto Colnago and Eduardo Ribeiro de Azevedo
Real-Time Monitoring Polymerization Reactions Using Dipolar Echoes in ^{1}H Time Domain NMR at a Low Magnetic Field
Reprinted from: *Molecules* **2022**, *27*, 566, doi:10.3390/molecules27020566 . . . . **209**

Anton Duchowny and Alina Adams
Compact NMR Spectroscopy for Low-Cost Identification and Quantification of PVC Plasticizers
Reprinted from: *Molecules* **2021**, *26*, 1221, doi:10.3390/molecules26051221 . . . . **223**

Grzegorz Stoch and Artur T. Krzyżak
Enhanced Resolution Analysis for Water Molecules in MCM-41 and SBA-15 in Low-Field T_2 Relaxometric Spectra
Reprinted from: *Molecules* **2021**, *26*, 2133, doi:10.3390/molecules26082133 . . . . **239**

Steffen Merz, Jie Wang, Petrik Galvosas and Josef Granwehr
MAS-NMR of [Pyr$_{13}$][Tf$_2$N] and [Pyr$_{16}$][Tf$_2$N] Ionic Liquids Confined to Carbon Black: Insights and Pitfalls
Reprinted from: *Molecules* **2021**, *26*, 6690, doi:10.3390/molecules26216690 . . . . **255**

Maiko Sasaki and Keiko Takahashi
Complete Assignment of the ^{1}H and ^{13}C NMR Spectra of Carthamin Potassium Salt Isolated from *Carthamus tinctorius* L.
Reprinted from: *Molecules* **2021**, *26*, 4953, doi:10.3390/molecules26164953 . . . . **273**

Nataliya E. Kuz'mina, Sergey V. Moiseev, Mikhail D. Khorolskiy and Anna I. Lutceva
Development and Validation of 2-Azaspiro [4,5] Decan-3-One (Impurity A) in Gabapentin Determination Method Using qNMR Spectroscopy
Reprinted from: *Molecules* **2021**, *26*, 1656, doi:10.3390/molecules26061656 . . . . **283**

Xudong Lv, Jeffrey Walton, Emanuel Druga, Raffi Nazaryan, Haiyan Mao and Alexander Pines et al.
Imaging Sequences for Hyperpolarized Solids
Reprinted from: *Molecules* **2020**, *26*, 133, doi:10.3390/molecules26010133 . . . . **293**

Alexander Adair, Sebastian Richard and Benedict Newling
Gas and Liquid Phase Imaging of Foam Flow Using Pure Phase Encode Magnetic Resonance Imaging
Reprinted from: *Molecules* **2020**, *26*, 28, doi:10.3390/molecules26010028 307

Sabina Haber-Pohlmeier, David Caterina, Bernhard Blümich and Andreas Pohlmeier
Magnetic Resonance Imaging of Water Content and Flow Processes in Natural Soils by Pulse Sequences with Ultrashort Detection
Reprinted from: *Molecules* **2021**, *26*, 5130, doi:10.3390/molecules26175130 317

Nitish Katoch, Bup-Kyung Choi, Ji-Ae Park, In-Ok Ko and Hyung-Joong Kim
Comparison of Five Conductivity Tensor Models and Image Reconstruction Methods Using MRI
Reprinted from: *Molecules* **2021**, *26*, 5499, doi:10.3390/molecules26185499 331

Xiaodong Li, Yanhong Sun, Lina Ma, Guifeng Liu and Zhenxin Wang
The Renal Clearable Magnetic Resonance Imaging Contrast Agents: State of the Art and Recent Advances
Reprinted from: *Molecules* **2020**, *25*, 5072, doi:10.3390/molecules25215072 349

About the Editor

Igor Serša

Igor Serša received a BSc degree in 1991, an MSc degree in 1994 (both at the Faculty of Natural Sciences and Technology, University of Ljubljana, Slovenia), and a PhD degree in 1996 (Faculty of Mathematics and Physics, University of Ljubljana, Slovenia). Since 1991, he has been employed at the Jožef Stefan Institute in Ljubljana, Slovenia, where he has been the head of the Magnetic Resonance Imaging Laboratory since 1999. Since 1994, he has also been affiliated with the Faculty of Mathematics and Physics of the University of Ljubljana, Slovenia, where he was promoted to full professor in 2017. In 1998–1999, he was a postdoctoral research fellow at the Mayo Clinic under Prof. Slobodan Macura, where he worked on the development of NMR pulse sequences with completely arbitrary regional volume excitation. His research interests are in magnetic resonance imaging and microscopy, with a focus on the development of new pulse sequences and their application in studies of various materials and processes. Thus, his research includes applications in various fields, from biology and medicine with extensive research on blood clots and their thrombolysis to wood science with the role of water in wood materials, food science with research on food processing, and porous materials, which he studies by diffusion. He is also one of the pioneers of current density imaging, a magnetic resonance imaging method that enables the imaging of electric currents in biological tissues and materials. He is the author of more than 130 scientific publications in peer-reviewed journals.

Preface to "Advances in NMR and MRI of Materials"

The development of science has led to the emergence of many new modern materials, which also require more advanced tools for their characterization and analysis. NMR and MRI are certainly among such tools, also due to their continuous development, which has made them more powerful, versatile, and sensitive. With these advances, these two techniques have been able to address many open problems associated with the emergence of new materials.

The aim of this Special Issue was to edit a collection of advanced NMR and MRI techniques and methods, together with a demonstration of their application to the target materials for which they were designed and optimized. In my humble opinion, I think this Special Issue has successfully accomplished that task. The Special Issue comprises an impressive collection of articles, which in this book are classified according to the following topics: MR methods in pharmaceutical research, NMR in cement research, MR methods in wood research, diffusion in materials, characterization of materials by NMR relaxometry, NMR spectroscopy of materials, and MRI of materials. I hope readers enjoy reading this Special Issue as much as I enjoyed editing it.

I would also like to thank all the authors who contributed to this Special Issue, all the reviewers for their time, effort, and constructive criticism, which helped to improve the quality of the published articles, and last but not least, the MDPI publisher and the editorial staff of the journal for their dedication and support in producing this Special Issue.

Igor Serša
Editor

Review

Solid-State NMR Spectroscopy: A Key Tool to Unravel the Supramolecular Structure of Drug Delivery Systems

Marianna Porcino [1,*], Xue Li [2], Ruxandra Gref [2] and Charlotte Martineau-Corcos [3,*]

1. CEMHTI UPR CNRS 3079, Université d'Orléans, 45071 Orléans, France
2. Institut des Sciences Moléculaires d'Orsay, UMR CNRS 8214, Paris-Sud University, Université Paris Saclay, 91400 Orsay, France; xue.li@universite-paris-saclay.fr (X.L.); ruxandra.gref@universite-paris-saclay.fr (R.G.)
3. CortecNet, 7 Avenue du Hoggar, 91940 Les Ulis, France
* Correspondence: marianna.porcino@cnrs-orleans.fr (M.P.); ccorcos@cortecnet.com (C.M.-C.)

Abstract: In the past decades, nanosized drug delivery systems (DDS) have been extensively developed and studied as a promising way to improve the performance of a drug and reduce its undesirable side effects. DDSs are usually very complex supramolecular assemblies made of a core that contains the active substance(s) and ensures a controlled release, which is surrounded by a corona that stabilizes the particles and ensures the delivery to the targeted cells. To optimize the design of engineered DDSs, it is essential to gain a comprehensive understanding of these core–shell assemblies at the atomic level. In this review, we illustrate how solid-state nuclear magnetic resonance (ssNMR) spectroscopy has become an essential tool in DDS design.

Keywords: solid-state NMR spectroscopy; porous material; drug delivery system; heteronuclei

1. Introduction

A nanosized drug delivery system (DDS) is defined as a formulation or a device that enables the introduction of a therapeutic substance in the body [1]. The major goals of DDS are to improve the efficacy and safety of a given active pharmaceutical ingredient (API) by controlling its rate, time, and place of release as well the compliance of the patient [2]. A plethora of nanosized DDSs were developed, including liposomes, polymeric and inorganic nanoparticles, dendrimers, carbon nanotubes, etc. However, there is no universal drug nanocarrier, and for each medical target, a new system has to be designed. An ideal DDS should have high colloidal stability, high drug loading capacity, be prepared without the need for toxic solvents, be inert, biocompatible, mechanically resistant, biodegradable, well tolerated by the patient, safe and simple to administer, and finally easy and cost-effective to fabricate and sterilize [3]. Most DDSs are composed of complex core–shell supramolecular structures where each component has different physicochemical characteristics and functions (Figure 1). The core of the system accommodates the drug, protects, and releases it in a controlled manner. It must ensure efficient drug release and degrade to avoid accumulation in the body. The corona plays a role in increasing the colloidal stability of the nanoparticles and governs their in vivo fate (confers "stealth" properties to evade the immune system and/or to ensure specific delivery to the biological target). To guide the design of complex core–shell particles, it is essential to have at hand analytical tools able to yield information at the atomic scale about the structure of the DDS, the host–drug interactions, dynamics, etc.

A variety of techniques have been used so far, such as thermal analysis [4] and nitrogen sorption [5], which can both give qualitative and quantitative information about the bulk structural and thermodynamic properties of the incorporated drugs. Unfortunately, they do not provide any insight into the microscopic properties of the embedded drugs and in particular about the crucial drug–drug and drug–host interactions. X-ray powder diffraction (XRPD), which is the usual method of choice for the investigation of structural properties, is also often inapplicable because of the lack of long-range order of complex DDSs. An alternative technique is solid-state nuclear magnetic resonance (ssNMR), which through spectroscopy and relaxation measurements offers valuable information about the structural and dynamical properties of the embedded compound and which is not limited to the materials or forms that exhibit long-range order [6]. ssNMR experiments probe the local environment of the nucleus and short-range order. This gives rise to various advantages, including the ability to provide detailed structural insights from polycrystalline samples and to examine local defects and disorders [7].

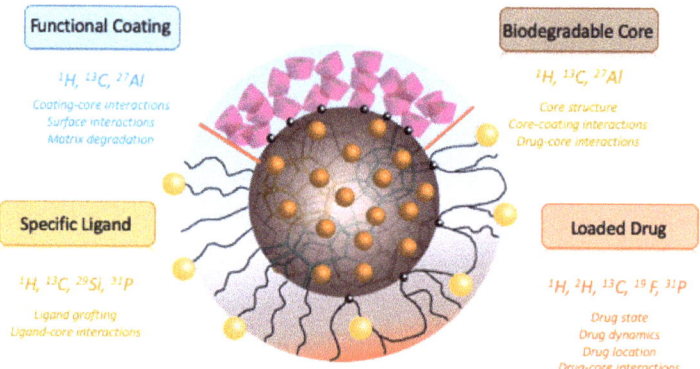

Figure 1. Scheme of a DDS nanoparticle containing a core, in which the drug is loaded, functional coating, and specific ligand. The nuclei of interest in ssNMR and the information that can be extracted from these measurements are indicated.

In the context of DDSs, because numerous NMR active nuclei are present in both the core and the corona, ssNMR spectroscopy can provide information in many aspects ranging from intramolecular and intermolecular interactions, localization, state and stability of the drug, core–shell interactions, the matrix degradation, delivery/release of the drug, etc. (Figure 1). These pieces of information can significantly contribute to guide the design and development of new DDSs. Moreover, ssNMR is non-destructive; i.e., the DDS can be studied intact without the need for dissolution/digestion (as it would be required for solution-state NMR, fluorescence spectroscopy or other strategies involving chemical modification).

In this review, we illustrate the versatility of ssNMR spectroscopy for the investigation of the supramolecular structure of DDSs. We focus this review on DDSs that are based on inorganic or hybrid organic–inorganic particles, i.e., rather than on organic particles such as liposomes or polymer-based ones. We first report ssNMR experiments based on one-dimensional (1D) NMR spectroscopy, then two-dimensional (2D) NMR starting from the simpler ^{1}H-^{1}H spin pair, then moving toward ^{1}H-X pairs and finally to more challenging pairs of heteronuclei. For each case, the advantages and limits of the ssNMR technique are described.

2. A Few Notes on ssNMR Spectroscopy

An ssNMR spectrum, similarly to a solution-state NMR spectrum, contains information about the chemical shift interaction, i.e., a line shift that gives indication about the neighboring environment of a nucleus, and, when it can be observed or detected, about scalar coupling, i.e., a line splitting, which represents the spin–spin interaction through chemical bonds and gives information about the number and nature of neighboring nuclei [8,9]. In the solid state, two other interactions are expressed in the ssNMR spectra: the dipolar interaction, that is a through-space interaction between nuclei, and the quadrupolar interaction for atoms with nuclear spin quantum numbers larger than $\frac{1}{2}$. The former is mostly responsible for an important homogeneous line broadening of ssNMR resonances as compared to solution NMR resonances (for which this interaction is averaged to zero due to fast molecular tumbling), the second can give rise to particularly broad NMR spectra. Magic-angle spinning (MAS), i.e., spinning the sample at an angle of 54.7° about the main—usually vertical—magnetic field is the most used method to remove the dipolar interaction, that is the major source of resolution loss for solids. The faster the spinning, the most efficient the removal of the interaction [10].

3. 1D ^{1}H ssNMR Spectroscopy

In the world of DDSs, mesoporous silica nanoparticles (MSNs) are among the most studied systems. Their long-range ordered pore structure with tailorable pore size (2–50 nm) and geometry facilitates a homogenous incorporation of guest molecules with different sizes and properties. Their surface can further be easily functionalized. Drug loading is based on the adsorptive properties of MSNs and both hydrophilic and hydrophobic cargos can be adsorbed into the pores. The release profile of drugs from MSNs mainly depends on its diffusion from the pores, which can be tailored by modifying the surface of the MSNs to suit the biological needs. The decisive factor responsible for controlling the release is the interaction between the surface groups on pores and the drug molecules. The two most widely explored materials for drug delivery are MCM-41 (Mobil Crystalline Materials) and SBA-15 (Santa Barbara Amorphous type material) [11,12]. MSNs owing to their modifiable surface chemistry can act as carriers for poorly soluble drugs and tackle their solubility issues. ^{1}H, ^{13}C, and ^{29}Si are the most used nuclei to study the interactions between silica molecules and drugs and the influence in the environment of silicon atoms in SBA-15 [13–16].

^{1}H is indeed the most obvious nucleus that can be used for the characterization of DDS as the majority of drugs are protonated, as are the drug carriers. 1D ^{1}H MAS NMR is a simple and fast experiment that gives information about the state of the drug in the DDS and the formation of hydrogen bonds (that leads to high-frequency proton shifts), thanks to a shift of the proton resonance. The spectra are known to be dominated by large homonuclear dipolar interactions; hence, techniques based on fast MAS are usually preferred to obtain high-resolution data [17]. Due to the small chemical shift dispersion of ^{1}H NMR spectra (below 20 ppm in most diamagnetic systems), it is preferable to perform measurements at high magnetic field to optimize the signal resolution.

Ukmar et al. [6] studied by ssNMR the composition and structure of nonfunctionalized and functionalized SBA-15 mesoporous silica matrices in which were loaded different amounts of indomethacin (IMC) molecules. From the ^{1}H MAS NMR spectra (Figure 2), a decrease in the signal intensity of the water resonance was detected when the amount of the drug was increased. Furthermore, the appearance of a new resonance revealed the formation of hydrogen bonds between the drug and the silica matrix. These were evidences for drugs in close contact in the pores with the silica matrix.

However, in DDSs, most systems are complex and lacking long-range order; therefore the resonances are strongly heterogeneous, and even using the fastest available MAS probes and highest static magnets is not enough to obtain satisfactory resolution. This is exemplified in Figure 3 for a model DDS system consisting of an aluminum-based nanoscale metal–organic framework (MOF). This MOF, MIL-100(Al) (MIL stands for Matériau Institut Lavoisier), was analyzed pure, loaded with a drug and with surface covered by β-cyclodextrin (CD) phosphates. The differences between all three samples are not easily seen on the 1D MAS (60 kHz, 20.0 T) ^{1}H NMR spectra, due to strong overlap between all the components, each of them containing a fairly large number of protons. Two solutions might be adopted to overcome this problem: performing a two-dimensional (2D) homonuclear experiment (illustrated in the next section) or the use of deuterated molecules.

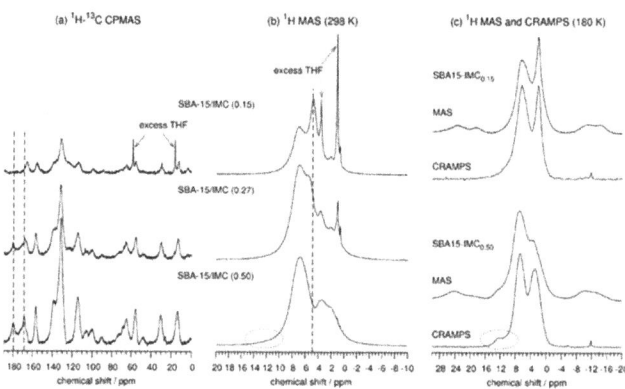

Figure 2. ^{1}H-^{13}C CPMAS (**a**), ^{1}H MAS (**b**), ^{1}H MAS, and CRAMPS (**c**) experiments of SBA-15 in which were loaded different amount of indomethacin [6]. (Reprinted with permission from T. Ukmar, T. Čendak, M. Mazaj, V. Kaučič, G. Mali, J. Phys. Chem. C 2012, 116, 2662. Copyright © 2012, American Chemical Society).

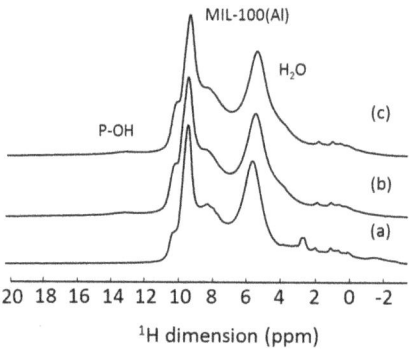

Figure 3. ^{1}H MAS NMR spectra of (**a**) nanoMIL-100(Al), (**b**) ATP-loaded nanoMIL-100(Al) and (**c**) CD-P coated ATP-loaded nanoMIL-100(Al).

4. ^{1}H-^{1}H 2D NMR

The amount of information from 1D NMR spectra is sometimes limited due to potential overlap on the resonances. Therefore, 2D NMR spectroscopy is often used for characterization in pharmaceutical analysis [17]. As an example, it was employed to study flurbiprofen incorporated in 200–400 nm silica capsules filled with Pluronic P123 (polyethylene oxide-polypropylene oxide-polyethylene oxide triblock copolymer). ^{1}H-^{1}H 2D single-quantum single-quantum (SQ-SQ) NMR experiments (Figure 4) revealed the close proximity between the protons of flurbiprofen molecules and those of polypropyleneoxide part of the P123 chains. This confirmed the solubilization of flurbiprofen inside the core of the micelles composed of poly(propylene oxide), and its absence from the shells made of poly(ethylene oxide) [18].

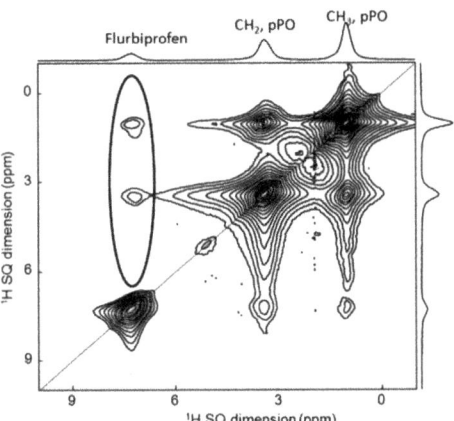

Figure 4. ^{1}H-^{1}H 2D SQ-SQ MAS NMR of flurbiprofen incorporated in silica capsules filled with Pluronic P123. As evidence (ringed), the cross-peaks presented between the protons of the drug and the ones of the DDS [18].

Another important class of porous nanosized DDS are MOFs [19], which are based on metal clusters linked to each other through organic linkers. They offer highly porous structures that can accommodate, transport, and release drugs. The most promising MOFs are based on non-toxic paramagnetic Fe^{3+} cations, which limits the NMR investigations [20]. Indeed, ssNMR offers more information when diamagnetic cations such as Ca^{2+}, Zn^{2+}, Al^{3+}, or Zr^{4+} are used. As an example, UiO-66(Zr), where UiO stands for the University of Oslo, has been tested as a carrier for caffeine [18]. The terephthalate linker can be functionalized with different polar/apolar groups, which can allow tuning of the host/guest interactions. ^{1}H and ^{13}C ssNMR experiments were performed to investigate these interactions. The interactions between the MOFs and the drug were studied by combining density functional theory (DFT) calculations and ^{1}H-^{1}H double-quantum single-quantum (DQ-SQ) MAS NMR experiments. To do so, different functionalized UiO-66(Zr) samples (-H, -NH$_2$, -2OH, -Br) loaded with caffeine were used (Figure 5). Notably, they show that the functional groups had little impact on the drug as no specific interaction between the caffeine and the functional group was found on the NMR spectra [21].

Coming back to the example shown in the previous section (MIL-100(Al) coated with cyclodextrin-phosphate and loaded with adenosine triphosphate (ATP), for which no resolution was obtained on the 1D MAS NMR spectrum, one can notice a slightly better resolution of the resonances on the 2D ^{1}H-^{1}H MAS NMR (Figure 6). In particular, a correlation peak between the phosphate group of the drug and the proton of the MOF linker is observed, confirming the incorporation of the drug in the pores of the MOF. However, the overlap is still very strong, and it is difficult from this spectrum to extract further unambiguous correlation patterns.

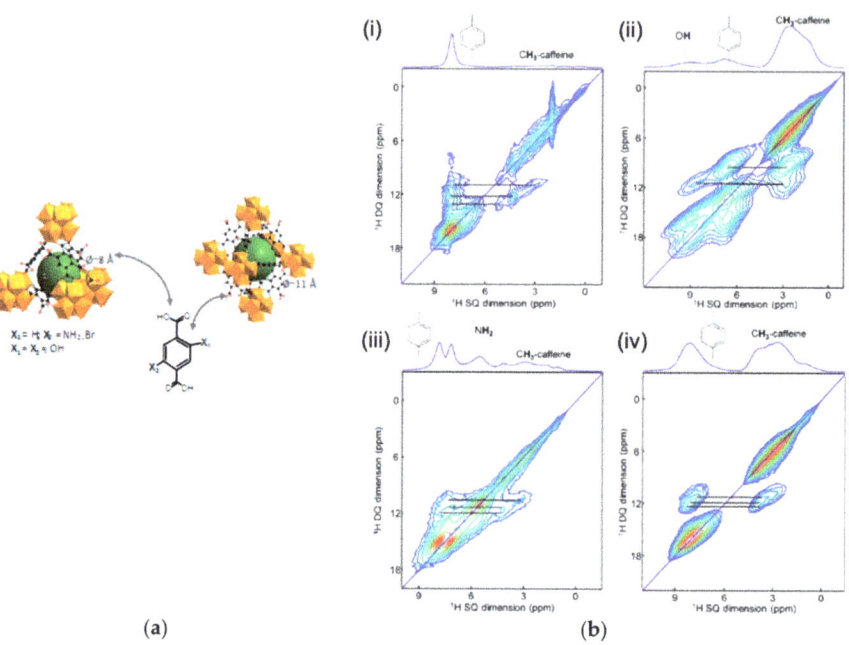

Figure 5. (**a**) Schematic view of the tetrahedral (left) and octahedral (center) cages of the dehydrated UiO-66(Zr). Zirconium polyhedra and carbon atoms are in orange and black, respectively. Hydrogen atoms are omitted for clarity. (**b**) ^{1}H-^{1}H DQ-SQ MAS NMR experiments on different functionalized UiO-66(Zr) samples (-H (**i**), -NH$_2$ (**ii**), -2OH (**iii**), -Br (**iv**)) loaded with caffeine. The lines indicate the correlation between the protons of the linker of the MOF and the ones of the CH$_3$ group of the caffeine [21].

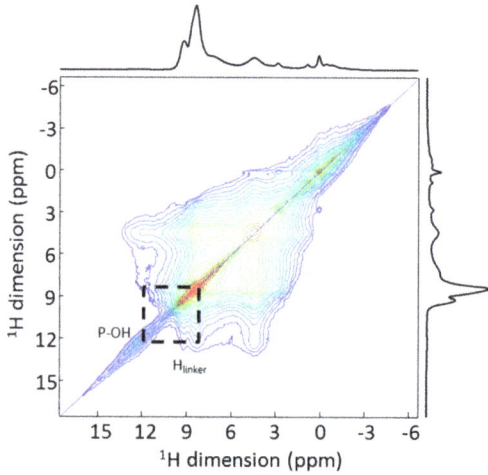

Figure 6. 2D ^{1}H-^{1}H spin diffusion NMR spectrum of CD-P coated ATP-loaded nanoMIL-100(Al). The dash lines indicate the spatial proximity between a phosphate proton and a proton from the linker of the MOF.

5. ^{1}H-X NMR

An alternative solution to address the challenges of analyzing the structure DDSs is to take advantage of the heteroatoms (i.e., non-hydrogen atoms) when they are present in the drug or in the host. Hereafter, we illustrate applications of ^{1}H-X ssNMR spectroscopy in this context.

^{13}C nucleus is the second most obvious nucleus to study by NMR, as most drugs but also numerous DDSs contain carbon atoms, either in the grafted molecules or in the host structure itself (e.g., organic DDSs, MOFs, or functionalized porous silica). Solid dispersions can be analyzed by ^{13}C ssNMR to probe the association between amorphous drug and polymers through differences in NMR spectra that are not visible in the PXRD pattern [22,23] or to provide a direct way to probe drug–carrier interactions [24–26] and analyze the polymorphic forms of drugs [27], or to distinguish between the free and bound steroid drug in a DDS [28]. For example, in solid dispersions formed by α-, β-, and γ-CD in polyethylene glycol (PEG) 6000 g/mol with or without the addition of 5% w/w indomethacin (IM), the ^{13}C cross-polarization under MAS (CPMAS) NMR spectra of the α- and β-CD solid dispersions gave spectra that were essentially superpositions of the spectra of the pure components of the system. On the contrary, for the γ-CD based dispersion, the spectral resolution was somewhat better, and therefore, several chemical shift data for C-1, C-4, and C-6 of the CDs were obtained. They concluded from these data that an inclusion complex might have been formed where the PEG molecules or their hydroxyl end groups interact with the CD cavity, chasing the bound water molecules from the cavities and changing the chemical shift in a way similar to that obtained when water was present. In the PEG/IM, PEG/α-CD/IM, and PEG/β-CD/IM samples as in pure IM spectra, these two peaks are of the same magnitude, while the C-10 peak has almost disappeared in the γ-CD dispersion (PEG/γ-CD/IM). For γ-CD, C-1 and C-4 signals showed a shift of 1–4 ppm downfield, while C-2,3,5 signals showed a shift of 1 ppm in the opposite direction compared to pure γ-CD. The C-6 carbon located at the exterior of the torus also showed a shift of 2 ppm upfield. These results indicate that IM may interact with γ-CD not only at the interior of the cavity but could also affect, directly or indirectly, the hydrogen bonds at the top of the torus where C-6 is located [29].

The surface of MCM-41 was modified with silane or other organic or amino groups [14–17] with the aim to better control the drug release. Azais et al. [30] reported a study on ibuprofen (Ibu), loaded in MCM-41 with pore size ranging from 35 to 116 Å (Figure 7a). Using ^{1}H, ^{13}C, and ^{29}Si ssNMR experiments recorded at room or low temperature (-50 °C), the authors clearly showed the local interactions between the drug and the MCM-41 host, which they could further relate to the drug release profile. Figure 7b displays the ^{13}C CPMAS NMR spectra of Ibu-116 (Ibu loaded in MCM-41 with pore size of 116 Å) and Ibu-35 (Ibu loaded in MCM-41 with pore size of 35 Å) recorded at -50 °C. The two spectra are very distinct; in particular, if the spectrum of Ibu-116 is close to the one of the crystalline Ibu, the one of Ibu-35 presents broader peaks. The pores with diameters of 116 Å are large enough to allow the nucleation of Ibu crystallites. In contrast, in narrow pores (35 Å), a vitrification process occurs as demonstrated in the ^{13}C NMR spectrum (Figure 7b), in which the quaternary carbons are now clearly detected [30].

β-IM was loaded in the cavities of both MIL-101(Al)-NH$_2$ and a mesoporous silica SBA-15 (Figure 8a). Detailed inspection of the ^{1}H-^{13}C CPMAS NMR spectrum of MIL-101(Al)-NH$_2$/IM (Figure 8b) leads to two interesting observations. First, unlike the spectrum of SBA-15/IM, the spectrum of MIL-101(Al)-NH$_2$/IM clearly confirms the presence of tetrahydrofuran (THF) within the pores. Second, the carboxylic carbon peak at 173 ppm (named L2, L3 in the spectrum) narrows substantially. This suggests that the metal–organic framework undergoes structural ordering when it is filled with IM molecules [31].

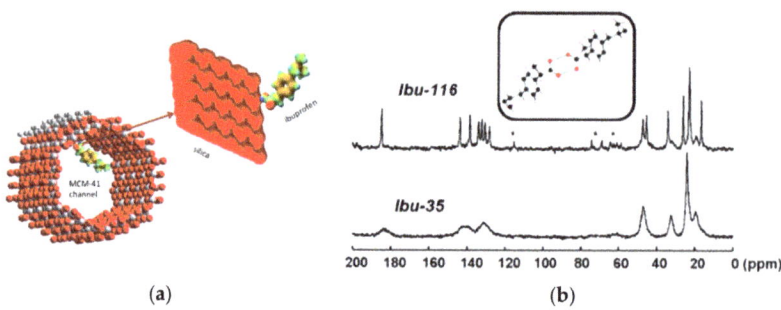

Figure 7. (a) Adsorption of a molecule of ibuprofen within the MCM-41 channels by electrostatic interactions; (b) ^{13}C CPMAS NMR spectra of Ibu-116 and Ibu-35 [30]. The stars indicate the presence of spinning sidebands. (Reprinted with permission from T. Azaïs, C. Tourné-Péteilh, F. Aussenac, N. Baccile, C. Coelho, J.-M. Devoisselle, F. Babonneau, Chem. Mater. **2006**, 18, 6382. Copyright © 2006, American Chemical Society).

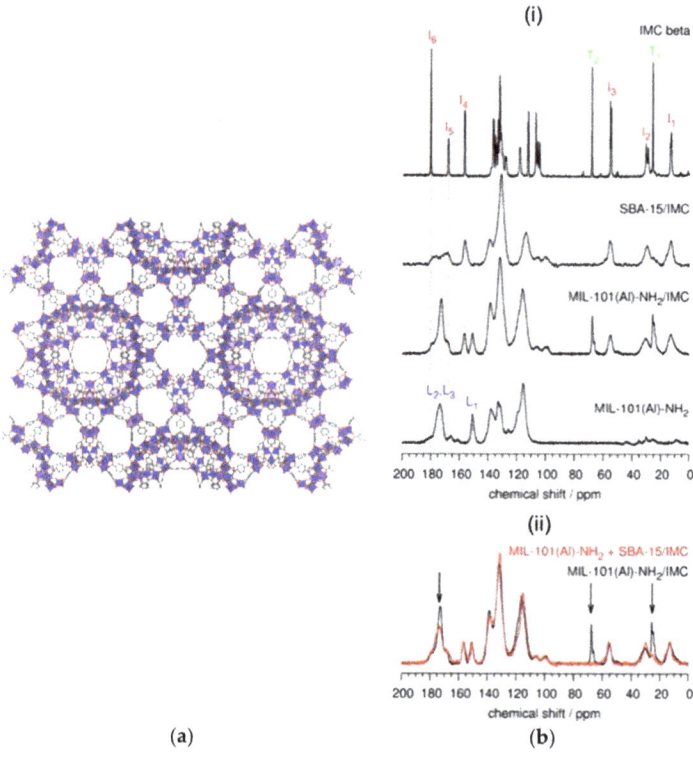

Figure 8. (a) Structure of MIL-101 viewed along the (101) direction. (b) (i) ^{1}H–^{13}C CPMAS NMR spectra of bulk crystalline β- IMC, SBA-15/IMC, loaded and empty MIL-101(Al)-NH$_2$. Peak labels correspond to selected carbon atoms within the IMC molecule (I1–I6), THF molecule (T1, T2), and BDC-NH$_2$ linker (L1–L3). Vertical dotted lines enable an easier comparison of signal positions. (ii) Comparison of the ^{1}H–^{13}C CPMAS NMR spectrum of MIL-101(Al)-NH$_2$/IMC with the sum of the spectra of SBA-15/IMC and MIL-101(Al)-NH$_2$. Arrows indicate details where the differences are the most pronounced [31]. (Reprinted with permission from T. Čendak, E. Žunkovič, T. Ukmar Godec, M. Mazaj, N. Zabukovec Logar, G. Mali, J. Phys. Chem. C **2014**, 118, 6140. Copyright © 2014, American Chemical Society).

Another interesting nucleus is ^{19}F, as about 25% of APIs present on the market contain a fluorine atom in their structure [32]. With a sensitivity close to that of proton, ^{19}F is highly attractive from an NMR point of view. It is usually present in little quantity in the API (often less than two or three fluorine groups), leading to ^{19}F MAS NMR spectra with a limited number of resonances. Furthermore, the chemical shift dispersion (but also the chemical shift anisotropy) is much larger than that of proton, leading to less signal overlap. It can provide information also on the molecular orientation inside a DDS [33].

Pham et al. [22] performed ^{1}H-^{19}F CP-HETCOR and Lee-Goldburg (LG) CP-HETCOR experiments (Figure 9) on acetaminophen amorphous dispersions in poly(vinyl pyrrolidone) (PVP) to confirm the formation of an amorphous glass solution. Note that to perform this type of experiment, a particular HFX triple resonance probe design is required [34]. Correlations are observed between the fluorine signal and both the types of protons (aromatic and aliphatic). The aliphatic ones can only be associated with the polymer. Increasing the contact time (2 ms), the correlation increases as expected from spin diffusion. To confirm spin diffusion effects, the authors performed LGCP-HETCOR experiments, which greatly reduces spin diffusion during the ^{1}H spin lock period (Figure 9b). As shown by the relative intensity of the correlations, the build-up of spin diffusion is eliminated; the remaining correlation between the fluorine signals and aliphatic protons can then be assigned to a direct through-space dipolar interaction.

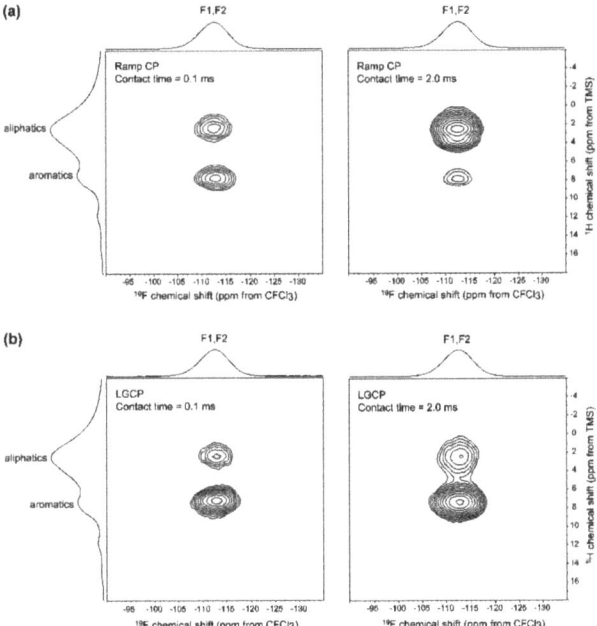

Figure 9. (a) ^{1}H-^{19}F CP-HETCOR spectra of a 30% w/w amorphous dispersion of diflunisal (VI) in PVP; (b) ^{1}H-^{19}F LGCP-HETCOR spectra obtained under the same conditions, except with the use of LGCP to suppress spin diffusion. The difference in relative correlation intensity with a 2.0 ms contact time between the two CP methods highlights the magnitude of the spin diffusion effects between VI and PVP in the dispersion. The F2 projections are the ^{19}F CP-MAS spectrum recorded at 15 kHz), and the F1 projections are the ^{1}H MAS spectrum recorded at 35 kHz) [22]. (Reprinted with permission from T. N. Pham, S. A. Watson, A. J. Edwards, M. Chavda, J. S. Clawson, M. Strohmeier, F. G. Vogt, Mol. Pharm. **2010**, 7, 1667. Copyright © 2010, American Chemical Society).

On a complex made by diflunisal and β-CD (Figure 10a), a 2D ^{1}H-^{19}F CP-HETCOR spectrum was performed to verify the incorporation of the drug in the cavities (Figure 10b). In the spectrum, the two components representing the included and free diflunisal can be clearly distinguished. A strong correlation arises between the aliphatic β-CD protons (3.5 ppm) and the more deshielded fluorine position at approximately −111 ppm. Within the included diflunisal, the expected correlation between aromatic proton positions and fluorine positions is more difficult to observe in the 2D contour plot because of a combination of the low concentration of this component and the loss of signal from spin diffusion to aliphatic β-CD protons. However, the presence of this expected correlation is evidenced by a deshielded shoulder between −105 and −110 ppm in the 1D row extraction shown for the free component [35].

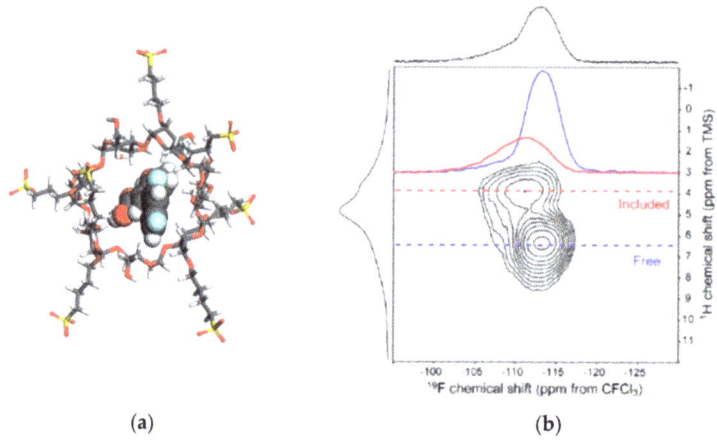

Figure 10. (a) Structure of diflunisal and β-CD complex; (b) ^{1}H−^{19}F CPHETCOR spectrum showing a mixture of bound and unbound diflunisal. In the F2 axis dimension is plotted the ^{19}F CP-MAS spectrum (at 14 kHz), and the ^{1}H MAS spectrum (at 35 kHz) is plotted along the F1 axis dimension. Extracted rows are shown [35]. (Reprinted with permission from F. G. Vogt, M. Strohmeier, Mol. Pharm. **2012**, 9, 3357. Copyright © 2012, American Chemical Society).

^{19}F nucleus was also used as a spy to get insights into the interactions of a drug (lansoprazole, LPZ, which contains a single CF$_3$ group) loaded in a CD-based MOF, namely γ-CD-MOF. ^{1}H-^{19}F-^{13}C double CP experiments provided the selection of the carbon atoms in the proximity of the fluorine atom. The resulting ^{13}C NMR spectrum is compared to the ^{1}H-^{13}C CP NMR spectrum, in which all ^{13}C atoms are present (since both the drug and the CD have protons to transfer magnetization to the carbon atoms); the spectra are normalized to the CD peak at 73 ppm. In the ^{19}F-^{13}C CPMAS, one can notice higher intensity, as expected, as these are the closest C atoms to the ^{19}F nuclei. There is also a significant signal for the ^{13}C nuclei of the CD, indicating its close spatial proximity to the drug. Among the ^{13}C of the CD, the one labeled C6 (which corresponds to the CH$_2$OH) has higher intensity than the other CD carbons. This indicates that the CF$_3$ group is in close contact to the CH$_2$OH; hence, it is located outside the CD. In combination with DFT simulation, the data indicated the formation of a 2:1 γ-CD:lansoprazole complex [36].

^{19}F nucleus was also used to try to distinguish between outer and inner surface interactions. This investigation was made on MIL-100(Al) nanoparticles, which were selected because they are the diamagnetic analogues of MIL-100(Fe). The nanoMIL-100(Al) was impregnated with two different F-labeled lipid conjugates: methyl perfluorooctanoate (FO), which is supposedly small enough to enter in the pores of the MOF, and 1-palmitoyl-2-(16-fluoropalmitoyl)-sn-glycero-3-phosphocholine (FP), which is supposedly too large to enter in the pores of the MOF (Figure 11a,b). ^{1}H-^{19}F CPMAS 2D correlation experiments were performed on the two fluorinated-lipid nanoMIL-100(Al) materials to provide unambiguous localization of the F-lipids inside and outside the MOF. Figure 12a shows the ^{19}F-^{1}H 2D NMR spectrum of FO@nanoMIL-100(Al). One can see correlation peaks of strong intensity between all fluorine atoms of the lipid and the trimesate protons. This unambiguously confirms the presence of the FO lipids in the bulk of the particles, i.e., in the pores of the MOF. In the ^{19}F-^{1}H 2D NMR spectrum of FP@nanoMIL-100(Al) (Figure 12b), cross-peaks of strong intensity are observed between the CFH$_2$ group and its neighboring CH$_2$ groups from the lipid. The cross-peak between the CFH$_2$ of the lipid and the proton of the trimesate has very low intensity compared to the ones of the first sample, confirming its localization on the surface of the particle, and not inside the pores. The fact that these cross-peaks are observed shows that the F-H$_{trimesate}$ distance is not very long, which in turn could indicate that the FP is folded on the nanoparticle surface (Figure 11d) and does not stand in a brush-like manner as was initially expected (Figure 11c) [37].

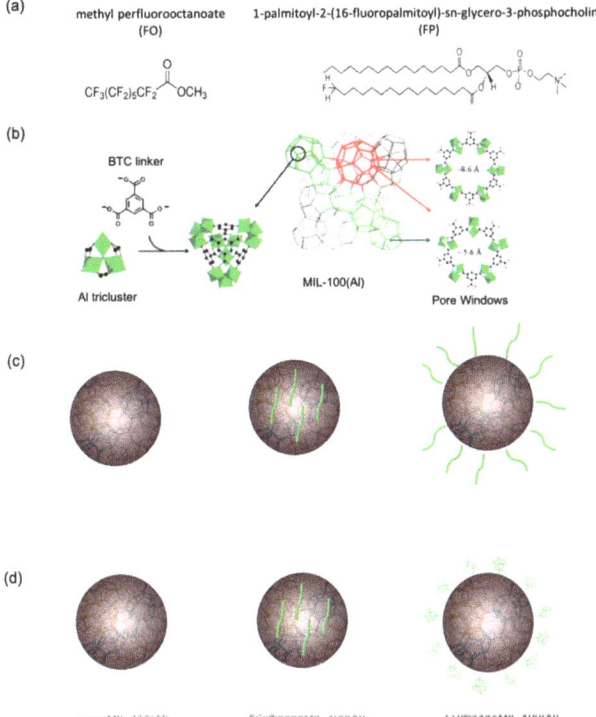

Figure 11. (**a**) Chemical structure of methyl perfluorooctanoate FO (middle) and 1-palmitoyl-2-(16-fluoropalmitoyl)-sn-glycero-3-phosphocholine FP (right). (**b**): Structure of MIL-100 (Al) based on the association of BTC linker and Al triclusters. The pore openings are shown on the right part. (**c**): Schematic structure of nanoMIL-100(Al) (left) and hypothesized localization of FO (middle) and FP (right) inside and outside the nanoMOFs, respectively. (**d**) Schematic structure of nanoMIL-100 (left) and localization of FO (middle) and FP (right) inside and outside the nanoMOFs, respectively, as deduced from the NMR data [37].

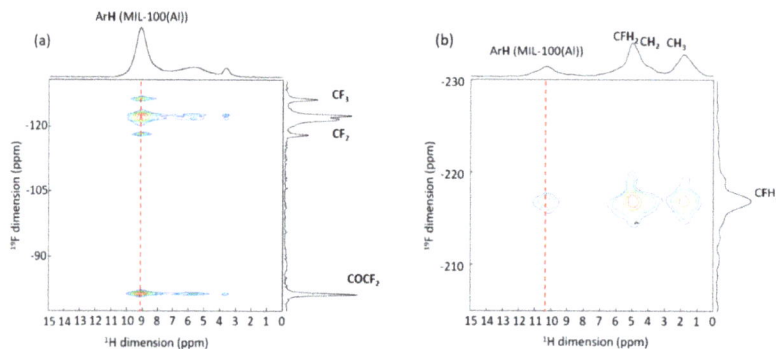

Figure 12. ^{19}F-^{1}H 2D CPMAS NMR spectra of (**a**) FO@nanoMIL-100(Al) and (**b**) FP@nanoMIL-100(Al). The lines are assigned. The red dash line indicates the spatial proximity between the ^{19}F resonances and the ^{1}H of the MOF [37].

For mesoporous silicas, ^{29}Si proves an interesting nucleus. It is quite useful to study the influence of mesoporous structure on the uptake of the drug [38], on the interactions drug–silica [39,40], on the drug delivery properties [41] and to study the different connectivity in the silica network [42] and explore the proton chemical environments around the silica [43].

Zeolites are also potential drug carriers; a ^{1}H–^{29}Si HETCOR spectrum of a zeolite beta-based drug formulation containing Ag and sulfadiazine (SD) gives details into the incorporation of the drug within the zeolite matrix. The strong correlation peak detected between the Si(OAl) sites and SD aromatic and NH protons evidences the localization of the drug near the Q^4 structures. A second correlation peak with lower intensity, is observed between the aromatic protons of SD and the signal, corresponding to [Si(1OH) + Si(1Al)] sites, at 103 ppm [44].

Numerous drugs contain a phosphorous atom in the form of phosphonate, phosphinate, or phosphate groups. ^{31}P NMR spectroscopy proves very sensitive to hydrogen bonds [45]. MCM-41 containing phosphorous atoms (P-MCM-41) was used as bioactive material. The ^{31}P MAS-NMR experiment (Figure 13) was performed to evaluate the amount of phosphorous and its impact in the structure of the material. In the spectrum, two groups of signals around 0 and -11 ppm are shown. Despite the low amount of phosphorus (<1%), they could be assigned, according to phosphate units PO$_4$ not bonded to silicon (called Q$_0$ species) or bonded to one silicon atom (called Q$_1$ species) through one P-O-Si bond. The relative intensities between both signals give a Q$_0$/Q$_1$m molar ratio = 1:2, which indicates that at least 66% of the P atoms are bonded to the silica framework [46].

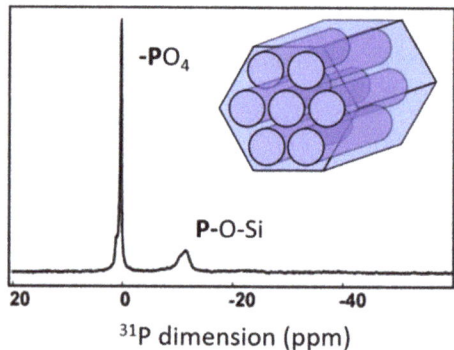

Figure 13. ^{31}P MAS-NMR spectrum of MCM-41 phosphate [46].

6. X-Y NMR

If all the examples reported in the previous sections show the potential of the ^{1}H nucleus in the study of DDS, this nucleus still sometimes leads to such complicated NMR spectra that it becomes difficult to extract information from them. In that case, although it is more demanding, it might be interesting to use pairs of heteronuclei X-Y.

6.1. ^{27}Al-^{31}P

These two nuclei happened to be present in Al-based MOFs loaded with phosphate drugs, or which had a surface covered with phosphate molecules (Figure 14a). These MOF carriers were selected because of the known affinity between aluminum and phosphorus (e.g., as in aluminophosphates), which could lead to strongly anchored drug/covering moieties. From an NMR point of view, the numerous methods have been developed for this pair of nuclei, and the required triple resonance ^{1}H-^{31}P-^{27}Al probes are nowadays available in most labs [47].

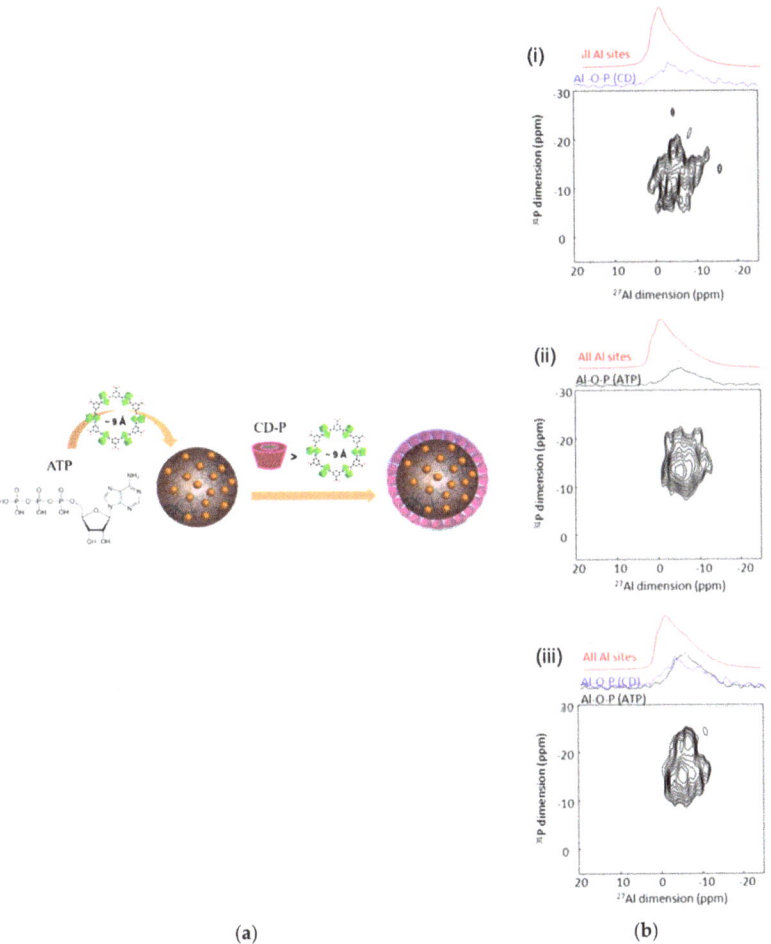

Figure 14. (a) Schematic representation of the highly porous MIL-100 (Al) nanoparticles loaded with ATP and then coated with CD-P; (b) ^{27}Al{^{31}P} MAS D-HMQC NMR spectra of CD-P coated (i), ATP loaded (ii), and CD-P coated ATP loaded nanoMIL100(Al) (iii). The top blue spectra are the full projections on the horizontal dimension for the surface sites, the black spectra are the full projection for the interphase sites, while the red spectra are the MAS NMR spectra shown for comparison [48].

In the nanoMIL100(Al) coated with CD-P, the surface aluminum species of the nanoparticles (NPs), i.e., those in interaction with the covering groups, could be detected by employing the through-space dipolar-based version of the solution-NMR experiments HMQC (D-HMQC) and keeping the recoupling time short enough to ensure that only the Al in very close proximity to the ^{31}P were selected. The 2D ^{27}Al-^{31}P MAS NMR spectrum of a CD-P-coated nanoMIL-100(Al) (Figure 14b(i)) showed a ^{27}Al NMR signal at −5 ppm, i.e., at a chemical shift that is close to the one observed in aluminophosphate species. This indicated the formation of an Al-O-P bond between the Al surface sites of the nanoMOF and the terminal phosphate groups of the CD-P, which very likely replace a water molecule. The same experiment was made for the ATP loaded nanoMIL-100 (Al) (Figure 14b(ii)). A six-fold coordinated ^{27}Al resonance around −7 ppm was identified and suggested a close proximity (formation of an Al-O-P chemical bond) between Al species of the MOF framework and the terminal phosphate of the drug. This ^{27}Al resonance is the signature of the grafted aluminum sites inside the pores of the MOF. Finally, the similarity of the spectra of ATP loaded nanoMIL-100(Al) and the target CD-P coated nanoMIL-100(Al) loaded with ATP (Figure 14b(iii)) clearly confirms that the CD-P coating on the external surface of the ATP loaded nanoMOF did not affect the Al-O-ATP bond formed inside the MOF cavities [48]. These experiments confirmed the assumption of strong affinity between aluminum and phosphorus species.

6.2. ^{27}Al-^{13}C

^{13}C-^{27}Al 2D MAS NMR experiments were performed to study drug carrier interactions and obtain information about the localization of the drug [49]. Oligomers based on CD cross-linked with citric acid (CD-CO) were shown to interact strongly with the anticancer drug, Doxorubicin (DOX) [50,51]. DOX was also shown to enter in the pores of nanoMIL-100 (Al) [46]. Therefore, this CD-CO water soluble oligomer was considered as a versatile coating to promote DOX incorporation and control its release. To study the interactions taking place in this system and have selectivity between the bulk and the surface, the coating was synthesized using a ^{13}C label citric acid (1,5-^{13}C$_2$ citric acid). The obtained CD-^{13}CO oligomer was used to cover the surface of the nanoMIL-100(Al) after loading of DOX drug in the pores (Figure 15) [49]. To understand the interaction between the CD-CO coating and the MOF nanoparticles, 2D ^{13}C-^{27}Al MAS correlation NMR experiments were performed (Figure 16), showing the spatial proximity between carbon and aluminum atoms. On the 2D NMR spectrum of CD-^{13}CO@nanoMIL-100(Al) (Figure 16b), correlation peaks of strong intensity were observed between the COO of the citric acid (^{13}C resonance at 180 ppm) and the surface Al sites, confirming that the CD-CO oligomer has high affinity with the NP surface. Note that the ^{13}C resonance at 175 ppm contains both the carboxylic group of the trimesate linker of the MOF (not labeled but present in large quantity) and the COO-CD of the CD-^{13}CO coating. The same experiment was performed on the DOX loaded CD-^{13}CO@nanoMIl-100(Al) (Figure 16c) and shows that the coating is still in strong interaction with the NP surface. One can notice a change of the relative intensity between the two carbon resonances. This indicates that in addition to going in the pores of the MOF, the DOX molecules also interact significantly with the CD-CO coating. Notably, since the intensity of ^{13}C resonance at 180 ppm has decreased, it very likely indicates that part of the ^{13}COO-Al bonds formed between the citric acid moieties and the surface Al sites have been broken, probably in favor of the interaction of the citric acid with the DOX drug.

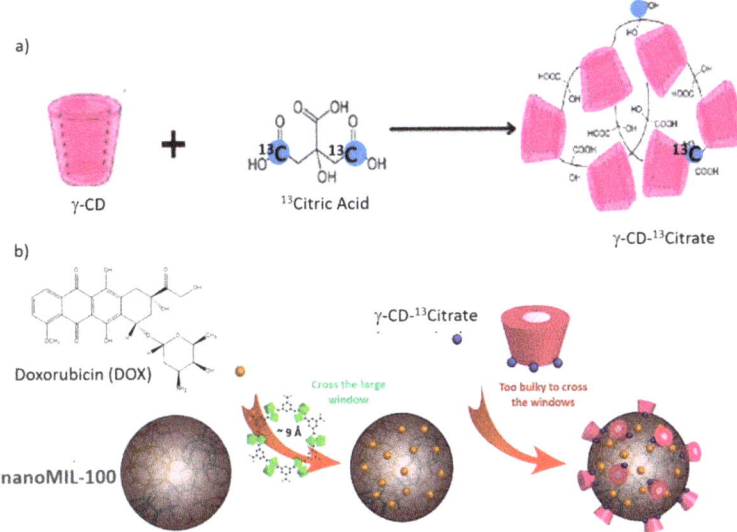

Figure 15. (**a**) Synthesis of γCD-^{13}citrate polymers (CD-^{13}CO); (**b**) Schematic representation of the highly porous MIL-100(Al) nanoparticles loaded with DOX and then coated with CD-^{13}CO [48].

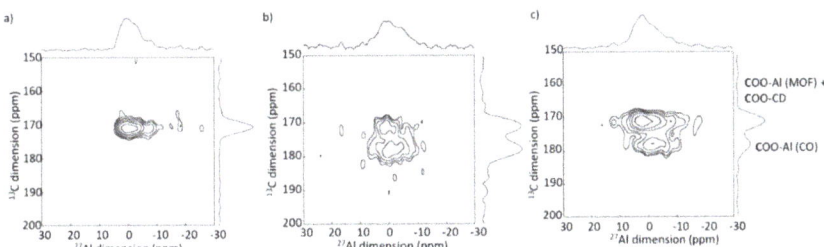

Figure 16. 2D ^{13}C-^{27}Al D-HMQC 2D of (**b**) CD-^{13}CO@nanoMIL-100(Al) and (**c**) DOX loaded CD-^{13}CO@nanoMIL-100(Al). The spectrum recorded on pure (**a**) nanoMIL-100(Al) is shown for comparison [49].

7. Conclusions

In this review, we have illustrated different approaches used during the years to analyze the supramolecular structures of DDS by ssNMR spectroscopy. This technique proves very useful to study the confinement of the drug in the nanoparticulate system, its interaction with the host matrix, and its release. Moreover, ssNMR spectroscopy gave invaluable information on the location of the shell, possible penetration inside the nanoparticles' pores, and its interaction with the core. The easiest and most analyzed nucleus is ^{1}H, but sometimes, the systems are so complex that the ^{1}H 1D NMR spectra result in a broad signal without any information available. Several strategies can be adopted in that case. The first one is to perform ^{1}H-^{1}H 2D NMR experiments, increasing the resolution and having more information thanks to the non-direct dimension. As an alternative, heteronuclei (X) NMR, such as ^{13}C, ^{29}Si, ^{27}Al, ^{31}P, and ^{19}F present on the drug or the delivery system can be used in conjunction with proton NMR in ^{1}H-X NMR experiments. Finally, when proton NMR is not informative at all, NMR of pairs of heteronuclei X-Y can be used. With this complete set of NMR tools and methods, the supramolecular structures of a large variety of DDSs, as illustrated in this review, have been studied. This deep characterization step is essential to guide the design of more performant DDSs. Although not shown here, ssNMR, in complement with solution state NMR, can also be used to study both the degradation of the particles and drug release.

Author Contributions: Conceptualization, M.P., X.L., R.G., C.M.-C.; writing—original draft preparation, M.P., C.M.-C.; writing—review and editing, X.L., R.G. All authors have read and agreed to the published version of the manuscript.

Funding: This work is supported by a public grant overseen by the French National Research Agency (ANR) as part of the "Investissements d'Avenir" program (Labex NanoSaclay, reference: ANR-10-LABX-0035).

Institutional Review Board Statement: Not applicable.

Informed Consent Statement: Not applicable.

Acknowledgments: M.P. thanks the Région Centre-Val de Loire for a PhD fellowship. R.G. is grateful for a support from the ANR-20-CE19-0020.

Conflicts of Interest: The authors declare no conflict of interest.

References

1. Mitchell, M.J.; Billinglsey, M.M.; Haley, R.M.; Wechsler, M.E.; Peppas, N.A.; Langer, R. Engineering precision nanoparticles for drug delivery. *Nat. Rev. Drug Discov.* **2021**, *20*, 101–124. [CrossRef]
2. Skorupska, E.; Jeziorna, A.; Kazmierski, S.; Potrzebowski, M.J. Recent progress in solid-state NMR studies of drugs confined within drug delivery systems. *Solid State Nucl. Magn. Reson.* **2014**, *57–58*, 2–16. [CrossRef]
3. Bhowmik, D.; Gopinath, H.; Kumar, B.P.; Duraivel, S.; Kumar, K.P.S. Controlled released drug delivery systems. *Pharma Innov.* **2012**, *1*, 24–32.
4. Loganathan, S.; Valapa, R.B.; Mishra, R.K.; Pugazhenthi, G.; Thomas, S. Thermogravimetric Analysis for Characterization of Nanomaterials. In *Thermal and Rheological Measurement Techniques for Nanomaterials Characterization*, 1st ed.; Thomas, S., Thomas, R., Zachariah, A., Mishra, R., Eds.; Elsevier: Amsterdam, The Netherlands, 2017; Volume 3, Chapter 4; pp. 67–108. [CrossRef]
5. Ambroz, F.; Macdonald, T.J.; Martis, V.; Parkin, I.V. Evaluation of the BET theory for the characterization of meso and microporous MOFs. *Small Methods* **2018**, *2*, 1800173. [CrossRef]
6. Ukmar, T.; Čendak, T.; Mazaj, M.; Kaučič, V.; Mali, G. Structural and Dynamical Properties of Indomethacin Molecules Embedded within the Mesopores of SBA-15: A Solid-State NMR View. *J. Phys. Chem. C* **2012**, *116*, 2662–2671. [CrossRef]
7. Wong, Y.T.A.; Martins, V.; Lucier, B.E.G.; Huang, Y. Solid-State NMR Spectroscopy: A Powerful Technique to Directly Study Small Gas Molecules Adsorbed in Metal–Organic Frameworks. *Chem. Eur. J.* **2019**, *25*, 1848–1853. [CrossRef]
8. Levitt, M.H. *Spin Dynamics*; Wiley: Chichester, UK, 2008.
9. Keeler, J. *Understanding NMR Spectroscopy*; Wiley: Chichester, UK, 2010.
10. Struppe, J.; Quinn, C.M.; Sarkar, S.; Gronenborn, A.M.; Polenova, T. Ultrafast ^{1}H MAS NMR Crystallography for Natural Abundance Pharmaceutical Compounds. *Mol. Pharm.* **2020**, *17*, 674–682. [CrossRef] [PubMed]
11. Narayan, R.; Nayak, U.; Raichur, A.; Garg, S. Mesoporous Silica Nanoparticles: A Comprehensive Review on Synthesis and Recent Advances. *Pharmaceutics* **2018**, *10*, 118. [CrossRef] [PubMed]
12. Tang, F.; Li, L.; Chen, D. Mesoporous silica nanoparticles: Synthesis, biocompatibility and drug delivery. *Adv. Mater.* **2012**, *24*, 1504–1534. [CrossRef]
13. Shen, S.; Chow, P.S.; Kim, S.; Zhu, K.; Tan, R.B.H. Synthesis of carboxyl-modified rod-like SBA-15 by rapid co-condensation. *J. Colloid Interface Sci.* **2008**, *321*, 365–372. [CrossRef]
14. Mellaerts, R.; Houthoofd, K.; Elen, K.; Chen, H.; Van Speybroeck, M.; Van Humbeeck, J.; Augustijns, P.; Mullens, J.; Van den Mooter, G.; Martens, J.A. Aging behavior of pharmaceutical formulations of itraconazole on SBA-15 ordered mesoporous silica carrier material. *Microporous Mesoporous Mater.* **2010**, *130*, 154–161. [CrossRef]
15. Aiello, D.; Folliet, N.; Laurent, G.; Testa, F.; Gervais, C.; Babonneau, F.; Azaïs, T. Solid state NMR characterization of phenylphosphonic acid encapsulated in SBA-15 and aminopropyl-modified SBA-15. *Microporous Mesoporous Mater.* **2013**, *166*, 109. [CrossRef]
16. Pérez-Quintanilla, D.; Gómez-Ruiz, S.; Žižak, Z.; Sierra, I.; Prashar, S.; del Hierro, I.; Fajardo, M.; Juranić, Z.D.; Kaluđerović, G.N. A new generation of anticancer drugs: Mesoporous materials modified with titanocene complexes. *Chem. Eur. J.* **2009**, *15*, 5588–5597. [CrossRef] [PubMed]
17. Brown, S.P. Applications of high-resolution 1H solid-state NMR. *Solid State Nucl. Magn. Reson.* **2012**, *41*, 1–27. [CrossRef] [PubMed]
18. Kerkhofs, S.; Saïdi, F.; Vandervoort, N.; Van den Mooter, G.; Martineau, C.; Taulelle, F.; Martens, J.A. Silica Capsules Enclosing P123 Triblock Copolymer Micelles for Flurbiprofen Storage and Release. *J. Mater. Chem. B* **2015**, *3*, 3054–3061. [CrossRef] [PubMed]

19. Horcajada, P.; Chalati, T.; Serre, C.; Gillet, B.; Sebrie, C.; Baati, T.; Eubank, J.F.; Heurtaux, D.; Clayette, P.; Kreuz, C.; et al. Porous metal-organic-framework nanoscale carriers as a potential platform for drug delivery and imaging. *Nat. Mater.* **2010**, *9*, 172–178. [CrossRef]
20. Mali, G. *Looking into Metal-Organic Frameworks with Solid-State NMR Spectroscopy*; InTech: London, UK, 2016.
21. Devautour-Vinot, S.; Martineau, C.; Diaby, S.; Ben-Yahia, M.; Miller, S.; Serre, C.; Horcajada, P.; Cunha, D.; Taulelle, F.; Maurin, G. Caffeine Confinement into a Series of Functionalized Porous Zirconium MOFs: A Joint Experimental/Modelling Exploration. *J. Phys. Chem. C* **2013**, *117*, 11694–11704. [CrossRef]
22. Pham, T.N.; Watson, S.A.; Edwards, A.J.; Chavda, M.; Clawson, J.S.; Strohmeier, M.; Vogt, F.G. Analysis of amorphous solid dispersions using 2D solid-state NMR and (1)H T(1) relaxation measurements. *Mol. Pharm.* **2010**, *7*, 1667–1691. [CrossRef]
23. Ito, A.; Watanabe, T.; Yada, S.; Hamaura, T.; Nakagami, H.; Higashib, K.; Moribe, K.; Yamamoto, K. Prediction of recrystallization behavior of troglitazone/polyvinylpyrrolidone solid dispersion by solid-state NMR. *Int. J. Pharm.* **2010**, *383*, 18–23. [CrossRef]
24. Liu, X.; Lu, X.; Su, Y.; Kun, E.; Zhang, F. Clay-Polymer Nanocomposites Prepared by Reactive Melt Extrusion for Sustained Drug Release. *Pharmaceutics* **2020**, *12*, 51. [CrossRef]
25. Li, S.; Lin, X.; Xu, K.; He, J.; Yang, H.; Li, H. Co-grinding Effect on Crystalline Zaltoprofen with β-cyclodextrin/Cucurbit[7]uril in Tablet Formulation. *Sci. Rep.* **2017**, *7*, 45984. [CrossRef]
26. Chierentin, L.; Garnero, C.; Chattah, A.K.; Delvadia, P.; Karnes, T.; Longhi, M.R.; Salgado, H.R. Influence of β-cyclodextrin on the Properties of Norfloxacin Form A. *AAPS PharmSciTech* **2014**, *16*, 683–691. [CrossRef]
27. Garnero, C.; Chattah, A.K.; Longhi, M. Improving furosemide polymorphs properties through supramolecular complexes of β-cyclodextrin. *J. Pharm. Biomed. Anal.* **2014**, *95*, 139–154. [CrossRef]
28. McCoy, C.F.; Apperley, D.C.; Malcolm, R.; Variano, B.; Sussman, H.; Louven, D.; Boyd, P.; Malcolm, R.K. Solid state 13C NMR spectroscopy provides direct evidence for reaction between ethinyl estradiol and a silicone elastomer vaginal ring drug delivery system. *Int. J. Pharm.* **2018**, *548*, 689. [CrossRef]
29. Wulff, M.; Aldén, M.; Tegenfeldt, J. Solid-State NMR Investigation of Indomethacin−Cyclodextrin Complexes in PEG 6000 Carrier. *Bioconjug. Chem.* **2002**, *13*, 240–248. [CrossRef] [PubMed]
30. Azaïs, T.; Tourné-Péteilh, C.; Aussenac, F.; Baccile, N.; Coelho, C.; Devoisselle, J.-M.; Babonneau, F. Solid-State NMR Study of Ibuprofen Confined in MCM-41 Material. *Chem. Mater.* **2006**, *18*, 6382–6390. [CrossRef]
31. Čendak, T.; Žunkovič, E.; Godec, T.U.; Mazaj, M.; Logar, N.Z.; Mali, G. Indomethacin Embedded into MIL-101 Frameworks: A Solid-State NMR Study. *J. Phys. Chem. C* **2014**, *118*, 6140–6150. [CrossRef]
32. Kirk, K.L. Fluorine in medicinal chemistry: Recent therapeutic applications of fluorinated small molecules. *J. Fluor. Chem.* **2006**, *127*, 1013–1029. [CrossRef]
33. Lau, S.; Stanhope, N.; Griffin, J.; Hughes, E.; Middleton, D.A. Drug orientations within statin-loaded lipoprotein nanoparticles by ^{19}F solid-state NMR. *Chem. Commun.* **2019**, *55*, 13287–13290. [CrossRef] [PubMed]
34. Bechmann, M.; Hain, K.; Marichal, C.; Sebald, A. X-[1H, 19F] triple resonance with a X-[1H] CP MAS probe and characterisation of a 29Si-19F spin pair. *Solid State Nucl. Magn. Reson.* **2003**, *23*, 50–61. [CrossRef]
35. Vogt, F.G.; Strohmeier, M. 2D Solid-State NMR Analysis of Inclusion in Drug—Cyclodextrin Complexes. *Mol. Pharm.* **2012**, *9*, 3357–3374. [CrossRef] [PubMed]
36. Li, X.; Porcino, M.; Martineau-Corcos, C.; Guo, T.; Xiong, T.; Zhu, W.; Patriarche, G.; Péchoux, C.; Perronne, B.; Hassan, A.; et al. Efficient incorporation and protection of lansoprazole in cyclodextrin metal-organic frameworks. *Int. J. Pharm.* **2020**, *585*, 119442. [CrossRef] [PubMed]
37. Porcino, M.; Li, X.; Gref, R.; Martineau-Corcos, C. Solid state NMR spectroscopy as a powerful tool to investigate the location of fluorinated lipids in highly porous hybrid organic-inorganic nanoparticles. *Magn. Reson. Chem.* **2021**. [CrossRef] [PubMed]
38. Manzano, M.; Aina, V.; Areán, C.O.; Balas, F.; Cauda, V.; Colilla, M.; Delgado, M.R.; Vallet-Regí, M. Studies on MCM-41 mesoporous silica for drug delivery: Effect of particle morphology and amine functionalization. *Chem. Eng. J.* **2008**, *137*, 30–37. [CrossRef]
39. Mellaerts, R.; Roeffaers, M.B.J.; Houthoofd, K.; Van Speybroeck, M.; De Cremer, G.; Jammaer, J.A.G.; Van den Mooter, G.; Augustijns, P.; Hofkens, J.; Martens, J.A. Molecular organization of hydrophobic molecules and co-adsorbed water in SBA-15 ordered mesoporous silica material. *Phys. Chem. Chem. Phys.* **2011**, *13*, 2706–2713. [CrossRef] [PubMed]
40. Azaïs, T.; Hartmeyer, G.; Quignard, S.; Laurent, G.; Tourné-Péteilh, C.; Devoisselle, J.-M.; Babonneau, F. Solid-state NMR characterization of drug-model molecules encapsulated in MCM-41 silica. *Pure Appl. Chem.* **2009**, *81*, 1345–1355. [CrossRef]
41. Gao, L.; Sun, J.; Zhang, L.; Wang, J.; Ren, B. Influence of alkyl chains as a strategy for controlling drug delivery different structured channels of mesoporous silicate on pattern. *Mater. Chem. Phys.* **2012**, *135*, 786–797. [CrossRef]
42. Möller, K.; Bein, T. Degradable Drug Carriers: Vanishing Mesoporous Silica Nanoparticles. *Chem. Mater.* **2019**, *31*, 4364–4378. [CrossRef]
43. Folliet, N.; Roiland, C.; Bégu, S.; Aubert, A.; Mineva, T.; Goursot, A.; Selvaraj, K.; Duma, L.; Tielens, F.; Mauri, F.; et al. Investigation of the interface in silica-encapsulated liposomes by combining solid state NMR and first principles calculations. *J. Am. Chem. Soc.* **2011**, *133*, 16815–16827. [CrossRef]
44. Shestakova, P.; Martineau, C.; Mavrodinova, V.; Popova, M. Solid state NMR characterization of zeolite Beta based drug formulations containing Ag and sulfadiazine. *RSC Adv.* **2015**, *5*, 81957–81964. [CrossRef]

45. Crutchfield, M.M.; Callis, C.F.; Irani, R.R.; Roth, G.C. Phosphorus Nuclear Magnetic Resonance Studies of Ortho and Condensed Phosphates. *Inorg. Chem.* **1962**, *4*, 813–817. [CrossRef]
46. Vallet-Regí, M.; Izquierdo-Barba, I.; Rámila, A.; Pérez-Pariente, J.; Babonneau, F.; González-Calbet, J.M. Phosphorous-doped MCM-41 as bioactive material. *Solid State Sci.* **2005**, *7*, 233–237. [CrossRef]
47. Nagashima, H.; Martineau-Corcos, C.; Tricot, G.; Trébosc, J.; Pourpoint, F.; Amoureux, J.-P.; Lafon, O. Recent Development in NMR Studies of Aluminophosphates. In *Annual Reports on NMR Spectroscopy*; Webb, G., Ed.; Academic Press: London, UK, 2018; Volume 94, Chapter 4.
48. Porcino, M.; Christodoulou, I.; Le Vuong, M.D.; Gref, R.; Martineau-Corcos, C. New insights on the supramolecular structure of highly porous core-shell drug nanocarriers using solid-state NMR spectroscopy. *RSC Adv.* **2019**, *4*, 32472–32475. [CrossRef]
49. Li, X.; Porcino, M.; Qiu, J.; Constantin, D.; Martineau-Corcos, C.; Gref, R. Doxorubicin-loaded metal-organic frameworks nanoparticles with engineered cyclodextrin coatings: Insights on drug location by solid state NMR spectroscopy. *Nanomaterials* **2021**, *11*, 945. [CrossRef] [PubMed]
50. Anand, R.; Malanga, M.; Manet, I.; Tuza, K.; Aykaç, A.; Ladavière, C.; Fenyvesi, E.; Vargas-Berenguel, A.; Gref, R.; Monti, S. Citric acid–γ-cyclodextrin crosslinked oligomers as carriers for doxorubicin delivery. *Photochem. Photobiol. Sci.* **2013**, *12*, 1841–1854. [CrossRef]
51. Anand, R.; Borghi, F.; Manoli, F.; Manet, I.; Agostoni, V.; Reschiglian, P.; Gref, R.; Monti, S. Host–Guest Interactions in Fe(III)-Trimesate MOF Nanoparticles Loaded with Doxorubicin. *J. Phys. Chem. B* **2014**, *118*, 8532–8539. [CrossRef]

Article

Magnetic Resonance Methods as a Prognostic Tool for the Biorelevant Behavior of Xanthan Tablets

Urša Mikac [1],* and Julijana Kristl [2]

[1] Jožef Stefan Institute, Jamova 39, 1000 Ljubljana, Slovenia
[2] Faculty of Pharmacy, University of Ljubljana, Aškerčeva 7, 1000 Ljubljana, Slovenia; Julijana.Kristl@ffa.uni-lj.si
* Correspondence: urska.mikac@ijs.si

Academic Editor: Clarisse Ribeiro
Received: 20 November 2020; Accepted: 10 December 2020; Published: 11 December 2020

Abstract: Hydrophilic matrix tablets with controlled drug release have been used extensively as one of the most successful oral drug delivery systems for optimizing therapeutic efficacy. In this work, magnetic resonance imaging (MRI) is used to study the influence of various pHs and mechanical stresses caused by medium flow (at rest, 80, or 150 mL/min) on swelling and on pentoxifylline release from xanthan (Xan) tablets. Moreover, a bimodal MRI system with simultaneous release testing enables measurements of hydrogel thickness and drug release, both under the same experimental conditions and at the same time. The results show that in water, the hydrogel structure is weaker and less resistant to erosion than the Xan structure in the acid medium. Different hydrogel structures affect drug release with erosion controlled release in water and diffusion controlled release in the acid medium. Mechanical stress simulating gastrointestinal contraction has no effect on the hard hydrogel in the acid medium where the release is independent of the tested stress, while it affects the release from the weak hydrogel in water with faster release under high stress. Our findings suggest that simultaneous MR imaging and drug release from matrix tablets together provide a valuable prognostic tool for prolonged drug delivery design.

Keywords: hydrophilic matrix tablets; magnetic resonance; hydrogel; drug release; biorelevant dynamic conditions

1. Introduction

Matrix tablets are considered as a dosage form that may maximize the bioavailability of drugs and enable good adjustment of the doses, patient friendly administration, and relatively low manufacturing cost and are, therefore, attractive delivery systems, which are not fully understood yet. The principal objective of dosage form design is to achieve a predictable release and therapeutic response of a drug included in a formulation that is capable of large-scale manufacture with reproducible product quality. The goal of drug administration is to achieve and maintain a plasma drug concentration within the therapeutic window. With conventional oral drug delivery systems, the drug level in the plasma rises after each administration of the tablet and then decreases until the next administration; therefore, frequent dosing is required. However, drug delivery is not easily controlled. In order to avoid the "peaks and valleys" of standard dosage forms, innovative drug delivery systems have been formulated by testing a wide array of hydrophilic polymers and production strategies to prolong drug release and for safe and efficient use. Additionally, matrix tablets with prolonged release are well accepted by patients due to their reduced frequency of administration.

Although the sustained release system was first described in 1952, intensive development is still taking place in this field [1]. Among different types of prolonged drug delivery systems, the most used ones are hydrophilic matrix tablets that are formulated by using a hydrophilic polymer as a material that directs the release kinetics. Basically, these tablets are composed of two major components, a polymer matrix carrier that swells upon exposure to the solvent or physiological environment and the embedded drug molecules that are gradually released [2]. Among hydrophilic polymers, non-ionic polymers of semi-synthetic origin such as hydrophilic derivatives of cellulose ethers [3–5] and high molecular weight polyethylene oxides (PEOs) [6] have been extensively studied. Polymers of a natural origin, such as agar–agar, alginate, carrageenan, or xanthan (Xan), are becoming more important in the development of matrix tablets [7,8]. Xan is a well-known and already widely used biopolymer produced by the bacterium Xanthomonas campestris biotechnologically [9]. It is a polysaccharide consisting of a cellulose backbone and trisaccharide side chains containing glucuronic acids and a pyruvate group, which are mainly responsible for its anionic polyelectrolyte character. The native ordered and rigid conformation of Xan chains has been reported to exist as a double-stranded helix [10]. In water, the rigid helix-coil structure transforms into the flexible coils, whose stability and physical properties are strongly influenced by the pH and the ionic environment [7,11].

To achieve the optimal performance of the hydrophilic matrix tablets, knowledge of the physicochemical, technological, and physiological parameters that influence the release kinetics is essential for their design [1,12]. One of the key factors for drug release kinetics are the hydrogel properties governed by the hydration of the matrix, mainly defined at the microscopic level by the pore size (i.e., mean linear distance between crossed polymer chains) and their distribution, which are determined by the polymer chain dynamics. These also define the hydrogel swelling kinetics and structure, thus the diffusion of drug through the hydrogel layer [13]. In addition to the hydrodynamic ratio of the drug volume and pore size in the hydrogel, the amount of free water in the network is also responsible for drug diffusion, since the drug has to dissolve before it can diffuse through the hydrogel. Therefore, detailed studies of the amount of free medium within hydrophilic network systems available for drug dissolution are essential to optimize and predict the release kinetics. Methods available for studying the type of water (free water among polymeric chains and water bound to the hydrophilic groups of the polymer) are differential dynamic calorimetry, nuclear magnetic resonance, and Fourier transform infrared spectroscopy [14–17]. To follow the swelling and the thickness of the hydrogel layers, which also determine drug release, methods such as rheology, gravimetric methods, texture analyzer, optical imaging, ultrasound, microcomputed tomography (micro CT), or magnetic resonance imaging (MRI) are used [4,5,16,18–28]. While the light-based techniques provide high spatial resolution and also enable the spectroscopic characterization of superficial layers of the tablet that are limited by the light penetration depth, MRI enables non-invasive tomographic characterization of the entire hydrogel structure, however with a comparatively larger voxel size. Therefore, experimental studies of tablet swelling that employ various complementary techniques to cover different spatio-temporal scales are commonly combined with a number of different mathematical modelling approaches [29–32]. On the other hand, also biorelevant conditions may not be neglected.

Drug formulations administered orally pass through a series of gastrointestinal (GI) compartments with varied contraction forces [33]. It is reasonable to know if the gastrointestinal contraction forces affect hydrogel structure and drug release during GI transit [34]. An approach to distinguish between the role of hydrodynamics and mechanical stresses similar to the contraction forces of the GI tract on drug release from modified release dosage forms was presented by Takieddin's group [35]. Measuring the effects of these forces in an in vitro setting paves the way for future improvements to drug delivery systems and methods that are more representative of in vivo conditions.

In this work, magnetic resonance imaging (MRI) is used to study hydrogel thickness and drug release from matrix tablets composed of Xan polymer and a non-ionic, highly water soluble drug, pentoxifylline (PF). More precisely, the swelling dynamics of the Xan matrix tablets during hydration and PF release influenced by pH, ionic strength, and mechanical stress caused by medium flow (without flow and with 80 or 150 mL/min flow) are analyzed. To follow the swelling and release under the same experimental conditions and simultaneously, an MRI flow-through system is designed.

For these two media differing in pH and ionic strength, the values where the largest differences in Xan swelling and drug release kinetics have been observed in previous studies [18,22] are used: pure water (i.e., with pH 5.7 and ionic strength of 0 M) and an acid medium (i.e., HCl with pH 1.2 and ionic strength of 0.28 M). To show the whole picture of the effectiveness of MR methods in the research on hydrophilic matrix tablets, the results of this study are combined with our previous MR studies of Xan tablets [18,22,36].

2. Hydrophilic Matrix Tablets

In their simplest form, hydrophilic matrix tablets are prepared by the compression of a powdered mixture of the drug, a hydrophilic polymer, and other excipients [37]. They do not decompose in contact with body fluids (medium), but a hydrated polymer layer is formed on the surface, which slows the further penetration of the medium and controls the release of the drug. During slow tablet swelling, water diffuses into the tablet, where it is locally taken up by the dry polymer matrix. The medium uptake results in a medium-mediated glassy-to-rubbery phase transition of the polymer matrix. The transition is microscopically manifested by disentangling of individual polymer chains and their cross-linking via medium-mediated intermolecular bridges, while macroscopically, the process is associated with the formation of a hydrogel layer around the tablet core in the glassy state that represents the drug reservoir [38]. The hydrogel layer slows down the penetration of the medium into the tablet and thus the dissolution and diffusion of the drug from the tablet. On the surface of the matrix, the polymer chains disentangle, erode and pass into the surrounding medium.

The tablet swelling is associated with the formation of four characteristic fronts established in the tablet's interior during its exposure to the medium [39]. These fronts are: the penetration front that is determined with the maximal reach of the diffusing medium into the tablet interior, i.e., the border between dry and hydrated polymer that is still in a glassy state; the swelling front at the interface between the hydrated glassy polymer and the polymer in a rubbery state (polymer that has taken up a sufficient amount of medium to lower the glass transition temperature T_g below the experimental temperature); the diffusion front at the interface between undissolved and dissolved drug in the hydrogel layer and the erosion front that contains completely swollen polymer matrix layers in a rubbery state and is in a contact with the bulk medium (Figure 1). As the tablet, immersed in the medium, is subjected to a time-dependent ingress of water molecules into its interior, the positions of these four fronts also become time dependent and determine the efficacy of a controlled drug release. Initially, when the swelling process predominates, the penetration and swelling fronts move toward the center of the tablet and the diffusion and erosion fronts outward. When the concentration of the medium exceeds the critical value, the polymer chains on the hydrogel surface begin to disentangle, and the diffusion and erosion fronts gradually move towards the center of the tablet until the entire tablet disintegrates.

Drug release from polymer matrix tablets is a very complex process determined by different factors, such as hydrogel layer properties and thickness, polymer-medium and polymer-drug interaction, drug solubility, etc. These factors result in different drug release mechanisms from the hydrophilic matrix tablets [40], such as swelling and erosion controlled release of the drug [1]. In swelling controlled systems, the drug diffuses through the hydrogel layer, while in the erosion controlled systems, the pores in the hydrogel layer are too small to enable drug diffusion. Often, however, the processes of diffusion and erosion occur simultaneously [25].

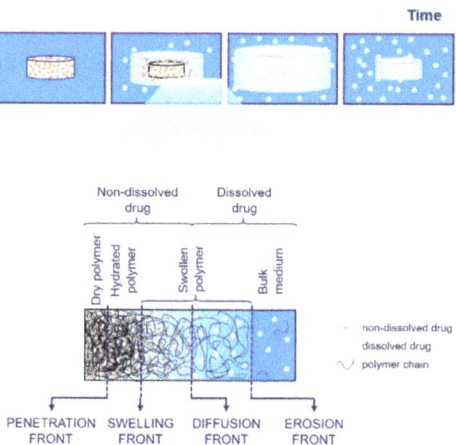

Figure 1. Schematic representation of swelling and drug release from the hydrophilic matrix tablets with different layers and fronts.

3. Results

The MRI studies using flow-through cells showed some different characteristics of swelling under mechanical stress than in conditions without it. The water profiles in the tablet (polymer matrix) as a function of position and swelling time were obtained through the magnitude of the MRI signal. The signal intensity of 2D MRI depends on the physical state of water in the sample and the chosen experimental conditions. Therefore, it varied through the sample (Figure 2a, upper row). In the MR image, the dry tablet core was black (zero signal), since there was only a small amount of water, and T_2 was too short to give any MRI signal. As the amount of water increased in the hydrogel layer, the signal intensity first increased (from dark grey to white) and, at very high water content, decreased again (from white to grey) due to the long T_1 and short repetition time used in the MRI sequence. Therefore, the brightness of the hydrogel was dependent on the Xan concentration. The signal intensity in the region of pure medium was grey under no flow conditions. The medium flow during MR signal acquisition caused the motional artefacts in the area where only the medium was present (Figure 2a, second and third rows: black regions above the grey hydrogel layer). On the other hand, no artefacts were observed in the hydrogel region. In order to confirm that the MRI signal in the hydrogel region was not affected by the flow, two consecutive images were compared, first without flow and immediately after with the flow. The same hydrogel thicknesses were determined from both images with and without the flow. At the end of the experiment at 150 mL/min flow, it was stopped, and another image without flow was acquired. In water, the MR signal intensity was uniform through the whole sample indicating that after 24 h, no hydrogel layer existed anymore, and disentangled polymer chains were uniformly distributed over the whole cell. The situation was different in acid medium where, after 24 h, the signal intensity still varied through the sample, showing that the hydrogel layer still existed even after 24 h, despite the strong flow.

Moving front positions and the hydrogel thicknesses for Xan tablets swelling in water and in the acid medium at no-flow and at flow rates of 80 mL/min and 150 mL/min were determined from 1D single point imaging (SPI), T_2 values, and 2D images and are shown in Figure 2b. The results show that the thickness of the hydrated hydrogel layers was influenced by the type of medium, flow rate, and swelling time. The hydrated hydrogel layer in water was 1.5 times thicker than in the acid medium. Within the same medium, the flow rate of 80 mL/min did not affect the hydrogel thickness. The situation was different in water at 150 mL/min, where the hydrogel was significantly thinner, and the tablet fully disintegrated at 15 h of swelling. On the other hand, the hydrogel thicknesses were the same for all tested mechanical stresses in the acid medium (Figure 2b).

The drug release profiles at different swelling times were obtained by medium withdrawal during the MRI measurements. Drug release kinetics at both flow rates were compared for Xan tablets swelling in both media (Figure 2b). In water, drug release was significantly faster at higher flow, while the same drug release kinetics were observed at both flow rates in the acid medium. The results of drug release kinetics were compared with Equation (1), and the results of the fitting obtained for each medium are shown in Table 1. In water, the value of the exponent $n \geq 1$ indicates that drug release was controlled by polymer erosion, and the kinetic constant k was higher at a higher flow rate where drug release was faster. The exponent $n = 0.6$ in the acid medium indicates that drug diffusion through the hydrogel layer was the prevailing mechanism for drug release in the acid medium.

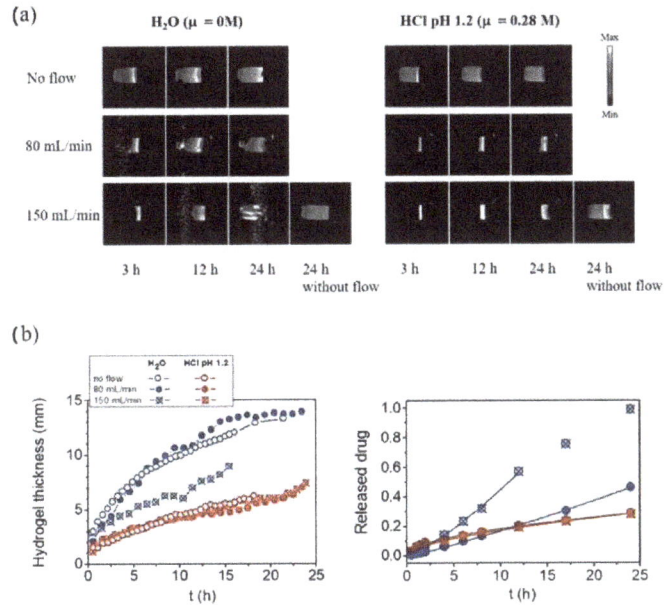

Figure 2. (a) 2D MR images of xanthan tablets during swelling at different flow rates in water and in the acid medium and (b) corresponding hydrogel layer thicknesses and fractions of the released pentoxifylline drug at different swelling times and mechanical stresses.

Table 1. Values of the fitting parameters determined from drug release data measured at two different flow rates and fit by Equation (1) in water and in the acid medium.

Flow Rate	H_2O ($\mu = 0$ M)		HCl pH 1.2 ($\mu = 0.28$ M)	
	k (h^{-n})	n	k (h^{-n})	n
80 mL/min	0.012	1	0.055	0.6
150 mL/min	0.025	1.2	0.048	0.6

Thus, the results indicate that the mechanical stress of the medium flow up to 150 mL/min reduced hydrogel thickness in water and had no effect on the hydrogel thickness in the acid medium, which represents the formation of a remarkably stronger hydrogel in the acid medium. The results agree with the drug release studies, which showed that the release of highly soluble PF drug was significantly increased at higher mechanical forces in water (the rate constant k increased from 0.012 h^{-1}

at 80 mL/min to 0.025 h$^{-1.2}$ at a flow rate of 150 mL/min) where the main release mechanism was polymer erosion, owing to the weaker hydrogel layer being more susceptible to mechanical forces. In the acid medium, where the main release mechanism was drug diffusion through the hydrogel, the effect of the mechanical stress on the rate constant was negligible.

4. Discussion

MR provides information about the hydrogel properties and enables following the moving fronts' (penetration, swelling, and erosion fronts) positions and the hydrogel's properties in situ. The spin-spin (T_2) and spin-lattice (T_1) relaxation times measured at the Larmor frequency in the MHz range together with the measurements of the frequency dependent T_1 in the kHz range using the fast field-cycling NMR relaxometry technique provide information about the molecular dynamics over a very wide frequency range. The ability to measure dynamics over a wide frequency range is very important in hydrogels where the molecular dynamics are particularly complicated and can range from the very fast free water dynamics, to slower bound water dynamics, as well as different types of polymer-chain dynamics (fast fluctuations of side groups, different types of backbone motions). The information about medium and polymer dynamics is important for the design of matrix tablets with the desired drug release kinetics since the dynamics determine the diffusion pathways for the drug in the hydrogel layer. By using the MR imaging techniques, the positions of the moving fronts and the hydrogel layer thickness together with polymer concentration profiles across the hydrogel layer during the swelling of the polymer tablets can be determined. MRI experiments using a flow through cell enable simultaneous measurements of the swelling and drug release kinetics, as well as determining the effect of mechanical stress caused by the flow on the hydrogel layer behavior. To better understand the hydrogel impact on drug release, a mathematical model that combines the polymer swelling kinetics and drug release can be applied.

The ability of MR in the research of hydrophilic matrix tablets and the information that the method can provide are shown in the case of xanthan matrix tablets (Figure 3). Different MR modalities were used to determine and understand the behavior of Xan tablets in media differing in pH and ionic strengths. Besides, the influence of the addition of the highly soluble model drug pentoxifylline on the hydrogel properties was also investigated (Figure 3). Xan is a natural polymer widely used in pharmacy. Due to its polyelectrolyte nature with a pKa of 3.1, its swelling depends on the pH and ionic strength of the medium. In our previous studies, six media that mimic gastric conditions were used [18,22]. It has been shown that the largest differences in Xan swelling and drug release are between a medium with pH 1.2 and pH 3.0; between pH 3.0 and water with pH 5.7, the differences are extremely small. No differences were observed between water and the medium with pH 7.4, so this medium was not included in further investigation with Xan. Therefore, the results of Xan where the largest differences were observed, HCl medium with pH 1.2 and ionic strength µ = 0.28 M and purified water with pH 5.7 and µ = 0 M, will be discussed. By measuring the spin-spin and spin-lattice relaxation times of hydrogels with a known Xan concentration, we found that the high frequency dynamics (measured by T_1 at 100 MHz) were the same in both media, while the dynamics at low frequencies depended on the medium properties [18,36]. The effect of medium pH and ionic strength resulted in slower medium and polymer-chain dynamics with a higher amount of free water available for drug dissolution in the hydrogels prepared with the acid medium than in the hydrogels prepared with water. No impact on the dynamics was observed after the addition of the PF drug in Xan hydrogels with a Xan to PF ratio of 1:1. To determine how the medium properties affect the swelling kinetics of the Xan tablets, MR imaging was used. By knowing how the NMR relaxations change with the polymer concentrations (concentration dependencies of T_1 and T_2 measured at 100 MHz), the MR imaging parameters (*TE* and *TR*), which provide the best contrast between the bulk medium, hydrogel, and dry tablet, can be determined. In addition, the polymer concentration profiles across the formed hydrogel can also be calculated for different swelling times. MRI showed that the pH and ionic strength of the media significantly influenced hydrogel layer thickness, i.e., by decreasing the pH or increasing the ionic strength of the medium, the hydrogel layer thickness decreased. Different hydrogel layer thicknesses were the result of different erosion front positions, while the positions of the penetration and swelling fronts were independent of the medium properties [18,22]. These can also be seen from the concentration profiles where the concentration of Xan in the acid medium decreased much faster at low polymer concentrations than in water (Figure 3). The influence of the mechanical stress on the hydrogel layer formation showed that the hydrogel layer was weaker in water, where the hydrogel layer was significantly thinner at a high flow rate, and the tablet disintegrated after 15 h (e.g., the hydrogel thickness was 7.4 mm at 80 mL/min and 5.3 mm at a 150 mL/min flow rate after 5 h and 10.7 mm at 80 mL/min and 6 mm at a 150 mL/min flow rate after 10 h of swelling), than in the acid medium where the flow up to 150 mL/min did not affect the hydrogel layer (e.g., the hydrogel thickness was 3.4 mm after 5 h and 4.4 mm after 10 h of swelling for both flow rates). The addition of PF drug in the Xan tablets showed that at a high drug amount (Xan:PF = 1:1), the hydrogel layer was thinner in media with a low pH or increased ionic strength than in the empty Xan tablets with no PF influence observed in water [22].

By using a mathematical model, which combines the polymer swelling kinetics and drug release that account for the superposition of Fickian diffusion and polymer erosion processes, the results of Xan swelling and the PF release kinetics can be better understood [22]. The obtained model parameters showed that in water, the fraction of dissolved drug was lower, and the main release mechanism was erosion; whereas in the acid medium, the amount of medium in the tablet that was available for PF dissolution was higher, and the main mechanism was drug diffusion. The parameters also clarify the reason for the unexpected behavior; i.e., the same hydrogel thickness and slightly faster drug release in water, on the one hand, and the thinner hydrogel and the same drug release kinetics in the acid medium for high (Xan:PF = 1:1) compared to low (Xan:PF = 3:1) drug loading, on the other hand. In water, more medium was available for drug dissolution and, consequently, for drug release in tablets with high drug loading, causing faster drug release despite the same hydrogel layer thickness. In acid medium, the smaller diffusion contribution led to a thinner hydrogel in the tablets with high drug loading; despite the slower drug diffusion, drug release was the same for both drug loadings owing to the higher amount of dissolved drug and the thinner hydrogel layer in tablets with high drug loading compared to the tablets with low drug loading.

Different MR measurements thus showed that more restricted mobility in the acid medium resulted in a thinner hydrogel layer, which was more resistant to erosion, with the drug release mainly governed by drug diffusion through the hydrogel layer. In water, the higher water and polymer-chain mobility resulted in a weaker and homogeneous hydrogel layer that was less resistant to erosion, leading to erosion controlled drug release. At low mechanical stress, the release was faster in the case of the diffusion controlled mechanism (acid medium) than in the erosion mechanism (water): at an 80 mL/min flow rate, the fraction of released drug was 0.06 in water and 0.11 in the acid medium after 4 h and 0.13 in water and 0.16 in the acid medium after 8 h of swelling. At longer times when the hydrogel was more diluted and, consequently, weaker, the erosion controlled release in water became faster than the diffusion controlled release in the acid medium (after 24 h, the fraction of released drug was 0.46 in water and 0.28 in the acid medium). When strong mechanical forces were applied, causing more pronounced hydrogel erosion in the weaker hydrogel in water, drug release was accelerated. This led to faster drug release in water than in the acid medium (e.g., at a 150 mL/min flow rate, the fraction of released drug was 0.15 in water and 0.11 in the acid medium after 4 h of swelling and 0.32 in water and 0.16 in the acid medium after 8 h of swelling), where the hydrogel was so robust that the flow up to 150 mL/min did not affect the hydrogel layer, and the main release mechanism remained drug diffusion.

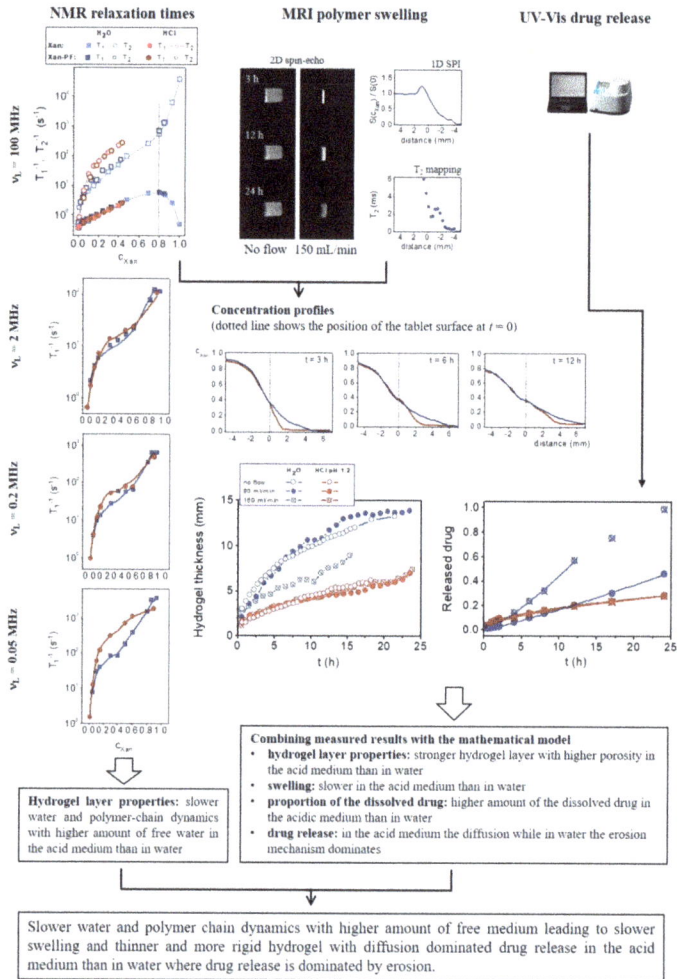

Figure 3. The use of MR methods in the study of the matrix tablets of the xanthan (Xan) polymer.

5. Materials and Methods

5.1. Materials

Xanthan with a MW of 2×10^6 was obtained from Sigma-Aldrich Chemie, Munich, Germany. A model drug, pentoxifylline (PF) (MW = 278.31) with a solubility in water at 25 °C of 77 mg/mL, was supplied by Krka, d.d. Novo mesto, Slovenia. For swelling and release experiments, two different media were used: purified H_2O with pH 5.7 and ionic strength of 0 M and HCl at pH 1.2 with increased ionic strength (11.7 g of NaCl per 1000 mL of HCl medium) resulted in ionic strength $\mu = 0.28$ M.

5.2. Preparation of Xanthan Matrix Tablets

Xan and the drug (PF) were mixed homogeneously using a laboratory model drum blender (Electric Inversina Tumbler Mixer, Paul Schatz principle, BioComponents Inversina 2L, Bioengineering AG, Wald, Switzerland), and cylindrical flat-faced tablets with composition of Xan:PF = 1:1 (200 mg of Xan and 200 mg of PF) were prepared by direct compression (SP 300, Kilian and Co., Cologne, Germany) to form tablets with a diameter of 12 mm and a crushing strength of 100 N ± 10 N (Tablet hardness tester, Vanderkamp, VK 200, USA).

5.3. MRI of Xanthan Tablets during Swelling

The MRI experiments were performed at room temperature with a superconducting 2.35 T (^{1}H NMR frequency of 100 MHz) horizontal bore magnet (Oxford Instruments, Oxon, U.K.) equipped with gradients and RF coils for MR macroscopy (Bruker, Ettlingen, Germany) using a TecMag Apollo (Tecmag, Houston, TX, USA) MRI spectrometer.

To follow the swelling and release of the drug from Xan tablets under the same experimental conditions and simultaneously, a flow-through system was designed. The tablet was inserted in a small MRI flow-through cell so that only one circular cylinder surface was exposed for the medium penetration. The flow-through cell was connected with the container with 900 mL of medium using plastic tubes. To determine drug release from the same Xan tablets and at the same time as MRI, five milliliter samples were withdrawn at predetermined time intervals. This means that the MRI scan and dissolution medium withdrawing were performed simultaneously. Gastrointestinal tract (GIT) mechanical stress was simulated with different flow rates of the dissolution medium, which was driven by a peristaltic pump (Anko, Bradenton, FL, USA) and controlled at two different flow rates: 80 ± 2 mL/min and 150 ± 2 mL/min. To determine the influence of the stress on hydrogel thickness and drug release, the experiments were also performed without flow. The first MR image was taken approximately 10 min after the tablet came in contact with the medium and then every 30 min for 24 h. The experiments were performed at room temperature ($\approx$22 °C).

To follow the moving (penetration, swelling, and erosion) front positions, two different MRI methods were used as described in our previous paper [18]: a 2D multi-echo pulse sequence to determine the erosion front and a 1D SPI T_2 mapping sequence to determine the position of the swelling and penetration fronts. Imaging parameters for the 2D multi-echo pulse sequence were: an echo time (*TE*) of 6.2 ms, a repetition time (*TR*) of 200 ms, a field of view of 50 mm with an in-plane resolution of 200 μm, and a slice thickness of 3 mm. For the 1D SPI, a single point on the free induction decay was sampled at the encoding time t_p of 0.17 ms after the radiofrequency detection pulse α of 20° with TR = 200 ms, and the inter-echo time was varied from 0.3 ms to 10 ms. The field of view was 45 mm with an in-plane resolution of 350 μm. The position of the erosion front was obtained from the one-dimensional signal intensity profiles of the 2D MR images; the position of the swelling front was determined from 1D SPI T_2 maps at T_2 = 2.7 ms; and the position of the penetration front was determined from the signal intensity profiles from SPI measurements at an inter-echo time of 0.3 ms (Figure 4).

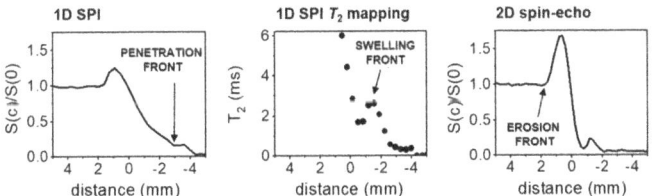

Figure 4. Determination of the moving fronts: the penetration front was determined from 1D SPI normalized signal intensity; the swelling front was determined from the T_2 profile; and the erosion front was obtained from the one-dimensional signal intensity profile along the horizontal direction of the 2D MR image. Zero on the x axis represents the surface of the tablet at the beginning of the experiment.

5.4. Drug Release from Xanthan Tablets

To determine drug release from the Xan tablets, the collected samples of the outflow medium from the MRI flow-through cell were filtered through a filter with 0.45 μm pores. Drug release was monitored as a function of time using the HP Agilent 8453 Diode Array UV-Vis Spectrophotometer, Waldbronn, Germany, to measure the absorbance of PF at 274 nm.

Drug release profiles were analyzed using the Korsmeyer–Peppas equation [41,42]:

$$\frac{M(t)}{M(\infty)} = k \cdot t^n, \qquad (1)$$

where $M(t)/M(\infty)$ is the fraction of the released drug at time t, k is the rate constant, and n is the diffusion exponent, which indicates the general release mechanism: $n = 0.5$ indicates Fickian diffusion controlled drug release (Case I), and $n = 1.0$ indicates Case II transport (erosion controlled drug release). Case I release occurs by molecular diffusion of the drug due to a chemical potential gradient, whereas the mechanism driving Case II drug release is the swelling or relaxation of polymeric chains. Values between $0.5 < n < 1.0$ indicate an anomalous (non-Fickian or both diffusion/erosion) controlled drug release. Equation (1) is the short time approximation, which is valid only up to the fraction of 0.6 of released drug.

6. Conclusions

NMR can be used to determine the structure of the hydrogel layer and MR imaging as a non-destructive and fast enough method, which allows monitoring the swelling of hydrophilic matrix tablets in situ. Here, matrix tablets composed of xanthan polymer and a non-ionic, highly water soluble drug pentoxifylline are investigated using the MRI flow-through cell with simultaneous drug release testing. Inclusion of in vitro mechanical stress simulating GI contraction forces during dissolution testing allows for better in vivo prediction.

The swelling dynamics of the Xan matrix tablets during hydration and PF release influenced by the medium properties and mechanical stress caused by medium flow (without flow and with 80 or 150 mL/min flow) were analyzed. The results of this study together with the previous MR studies of Xan tablets show the more restricted mobility of Xan polymer chains and water molecules in the acid medium than in water, which results in a thinner hydrogel layer more resistant to erosion. Xan matrix in the acid medium releases PF mainly by drug diffusion through the strong hydrogel layer. Xan in water swells faster due to the higher water and polymer-chain mobility, resulting in a weaker hydrogel, less resistant to erosion, which causes the erosion controlled drug release. High mechanical stress affects the release from the soft hydrogel in water, while the release of the harder hydrogel in the acid medium remains the same, independent of mechanical stress.

The combination of the obtained results together with the results of other analytical methods and applied mathematical models enables the understanding of polymer systems at the molecular, microscopic, and macroscopic levels and therefore contributes to the development of efficient systems with the desired drug release kinetics.

Author Contributions: The manuscript was written through the contributions of all of the authors. Both authors have read and agreed to the published version of the manuscript.

Funding: This research was funded by the Slovenian Research Agency through the Research Core Funding Nos. P1-0189, P1-0060, and Project J1-9194.

Acknowledgments: The authors thank Kanza Awais for proofreading an earlier version of the manuscript.

Conflicts of Interest: The authors declare no conflict of interest.

References

1. Maderuelo, C.; Zarzuelo, A.; Lanao, J.M. Critical factors in the release of drugs from sustained release hydrophilic matrices. *J. Control. Release* **2011**, *154*, 2–19. [CrossRef] [PubMed]
2. Harland, R.S.; Gazzaniga, A.; Sangalli, M.E.; Colombo, P.; Peppas, N.A. Drug Polymer Matrix Swelling and Dissolution. *Pharm. Res.* **1988**, *5*, 488–494. [CrossRef] [PubMed]
3. Doelker, E. Cellulose Derivatives. *Adv. Polym. Sci.* **1993**, *107*, 199–265.

4. Baumgartner, S.; Lahajnar, G.; Sepe, A.; Kristl, J. Quantitative evaluation of polymer concentration profile during swelling of hydrophilic matrix tablets using H-1 NMR and MRI methods. *Eur. J. Pharm. Biopharm.* **2005**, *59*, 299–306. [CrossRef] [PubMed]
5. Fyfe, C.A.; Blazek, A.I. Investigation of hydrogel formation from hydroxypropylmethylcellulose (HPMC) by NMR spectroscopy and NMR imaging techniques. *Macromolecules* **1997**, *30*, 6230–6237. [CrossRef]
6. Maggi, L.; Bruni, R.; Conte, U. High molecular weight polyethylene oxides (PEOs) as an alternative to HPMC in controlled release dosage forms. *Int. J. Pharm.* **2000**, *195*, 229–238. [CrossRef]
7. Coviello, T.; Alhaique, F.; Dorigo, A.; Matricardi, P.; Grassi, M. Two galactomannans and scleroglucan as matrices for drug delivery: Preparation and release studies. *Eur. J. Pharm. Biopharm.* **2007**, *66*, 200–209. [CrossRef]
8. Baumgartner, S.; Pavli, M.; Kristl, J. Effect of calcium ions on the gelling and drug release characteristics of xanthan matrix tablets. *Eur. J. Pharm. Biopharm.* **2008**, *69*, 698–707. [CrossRef]
9. Garcia-Ochoa, F.; Santos, V.E.; Casas, J.A.; Gomez, E. Xanthan gum: Production, recovery, and properties. *Biotechnol. Adv.* **2000**, *18*, 549–579. [CrossRef]
10. Capron, I.; Brigand, G.; Muller, G. About the native and renatured conformation of xanthan exopolysaccharide. *Polymer* **1997**, *38*, 5289–5295. [CrossRef]
11. Talukdar, M.M.; Michoel, A.; Rombaut, P.; Kinget, R. Comparative study on xanthan gum and hydroxypropylmethyl cellulose as matrices for controlled-release drug delivery I. Compaction and in vitro drug release behaviour. *Int. J. Pharm.* **1996**, *129*, 233–241. [CrossRef]
12. Ghori, M.U.; Conway, B.R. Hydrophilic Matrices for Oral Control Drug Delivery. *Am. J. Pharmacol. Sci.* **2015**, *3*, 103–109. [CrossRef]
13. Koetting, M.C.; Peters, J.T.; Steichen, S.D.; Peppas, N.A. Stimulus-responsive hydrogels: Theory, modern advances, and applications. *Mat. Sci. Eng. R* **2015**, *93*, 1–49. [CrossRef] [PubMed]
14. Higuchi, A.; Iijima, T. Dsc Investigation of the States of Water in Polyvinyl Alcohol-Co-Itaconic Acid) Membranes. *Polymer* **1985**, *26*, 1833–1837. [CrossRef]
15. Gun'ko, V.M.; Savina, I.N.; Mikhalovsky, S.V. Properties of Water Bound in Hydrogels. *Gels* **2017**, *3*, 37. [CrossRef]
16. Wray, P.S.; Clarke, G.S.; Kazarian, S.G. Application of FTIR Spectroscopic Imaging to Study the Effects of Modifying the pH Microenvironment on the Dissolution of Ibuprofen from HPMC Matrices. *J. Pharm. Sci.* **2011**, *100*, 4745–4755. [CrossRef]
17. Saalwachter, K.; Chasse, W.; Sommer, J.U. Structure and swelling of polymer networks: Insights from NMR. *Soft Matter* **2013**, *9*, 6587–6593. [CrossRef]
18. Mikac, U.; Sepe, A.; Kristl, J.; Baumgartner, S. A new approach combining different MRI methods to provide detailed view on swelling dynamics of xanthan tablets influencing drug release at different pH and ionic strength. *J. Control. Release* **2010**, *145*, 247–256. [CrossRef]
19. Nott, K.P. Magnetic resonance imaging of tablet dissolution. *Eur. J. Pharm. Biopharm.* **2010**, *74*, 78–83. [CrossRef]
20. Knoos, P.; Topgaard, D.; Wahlgren, M.; Ulvenlund, S.; Piculell, L. Using NMR Chemical Shift Imaging To Monitor Swelling and Molecular Transport in Drug-Loaded Tablets of Hydrophobically Modified Poly(acrylic acid): Methodology and Effects of Polymer (In)solubility. *Langmuir* **2013**, *29*, 13898–13908. [CrossRef]
21. Mikac, U.; Kristl, J.; Baumgartner, S. Using quantitative magnetic resonance methods to understand better the gel-layer formation on polymer-matrix tablets. *Expert Opin. Drug Deliv.* **2011**, *8*, 677–692. [CrossRef] [PubMed]
22. Mikac, U.; Sepe, A.; Baumgartner, S.; Kristl, J. The Influence of High Drug Loading in Xanthan Tablets and Media with Different Physiological pH and Ionic Strength on Swelling and Release. *Mol. Pharm.* **2016**, *13*, 1147–1157. [CrossRef] [PubMed]
23. Zhang, Q.L.; Gladden, L.; Avalle, P.; Mantle, M. In vitro quantitative H-1 and F-19 nuclear magnetic resonance spectroscopy and imaging studies of fluvastatin (TM) in Lescol (R) XL tablets in a USP-IV dissolution cell. *J. Control. Release* **2011**, *156*, 345–354. [CrossRef] [PubMed]
24. Kulinowski, P.; Mlynarczyk, A.; Jasinski, K.; Talik, P.; Gruwel, M.L.H.; Tomanek, B.; Weglarz, W.P.; Dorozynski, P. Magnetic Resonance Microscopy for Assessment of Morphological Changes in Hydrating Hydroxypropylmethylcellulose Matrix Tablets In Situ-Is it Possible to Detect Phenomena Related to Drug Dissolution Within the Hydrated Matrices? *Pharm. Res.* **2014**, *31*, 2383–2392. [CrossRef] [PubMed]

25. Gao, P.; Skoug, J.W.; Nixon, P.R.; Ju, T.R.; Stemm, N.L.; Sung, K.C. Swelling of hydroxypropyl methylcellulose matrix tablets. 2. Mechanistic study of the influence of formulation variables on matrix performance and drug release. *J. Pharm. Sci.* **1996**, *85*, 732–740. [CrossRef]
26. Lamberti, G.; Cascone, S.; Cafaro, M.M.; Titomanlio, G.; d'Amore, M.; Barba, A.A. Measurements of water content in hydroxypropyl-methyl-cellulose based hydrogels via texture analysis. *Carbohyd. Polym.* **2013**, *92*, 765–768. [CrossRef]
27. Konrad, R.; Christ, A.; Zessin, G.; Cobet, U. The use of ultrasound and penetrometer to characterize the advancement of swelling and eroding fronts in HPMC matrices. *Int. J. Pharm.* **1998**, *163*, 123–131. [CrossRef]
28. Zahoor, F.D.; Mader, K.T.; Timmins, P.; Brown, J.; Sammon, C. Investigation of Within-Tablet Dynamics for Extended Release of a Poorly Soluble Basic Drug from Hydrophilic Matrix Tablets Using ATR-FTIR Imaging. *Mol. Pharm.* **2020**, *17*, 1090–1099. [CrossRef]
29. Siepmann, J.; Siepmann, F. Mathematical modeling of drug delivery. *Int. J. Pharmaceut.* **2008**, *364*, 328–343. [CrossRef]
30. Kaunisto, E.; Abrahmsen-Alami, S.; Borgquist, P.; Larsson, A.; Nilsson, B.; Axelsson, A. A mechanistic modelling approach to polymer dissolution using magnetic resonance microimaging. *J. Control. Release* **2010**, *147*, 232–241. [CrossRef]
31. Lamberti, G.; Galdi, I.; Barba, A.A. Controlled release from hydrogel-based solid matrices. A model accounting for water up-take, swelling and erosion. *Int. J. Pharm.* **2011**, *407*, 78–86. [CrossRef] [PubMed]
32. Peppas, N.A.; Narasimhan, B. Mathematical models in drug delivery: How modeling has shaped the way we design new drug delivery systems. *J. Control. Release* **2014**, *190*, 75–81. [CrossRef] [PubMed]
33. Kamba, M.; Seta, Y.; Kusai, A.; Nishimura, K. Comparison of the mechanical destructive force in the small intestine of dog and human. *Int. J. Pharm.* **2002**, *237*, 139–149. [CrossRef]
34. Mohylyuk, V.; Goldoozian, S.; Andrews, G.P.; Dashevskiy, A. IVIVC for Extended Release Hydrophilic Matrix Tablets in Consideration of Biorelevant Mechanical Stress. *Pharm. Res.* **2020**, *37*, 227. [CrossRef] [PubMed]
35. Takieddin, M.; Fassihi, R. A Novel Approach in Distinguishing between Role of Hydrodynamics and Mechanical Stresses Similar to Contraction Forces of GI Tract on Drug Release from Modified Release Dosage Forms. *AAPS Pharmscitech* **2015**, *16*, 278–283. [CrossRef] [PubMed]
36. Mikac, U.; Sepe, A.; Gradisek, A.; Kristl, J.; Apih, T. Dynamics of water and xanthan chains in hydrogels studied by NMR relaxometry and their influence on drug release. *Int. J. Pharm.* **2019**, *563*, 373–383. [CrossRef] [PubMed]
37. Melia, C.D. Hydrophilic Matrix Sustained-Release Systems Based on Polysaccharide Carriers. *Crit. Rev. Ther. Drug* **1991**, *8*, 395–421.
38. Gupta, P.; Vermani, K.; Garg, S. Hydrogels: From controlled release to pH-responsive drug delivery. *Drug Discov. Today* **2002**, *7*, 569–579. [CrossRef]
39. Ferrero, C.; Massuelle, D.; Jeannerat, D.; Doelker, E. Towards elucidation of the drug release mechanism from compressed hydrophilic matrices made of cellulose ethers. I. Pulse-field-gradient spin-echo NMR study of sodium salicylate diffusivity in swollen hydrogels with respect to polymer matrix physical structure. *J. Control. Release* **2008**, *128*, 71–79.
40. Lowman, A.M.; Peppas, N.A. Hydrogels. In *Encyclopedia of Controlled Drug Delivery*; Mathiowitz, E., Ed.; Wiley: Hoboken, NJ, USA, 2000; pp. 397–417.
41. Ritger, P.L.; Peppas, N.A. A simple equation for description of solute release I. Fickian and non-Fickian release from non-swellable devices in the form of slabs, spheres, cylinders or discs. *J. Control. Release* **1987**, *5*, 23–36. [CrossRef]
42. Korsmeyer, R.W.; Peppas, N.A. Effect of the Morphology of Hydrophilic Polymeric Matrices on the Diffusion and Release of Water-Soluble Drugs. *J. Membr. Sci.* **1981**, *9*, 211–227. [CrossRef]

Sample Availability: Samples of the compounds xanthan and pentoxifylline are available from the authors.

Publisher's Note: MDPI stays neutral with regard to jurisdictional claims in published maps and institutional affiliations.

© 2020 by the authors. Licensee MDPI, Basel, Switzerland. This article is an open access article distributed under the terms and conditions of the Creative Commons Attribution (CC BY) license (http://creativecommons.org/licenses/by/4.0/).

Article

Identification and Quantification by NMR Spectroscopy of the 22R and 22S Epimers in Budesonide Pharmaceutical Forms

Natalia E. Kuz'mina [1,*], Sergey V. Moiseev [1], Elena Y. Severinova [1], Evgenii A. Stepanov [2] and Natalia D. Bunyatyan [1,3]

[1] Scientific Centre for Expert Evaluation, Medicinal Products of the Ministry of Health of the Russian Federation, Federal State Budgetary Institution, 8/2 Petrovsky Blvd, 127051 Moscow, Russia; moiseevsv@expmed.ru (S.V.M.); severinova@expmed.ru (E.Y.S.); bunyatyan@expmed.ru (N.D.B.)

[2] Department of Chemistry and Chemistry Education, Charles University, Ovocný trh 560/5, 116 36 Prague, Czech Republic; orexrom@gmail.com

[3] Department of Pharmaceutical Technology and Pharmacology, I.M. Sechenov First Moscow State Medical University (Sechenov University), 8, Bldg. 2 St. Trubetskaya, 119991 Moscow, Russia

* Correspondence: kuzminan@expmed.ru

Abstract: The authors developed four variants of the qNMR technique (^{1}H or ^{13}C nucleus, DMSO-d6 or CDCl$_3$ solvent) for identification and quantification by NMR of 22R and 22S epimers in budesonide active pharmaceutical ingredient and budesonide drugs (sprays, capsules, tablets). The choice of the qNMR technique version depends on the drug excipients. The correlation of ^{1}H and ^{13}C spectra signals to molecules of different budesonide epimers was carried out on the basis of a comprehensive analysis of experimental spectral NMR data (^{1}H-^{1}H gCOSY, ^{1}H-^{13}C gHSQC, ^{1}H-^{13}C gHMBC, ^{1}H-^{1}H ROESY). This technique makes it possible to identify budesonide epimers and determine their weight ratio directly, without constructing a calibration curve and using any standards. The results of measuring the 22S epimer content by qNMR are comparable with the results of measurements using the reference HPLC method.

Keywords: budesonide; 22R and 22S epimers; identification; quantification; qNMR; HPLC

1. Introduction

Budesonide [Bud; 22(R,S)-(11β,16α)-16,17-Butylidenebis(oxy)-11,21-dihydroxypregna-1,4-diene-3,20-dione] is a synthetic compound of the glucocorticoid family with anti-inflammatory, anti-allergic, and immunosuppressive effects. Bud is actively used for the topical treatment of asthma, rhinitis, and inflammatory bowel disease [1–5] and included in the WHO list of essential medicines.

Bud is a racemic mixture of two epimers (22R and 22S, Figure 1). The epimers ratio in the mixture is determined by the synthesis method [6]. Although they have similar qualitative pharmacological effects, the Bud-22R is several times more potent than Bud-22S [7,8]. Therefore, the content of the less active epimer in the Bud active pharmaceutical ingredient (API) and Bud drug products is strictly normalized.

The identification and quantification of the Bud-22R and Bud-22S are carried out by capillary gas chromatography [6], high performance liquid chromatography (HPLC) [9,10], and sensitive ultra-high-performance liquid chromatography–tandem mass spectrometry method (HPLC-MS) [11,12]. These methods identify Bud epimers indirectly by comparing test samples with reference standards. Quantitative measurements by GC, HPLC, and HPLC-MS methods are relative and include the step of building a calibration function using a reference standard of the measured compound. It is important to use absolute and direct methods to identify and quantify Bud epimers. Absolute and direct methods (for example, qNMR) do not require the use of reference standards and the construction of calibration functions. The aim of this article is to develop the technique of the identification and quantification using qNMR of Bud-22R and Bud-22S in APIs and Bud drugs. The developed technique will allow selective identification of Bud-epimers and quantitative evaluation of its weight ratio directly by recording the characteristic signals of Bud-22R and Bud-22S in the NMR spectra and measuring their integral intensities.

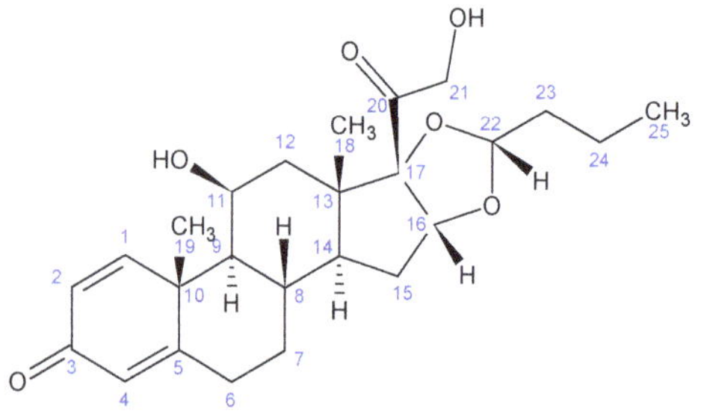

Figure 1. Chemical structures of Bud-22R and Bud-22S.

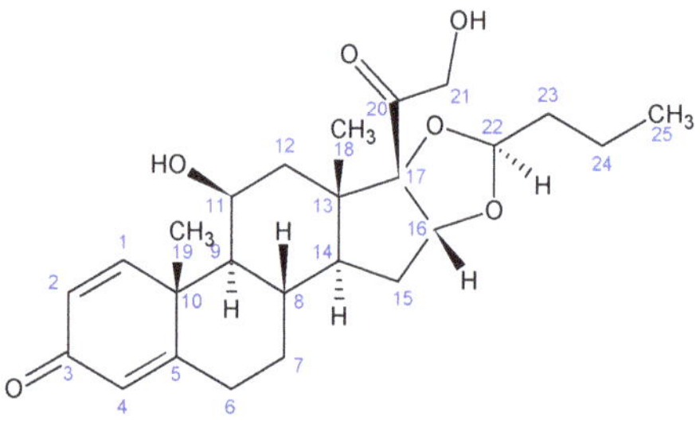

2. Results and Discussion

The simplest option for structural interpretation is the Bud-API spectrum, since it does not contain excipient signals. The comprehensive analysis of spectral data from 2D experiments (^{1}H-^{1}H gCOSY, ^{1}H-^{13}C gHSQC, ^{1}H-^{13}C gHMBC, ^{1}H-^{1}H ROESY) allowed us to correlate the ^{1}H and ^{13}C signals to different epimer molecules (Table 1).

Table 1. Spectral characteristics of 22R-Bud and 22S-Bud.

No.	22R		22S	
	δ, ppm		δ, ppm	
	^{1}H	^{13}C	^{1}H	^{13}C
	DMSO-d6			
1	7.31 d (J = 10.0)	156.40	7.30 d (J = 10.0)	156.43
2	6.16 dd (J = 10.0; 1.9)	127.11	6.16 d (J = 10.0; 1.9)	127.08
3		185.08		185.06
4	5.91 br.s	121.67	5.91 br.s	121.62
5		170.09		170.16
6	2.29 m; 2.52 m	31.17	2.29 m; 2.52 m	31.15
7	1.07 dd (J = 12.3; 4.7); 2.00 m	33.84	1.11 dd (J = 12.3; 4.7); 1.96 m	33.51
8	2.07 m	29.97	2.01 m	30.58
9	0.99 dd (J = 11.2; 3.5)	55.01	0.94 dd (J = 11.2; 3.5)	54.99
10		43.64		43.66
11	4.30 m	68.17	4.28 m	68.13
12	1.73 m	39.34	1.78 m	39.57
13		45.14		46.26
14	1.51 m	49.39	1.52 m	51.96
15	1.52 m; 1.59 m	32.93	1.58 m; 1.72 m	32.38
16	4.75 d (J = 4.3)	80.83	5.05 d (J = 7.3)	81.90
17		97.17		97.92
18	0.81 s	16.84	0.85 s	17.50
19	1.38 s	20.76	1.37 s	20.74
20		209.11		207.71
21	4.13 d (J = 19.4); 4.39 d (J = 19.4)	66.00	4.06 d (J = 19.2); 4.45 d (J = 19.2)	65.60
22	4.52 t (J = 4.5)	103.42	5.17 t (J = 4.8)	107.04
23	1.53 m	34.46	1.39 m	36.50
24	1.33 m	16.42	1.26 m	16.75

Table 1. Cont.

No.	22R		22S	
	δ, ppm		δ, ppm	
	^{1}H	^{13}C	^{1}H	^{13}C
25	0.85 t (J = 7.4)	13.79	0.85 t (J = 7.4)	13.79
11-OH	4.74 br.s		4.74 br.s	
		CDCl$_3$		
1	7.25 d (J = 10.1)	156.01	7.24 d (J = 10.1)	156.04
2	6.28 dd (J = 10.1; 1.8)	128.14	6.27 dd (J = 10.1; 1.8)	128.14
3		186.63		186.58
4	6.03 br.s	122.71	6.02 br.s	122.71
5		169.88		169.75
6	2.35 ddd (J = 13.7; 4.5; 1.8) 2.56 ddd (J = 13.7; 13.5; 5.5)	32.02	2.35 ddd (J = 13.7; 4.5; 1.8) 2.56 ddd (J = 13.7; 13.5; 5.5)	32.00
7	1.17 m; 2.07 m	34.14	1.17 m; 2.07 m	34.11
8	2.16 m	30.54	2.11 m	31.19
9	1.12 m	55.31	1.12 m	55.41
10		44.14		44.14
11	4.50 br.d (J = 3.3)	70.16	4.49 br.d (J = 3.3)	70.08
12	1.63 m; 2.07 m	41.17	1.63 m; 2.07 m	41.51
13		46.09		47.51
14	1.61 m	49.90	1.57 m	52.92
15	1.61 m; 1.78 m	33.58	1.75 m; 1.82 m	33.13
16	4.90 d (J = 4.7)	82.26	5.17 d (J = 6.8)	83.53
17		97.31		97.99
18	0.92 s	17.56	0.98 s	17.85
19	1.44 s	21.23	1.45 s	21.22
20		210.26		209.17
21	4.24 d (J = 19.8); 4.50 d (J = 19.8)	67.41	4.19 d (J = 19.8); 4.61 d (J = 19.8)	67.31
22	4.54 t (J = 4.5)	104.80	5.16 t (J = 5.1)	108.54
23	1.62 m	35.13	1.48 m	37.22
24	1.39 m	17.25	1.35 m	17.56
25	0.92 t (J = 7.5)	14.09	0.90 t (J = 7.5)	14.06

The C22-H bond direction (S or R) in each of the two epimers was determined by the technique ^{1}H-^{1}H ROESY. Only Bud-22R has protons C16-H and C22-H on the same side of the 1,3-dioxolane ring (Figure 1). This is the reason for the appearance of cross-peaks between these valence unbound protons in the ROESY spectrum. Figure 2 shows a fragment of the ROESY spectrum of Bud-API in DMSO-d6, containing the C16-H and C22-H proton signals (δ 4.75 and 4.52 ppm for one epimer and 5.05 and 5.17 ppm for the other). Only the proton pair 4.75–4.52 ppm had cross-peaks. This fact indicates that protons 4.75 and 4.52 ppm belong to the Bud-22R. The proton pair 5.05–5.17 ppm is part of the Bud-22S.

It should be noted that the Bud NMR spectral data presented in the literature [13,14] lack structural correlation of Bud NMR spectra signals to specific 22R and 22S epimers.

The spectra analysis of Bud-API solutions in DMSO-d6 and CDCl$_3$ (Figures 3–6, Table 1) allowed to determine isolate signals for each epimer (characteristic signals). There are following characteristic signals for Bud-22R:

(1) ^{1}H (DMSO-d6), δ, ppm: 4.13 d (C21-H), 4.39 d (C21-H), 4.52 t (C22-H);
(2) ^{1}H (CDCl$_3$), δ, ppm: 4.24 d (C21-H), 4.54 t (C22-H), 4.89 d (C16-H);
(3) ^{13}C (DMSO-d6), δ, ppm: 66.00 (C21), 80.83 (C16), 97.17(C17); 103.42 (C22);
(4) ^{13}C (CDCl$_3$), δ, ppm: 46.09 (C13); 49.90 (C14), 82.26 (C16), 97.31 (C17), 104.80 (C22).

There are the following characteristic signals for Bud-22S:

(1) ^{1}H (DMSO-d6), δ, ppm: 4.06 d (C21-H), 4.45 d (C21-H), 5.05 d (C16-H), 5.17 t (C22-H);
(2) ^{1}H (CDCl$_3$), δ, ppm: 4.19 d (C21-H), 4.61 d (C21-H);
(3) ^{13}C (DMSO-d6), δ, ppm: 65.60 (C21), 81.90 (C16), 97.92(C17); 107.04 (C22);
(4) ^{13}C (CDCl$_3$), δ, ppm: 47.51 (C13); 52.92 (C14), 83.52 (C16), 97.99 (C17), 108.54 (C22).

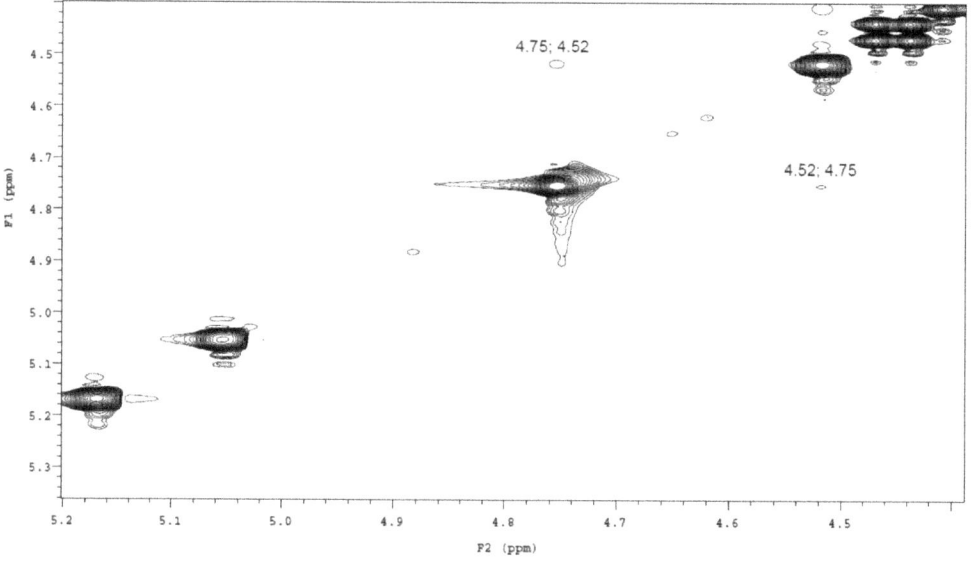

Figure 2. ^{1}H-^{1}H ROESY spectrum fragment of the Bud-API solution in DMSO-d6.

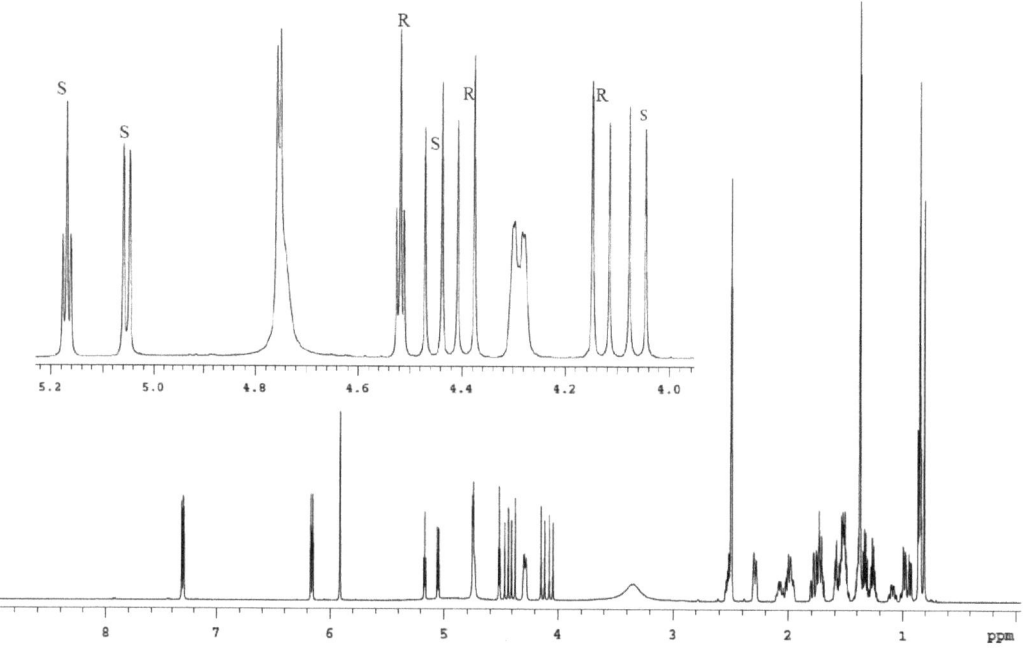

Figure 3. ^{1}H spectrum of the Bud-API solution in DMSO-d6.

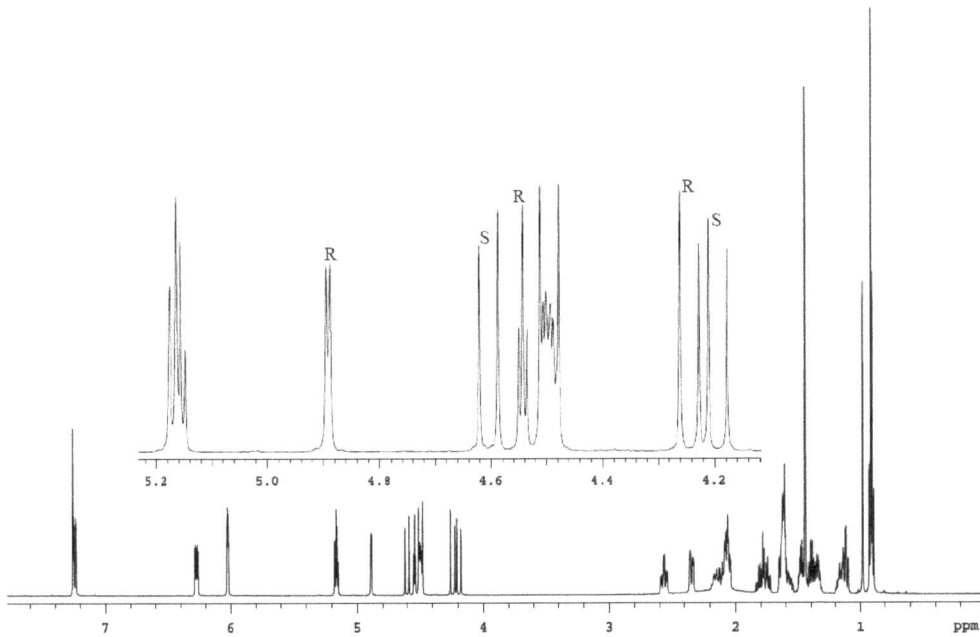

Figure 4. ^{1}H spectrum of the Bud-API solution in CDCl$_3$.

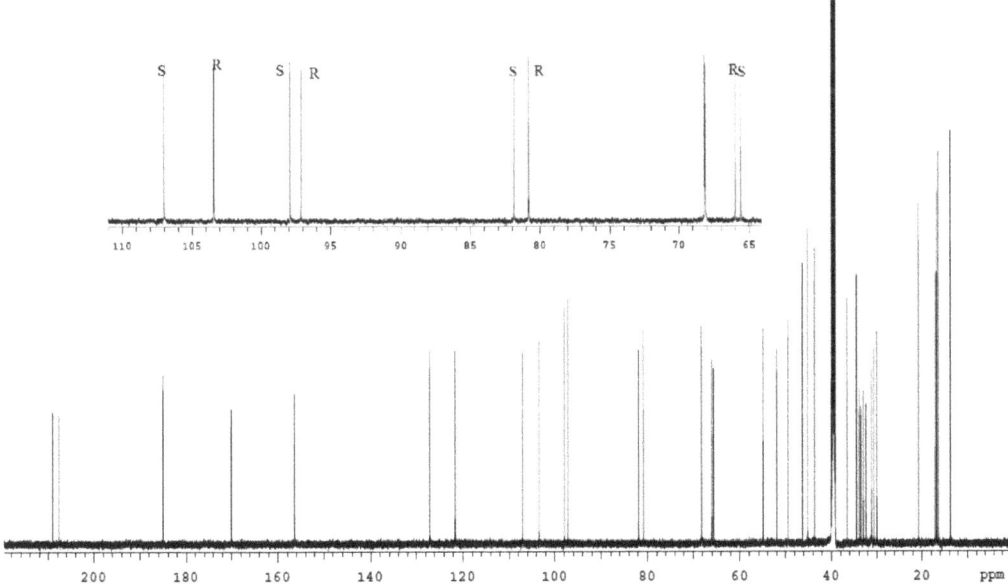

Figure 5. ^{13}C spectrum of the Bud-API solution in DMSO-d6.

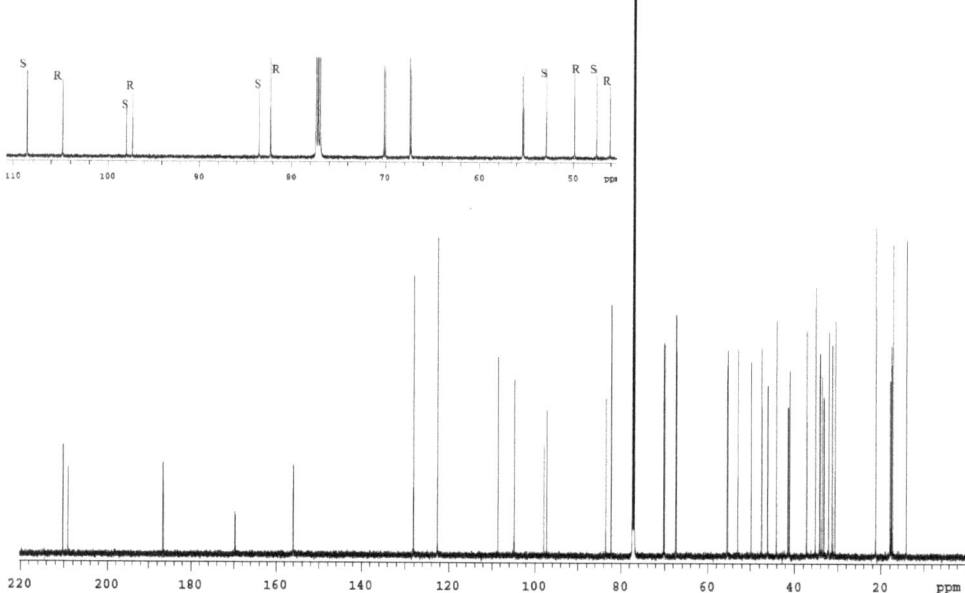

Figure 6. ^{13}C spectrum of the Bud-API solution in CDCl$_3$.

It should be noted that the use of DMSO-d6 provides a better separation of the characteristic signals of the Bud-22R and Bud-22S epimers in the proton spectrum. CDCl$_3$ provides better ^{13}C spectrum resolution.

The characteristic signals can be spectral markers of these epimers in the analyzed sample. Their normalized integral intensities are equal to the fraction of each epimer in the racemate mixture. It should be noted that qNMR is considered in the literature as an absolute and direct method for measuring the molar ratio of the analytes in a test sample, as well as the weight content of one component relative to another component, because the functional relationships between the analytes and the measurands (integrated intensities) are well-known: the molar ratio of the components in a mixture is equal to the ratio of the normalized integrated intensities of the signals of these components. Uncertainty of the measuring result by qNMR relies only on the uncertainty of the integral intensities ratio measurement [15]. The results of measurements by HPLC (pharmacopeial method) are relative and indirect by nature. Determination of Bud-22R and Bud-22S epimers by HPLC requires generation of a calibration curve using their pharmacopeial reference standards (the relative nature of measurements). The measurement by the HPLC method has a combined uncertainty (the indirect nature of measurements). Sources of the total uncertainty are the peak area measurement in the chromatogram, weighing of the test and standard samples, and solvent volume measurements. Therefore, the accuracy of measurement of Bud epimeric composition by direct and absolute method qNMR is higher than by indirect and relative method HPLC. Moreover, both normalized integral intensities of a selected individual pair of 22R and 22S epimeric signals and the average value of pairwise normalized integral intensities of all observed pairs of characteristic signals can be taken as a result of measuring the epimeric composition of the Bud sample. Averaging the measurement results reduces its uncertainty. In chromatographic methods, averaging is only possible with a series of experiments.

Bud drugs of different manufacturers have in their content a nonequal set of excipients. The solubility of excipients influences the choice of solvent (DMSO-d6 or CDCl$_3$) For example, a nasal spray is an aqueous suspension of Bud. The excipitents of this suspension are DMSO-soluble sodium methylparaben, carboxymethylcellulose and sodium carmellose, polysorbate 80, sucrose, polypropylene glycol and disodium edetate. Obviously, it is appropriate to use CDCl$_3$ rather than DMSO-d6 when analyzing this drug. The sample extraction with chloroform will concentrate Bud and remove excipients that do not pass into the extractant. In the ^{1}H (CDCl$_3$) spectrum of the Bud nasal spray, all characteristic signals of the Bud-22R and Bud-22S are observed (Figure 7a). For quantitative measurements, it is reasonable to use the most isolated signals 4.89 d (22R) and 4.61 d (22S). In the ^{13}C (CDCl$_3$) spectrum of this preparation, all characteristic signals are also present (Figure 7b).

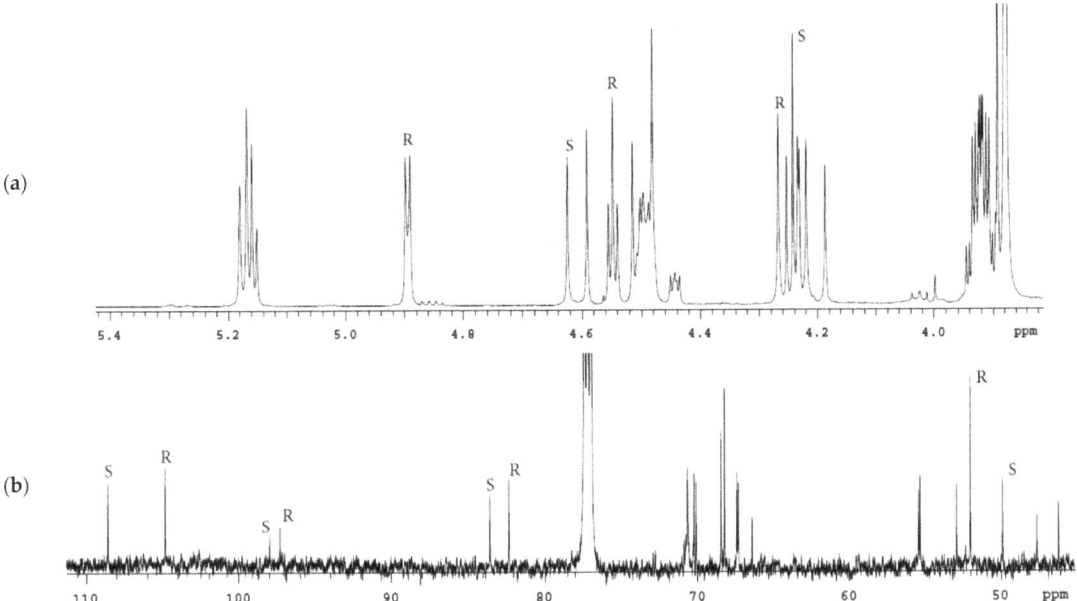

Figure 7. ^{1}H (a) and ^{13}C (b) spectra fragments of the Bud nasal spray solution in CDCl$_3$ with characteristic signals of 22R and 22S epimers.

Bud capsules contain chloroform-insoluble lactose monohydrate; therefore, it is also advisable to use CDCl$_3$ for this drug. The characteristic signals ^{1}H and ^{13}C of the Bud-22R and Bud-22S for capsule solutions in CDCl$_3$ are shown in Figure 8.

Bud tablets contain excipients with different solubility in DMSO-d6 and CDCl$_3$: stearic acid, soy lecithin, cellulose, hydroxypropylcellulose, lactose monohydrate, and magnesium stearate. The characteristic signals of Bud epimers partially overlap with the signals of excipients in ^{1}H spectra of Bud tablets solution in DMSO-d6 and CDCl$_3$ (Figure 9). For this reason, precise quantitative measurements are not possible. When selecting the ^{13}C nucleus, isolated characteristic signals are observed for each solvent (DMSO-d6 and CDCl$_3$; Figure 10).

Table 2 shows the results of quantitative measurements of the 22S and 22R epimers content in the Bud-API and Bud drugs, obtained using different versions of the developed technique.

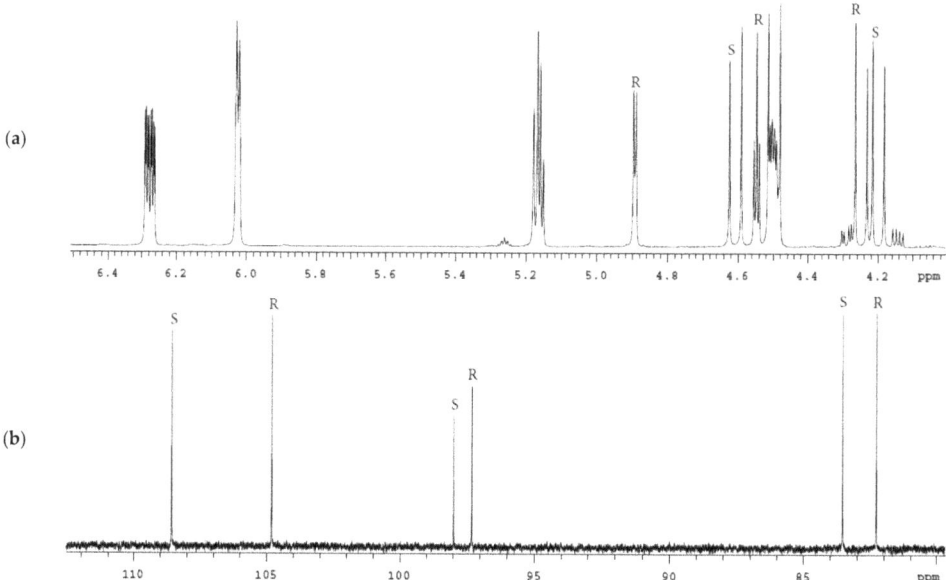

Figure 8. ^{1}H (**a**) and ^{13}C (**b**) spectra fragments of the Bud capsules solution in CDCl$_3$ with characteristic signals of 22R and 22S epimers.

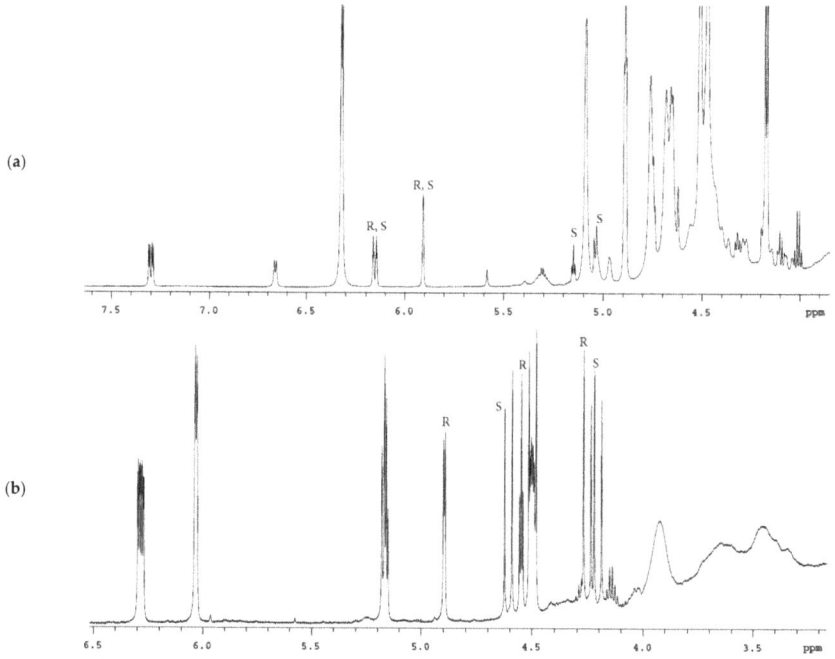

Figure 9. ^{1}H spectra fragments of the Bud tablets solutions in DMSO-d6 (**a**) and CDCl$_3$ (**b**) with characteristic signals of 22R and 22S epimers.

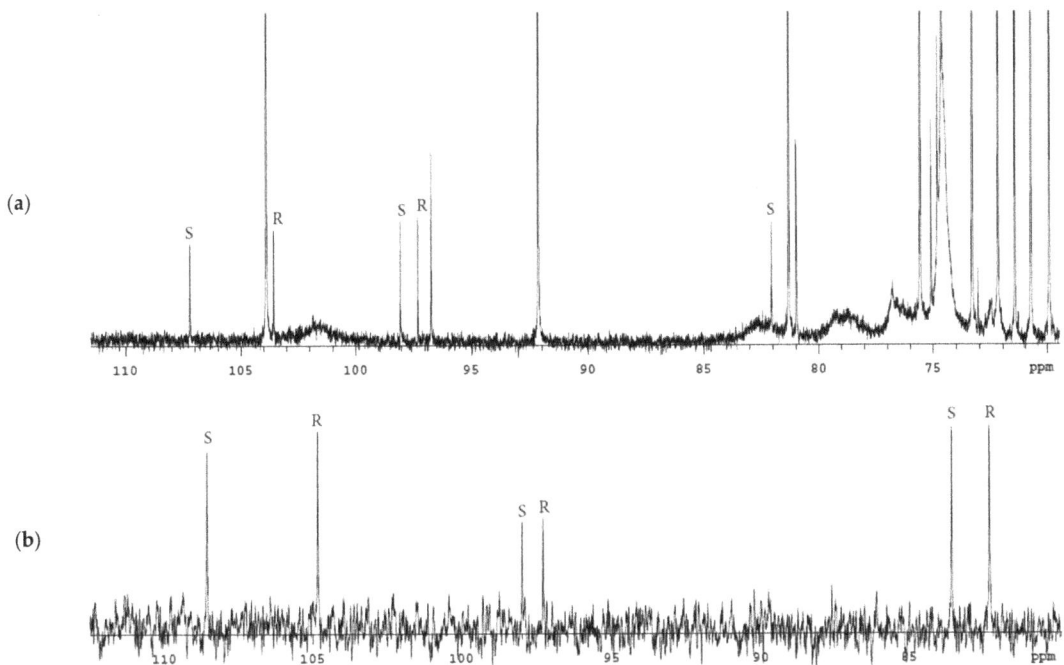

Figure 10. ^{13}C spectra fragments of the Bud tablets solutions in DMSO-d6 (**a**) and CDCl$_3$ (**b**) with characteristic signals of 22R and 22S epimers.

Table 2. The quantitative measurements results of the content of 22S and 22R epimers in the Bud-API and Bud drugs.

Bud	Content of 22S (22R), %				Mean Volume
	DMSO-d6		CDCl$_3$		
	^{1}H	^{13}C	^{1}H	^{13}C	
API	47.45 (52.55)	47.60 (52.40)	47.47 (52.53)	47.56 (52.44)	47.52 (52.48)
Nasal spray	-	-	46.67 (53.33)	46.53 (53.47)	46.60 (53.40)
Capsules	-	-	47.82 (52.18)	47.67 (52.33)	47.75 (52.25)
Tablets	-	48.60 (51.40)	-	48.81 (51.19)	48.71 (51.29)

In the ^{13}C spectra, characteristic signals of Bud-22R, Bud-22S, and excipients are located at a considerable distance from each other. Therefore, the signal ^{13}C can be integrated using the general rule for choosing the integration limit (the integration limit is equal to 64 times the half-width of a Lorentzian shape NMR signal [15]). In the ^{1}H spectra of Bud drugs, there is a partial overlap of the signals of Bud and excipients in this frequency range. Therefore, the integration limit of the Bud epimer signals in the ^{1}H spectra were narrowed to 20 times the half-width of a Lorentzian shape. It should be noted that variation in the solvent and nucleus does not affect the result of quantitative measurement of the Bud epimers content. For example, the RSD of the measurement results of Bud-22S content in Bud-API is 0.15% (mean volume is 47.52%).

The results of measurement of 22S epimer content using qNMR are comparable with the results of the HPLC reference method. Thus, the content of Bud-22S in the API and nasal spray, measured by HPLC, was 47.3 and 46.8% (47.52 and 46.60% by qNMR). The similarity of the measurement results, obtained by qNMR and HPLC methods, is an additional proof of the correctness of the proposed technique.

3. Materials and Methods

3.1. Materials

The following materials were used in the qNMR technique development: Bud-API by Farmabios S.p.A., Italy (A), nasal spray «Tafen Nasal» by Lek d.d., Slovenia (B), Bud capsules «Respinid» by Sava Healthcare Limited, India (C), tablets «Kortiment» by Cosmo S.p.A., Italy (D). Deuterated dimethylsulfoxide (DMSO-d6, 99.90% D) and chloroform (99.8% D) by Cambridge Isotope Laboratories, Inc. (St. Louis, MO, USA) were used in the NMR experiments.

HPLC measurements were carried out using the certified reference standard for Bud, manufactured by the European Pharmacopoeia, glacial acetic acid, potassium hydroxide (Sigma-Aldrich, Saint Louis, MO, USA). HPLC grade acetonitrile was purchased from Fisher Scientific (Fairlawn, NJ, USA). HPLC ready 18 MΩ water was obtained, in-house, from a Milli-Q Integral 3 water purification system, Merck Millipore Corp. (Burlington, MA, USA). Duran filter funnels (porosity 3) were used for filtration.

3.2. NMR Spectroscopy Method

3.2.1. Sample Preparation

API: About 20 mg of the Bud-API (exact mount is optional) were placed in an NMR tube, 0.5 mL of solvent (DMSO-d6 or CDCl$_3$) was added, shaken vigorously until the sample was completely dissolved.

Nasal spray: The contents of 1 vial was transferred to a separating funnel, 2 mL of CDCl$_3$ were added and thoroughly shaken for 5 min; then, the bottom organic layer was separated and transferred to the NMR tube.

Capsules: 10 mL of CHCl$_3$ were added to the contents of 30 capsules, thoroughly mixed and filtered; then, the filtrate was centrifuged. The supernatant was separated and dried by air. The resulting dry residue was dissolved in 0.6 mL CDCl$_3$ and transferred to an NMR tube.

Tablets DMSO-d6: 3 mL of DMSO-d6 were added to the two powdered tablets, thoroughly mixed and filtered; then, the filtrate was centrifuged. A total of 0.6 mL of the supernatant was separated and transferred to the NMR tube.

Tablets, CDCl$_3$: 10 mL of CHCl$_3$ were added to the 2 tablets crushed into a powder, thoroughly mixed and filtered; then, the filtrate was centrifuged. The supernatant was separated and dried by air. The resulting dry residue was dissolved in 0.6 mL of CDCl$_3$ and transferred to an NMR tube.

3.2.2. Instrumentation and Experiment Conditions

NMR spectra were collected on the Agilent DD2 NMR System 600 NMR spectrometer equipped with a 5 mm broadband probe and a gradient coil (VNMRJ 4.2 software). Parameters of 1D experiments: temperature—27 °C; spectral width—6009.6 Hz (^{1}H) and 37,878.8 Hz (^{13}C); observed pulse 90° (^{1}H) and 45° (^{13}C); acquisition time—5.325 s (^{1}H) and 0.865 s (^{13}C); relaxation delay—10 s (^{1}H) and 1 s (^{13}C); number of scans—256 (^{1}H) and 10,000 (^{13}C); the number of analog-to-digital conversion points—64 K; exponential multiplication—0.3 Hz (^{1}H) and 3 Hz (^{13}C); zero filling—64 K; automatic linear correction of the spectrum baseline, manual phase adjustment, calibration of the δ scale under DMSO (δ = 2.50 ppm for ^{1}H and 39.52 ppm for ^{13}C) or CHCl$_3$ (δ = 7.26 ppm for ^{1}H and 77.16 ppm for ^{13}C) [16]. The manual mode was also used for the signal integration. The integration limit was equal to 20 (^{1}H) and 64 (^{13}C) times the half-width of a Lorentzian shape NMR signal. The relaxation delay value was estimated using an inversion-recovery experiment: T1 is equal to 1.55 s. The ROESY experimental parameters: the relaxation time—1 s; the number of free induction signal accumulation per increment—16; the number of analog-to-digital conversion points—2K × 256; the mixing time—0.2 s; the pulse duration—0.15 s.

3.3. Reference Measurement with HPLC Method

3.3.1. Preparation of Solution

System suitability test solution, buffer solution, test solution of samples A–D, reference solutions, and mobile phase were prepared according to USP methods [9,10].

3.3.2. Instrumentation and Chromatographic Conditions

The HPLC system consists of an Agilent Infinity 1260 series (Agilent Technologies, Wilmington, DE). Data collection and analysis were performed using ChemStation software. Chromatographic conditions: column—Zorbax RX-C-18 250 mm × 4.6 mm × 5 μm (Agilent Technologies, Santa-Clara, CA, USA); column temperature—50 °C; mobile phase—acetonitrile and buffer pH 3.9 (45:55) for sample A and acetonitrile and water (70:30) for sample B; flow rate—1 mL/min; detector—UV 240 nm for sample A and 245 nm for sample B; injection volume—20 μL for sample A and 50 μL for sample B; run time—no less 40 min.

4. Conclusions

Different versions of the qNMR technique for identification and quantification Bud-22R and Bud-22S epimers (^{1}H or ^{13}C core, DMSO-d6 or CDCl$_3$ solvent) were developed for Bud APIs and Bud drugs. This technique does not need Bud-epimers reference standards. The choice of the qNMR technique version depends on the drug excipients in Bud drugs. Application of this technique will reduce the uncertainty of the measurement result, since the experimental procedure does not contain the stages of taking accurate weights, volumes, and constructing a calibration curve. This technique can be used for carrying out GP APIs and drug analyses.

Author Contributions: Conceptualization, N.E.K. and S.V.M.; data curation, N.E.K. and S.V.M.; formal analysis, N.E.K., S.V.M. and E.Y.S.; investigation, N.E.K. and S.V.M.; sample preparation, E.Y.S.; methodology, N.E.K. and S.V.M.; project administration, N.D.B.; supervision, N.D.B.; visualization, S.V.M. and E.A.S.; writing—original draft, N.E.K.; writing—review and editing, S.V.M., E.A.S. and N.D.B. All authors have read and agreed to the published version of the manuscript.

Funding: This research was funded by Ministry of Health of Russia [research project No. 056-00005-21-02, R&D public accounting No. 121022400083-1]. The APC was funded by the Scientific Centre for Expert Evaluation of Medicinal Products.

Institutional Review Board Statement: Not applicable.

Informed Consent Statement: Not applicable.

Data Availability Statement: The data that support the findings of this study are available from the corresponding author upon reasonable request.

Conflicts of Interest: The authors declare no conflict of interest. The funders had no role in the design of the study; in the collection, analyses, or interpretation of data; in the writing of the manuscript, or in the decision to publish the results.

Sample Availability: Not applicable.

References

1. Vandevyver, S.; Dejager, L. New insights into the anti-inflammatory mechanisms of glucocorticoids: An emerging role for glucocorticoid-receptor-mediated transactivation. *Endocrinology* **2013**, *154*, 993–1007. [CrossRef] [PubMed]
2. Varga, G.; Ehrchen, J. Immune suppression via glucocorticoid-stimulated monocytes: A novel mechanism to cope with inflammation. *J. Immunol.* **2014**, *193*, 1090–1099. [CrossRef] [PubMed]
3. Miehlke, S.; Madisch, A. Budesonide is more effective than mesalamine or placebo in short-term treatment of collagenous colitis. *Gastroenterology* **2014**, *146*, 1222–1230. [CrossRef] [PubMed]
4. Miehlke, S.; Aust, D. Efficacy and Safety of Budesonide, vs. Mesalazine or Placebo, as Induction Therapy for Lymphocytic Colitis. *Gastroenterology* **2018**, *155*, 1795–1804. [CrossRef] [PubMed]

5. Maeda, K.; Yamaguchi, M. Utility and effectiveness of Symbicort® Turbuhaler® (oral inhalation containing budesonide and formoterol) in a patient with severe asthma after permanent tracheostomy. *J. Pharm. Health Care Sci.* **2018**, *4*, 24–28. [CrossRef] [PubMed]
6. Krzek, J.; Czekaj, J.S. Direct separation, identification and quantification of epimers 22R and 22S of budesonide by capillary gas chromatography on a short analytical column with Rtx®-5 stationary phase. *J. Chromatogr. B* **2004**, *803*, 191–200. [CrossRef] [PubMed]
7. Cortijo, J.; Urbieta, E. Biotransformation in vitro of the 22R and 22S epimers of budesonide by human liver, bronchus, colonic mucosa and skin. *Fundam. Clin. Pharm.* **2008**, *15*, 47–54. [CrossRef] [PubMed]
8. Szefler, S.J. Pharmacodynamics and pharmacokinetics of budesonide: A new nebulized corticosteroid. *J. Allergy Clin. Immunol.* **1999**, *104*, S175–S183. [CrossRef]
9. USP43-NF38. Budesonide. *604*. Available online: https://online.uspnf.com/ (accessed on 8 November 2021).
10. USP43-NF38. Budesonide Nasal Spray. 2S. Available online: https://online.uspnf.com/ (accessed on 8 November 2020).
11. Li, Y.N.; Tattam, B. Seale Determination of epimers 22R and 22S of budesonide in human plasma by high-performance liquid chromatography-atmospheric pressure chemical ionization mass spectrometry. *J. Chromatogr. B* **1996**, *683*, 259–268. [CrossRef]
12. Lu, Y.; Sun, Z. Simultaneous quantification of 22R and 22S epimers of budesonide in human plasma by ultra-high-performance liquid chromatography–tandem mass spectrometry: Application in a stereoselective pharmacokinetic study. *J. Chromatogr. B* **2013**, *921–922*, 27–34. [CrossRef] [PubMed]
13. Yan, Y.; Wang, P. Synthesis of budesonide conjugates and their anti-inflammatory effects: A preliminary study. *Drug Des. Dev. Ther.* **2019**, *13*, 681–694. [CrossRef] [PubMed]
14. Bhutnar, A.; Khapare, S. Isolation and Characterization of Photodegradation Impurity in Budesonide Drug Product Using LC-MS and NMR Spectroscopy. *Am. J. Anal. Chem.* **2017**, *8*, 449–461. [CrossRef]
15. Malz, F.; Jancke, H. Validation of quantitative nuclear magnetic resonance. *J. Pharm. Biomed.* **2005**, *38*, 813–823. [CrossRef] [PubMed]
16. Gottlieb, H.E.; Kotlyar, V. NMR Chemical Shifts of Common Laboratory Solvents as Trace Impurities. *J. Org. Chem.* **1997**, *62*, 7512–7515. [CrossRef] [PubMed]

Article

Spin-Space-Encoding Magnetic Resonance Imaging: A New Application for Rapid and Sensitive Monitoring of Dynamic Swelling of Confined Hydrogels

Rui Wang, Jiaxiang Xin, Zhengxiao Ji, Mengni Zhu, Yihua Yu and Min Xu *

Shanghai Key Laboratory of Magnetic Resonance, School of Physics and Electronic Science, East China Normal University, Shanghai 200241, China; 51184700048@stu.ecnu.edu.cn (R.W.); 52174701014@stu.ecnu.edu.cn (J.X.); 51214700058@stu.ecnu.edu.cn (Z.J.); 52214700018@stu.ecnu.edu.cn (M.Z.); yhyu@phy.ecnu.edu.cn (Y.Y.)
* Correspondence: xumin@phy.ecnu.edu.cn

Abstract: An NMR method based on the gradient-based broadening fingerprint using line shape enhancement (PROFILE) is put forward to precisely and sensitively study hydrogel swelling under restricted conditions. This approach achieves a match between the resonance frequency and spatial position of the sample. A three-component hydrogel with salt ions was designed and synthesized to show the monitoring more clearly. The relationship between the hydrogel swelling and the frequency signal is revealed through the one-dimensional imaging. This method enables real-time monitoring and avoids changing the swelling environment of the hydrogel during contact. The accuracy of this method may reach the micron order. This finding provides an approach to the rapid and non-destructive detection of swelling, especially one-dimensional swelling, and may show the material exchange between the hydrogel and swelling medium.

Keywords: gradient broadening; profile; hydrogel; swelling

1. Introduction

At present, nuclear magnetic resonance (NMR), including magnetic resonance imaging (MRI), has become one of the most important technologies in scientific research, and is extensively used in chemistry, biology, materials science and medical imaging [1]. NMR spectroscopy is widely used to identify the structure of organic natural products [2], to identify the chemical composition distribution and fracture mode of random and block copolymers [3], to screen drug targets corresponding to proteins [4], to predict protein torsion angle by chemical shift [5], to characterize the technology and water content in food processing [6], to measure the diffusion of oil in rubber [7], and to determine the internal microstructure of wood [8]. Although MRI is widely accepted as a powerful medical imaging modality, it has also been recently used for analytical measurements in materials systems [9,10]. Under MRI mode, the spin position is searched and radiofrequency (RF) pulses are scanned in the presence of a magnetic field gradient. When the spin signal is properly reunited through the gradient field, a time-domain signal is produced to reflect the spatial distribution of the spin of the whole sample at that amplitude. High-resolution spectral data and high-definition imaging can be provided, even in the presence of an obviously nonuniform magnetic field. In addition, spectral imaging can be measured without complex conditions by relying on multidimensional experiments. With the help of the auxiliary field gradient, the spin coordinates are converted into offsets in a one-to-one manner and mapped into frequencies. Although the resolution of the NMR peaks is sacrificed, the spatial positions are imprinted on the NMR line shapes [10].

A hydrogel is a hydrophilic polymer material with a three-dimensional polymer network, and chemically synthesized hydrogels are polymerized by monomers containing functional groups, such as hydroxyl and carbon double bonds [11]. The hydrophilicity, thermal stability, mechanical stability, biocompatibility, and responsiveness the material possesses are different in the swelling state. These properties make it useful for a number of applications in chemical engineering, drug delivery, tissue engineering, food and agriculture [12]. For example, silicone hydrogel is used in contact lenses. Polyacrylamide hydrogel is used as an absorbent for wound dressing [13]. PVA hydrogel is used in drug delivery applications and as a soft tissue replacement. [14]. The hydrogel swelling behavior in fluids is a special characteristic, and the diffusion of loose macromolecules in the swelling state plays an important role in these applications [15]. Polymer swelling includes the penetration of small molecules into the interior of larger polymer molecules, which leads to changes in polymer volume [16]. When the molecular weight and the degree of polymer crosslinking are higher, such as in chemicals with crosslinked hydrogels in contact with water or other solvents, the solvent cannot disperse further because the chemical bonds in the polymer chains have a binding effect on the small molecules of the solvent. Therefore, the polymer remains in the swelling process [17]. The degree and speed of swelling are important indices for investigation. For example, the swelling speed of resin absorbing water in diapers needs to be large enough to ensure there is no side leakage. However, the gasket of the mechanical seal rubber needs only a tiny degree of expansion or even no expansion to ensure dimensional stability.

The most common methods in the swelling test include the mass method, volume method, and length method. Taking the length method as an example, the swelling degree can be obtained by evaluating the ratio of the change in the height (or length) of a hydrogel sample before and after swelling:

$$Q = (h - h_0)/h_0 \qquad (1)$$

where Q is the degree of swelling, h_0 is the height before swelling, and h is the height after swelling. Similarly, the principles of the mass method and volumetric method are similar to those of the length method (Supplementary S1). There are also some tests based on altered masses, such as centrifugation, filtration, tea bagging, and Prudential dextrin methods [18].

Although these traditional methods can easily test the degree of swelling, crosslinking density, and swelling curve of the polymer, they still have some disadvantages, such as low accuracy, cumbersome operation, and lack of continuous measurement. Usually, during the measuring process, the samples are always separated from the solvent for weighing or measurement. This process changes the gel's environment and is very time-consuming. The low mechanical strength of some highly swellable hydrogels may result in their structural destruction during the weight measurement, which may result in significant errors. Thus, these methods are unable to precisely monitor the dynamic swelling process of specimens without contact [19]. In addition, the relationship between swelling rates and experimental temperature, time, and various other factors cannot be used to monitor these sensitive details [20].

To solve these problems, some improved methods have been put forward, such as the linear variable differential transformer (LVDT) [21], fluid-dynamic gauging [22] and nonintrusive inductance swelling instruments using a dynamic mechanical analyzer [23]. A customized sample cell and time-varying attenuated total reflection Fourier transform infrared spectroscopy were used to track the swelling processes in polymer films [24]. On the basis of the length method, researchers from the University of Cambridge used an opposed laser displacement sensor to measure the swelling, but the actual measurement range was limited by the length of the laser beam and severe warping [25]. The most recent method, proposed by Tang and colleagues, employed an aggregation-induced emission approach to suggest a new technique for the measurement of swelling properties in hydrogels [26]. The one-dimensional swelling of NMR has been less reported in examinations of the swelling of hydrogels. Lee reported that NMR imaging requires a three-dimensional gradient [27]. The proposed technique significantly enhanced the traditional method of swelling measurement during observations of the swelling process. However, the initial weight and size of the hydrogel should be carefully selected to avoid over-swelling. Due to the difficulties in managing the subtle changes in the hydrogel during the swelling process, it is not easy to investigate the influence of various other factors, such as temperature and time, on swelling.

To meet the demand for continuous noncontact measurements and detect the restricted swelling behavior, we present an NMR technique based on spin-space encoding and the gradient-broadening method (PROFILE), which can lead to the real-time detection of swelling behavior at the micron scale. By applying PROFILE, the relationship between proton density and signal intensity was obtained using one-dimensional imaging, and the height of the whole gel sample could be expanded in the frequency domain. Thus, the change in the height of the swelling gel corresponds to the change in frequency.

A three-component hydrogel containing barium chloride was designed and successfully synthesized to demonstrate the detection process. By using very dilute Na_2SO_4 solution as the solvent, the barium ions in the hydrogel react with Na_2SO_4, creating a precipitation layer on the interface between the gel and liquid. Because the precipitation layer contains far fewer hydrogen protons than the hydrogel and the solvent, the NMR signal intensity sharply decreases at the interface, forming a deep trough in the NMR spectra, which makes the interface position and the corresponding frequency in the spectra more obvious. With the swelling of the hydrogel, the interface moves and the frequency of the trough also moves. By continuously tracing the trough (the NMR spectra can be obtained by a single scan of up to 4 s), the position of the interface was obtained; therefore, the change in the length of the swelled hydrogel can be detected in real-time. The accuracy of this method could be on the micron order, which is much better than that of traditional methods. This method provides a new strategy for investigating the degree of swelling and swelling dynamics of hydrogels or other crosslinked polymer systems, especially tiny swelling systems. It may also have many potential applications in other correlating fields.

2. Result and Discussion

2.1. The Synthesis and Characterization of Hydrogels

In the absence of Ba^{2+}, the PMAB gel can be synthesized by following the mechanism of radical polymerization, with PEGA, MEA, and AM as polymeric monomers in an aqueous solution and APS as initiators. The ^{1}H-NMR spectra of the monomer and the synthesized gel are given in S2–S5. The monomer showed obvious double-bond peaks in the range of δ = 5.5 − 6.5, while the double bond disappeared in the polymer, and the state changed from liquid to solid, indicating that the gel was successfully synthesized. Although the three-component gel was successfully synthesized, the polymerization effect cannot be achieved in the presence of higher concentrations of ions, and the resulting gel is extremely uneven. There are two possible reasons for this phenomenon: one is ion inhibition and the other is the impact of APS. In the presence of higher concentrations of inorganic metal ions, the ions will have an effect on the initiator, which, in turn, will inhibit polymerization. To verify this, we attempted to reduce the ion concentration, but the lower ion concentration still could not form the gel. The second reason is that the persulfate in the APS at the early stage of the reaction initiates the reaction, producing a primary radical with a sulfate-like structure, which binds to Ba^{2+} and loses the initiation effect. Therefore, this reaction used azodiisobutyronitrile (AIBN) instead of APS as the initiator and used a smaller amount of the initiator at a lower temperature. This was carefully protected from light agitation, and gradient heating was used at later stages of gel formation to reduce bursting and bubble formation. The dosage ratio of the initiator and monomer, adjusted to the substance, was n(AIBN):n(MEA):n(PEGA):n(AM) = 1:133:100:270 (optimal ratio), which can be completely dissolved into a clear system before polymerization, and can be polymerized at a suitable rate to obtain a homogeneous gel without bubbles.

2.2. Gradient-Based Broadening Fingerprint Obtained by Line Shape Enhancement (PROFILE)

The resonance frequency ω_0 of protons in the static magnetic field B_0 can be expressed as:

$$\omega_0 = \gamma B_0 \quad (2)$$

where γ is the ^1H gyromagnetic ratio. The resonance frequency ω_0 of the protons is constant and independent of the spatial position of the protons.

When the gradient field is applied in the direction of B_0, the effective field B_{eff} of the protons can be written as:

$$B_{eff} = B_0 + G_z z \quad (3)$$

where G_z is the amplitude of the gradient pulse; and z is the spatial position of the protons. Here, the effective field B_{eff} is not uniform and changes linearly with the space position z. The effective resonance frequency of protons ω_{eff} can be modified as:

$$\omega_{eff} = \gamma(B_0 + G_z z) \quad (4)$$

Therefore, the proton signal will be detected with a wide bandwidth while the gradient field is active and transformed in the magnitude mode; the NMR spectrum is a square-shaped spread. In the spectrum, the intensity of the spread signal is correlated with the spatial distribution of the proton density. The broadening range Δf of the signal can be written as:

$$\Delta f = \gamma \times G_z \times \Delta z \quad (5)$$

Figure 1 shows the 1D NMR spin-echo imaging sequence, which is used to measure the profile of an object. The radiofrequency (RF) pulses 90_y and 180_x were used to excite and refocus the NMR signals, respectively. The gradient pulse g_1 was used to dephase the NMR signal, and the magnetic field gradient g_2 was turned on for spatial encoding during acquisition. The profile of the object can be obtained by a Fourier transform of the temporal signal. The following sequence parameters were used in the studies: echo time (TE) = 8 ms; average number = 4; repetition time (TR) = 4 s; acquisition time = 6 ms; d_2 = 4 ms; and d_3 = 1 ms. A 10% maximum gradient strength G_{max} was used in all experiments.

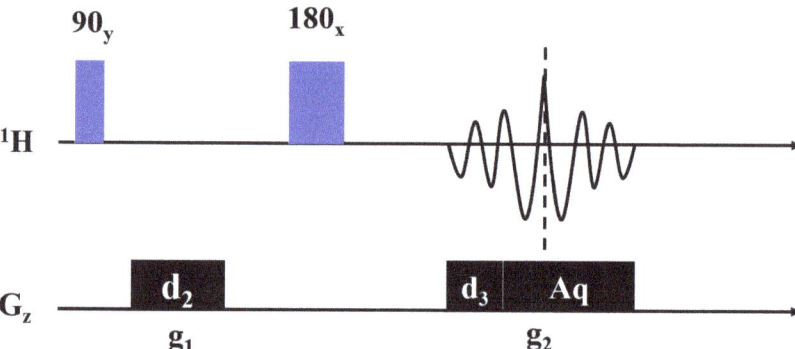

Figure 1. 1D NMR spin-echo imaging sequence. The blue- and black-filled rectangles represent the hard pulses and gradient pulses, respectively. (90_y: 90 degree hard pulse in y direction, 180_x: 180 degree hard pulse in x direction, d_2: the duration of gradient pulse g_1, d_3: the duration of gradient pulse g_2, Aq: acquisition time)In order to collect a complete echo signal and avoid redundant noise, the durations can be set to: $d_2 = (d_3 + aq)/2$.

The Shigemi tube (see Figure 2a,c), constructed as two solid glass sections with no NMR signal at the top and bottom, was used to measure the maximum gradient field strength. Using deionized water as the standard sample, by applying PROFILE, the square-shaped spectrum was obtained and is shown in Figure 2b,d. For each experiment, the bottom interface was placed in the center of the probe, following Bruker's NMR tube measuring cylinder. When water is added, the lower part of the tube is water and the upper part is solid (no proton). Thus, in the corresponding spectrum, the left part is water, and the right part is solid (no signal). The spectra of Sample 1 (220 µL; 2 cm high) are shown in Figure 2b; the frequency length Δf_1 of the profile at half the height of the gradient profile was about 45,780 Hz. According to Equation (5), we have

$$\Delta f_1 = \gamma \cdot G \cdot z_1 = 42.58 \times 10^6 \times 10\% \times G_{max} \times 2 \times 10^{-2} \quad (6)$$

where z_1 is the height of the water in Sample 1. The gradient that was applied is 10 percent of the maximum gradient.

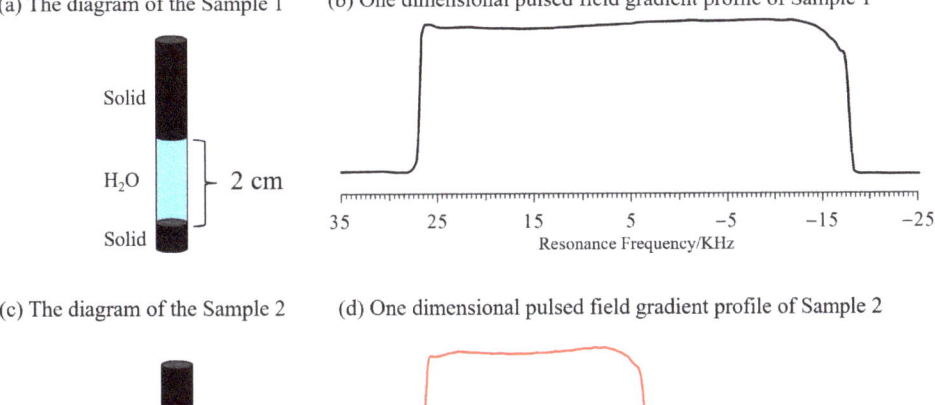

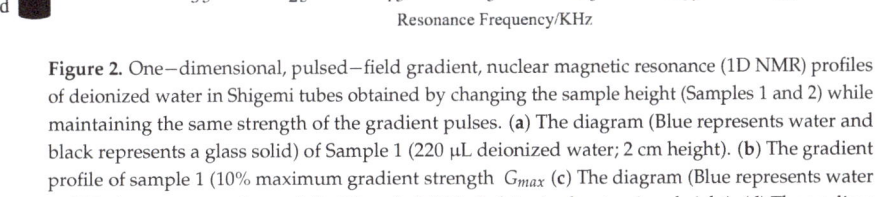

Figure 2. One–dimensional, pulsed–field gradient, nuclear magnetic resonance (1D NMR) profiles of deionized water in Shigemi tubes obtained by changing the sample height (Samples 1 and 2) while maintaining the same strength of the gradient pulses. (**a**) The diagram (Blue represents water and black represents a glass solid) of Sample 1 (220 µL deionized water; 2 cm height). (**b**) The gradient profile of sample 1 (10% maximum gradient strength G_{max} (**c**) The diagram (Blue represents water and black represents a glass solid) of Sample 2 (110 µL deionized water; 1 cm height). (**d**) The gradient profile of sample 2 (10% maximum gradient strength). All profiles were obtained with 1 scan.

From Equation (6), the maximum gradient strength G_{max} could be calculated as 53.8 Gauss/cm.

Similarly, the spectra of Sample 2 (110 µL; 1 cm high) are shown in Figure 2d; the frequency length f_2 of the profile at half the height of the gradient profile was about 22,800 Hz, which is exactly half that of Δf_1. According to Equation (5), we also have

$$\Delta f_2 = \gamma \cdot G \cdot z_2 = 42.58 \times 10^6 \times 10\% \times G_{max} \times 1 \times 10^{-2} \quad (7)$$

where z_2 is the height of the water in Sample 2. The gradient that was applied is 10 percent of the maximum gradient.

From Equation (7), the G_{max} could be calculated as 53.7 Gauss/cm, which is very similar to the result of Sample 1, and both results are in agreement with the data in the Bruker protocol. This result shows the reliability of this method.

2.3. Swelling Behavior of PMAB

After verifying the feasibility of the method, the swelling behavior of PMAB gel was studied using the PROFILE method. The interface was placed in the center of the probe, as described in the previous experiment. The schematic diagram is shown in Figure 3.

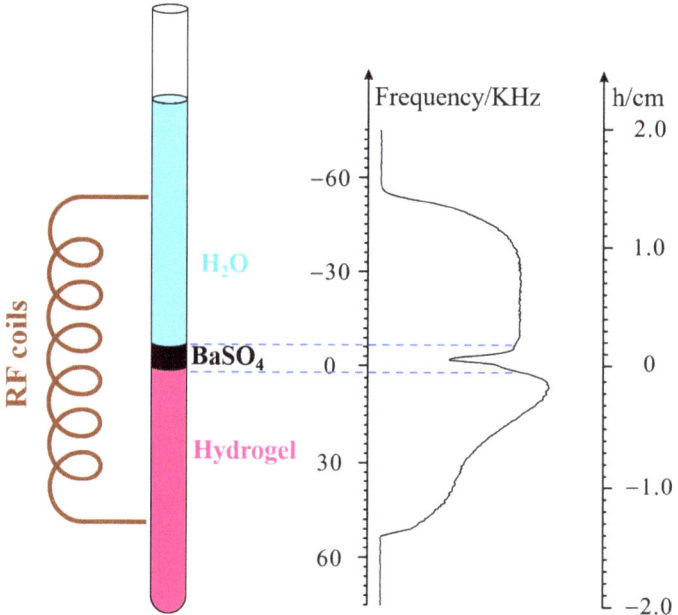

Figure 3. The spatial position of the NMR tube corresponds to the gradient profile. The brown represents the RF coil, and in the left NMR tube schematic diagram, the gel represents pink, the barium sulfate precipitate is black, and the water is blue from the bottom up.

The NMR−spectra−based PROFILE can be obtained by a single scan (a scan takes up to 4 s) and is not sensitive to the field inhomogeneity. Each scan obtains a spectrum. Thus, the dynamic swelling behavior can be studied by continuous NMR scans or according to a designed time schedule. When there is only gel before adding the solvent, the spectra have a square-shaped profile (see Supplementary S7).

As mentioned above, the PMAB gel contains Ba^{2+}, and diluted Na_2SO_4 solution was used as a swelling solvent. Both Ba^{2+} and Na^+ are diamagnetic metal ions and have little effect on the NMR signal. The PMAB gel was directly synthesized in the NMR tube. When the Na_2SO_4 solution was added to the gel, barium sulfate was formed, and a precipitation layer was formed at the interface between the gel and solvent. Because barium sulfate has no hydrogen, the precipitation layer contains much less hydrogen than both the gel and solvent. When PROFILE pulse sequences are performed, the profile of the system can be reflected in the frequency domain. A deep trough appears at the position corresponding to the precipitation layer, which can serve as a sign to trace the swelling of the gel. As the swelling continues, the precipitation layer moves upwards, and the corresponding trough moves downwards.

If there is only a gel before the addition of solvent, the spectral line shows an approximate rectangle. The slight distortion on the right side may be caused by the slight surface irregularities in the gel (see Supplementary S7). Figure 4 shows the spectra of swelling PMAB at different time points. The experimental parameters were the same as those in the verifying experiments. Compared with the spectra of pure gel, when the solvent was added, the profile of the spectra became a large square, with a deep trough at the interface. It is clear that as the swelling continued, the trough shifted toward the negative direction, which means that the PMAB gel swelled, and the interface moved upwards. In the verifying experiments, the relationship between the length change (Δh) and the frequency change (Δf) was calculated to be 4.36 μm/10 Hz.

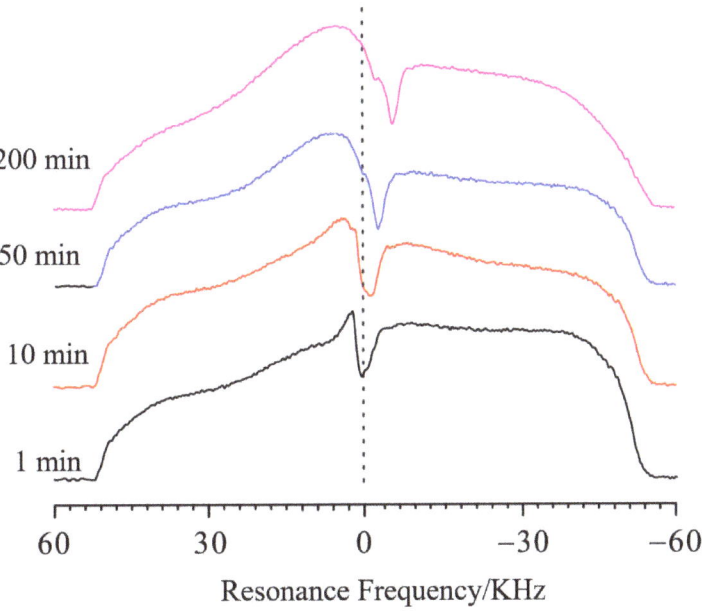

Figure 4. The NMR profile spectra of PMAB gels at different swelling times (25 °C). The black line represents 1 min, the red line represents 10 min, the blue line represents 50 min, and the pink line represents 200 min.

Each profile spectrum on the left side is aligned and the frequency value is 52,400 Hz.

Then, the frequency value of the trough of the PMAB swelling spectrum after 1 min is 2759 Hz; the f values for 10 min, 50 min and 200 min of swelling are 1254 Hz, 167 Hz and −2174 Hz, respectively. If we set the frequency value of the trough of the PMAB swelling spectrum at 1 min as the standard, the Δf values (compared to 1 min) for swelling periods of 10 min, 50 min and 200 min are −1505 Hz, −2592 Hz and −4933 Hz, respectively. Thus, the Δh at different time points can be calculated as 656 μm, 1130 μm and 2151 μm, respectively, and the swelling at different time points can be obtained. The swelling degree of PMAB is quite low, making it difficult to detect with other traditional methods.

To obtain a higher resolution for the spectral characterization of hydrogen and investigate the material transfer, the PGAD gel was designed to be partially deuterated (to lock and shim the field); thus, the intensity of the spectra of the gel part is lower than that of the solvent part. When swelling began, the water diffused into the gel and made a ridge at the top of the gel. As the swelling continued, the ridge became broader, as the water diffused deeper into the gel. This shows that this method could be used to study substance exchange.

The influence of temperature on the swelling was also studied. Figure 5 shows the NMR profile spectra of PMAB gels swelling for 80 min at different temperatures. The frequency values f of that swelling for 10 °C, 25 °C and 40 °C were 1288 Hz, −1889 Hz and −5238 Hz, respectively. Compared to 10 °C, the Δf values of that swelling for 25 °C and 40 °C were −3177 Hz and −6526 Hz, respectively. Thus, compared to 10 °C, the Δh at different temperatures can be calculated as 1385 μm and 2845 μm, respectively, and the difference in swelling height at different temperatures can be obtained.

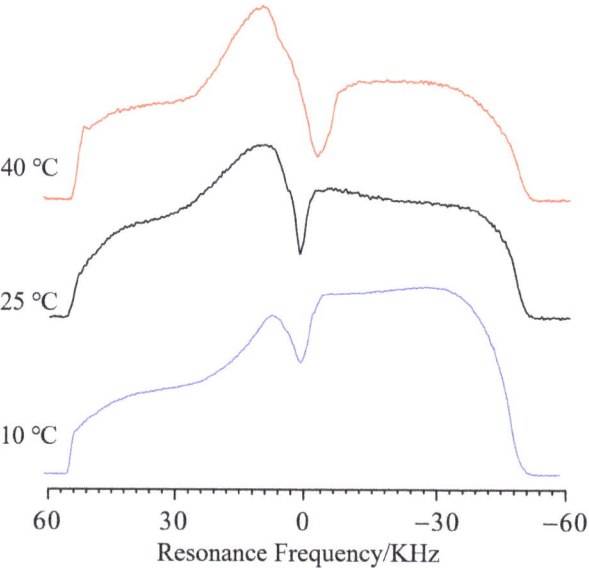

Figure 5. The NMR profile spectra of PMAB gels swelling at different temperatures but for the same amount of time (80 min). The blue line represents 10 °C, the black line represents 25 °C, and the red line represents 40 °C.

When the temperature was higher, the Δf became higher for the same amount of swelling time, which means that the swelling speed is higher at higher temperatures. As the temperature rises, the gradual bulge on the left side of the curve indicates that the upper part of the water enters the gel faster, and the hydrogen signal further increases. A faster decline on the right side (height different from that of the baseline) also indicates that deuterium water from the gel enters the upper water layer.

By transferring the Δf into Δh, the relationship between Δh and swelling time can be observed, as shown in Figure 6, and the plot can be simulated by a logarithmic function:

$$\Delta h(t) = a \cdot \log(t) + b \cdot t \tag{8}$$

where Δh represents the height compared to the initial increase in the gel, t stands for time, and the units are microns and minutes, respectively. When the temperature is 25 °C, the resulting formula is $\Delta h(t) = 121.4 \cdot \log(t) + 2.338 \cdot t$ and the fitting degree R is greater than 0.99. At other temperatures, the experimental data can also be simulated by Equation (8), as shown in Figure 6, and the simulated parameters are shown in Table 1.

Table 1. Influencing factors during swelling and fitting (with 95% confidence bounds) at different temperatures.

T/°C	Fitting Formula	$\Delta h(t) = a \cdot \log(t) + b \cdot t$		
		a	b	R
10		56.33 (±3.03)	1.142 (±0.058)	0.9961
25		121.4 (±15.1)	2.338 (±0.273)	0.9920
40		164.3 (±25.5)	4.159 (±0.585)	0.9753

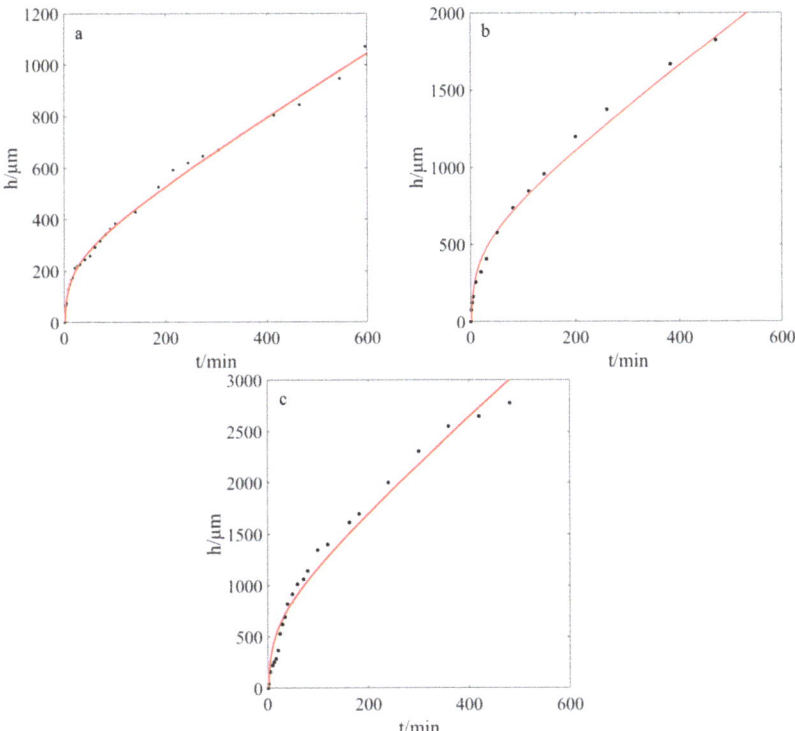

Figure 6. The relationship between gel height and swelling time at different temperatures. (**a**): 10 °C (**b**): 25 °C (**c**): 40 °C (the black dots are experiment data, and the red curves are simulated curves).

The swelling process can be roughly divided into two phases: one segment is more of a logarithmic function and the remaining segment is more of a linear function. For example, at 40 °C, the portion that is closer to 200 min is closer to logarithmic expansion, and at 25 °C, the portion that is closer to 150 min is closer to logarithmic expansion. Finally, at 10 °C, the fraction that is closer to 100 min is closer to logarithmic expansion. The simulated function is similar to that previously reported by Luo [28].

3. Materials and Methods

3.1. Materials

2-Methoxyethyl acrylate (MEA, 98%), polyethylene glycol acrylate (PEGA, Mn = 480), and D_2O (99.9% deuterated) were purchased from Sigma Aldrich, USA; acrylamide (AM, 99%) was purchased from Greagent; azodiisobutyronitrile (AIBN) (98%), and $BaCl_2·2H_2O$ (99%) were purchased from Adamas; and ammonium persulfate (APS, 98%) was purchased from Sinopharm Chemical Reagent Co., Ltd. (Shanghai, China). Other reagents were of analytical grade purity. The above reagents were used directly without treatment.

3.2. Experiment

3.2.1. The Synthesis of PEGA-MEA-AM Multi Component Gel

2-Methoxyethyl acrylate (1.20 g, 9.22 mmol) was added to the solvent of PEGA (3.0 g, 6.25 mmol), dissolved in water (10.40 g, 577.78 mmol) and mixed for 30 min to ensure that the compounds were completely dissolved. Acrylamide and a small amount of APS were then added to the solution and the mixture was stirred at 60 °C for 6 h. The hydrogel was cooled to room temperature. The synthetic pathway is shown in Figure 7.

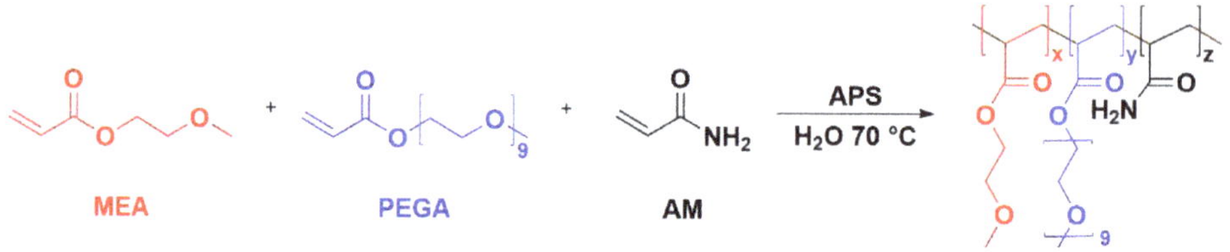

Figure 7. The synthesis of PEGA-MEA-AM multi-component gel. Red molecules represent the MEA fraction, blue molecules represent the PEGA fraction, and black molecules represent the AM fraction.

3.2.2. The Synthesis of PEGA-MEA-AM Gel with Salt

A small amount of 2-methoxyethyl acrylate (1.20 g, 9.22 mmol) was added to AIBN; the mixture was stirred at 15 °C for 30 min in the absence of light until the white solid was totally dissolved. PEGA (3.0 g, 6.25 mmol) was stirred in water (1.44 g, 80.00 mmol) for 30 min to completely dissolve the monomers. The mixture of MEA and AIBN was then added to the solution of PEGA. Next, $BaCl_2 \cdot 2H_2O$ (2.44 g, 10 mmol) was dissolved in water (7 g, 388.89 mmol), and AM (1.2 g, 16.88 mmol) was dissolved in water (2.0 g, 111.11 mmol). These were successively added to the mixture of MEA, AIBN and PEGA. The above solution was injected into the NMR tube and heated at 70 °C, then heated in a water bath by gradient heating for 8 h while remaining upright. The hydrogel was then allowed to cool to room temperature.

3.2.3. The Synthesis of Deuterated Hydrogel

On the basis of the above reaction conditions and method, the deuterated hydrogel can be obtained by changing the water in the reaction into a mixture of deuterated water and water at a certain ratio. However, only some of the water was changed to a mixture of water (1.08 g, 60.00 mmol) and deuterated water (0.4 g, 60.00 mmol). The resulting hydrogel was named PMAB gel.

3.3. Characterization

One-dimensional ^1H and ^{13}C NMR spectra of the monomers and hydrogel samples were acquired with a Bruker 500 MHz AVANCE III NMR spectrometer. All the experiments were performed at 298 K and had no spin. ^1H NMR experiments were acquired using the Bruker sequence "zg". The acquisition parameters were as follows: time domain (number of data points), 39,998; dummy scans, 0; number of scans, 16; acquisition time, 1.99 s; delay time, 5 s; pulse width, 10 μs; spectral width, 19.99 ppm (10,000 Hz); FID resolution 0.500025 Hz; and digitization mode, digital. The total acquisition time was 1 min and 20 s. ^{13}C NMR experiments were performed by optimizing a sequence that was modified to reduce the ringing effect and completely avoid ^1H-^{13}C coupling and NOE during relaxation. After measuring each carbon T_1, a delay time (D_1) equal to $5 \times T_1MAX$ (the longest relaxation time) was set to assure the complete relaxation of ^{13}C nuclei. All these experimental conditions were used to make the carbon integral suitable for quantitative purposes. The experiments were acquired using the Bruker sequence "zgdc" (details are provided in the Supporting Information), and the acquisition parameters were consequently modified and set as follows: time domain (number of data points), 22,722; dummy scans, 0; number of scans, 64; acquisition time, 0.98 s; delay time, 10 s; spectral width, 301.16 ppm (37,878.789 Hz); FID resolution, 3.334 Hz; and digitization mode, digital.

4. Conclusions

In summary, a non-destructive testing method based on the NMR gradient-broadening profile was put forward. Using this method, gel swelling can be continuously observed at the micron level with high sensitivity and spatial resolution, and without requiring contact. By using PROFILE, the sample profile can be extended to the frequency domain.

Under these experimental conditions, a 20 mm sample was extended to 45,780 Hz. Due to the high accuracy of NMR in the frequency domain (normally on the order of Hz), the hydrogel detection accuracy of this method should be at the micron scale, which is much higher than that of most other methods. A three-component hydrogel containing salt ions was synthesized. By choosing the proper swelling medium, the produced precipitate can be used to show the position of the interface between gel and liquid. By performing continuous NMR scans, the swelling behavior can be studied in real time. Thus, this method has obvious advantages when small swelling occurs and in real-time studies. Although the PROFILE was confirmed to be a powerful technique, it still has some restrictions. The RF coil length of the NMR instrument is limited and the maximal swelling height should not exceed that of the coil; thus, the initial sample length should be properly selected.

As an additional advantage, the diffusion of water into the gel was observed, which provides further possibilities to use this method in gel studies, such as in studies of the material exchange between the gel and solvent, dynamics of the solvent molecules, etc. A related investigation is ongoing.

Supplementary Materials: The following supporting information can be downloaded at: https://www.mdpi.com/article/10.3390/molecules28073116/s1, Supplementary S1, The principle of the volumetric method. Supplementary S2, The ^{1}H NMR spectrum of PEGA: Figure S1. The ^{1}H NMR spectrum of PEGA. Supplementary S3, The ^{1}H NMR spectrum of AM: Figure S2. The ^{1}H NMR spectrum of AM. Supplementary S4, The ^{1}H NMR spectrum of MEA: Figure S3. The ^{1}H NMR spectrum of MEA. Supplementary S5, The ^{1}H NMR spectrum of PMAB gel: Figure S4. The 1H NMR spectrum of PMAB gel. Supplementary S6, The ^{13}C and DEPT-135° NMR spectrum of gel: Figure S5. The ^{13}C and DEPT-135° NMR spectra of PMAB gel. Supplementary S7, The NMR profile sperctra of pure PMAB gel without solvent: Figure S6. The NMR profile spectra of pure PMAB gel without solvent.

Author Contributions: R.W., Y.Y. and M.X. performed the organic synthesis experiments and analyzed the spectroscopic data; R.W., J.X., Z.J. and M.Z. wrote the manuscript. M.X. edited and reviewed the article All authors have read and agreed to the published version of the manuscript.

Funding: This research received no external funding.

Institutional Review Board Statement: Not applicable.

Informed Consent Statement: Not applicable.

Data Availability Statement: The original data presented in this study are available from the authors.

Conflicts of Interest: The authors declare no conflict of interest.

Sample Availability: Samples of the compounds are not available from the authors.

References

1. Larive, C.K.; Larsen, S.C. NMR Developments and Applications. *Anal. Chem.* **2017**, *89*, 1391. [CrossRef] [PubMed]
2. Newton, C.G.; Drew, S.L.; Lawrence, A.L.; Willis, A.C.; Paddon-Row, M.N.; Sherburn, M.S. Pseudopterosin synthesis from a chiral cross-conjugated hydrocarbon through a series of cycloadditions. *Nat. Chem.* **2015**, *7*, 82–86. [CrossRef] [PubMed]
3. Wolf, A.; Desport, J.S.; Dieden, R.; Frache, G.; Weydert, M.; Poorters, L.; Schmidt, D.F.; Verge, P. Sequence-Controlled α-Methylstyrene/Styrene Copolymers: Syntheses and Sequence Distribution Resolution. *Macromolecules* **2020**, *53*, 8032–8040. [CrossRef]
4. Shen, Y.; Delaglio, F.; Cornilescu, G.; Bax, A. TALOS+: A Hybrid Method for Predicting Protein Backbone Torsion Angles from NMR Chemical Shifts. *J. Biomol. NMR* **2009**, *44*, 213. [CrossRef] [PubMed]
5. Mateus, N.; Silva, A.M.; Santos-Buelga, C.; Rivas-Gonzalo, J.C.; de Freitas, V. Identification of Anthocyanin-Flavanol Pigments in Red Wines by NMR and Mass Spectrometry. *J. Agric. Food Chem.* **2002**, *50*, 2110–2116. [CrossRef]
6. Parenti, O.; Guerrini, L.; Zanoni, B.; Marchini, M.; Tuccio, M.G.; Carini, E. Use of the 1H NMR technique to describe the kneading step of wholewheat dough: The effect of kneading time and total water content. *Food Chem.* **2020**, *338*, 128120. [CrossRef]
7. Meerwall, E.V.; Ferguson, R.D. Pulsed-Field Gradient NMR Measurements of Diffusion of Oil in Rubber. *J. Appl. Polym. Sci.* **2010**, *23*, 877–885. [CrossRef]
8. Toumpanaki, E.; Shah, D.U.; Eichhorn, S.J. Beyond What Meets the Eye: Imaging and Imagining Wood Mechanical–Structural Properties. *Adv. Mater.* **2020**, *33*, 2001613. [CrossRef]

9. Serša, I. Magnetic resonance microscopy of samples with translational symmetry with FOVs smaller than sample size. *Sci. Rep.* **2021**, *11*, 541. [CrossRef]
10. Balcom, B.J.; Fischer, A.E.; Carpenter, T.A.; Hall, L.D. Diffusion in aqueous gels. mutual diffusion coefficients measured by one-dimensional nuclear magnetic resonance imaging. *J. Am. Chem. Soc.* **1993**, *115*, 3300–3305. [CrossRef]
11. Piechocki, K.; Kozanecki, M.; Saramak, J. Water structure and hydration of polymer network in PMEO2MA hydrogels. *Polymer* **2020**, *210*, 122974. [CrossRef]
12. El-Husseiny, H.M.; Mady, E.A.; Hamabe, L.; Abugomaa, A.; Shimada, K.; Yoshida, T.; Tanaka, T.; Yokoi, A.; Elbadawy, M.; Tanaka, R. Smart/stimuli-responsive hydrogels: Cutting-edge platforms for tissue engineering and other biomedical applications. *Mater. Today Bio* **2022**, *13*, 100186. [CrossRef] [PubMed]
13. Tang, S.; Gong, Z.; Wang, Z.; Gao, X.; Zhang, X. Multifunctional hydrogels for wound dressings using xanthan gum and polyacrylamide. *Int. J. Biol. Macromol.* **2022**, *217*, 944–955. [CrossRef] [PubMed]
14. Gibas, I.; Janik, H. Synthetic polymer hydrogels for biomedical applications. *Chem. Chem. Technol.* **2010**, *4*, 297–298. [CrossRef]
15. Nandi, S.; Winter, H.H. Swelling behavior of partially cross-linked polymers: A ternary system. *Macromolecules* **2005**, *38*, 4447–4455. [CrossRef]
16. Madduma-Bandarage US, K.; Madihally, S.V. Synthetic hydrogels: Synthesis, novel trends, and applications. *J. Appl. Polym. Sci.* **2021**, *138*, 50376. [CrossRef]
17. Pekcan, Ö.; Uğur, Ş. Molecular weight effect on polymer dissolution: A steady state fluorescence study. *Polymer* **2002**, *43*, 1937–1941. [CrossRef]
18. Huang, F.; Zhang, X.; Tang, B.Z. Stimuli-responsive materials: A web themed collection. *Mater. Chem. Front.* **2019**, *3*, 10–11. [CrossRef]
19. Dehbari, N.; Tavakoli, J.; Khatrao, S.S.; Tang, Y. In situ polymerized hyperbranched polymer reinforced poly (acrylic acid) hydrogels. *Mater. Chem. Front.* **2017**, *1*, 1995–2004. [CrossRef]
20. Leslie, K.A.; Doane-Solomon, R.; Arora, S. Gel rupture during dynamic swelling. *Soft Matter* **2021**, *17*, 1513–1520. [CrossRef]
21. Kanchanavasita, W.; Pearson, G.J.; Anstice, H.M. Influence of humidity on dimensional stability of a range of ion-leachable cements. *Biomaterials* **1995**, *16*, 921–929. [CrossRef] [PubMed]
22. Tsai, J.H.; Cuckston, G.L.; Hallmark, B.; Wilson, D.I. Fluid-dynamic gauging for studying the initial swelling of soft solid layers. *AIChE J.* **2019**, *65*, e16664. [CrossRef]
23. Bailet, G.; Denis, A.; Bourgoing, A.; Laux, C.O.; Magin, T.E. Nonintrusive Instrument for Thermal Protection System to Measure Recession and Swelling. *J. Spacecr. Rocket.* **2022**, *59*, 6–18. [CrossRef]
24. Wiehemeier, L.; Cors, M.; Wrede, O.; Oberdisse, J.; Hellweg, T.; Kottke, T. Swelling behaviour of core–shell microgels in H_2O, analysed by temperature-dependent FTIR spectroscopy. *Phys. Chem. Chem. Phys.* **2019**, *21*, 572–580. [CrossRef]
25. Maleki, A.; Beheshti, N.; Zhu, K.; Kjøniksen, A.L.; Nyström, B. Shrinking of Chemically Cross-Linked Polymer Networks in the Postgel Region. *Polym. Bull.* **2007**, *58*, 435–445. [CrossRef]
26. Tavakoli, J.; Zhang, H.P.; Tang, B.Z.; Tang, Y. Aggregation-induced emission lights up the swelling process: A new technique for swelling characterisation of hydrogels. *Mater. Chem. Front.* **2019**, *3*, 664–667. [CrossRef]
27. Lee, D.H.; Ko, R.K.; Cho, Z.H.; Kim, S.S. Applications of NMR imaging to the time-dependent swelling effect in polymers. *Bio-Med. Mater. Eng.* **1996**, *6*, 313–322. [CrossRef]
28. Luo, W.; Liu, Y.; Bu, T. Determination on swelling degree of crosslinking polymer and its empirical characteristic function. *Exp. Technol. Manag.* **2012**, *29*, 45–52.

Disclaimer/Publisher's Note: The statements, opinions and data contained in all publications are solely those of the individual author(s) and contributor(s) and not of MDPI and/or the editor(s). MDPI and/or the editor(s) disclaim responsibility for any injury to people or property resulting from any ideas, methods, instructions or products referred to in the content.

Article

The Effect of an Accelerator on Cement Paste Capillary Pores: NMR Relaxometry Investigations

Ioan Ardelean

Physics and Chemistry Department, Technical University of Cluj-Napoca, 400114 Cluj-Napoca, Romania; ioan.ardelean@phys.utcluj.ro

Abstract: Nuclear Magnetic Resonance (NMR) relaxometry is a valuable tool for investigating cement-based materials. It allows monitoring of pore evolution and water consumption even during the hydration process. The approach relies on the proportionality between the relaxation time and the pore size. Note, however, that this approach inherently assumes that the pores are saturated with water during the hydration process. In the present work, this assumption is eliminated, and the pore evolution is discussed on a more general basis. The new approach is implemented here to extract information on surface evolution of capillary pores in a simple cement paste and a cement paste containing calcium nitrate as accelerator. The experiments revealed an increase of the pore surface even during the dormant stage for both samples with a faster evolution in the presence of the accelerator. Moreover, water consumption arises from the beginning of the hydration process for the sample containing the accelerator while no water is consumed during dormant stage in the case of simple cement paste. It was also observed that the pore volume fractal dimension is higher in the case of cement paste containing the accelerator.

Keywords: NMR relaxometry; cement hydration; accelerators; pore evolution; partially saturated; fractal dimension

Citation: Ardelean, I. The Effect of an Accelerator on Cement Paste Capillary Pores: NMR Relaxometry Investigations. *Molecules* **2021**, *26*, 5328. https://doi.org/10.3390/molecules26175328

Academic Editor: Igor Serša

Received: 30 July 2021
Accepted: 28 August 2021
Published: 2 September 2021

Publisher's Note: MDPI stays neutral with regard to jurisdictional claims in published maps and institutional affiliations.

Copyright: © 2021 by the author. Licensee MDPI, Basel, Switzerland. This article is an open access article distributed under the terms and conditions of the Creative Commons Attribution (CC BY) license (https://creativecommons.org/licenses/by/4.0/).

1. Introduction

Reducing the carbon footprint associated with cement production is an important objective nowadays, and it can be achieved by a better exploitation of the cement composites, for instance, using 3D printing technology [1]. Building without formworks has the advantage of saving cost, time and materials associated with formwork construction. However, it also implies some significant materials engineering challenges to substitute all the requirements which are typically fulfilled by the formwork. One of the most important characteristics of concrete extrusion, opposite to castable concrete, is that it requires fast-setting and low slump [2]. These requirements must be fulfilled because the material is unsupported after leaving the print nozzle. That is why the cement-based materials for 3D printing applications are designed to exhibit fast build-up process. Moreover, it is necessary that the mix energy used to break the bounds inside the material should be low to enable delivery by normal pumps [1].

The speed of strength development of cement mixtures is controlled by adding accelerators [3–5] or retarders [5–7] but also admixtures such as silica fume [8] or carbon nanotubes [9]. Note, however, that choosing the correct type and amount of accelerator, retarder or admixture is a difficult task. On the one hand, the rapid strength development allows building of more layers on top of one another, and on the other hand, it reduces the building time. Moreover, long setting times would be necessary to keep the surface of the layers chemically active to form interfaces between layers of which behavior is close to the bulk material [2,5,10,11]. Consequently, hydration kinetics must be controlled in an accurate manner so that the material does not set during the printing process but, instead, right after deposition to support its own weight and that of subsequently deposited layers of material. There is a so-called "open time" [5,10] during which a specified volume of material must be extruded in the 3D printing process. In practice, the open time overlaps with the dormant stage of the hydration process [5]. The open time is influenced by the cement sample constituents and the temperature [12]. Controlling the open time for cement mixtures is essential for successful 3D printing applications. That is why the development of new approaches to monitor the hydration process is an important issue.

Low-field nuclear magnetic resonance (NMR) relaxometry techniques are valuable instruments for monitoring the cement hydration under the influence of different additives and admixtures [6–8,13–19]. They are completely noninvasive, do not require any special sample preparation and can be applied even during the hydration process. By using NMR relaxometry techniques it is possible to monitor the pore evolution and water consumption during the hydration process. Note, however, that when applying NMR relaxometry to cement-based materials, during the hydration, it is arbitrarily assumed that the pores are fully saturated with liquid, an assumption which was never demonstrated but continues to be used in the literature. Here, this assumption will be disregarded, and the data will be analyzed in a more general manner. The results of the new approach will be compared with those based on the pore volume fractal dimension analysis [20–24].

2. Theoretical Background
2.1. The Process of Cement Hydration

The cement hydration process starts immediately after mixing the cement grains with water molecules [25]. It produces not only a simple neutral colloidal gel where the cement grains are dispersed in water, but also some internal organization arises [25]. This is because some amount of water almost instantly combines with the cement grains producing micro-organized systems such as flocculation of cement grains, chemical reactions, ettringitic pores and so on [25]. Water inside these pores was called "embedded water" and is characterized by a shorter transverse relaxation time in NMR relaxometry [19]. The remaining water, filling in the empty space between these microstructures, represents the "capillary water" and has a longer relaxation time [19]. Note that in the present work, the cement chemistry abbreviations are used, where $C = CaO$, $S = SiO_2$, $A = Al_2O_3$, $F = Fe_2O_3$ and $H = H_2O$ [25].

It is customary to separate the hydration process of cement paste into five stages: the initial stage, the dormant stage, the hardening stage, the cooling stage and the densification stage [12,19,26]. These hydration stages were extensively discussed in the literature both with respect of their duration and the influences introduced by different experimental parameters [6,19,25]. Here, we will only shortly describe them to understand the pore development. Thus, during the initial stage (less than 15 min) the C3A component of the clinker reacts with water and releases heat. The ettringite formation starts immediately creating a layer around the cement grains that isolates the paramagnetic relaxation centers (Fe^{3+}) on the surface of cement grains from the bulk water thus reducing the relaxation rate [6,15,27]. During the dormant stage (between 15 min and 2 h), the silicates (C3S and C2S) dissolve in water and the calcium and hydroxide ions are slowly released into the solution. No changes in the porosity and no increase of the ettringite layer are expected during this stage. During the hardening stage (between 2 h and 12 h), the hydroxide and calcium ions reach a critical concentration and the calcium silicate hydrate (C-S-H) and calcium hydroxide (CH) begin the crystallization process. Furthermore, during this period the development of the ettringite layer continues. In the cooling period (between 12 h and 20 h), the reaction of C3S is much slower because the C-S-H and CH restricts the contact between water and unhydrated cement grains. However, the porosity reduces, and the relaxation time decreases accordingly. The densification stage lasts from 20 h to the end of the cement hydration (conventionally considered 28 days). During this period C-S-H and CH form a solid mass; this produces an increase in the strength and durability of cement paste and, at the same time, a decrease in the permeability. The slow formation of hydrate products occurs and continues providing that water and unhydrated silicates are present.

2.2. NMR Relaxation in Partially Saturated Pores

In the NMR relaxometry of porous materials it is routinely considered that the pores are saturated with the filling liquid and the observed transverse relaxation rate of confined molecules is a weighted average between the bulk relaxation rate and the surface relaxation rate of molecules confined inside a thin layer of few molecular diameters, uniformly covering the internal surface of the pores [16,28]. However, there are many situations when the pores are only partially saturated with the liquid. In that case, the relaxation rate depends on the pore filling and the liquid distribution on the pore surface [16,28–30]. Assuming that the confined molecules wet the surface of the pores, the relaxation rate can be expressed as a weighted average between the relaxation rate of the remaining bulk-like liquid and the surface relaxation rate. In the case of cement-based materials, the bulk-like contribution can be neglected, and the relaxation rate can be approximated as

$$\frac{1}{T_2} = \rho \frac{S_p}{V_l}. \tag{1}$$

where S_p is the pore surface and V_l is the liquid volume inside the pores, under partially saturated conditions. The constant ρ is called relaxivity and depends on pore surface properties, filling molecules and the intensity of the magnetic field of the experiment [15,18,27,30]. In the case of saturated pores, $V_l = V_p$, where V_p is the pore volume. Provided that relaxivity of the surface is known, for saturated pores it is possible to determine the pore size distribution from relaxation time distribution measurements. Note that in all the investigations reported in the literature related to relaxation studies on cement hydration, it is a priori assumed that $V_l = V_p$. In the present work, this assumption is eliminated, and the data are evaluated based on Equation (1) where only the volume of the confined liquid is considered. Consequently, by representing the ratio V_l/T_2 as a function of hydration time, information on surface evolution during the hydration can be extracted. This approach will be exploited here to monitor the surface evolution of the capillary pores in cement paste during the first hours of hydration (dormant and hardening stage).

2.3. The Transverse Relaxation Time and the Fractal Dimension

Starting with the introduction of fractals by Mandelbrot in 1977 [31], the geometrical structure of pores and the pore surface could be described based on fractal dimension [20–22]. The concept of fractal dimension can be used also for analyzing a volume distribution of pores [22]. Thus, a uniform pore size distribution corresponds to the topological dimension of three while a variation in the pore size distribution can be described by a fractal volume dimension $D_f < 3$. From the practical side, it was demonstrated that, in the case of fiber recycled concrete, there is a linear relationship between the pore volume fractal dimension and the strength [21].

The basic theory that relates the fractal geometry of the porous structure to the NMR relaxation data is comprehensively described by Zhang and Weller [22]. They have shown that the transverse relaxation time of molecules confined inside porous structures can be related to the pore volume fractal dimension, D_f, in accordance with the work in [22]:

$$\log(V_c) = \left(3 - D_f\right) \log(T_2) - \left(3 - D_f\right) \log\left(T_2^{\max D}\right) \tag{2}$$

where V_c is the cumulative volume fraction of the wetting fluid in the pore space, with the relaxation time smaller than T_2. It is defined as the ratio between the volume of pores characterized by a relaxation time smaller than T_2 and the total pore volume. The cumulative volume can be calculated from NMR relaxation time distribution by dividing the area under the curve, obtained for relaxation times smaller than T_2, to the total area of the distribution [22]. $T_2^{\max D}$ represents the maximum detectable relaxation time in the relaxation time distribution. Note that the above formula was derived under the condition $T_2 \gg T_2^{\min D}$, where $T_2^{\min D}$ is the minimum relaxation time detected in the distribution. That is why fitting of cumulative volumes with a linear curve to extract the slope and thus to determine the pore volume fractal dimension is only possible under such circumstances [22]. Consequently, the data will be fitted here only for relaxation values close to the T_2^{maxD}. The above Equation (2) originates in the assumption that the pore dimension is proportional with the relaxation time. In the case of partially saturated pores, the above equation is still valid, but the probed dimension refers there to the liquid volume inside the pore space.

3. Results and Discussion

The samples under investigation were the simple cement paste (CP) prepared with a water-to-cement ratio of 0.4 and a cement paste additionally containing 3% by cement weight of $Ca(NO_3)_2$, as accelerator. The echo trains recorded in the Carr–Purcell–Meiboom–Gill (CPMG) [32,33] experiments, performed at 15 min intervals, during the first 6 h of hydration are shown in Figure 1. Comparing the time evolution of the two samples it is observed a faster increase in the slope of the echo trains recorded for the sample containing the accelerator (Figure 1b). This effect arises as calcium nitrate increases the concentration of calcium ions leading to a faster supersaturation with respect to the silicate hydrates [5]. This in turn produces a faster development of the C-S-H phase and faster consumption of the capillary water in the hydration process (see Section 2.1 above).

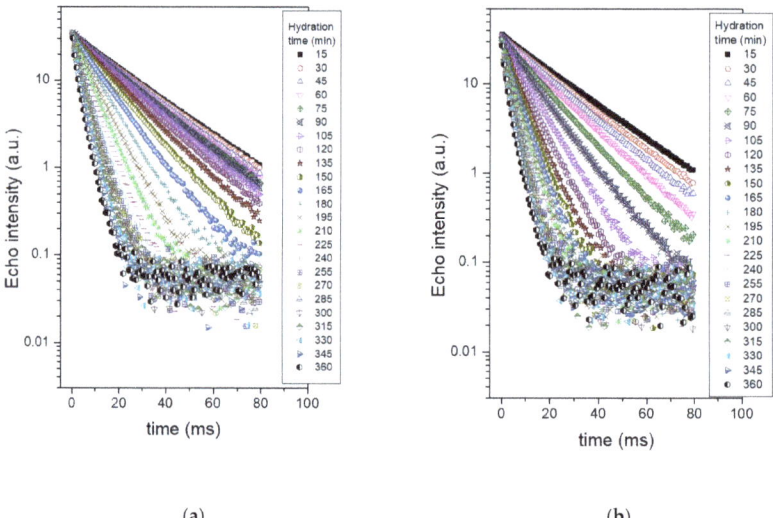

Figure 1. CPMG echo trains recorded for the two samples, at different hydration times, as indicated in the legend: (**a**) Cement paste sample prepared at a water-to-cement ratio of 0.4; (**b**) Cement paste prepared at 0.4 water-to-cement ratio which additionally contains 3% accelerator ($Ca(NO_3)_2$).

To monitor the effects of water consumption and the evolution of capillary pores the curves depicted in Figure 1 can be analyzed using a numerical Laplace inversion [34,35]. The numerical analysis provides the relaxation time distributions shown in Figure 2. One can observe three peaks corresponding to different water reservoirs inside the sample: two peaks of smaller area and one peak of larger area. The first and the second peak (from the left) can be attributed to the water inside intra- and inter C-S-H pores, respectively [14–16,27]. They arise immediately after mixing the cement grains with water molecules and remain constant during the dormancy stage. The shift in the position of the second peak to smaller values can be attributed to a denser inter C-S-H phase [19]. Beginning with the acceleration stage, the area of the two peaks starts to increase showing that more and more intra and inter C-S-H pores are formed inside the sample.

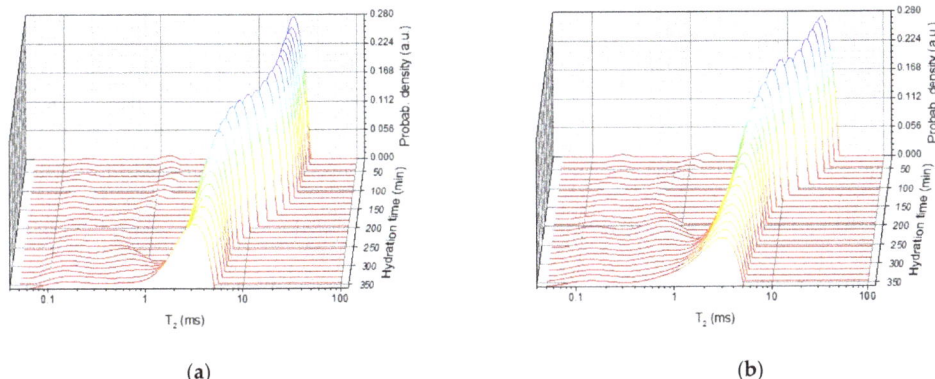

Figure 2. Relaxation time distributions for the two samples, at different hydration times, as indicated in the legend: (**a**) Cement paste sample prepared at a water-to-cement ratio of 0.4; (**b**) Cement paste prepared at 0.4 water-to-cement ratio which additionally contains 3% accelerator (Ca(NO$_3$)$_2$), by cement mass.

The third peak (the largest one) in the relaxation time distributions, shown in Figure 2, corresponds to the capillary water contained between the cement grains during the first stage of hydration and will be monitored here in more detail. The area of the peak is proportional with the amount of water inside the capillary pores. Figure 3 shows the dependence of the peak area (Figure 3a) and of the position of the peak maximum (Figure 3b) on the hydration time in the case of the simple cement paste (CP) and the cement paste containing the accelerator (CP + 3% Ca(NO$_3$)$_2$). One can observe faster decay of the peak area and of T_2^{max} in the case of sample containing the accelerator as compared with the simple cement paste. The faster evolution demonstrates accelerated hydration dynamics introduced by the calcium nitrate. This observation is consistent with the previous reports that calcium nitrate increases the concentration of calcium ions, leading to a faster supersaturation with respect to the silicate hydrates and thus producing a faster development of the C-S-H phase and faster consumption of the capillary water in the hydration process [5].

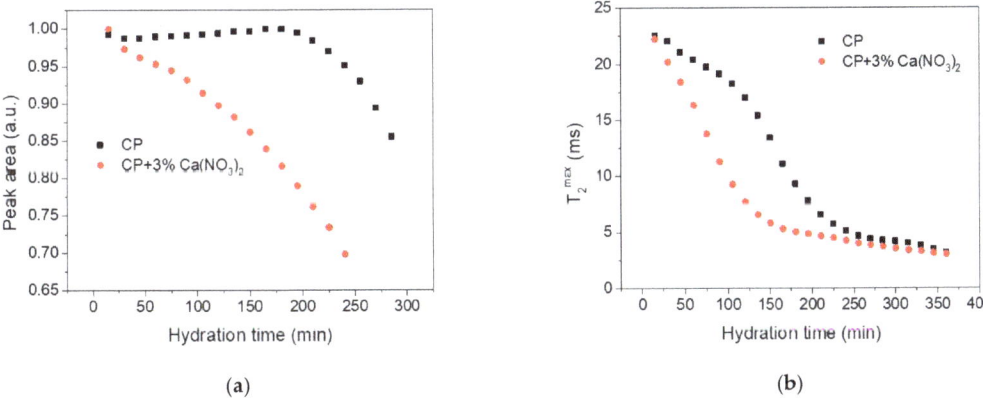

Figure 3. (**a**) The peak area evolution during hydration for the capillary pores of the two samples; (**b**) Evolution of the relaxation time corresponding to the peak maximum during the hydration.

Comparing the area evolution of the capillary water peak for the two samples (Figure 3a) it is observed that the area of CP peak remains constant up to 3 h of hydration, but a continuous decrease arises in the case of cement paste containing the accelerator (CP + 3% Ca(NO$_3$)$_2$).

The relatively constant peak area of the CP capillary pores indicates that the pore volume does not change during the dormancy stage and water is not consumed to form hydration products. Consequently, the reduction of the relaxation time observed in Figure 3b during the dormant stage can only be attributed to the changes in the pore surface. To demonstrate this conclusion, we notice from Equation (1) that if we represent the ratio V_l/T_2^{max} as a function of hydration time, this ratio will describe the pore surface evolution. Note that V_l is proportional with the peak area, consequently $Area/T_2^{max}$ is represented in Figure 4 as a function of hydration time. According to Equation (1), this representation is independent of the assumption that the pores are saturated with water and provides information of pore surface evolution during the hydration. As one can observe, the surface of capillary pores increases for both samples, but the process is faster for the sample containing the accelerator (circles).

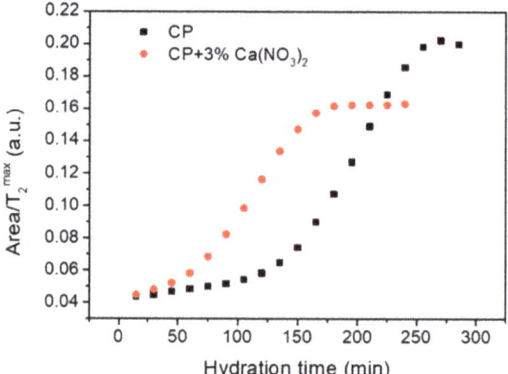

Figure 4. Ratio between the peak area and the peak maximum in the relaxation time distributions shown in Figure 2 for the capillary water component of the two samples, as indicated in the legend. In the case of sample containing the accelerator, the evaluation was restricted to shorter relaxation times due to the overlapping of the capillary peak with the signal corresponding to the C-S-H pores.

The changes in the pore morphology can be also described by monitoring the volume fractal dimension D_f. This quantity can be evaluated based on the log-log representation suggested by Equation (2). The $\log(V_c)$ versus $\log(T_2)$ curves are shown in Figure 5 for the two samples during the first six hours of hydration. A decrease in the slope for both samples is obtained during the hydration which is equivalent, based on Equation (2), with an increase in the fractal dimension. As can be observed from the figure, the fractal dimension varies from 2.237 to 2.663 in the case of simple cement paste (Figure 5a), and from 2.356 to 2.723 in the case of cement paste containing the accelerator (Figure 5b). This variation in fractal dimension is continuous (see the slopes in Figure 5c,d, respectively) and correlated with the change in the surface size revealed in Figure 4. Note that, the increase in the fractal dimension of cement based materials was associated with an increase in their strength [21]. Here, the increase in fractal dimension could be again correlated with the increase in strength during the hydration. However, establishing a direct relationship between compressive strength and pore volume fractal dimension, determined by NMR, requires supplemental investigations.

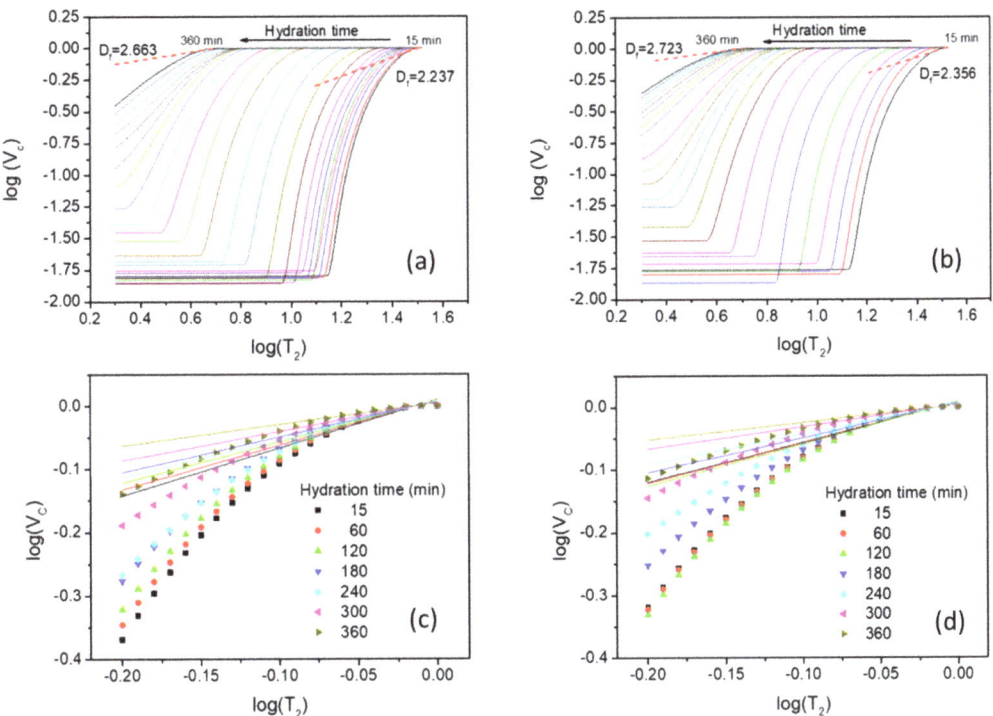

Figure 5. Versus $\log(T_2)$ at different hydration times for the capillary water peak in Figure 2 in the case of cement paste sample (**a**) and the cement paste containing the accelerator (**b**). T_2 was considered here in milliseconds. The linear fits with Equation (2) for the two samples were performed after shifting the data on the abscissa axis, for a direct comparison. The same region, between −0.08 and 0.00 on the abscissa axis, was used for all fits. A continuous change in the slope is observed both for the simple cement paste (**c**) and the sample containing the accelerator (**d**). The coefficients of determination R^2 in all fits were bigger than 0.93, indicating a good linear approximation.

4. Materials and Methods

4.1. Sample Preparation

Two samples, with the same water-to-cement ratio of 0.4, were prepared and comparatively investigated in the present study. One sample is a pure cement paste (CP) obtained by mixing Portland cement with water and the other additionally contains 3% $Ca(NO_3)_2$ by cement mass. The two samples were prepared with white Portland cement CEM I 52.5 R (Holcim, Bucuresti, Romania), fulfilling the European Standard BS EN 197-1. The white cement was chosen here on purpose due to its low content of iron oxide (<0.5%), in order to reduce internal gradients that can be induced by susceptibility difference between the solid matrix and filling liquid [13,36]. The $Ca(NO_3)_2$ accelerator was acquired from NORDIC Chemicals SRL, Cluj-Napoca, Romania. Before mixing with cement powder, the accelerator was dissolved in water. The ingredients were then mixed for 5 min using a mixer, at 100 rpm. The resulting paste was poured into 10 mm glass tubes and then introduced inside the probe head. The tubes were sealed to prevent water evaporation. The first NMR measurements were always performed at 15 min counting from the initiation of the mixing process and the last after 6 h of hydration.

4.2. NMR Measurements

Transverse relaxation measurements of fluids confined inside the cement paste pores were performed using the well-known Carr–Purcell–Meiboom–Gill (CPMG) technique [32,33].

In the CPMG pulse sequence, an initial 90° radiofrequency pulse around the y-axis is followed by a train of 180° pulses around the x-axis at time instants $\tau, 3\tau, 5\tau, \ldots$ The attenuation of the echo train recorded at the time instants $2\tau, 4\tau, 6\tau, \ldots$ contains information about the transverse relaxation time distribution inside the sample. If the sample is heterogeneous and the echo train attenuation is multiexponential, a numerical Laplace inversion [34,35] of the echo train provides the relaxation time distribution. In the case of liquids confined inside porous media, the relaxation time distribution mimics the pore size distribution and a quantitative description of the pore sizes can be obtained provided that the relaxivity of the pore surface is known from independent measurements. The main advantage of such a multiple echo technique for the determination of the transverse relaxation time is that it allows fast multiple accumulations of the echo train signal—an important issue in increasing the detection sensitivity. Furthermore, due to the short echo times implemented it reduces diffusion effects on transverse relaxation measurements [13,36].

The experiments were performed using a Bruker Minispec MQ20 instrument (Bruker BioSpin GmbH, Rheinstetten, Germany), operating at a proton resonance frequency of 20 MHz. The CPMG echo trains consisting of 1000 echoes were recorded after 32 scans, with an echo time of 80 µs and a recycle delay of 0.5 s. With these parameters, the recording duration of one echo train was short enough to prevent sample changes during the experiment. The measurements were performed at 35 °C, the working temperature of the Bruker Minispec MQ20 instrument, without using the external temperature control unit.

5. Conclusions

Monitoring the evolution of NMR relaxation time under the influence of different parameters can be used as a tool in determining the pore evolution of cement-based materials. Here, the influence of an accelerator on the surface evolution of capillary pores was studied using a low-field NMR instrument. The approach employed here removes the generally used assumption that the capillary pores are saturated with water during the early hydration. Based on the new approach, it was shown that the surface of capillary pores increases during the early hydration (less than 6 h) even during the dormant stage and this effect is higher in the presence of an accelerator. However, in the case of simple cement paste, the pore volume sems to be constant during the dormant stage, and a clear identification of the dormant stage is possible by plotting the area of the peak versus hydration time. In the presence of an accelerator there is a continuous consumption of water in capillary pores, and one cannot clearly identify the dormant stage, at least for the accelerator content and hydration temperature used in the experiments. The volume fractal dimension of capillary pores increases during the hydration in the case of both samples, with higher values in the case of cement paste containing the accelerator. This indicates a more uniform pore distribution in the case of the sample containing the accelerator.

Funding: This work was supported by a grant of the Romanian Ministry of Education and Research, CNCS–UEFISCDI, project number PN-III-P4-ID-PCE-2020-0533, within PNCDI III.

Institutional Review Board Statement: Not applicable.

Informed Consent Statement: Not applicable.

Data Availability Statement: Not applicable.

Conflicts of Interest: The author declares no conflict of interest.

Sample Availability: Samples of the compounds are not available.

References

1. Buswell, R.A.; De Silva, W.R.L.; Jones, S.Z.; Dirrenberger, J. 3D printing using concrete extrusion: A roadmap for research. *Cem. Concr. Res.* **2018**, *112*, 37–49. [CrossRef]
2. Bos, F.; Wolfs, R.; Ahmed, Z.; Salet, T. Additive manufacturing of concrete in construction: Potentials and challenges of 3D concrete printing. *Virtual Phys. Prototyp.* **2016**, *11*, 209–225. [CrossRef]

3. Aïtcin, P.C. Accelerators. In *Science and Technology of Concrete Admixtures*; Woodhead Publishing: Sawston, UK, 2016; pp. 405–413. [CrossRef]
4. Myrdal, R. *Accelerating Admixtures for Concrete, State of the Art*; SINTEF: Trondheim, Norway, 2007; ISBN 978-82-536-0989-8.
5. Marchon, D.; Kawashima, S.; Bessaies-bey, H.; Mantellato, S.; Ng, S. Hydration and rheology control of concrete for digital fabrication: Potential admixtures and cement chemistry. *Cem. Concr. Res.* **2018**, *112*, 96–110. [CrossRef]
6. Pop, A.; Badea, C.; Ardelean, I. The effects of different superplasticizers and water-to-cement ratios on the hydration of gray cement using T2-NMR. *Appl. Magn. Reson.* **2013**, *44*, 1223–1234. [CrossRef]
7. Pop, A.; Bede, A.; Dudescu, M.C.; Popa, F.; Ardelean, I. Monitoring the influence of aminosilane on cement hydration via low-field NMR relaxometry. *Appl. Magn. Reson.* **2016**, *47*, 191–199. [CrossRef]
8. Crețu, A.; Mattea, C.; Stapf, S.; Ardelean, I. The effect of silica fume and organosilane addition on the porosity of cement paste. *Molecules* **2020**, *25*, 1762. [CrossRef]
9. Irshidat, M.R.; Al-Nuaimi, N.; Rabie, M. Influence of carbon nanotubes on phase composition, thermal and post-heating behavior of cementitious composites. *Molecules* **2021**, *26*, 850. [CrossRef]
10. Le, T.T.; Austin, S.A.; Lim, S.; Buswell, R.A.; Gibb, A.G.F.; Thorpe, T. Mix design and fresh properties for high-performance printing concrete. *Mater. Struct.* **2012**, *45*, 1221–1232. [CrossRef]
11. Moeini, M.A.; Hosseinpoor, M.; Yahia, A. Effectiveness of the rheometric methods to evaluate the build-up of cementitious mortars used for 3D printing. *Constr. Build. Mater.* **2020**, *257*, 119551. [CrossRef]
12. Badea, C.; Bede, A.; Ardelean, I. The effect of silica fume on early hydration of white Portland cement via fast field cycling-NMR relaxometry. In Proceedings of the AIP Conference, Yogyakarta, Indonesia, 9–10 November 2017; Volume 1917, p. 040001-5.
13. Stepišnik, J.; Ardelean, I. Usage of internal magnetic fields to study the early hydration process of cement paste by MGSE method. *J. Magn. Reson.* **2016**, *272*, 100–107. [CrossRef]
14. McDonald, P.J.; Rodin, V.; Valori, A. Characterisation of intra- and inter-C–S–H gel pore water in white cement based on an analysis of NMR signal amplitudes as a function of water content. *Cem. Concr. Res.* **2010**, *40*, 1656–1663. [CrossRef]
15. Korb, J.-P.; Monteilhet, L.; McDonald, P.J.; Mitchell, J. Microstructure and texture of hydrated cement-based materials: A proton field cycling relaxometry approach. *Cem. Concr. Res.* **2007**, *37*, 295–302. [CrossRef]
16. Bede, A.; Scurtu, A.; Ardelean, I. NMR relaxation of molecules confined inside the cement paste pores under partially saturated conditions. *Cem. Concr. Res.* **2016**, *89*, 56–62. [CrossRef]
17. Badea, C.; Pop, A.; Mattea, C.; Stapf, S.; Ardelean, I. The effect of curing temperature on early hydration of gray cement via Fast Field Cycling-NMR relaxometry. *Appl. Magn. Reson.* **2014**, *45*, 1299–1309. [CrossRef]
18. Ardelean, I. Applications of Field-Cycling NMR Relaxometry to Cement Materials. In *Field-Cycling NMR Relaxometry: Instrumentation, Model Theories and Applications*; Kimmich, R., Ed.; The Royal Society of Chemistry: Cambridge, UK, 2019; Chapter 19; pp. 462–489, ISBN 978-1-78801-154-9.
19. Faure, P.F.; Rodts, S. Proton NMR relaxation as a probe for setting cement pastes. *Magn. Reson. Imaging* **2008**, *26*, 1183–1196. [CrossRef]
20. Issa, M.A.; Issa, M.A.; Islam, M.S.; Chudnovsky, A. Fractal dimension-a measure of fracture roughness and toughness of concrete. *Eng. Fract. Mech.* **2003**, *70*, 125–137. [CrossRef]
21. Zhou, J.; Kang, T.; Wang, F. Pore structure and strength of waste fiber recycled concrete. *J. Eng. Fiber. Fabr.* **2019**, *14*, 1558925019874701. [CrossRef]
22. Zhang, Z.; Weller, A. Fractal dimension of pore-space geometry of an eocene sandstone formation. *Geophysics* **2014**, *79*, D377–D387. [CrossRef]
23. Hansen, J.P.; Skjeltorp, A.T. Fractal pore space and rock permeability implications. *Phys. Rev. B* **1988**, *38*, 2635–2638. [CrossRef]
24. Lü, Q.; Qiu, Q.; Zheng, J.; Wang, J.; Zeng, Q. Fractal dimension of concrete incorporating silica fume and its correlations to pore structure, strength and permeability. *Constr. Build. Mater.* **2019**, *228*, 116986. [CrossRef]
25. Taylor, P.C.; Kosmatka, S.H.; Voigt, G.F.; Ayers, M.E.; Davis, A.; Fick, G.J.; Gajda, J.; Grove, J.; Harrington, D.; Kerkhoff, B.; et al. *Integrated Materials and Construction Practices for Concrete Pavement: A State of the Practice Manual*; Federal Highway Administration: Washington, DC, USA, 2006.
26. Bhardwaj, A.; Jones, S.Z.; Kalantar, N.; Pei, Z.; Vickers, J.; Wangler, T.; Zavattieri, P.; Zou, N. Additive Manufacturing Processes for Infrastructure Construction: A Review. *J. Manuf. Sci. Eng.* **2019**, *141*, 091010. [CrossRef]
27. Korb, J.P. Microstructure and texture of cementitious porous materials. *Magn. Reson. Imaging* **2007**, *25*, 466–469. [CrossRef]
28. Gallego-Gómez, F.; Cadar, C.; López, C.; Ardelean, I. Imbibition and dewetting of silica colloidal crystals: An NMR relaxometry study. *J. Colloid Interface Sci.* **2020**, *561*, 741–748. [CrossRef]
29. Simina, M.; Nechifor, R.; Ardelean, I. Saturation-dependent nuclear magnetic resonance relaxation of fluids confined inside porous media with micrometer-sized pores. *Magn. Reson. Chem.* **2011**, *49*, 314–319. [CrossRef]
30. Cadar, C.; Cotet, C.; Baia, L.; Barbu-Tudoran, L.; Ardelean, I. Probing into the mesoporous structure of carbon xerogels via the low-field NMR relaxometry of water and cyclohexane molecules. *Microporous Mesoporous Mater.* **2017**, *251*, 19–25. [CrossRef]
31. Mandelbrot, B.B. *Fractals: Form, Chance, and Dimension*; W. H. Freeman & Company: San Francisco, CA, USA, 1977; ISBN 9780716704737.
32. Carr, H.Y.; Purcell, E.M. Effects of diffusion on free precession in nuclear magnetic resonance experiments. *Phys. Rev.* **1954**, *94*, 630–638. [CrossRef]

33. Meiboom, S.; Gill, D. Modified spin-echo method for measuring nuclear relaxation times. *Rev. Sci. Instrum.* **1958**, *29*, 688–691. [CrossRef]
34. Provencher, S.W. CONTIN: A general purpose constrained regularization program for inverting noisy linear algebraic and integral equations. *Comput. Phys. Commun.* **1982**, *27*, 229–242. [CrossRef]
35. Venkataramanan, L.; Song, Y.; Hürlimann, M.D. Solving Fredholm integrals of the first kind with tensor product structure in 2 and 2.5 dimensions. *IEEE Trans. Signal Process.* **2002**, *50*, 1017–1026. [CrossRef]
36. Pop, A.; Ardelean, I. Monitoring the size evolution of capillary pores in cement paste during the early hydration via diffusion in internal gradients. *Cem. Concr. Res.* **2015**, *77*, 76–81. [CrossRef]

Article

The Role of Titanium Dioxide on the Hydration of Portland Cement: A Combined NMR and Ultrasonic Study

George Diamantopoulos [1,2], Marios Katsiotis [2], Michael Fardis [2], Ioannis Karatasios [2], Saeed Alhassan [3], Marina Karagianni [2], George Papavassiliou [2] and Jamal Hassan [1,*]

1   Department of Physics, Khalifa University, Abu Dhabi 127788, UAE; g.diamantopoulos@inn.demokritos.gr
2   Institute of Nanoscience and Nanotechnology, NCSR Demokritos, 15310 Aghia Paraskevi, Attikis, Greece; mikappa1@gmail.com (M.K.); m.fardis@inn.demokritos.gr (M.F.); i.karatasios@inn.demokritos.gr (I.K.); m.karagianni@inn.demokritos.gr (M.K.); g.papavassiliou@inn.demokritos.gr (G.P.)
3   Department of Chemical Engineering, Khalifa University, Abu Dhabi 127788, UAE; saeed.alkhazraji@ku.ac.ae
*   Correspondence: jamal.hassan@ku.ac.ae

Academic Editor: Igor Serša
Received: 30 September 2020; Accepted: 9 November 2020; Published: 17 November 2020

Abstract: Titanium dioxide (TiO_2) is an excellent photocatalytic material that imparts biocidal, self-cleaning and smog-abating functionalities when added to cement-based materials. The presence of TiO_2 influences the hydration process of cement and the development of its internal structure. In this article, the hydration process and development of a pore network of cement pastes containing different ratios of TiO_2 were studied using two noninvasive techniques (ultrasonic and NMR). Ultrasonic results show that the addition of TiO_2 enhances the mechanical properties of cement paste during early-age hydration, while an opposite behavior is observed at later hydration stages. Calorimetry and NMR spin–lattice relaxation time T_1 results indicated an enhancement of the early hydration reaction. Two pore size distributions were identified to evolve separately from each other during hydration: small gel pores exhibiting short T_1 values and large capillary pores with long T_1 values. During early hydration times, TiO_2 is shown to accelerate the formation of cement gel and reduce capillary porosity. At late hydration times, TiO_2 appears to hamper hydration, presumably by hindering the transfer of water molecules to access unhydrated cement grains. The percolation thresholds were calculated from both NMR and ultrasonic data with a good agreement between both results.

Keywords: cement hydration; titanium dioxide TiO_2; NMR; ultrasonic; calorimetry

1. Introduction

Titanium dioxide (TiO_2) has been studied for potential applications, notably as a white pigment and in hydrolysis [1] and electricity production [2], as well as an additive in construction materials (cement, concrete, tiles and windows) for its sterilizing, deodorizing and antifouling properties [3–8]. TiO_2 integrated into construction materials effectively decomposes or deactivates volatile organic compounds, removing bacteria and other harmful agents. Cement-based, self-cleaning construction materials can play a major role in achieving clean air conditions in modern urban environments. Accordingly, it is of critical importance to characterize the structural properties and hydration kinetics of these materials to improve their mechanical performance.

Different, nondestructive methods are used to investigate the hydration of cement-based materials. The authors of [9,10] provide an overall view of these techniques. NMR has the advantage of nuclear-spin selectivity, where only one nuclear-spin isotope is detected at a time. The resulting resonances provide information on the local structure and dynamic effects. NMR allows studying the development of cement microstructures and kinetics in real-time during cement hydration, even at the earliest hydration times. Until now, several NMR techniques have been reported to provide valuable information on the porosity, pore size distribution and hydration kinetics of cement pastes. Such techniques include NMR cryoporometry [11,12], imaging [13–15] and diffusion studies [16–19]. The most widely used technique to study the porosity and hydration of cement pastes is proton (^{1}H) NMR relaxometry [20–23]. In this method, the molecular motion and chemical and physical environments of water molecules are probed continuously by measuring the ^{1}H nuclear-spin–lattice and spin–spin relaxation times of hydrogen nuclei. Ultrasonic wave velocity measurements similarly provide an excellent nondestructive tool to continuously monitor the evolution of the solid matrix as cement hydrates from the initial fluid state to the final solid configuration [24,25]. Measuring the speed of ultrasonic waves propagating through the cement slurry allows for in situ monitoring of hydration dynamics and determination of elastic properties [26–28].

When additives with hydrophilic properties, such as TiO_2, are added to a cement paste, the hydration process and pore size development are affected in multiple ways [29,30]. It is known that the surface of fine fillers provides additional sites for the nucleation of C–S–H, accelerating the hydration reaction by reducing the energy barrier [31]. The effectiveness of this catalytic effect depends on the dosage and fineness of the nanoparticles. For TiO_2 nanoparticles, their addition to cement can result in decreased water permeability and improved durability properties, such as chloride penetration and capillary adsorption [29]. It has been reported that the addition of TiO_2 nanoparticles to ordinary Portland cement results in an accelerating effect of early cement hydration, directly proportional to particle/agglomerate size [32]. TiO_2 nanoparticles acquire a negatively charged surface in the early-stage "ionic soup" of hydrating cement, balanced by increasing Ca^{+2} concentrations. Agglomeration is promoted via ion-ion correlations, also observed in C–S–H gel particles [33]. It is known that adding TiO_2 to cementitious materials enhances the mechanical properties of the cement in the early age of hydration. Enhanced mechanical properties for cementitious materials doped with TiO_2 are also reported at the late stage of hydration [32,34–39]. In a recent review by Rashad [40], the optimum percentages of TiO_2 in cementitious materials at which mechanical properties enhanced were summarized as 1–4% for concrete, 2–10% in mortars and up to 10% in pastes. Other studies reported opposite results, where the mechanical properties of cementitious materials doped with TiO_2 decreased at later stages of hydration [41–43]. The effect of adding TiO_2 in cement hydration and hardening, as well as the effect on the mechanical properties at late hydration times, are not fully clarified. This is considerably important for the development of new construction materials that incorporate TiO_2 for its sterilizing effects but less likely for its photocatalytic properties where a more practical approach would be the application of coatings to the exterior surface of the structures.

In this article, results from both NMR and ultrasonic velocity measurements on the effect of TiO_2 on the hydration of Portland cement are presented. NMR ^{1}H spin–lattice relaxation and diffusion data are recorded. These measurements monitor the dynamics of water molecules confined in cement pores and their interaction with the pore surface.

2. Experimental

2.1. Ultrasonic Section

In bulk solid materials, ultrasonic waves propagate mainly in two modes: shear and longitudinal waves. In cement, shear waves are expected to propagate only after C–S–H cement gel first percolates (i.e., at cement setting time) [24]. On the other hand, longitudinal waves propagate through the initial suspension. As the hydration procedure continues, the system becomes increasingly rigid and develops a solid matrix of pores filled with water. The longitudinal V_L and shear V_s wave velocities increase as both the bulk and shear moduli increase rapidly in value. V_L is related to the constrained modulus [44],

$M = \rho V_L^2$, where ρ is the density of the sample. By monitoring the evolution of V_L during the hydration process and the evolution of the cement paste density, it is possible to deduce the evolution of the M modulus of the cement paste as a function of the hydration time. During the propagation of the signal through a material, its spectrum deforms and experiences high damping [25,26,45,46].

2.2. NMR Section: Spin-Lattice Relaxation Time

In this study, ^{1}H-NMR spin–lattice relaxation time T_1 measurements are used to obtain information on the hydration process in a nondestructive manner. T_1 is determined by the water/solid interface of the cement system and the development of the pore network. The relaxation rate, $1/T_1$, of mobile water molecules increases near a liquid-solid interface due to the exchange between the free and bonded water and the presence of paramagnetic sites on the solid surface [13]. In the fast-exchange approximation:

$$\frac{1}{T_1} = \frac{1}{T_{1b}}(p) + \frac{1}{T_{1f}}(1-p) \tag{1}$$

where the subscripts b and f refer to the "bonded" water near the pore surface and "free" water, respectively; p is the fraction of bonded water molecules at pore surfaces [22,47]. For spherical pores with mean radius r, assuming bonded water molecules form a layer of thickness ε [23],

$$p = \frac{3\varepsilon}{r} = \frac{\varepsilon S}{V} \tag{2}$$

where S/V is the pore surface area to volume ratio. As $\frac{1}{T_{1f}} \ll \frac{1}{T_{1b}}$, due to the presence of paramagnetic sites on the solid surface, the overall relaxation rate depends linearly on the S/V of the pores.

$$\frac{1}{T_1} = \frac{1}{T_{1b}}\left(\frac{pS}{V}\right) \tag{3}$$

In cement and other complex porous materials, T_1 spreads over a wide distribution of relaxation times due to the existence of pore sizes ranging from nanometers to micrometers. During cement hydration, pore networks develop within the cement matrix. The nuclear magnetization in a saturation recovery experiment can be expressed as [23]:

$$R(t) = \frac{M_0 - M(t)}{M_0} \int_0^\infty g(T_1) \exp\left(-\frac{t}{T_1}\right) dT_1 \tag{4}$$

where $R(t)$ is the proton magnetization recovery function, M_0 is the magnitude of the magnetization at equilibrium and $M(t)$ is the observed magnetization as a function of time t. Here, $g(T_1)$ is the T_1 distribution function, which can be resolved by means of an inverse Laplace transform [23], unveiling important information on the porous microstructure in the hardened material.

There are mainly three different "water groups" in hydrating cement pastes, which can be monitored by T_1 ^{1}H-NMR relaxometry. First, water chemically bound to OH groups (portlandite, gypsum and ettringite) exhibits a restricted motion, characterized by long T_1 (> 100 ms) and very short spin-spin T_2 (≈10 µs) relaxation times. In this study, this water group is intentionally excluded from the data acquisition by setting the experimental time window for the NMR measurements accordingly. Second, mobile water is incorporated into the C–S–H phase and located in the restricted volume of the gel pores. The relaxation is dominated by the pore–surface interactions, resulting in short T_1 and T_2 values (0.5–1.0 ms). Monitoring the gel pore water is of primary importance, as it controls the viscoelastic response of C–S–H gel to mechanical loading and relative humidity changes (drying shrinkage). Third, water is trapped inside capillary pores (3–50 nm) and microcracks of the hydrating cement pastes, with considerably higher relaxation times (~5–10 ms) [48,49], albeit lower compared to bulk water (~2 s).

2.3. NMR Section: Spin-Spin Relaxation Time and Diffusion Measurements

Measuring the water self-diffusion coefficient (D) is important in studying cement, as it is directly related to hydration and the development of the gel matrix. D is directly connected to water permeability and thus to the durability and aging properties of cement. The conventional NMR method for measuring D is to monitor the ^{1}H-NMR spin echo decay in a constant linear magnetic field gradient. For the isotropic diffusion, the NMR data can be fitted to the relation [19]:

$$M(2\tau) = M_o \exp\left(\frac{2\tau}{T_2} - \frac{2}{3}\gamma^2 D G^2 \tau^3\right) \tag{5}$$

where $\gamma = 26.7522 \times 10^7$ rad s^{-1}T^{-1} is the gyromagnetic ratio for proton, and G is the magnetic field gradient. The linearly exponential part of the decay corresponds to T_2 and the cubic exponential decay corresponds to dephasing due to the presence of the magnetic field gradient. However, water molecules in porous systems do not diffuse freely due to pore confinements. Therefore, the dephasing part of the spin echo deviates from the previous equation [18,19,50]. To describe the dephasing behavior, two length scales need to be compared [50,51]: structural length l_S (=V/S, for spherical pores) and dephasing length l_G, i.e., the distance a particle must travel to dephase by a full cycle in the magnetic field gradient. If the diffusion length $l_D (= \sqrt{6D\tau}) < l_S$ or $< l_G$, water molecules diffuse freely in the porous matrix and Equation (5) is valid. However, if $l_S < l_D$, the magnetization decay is in the so-called motional averaging regime [50,51], and the dephasing part of the spin echo decay is characterized by a single exponential law. For the case of spherical pores [52,53]:

$$M(2\tau) = M_o \exp\left[-\left(\frac{8}{175}\frac{\gamma^2 G^2 R^4}{D}\right) 2\tau\right] \tag{6}$$

On the other hand, if $l_G < l_S$ and $< l_D$, the magnetization decay is in the so-called localization regime [50], and the following expression applies:

$$M(2\tau) = M_o \exp\left[-1.02\left((\gamma G)^{\frac{2}{3}} D^{\frac{1}{3}}\right) 2\tau\right] \tag{7}$$

The above is applicable when the magnetic field gradient is very strong and water molecules have already dephased significantly before they reach the pore walls.

2.4. Materials

White cement (CEM II-42.5) was provided by Lafarge–Heracles (Greece) and TiO$_2$ (P-25 Aeroxide) was purchased from Degussa. White cement was selected because of its low iron oxide content, which causes line shape broadening in NMR experiments due to magnetic susceptibility effects. According to the manufacturer, P-25 TiO$_2$ contains 70% anatase and 30% rutile (w/w), with average grain sizes of 21 nm and a specific surface area of 50 m^2/g. Four cement paste mixtures were prepared, namely C100, C97T3, C93T7 and C85T15. The number at the end of the doped sample names refers to the weight percentage of TiO$_2$ that replaced equal amounts of cement (3, 7 and 15% w/w, respectively). The water-to-cement ratio (w/c) was kept constant for all samples at w/c = 0.40. The percentages were chosen based on expansion measurements by Flow Table. The sample with 15% titania (C85T15) has a w/c ratio of 0.40, which is also its normal plasticity water, thus ensuring its workability. All other percentages are roughly selected by dividing the percentage of titania. Regarding sample preparation, cement and TiO$_2$ were initially mixed together at the appropriate weight ratio of each specimen and stirred at a low mixer speed for 10 min to ensure excellent dispersion of TiO$_2$ in cement. Mixing with distilled water was performed according to the procedure described in the EN 196-1 standard [54]. For the ultrasonic experiments, each sample was cast in a Plexiglas cubic mold (10 × 10 × 10 cm) immediately after mixing with water. The thickness of the Plexiglas wall at the position of contact with the ultrasonic transducers was 0.25 cm. Care was taken to ensure constant pressure on the transducers attached on opposite sides of the mold by the use of springs. The quality of the contact was assured by the application of lubricant between the transducers and the Plexiglas mold. Samples were saturated with distilled water after 6 h of casting and the open top of the molds was membrane-sealed. With this procedure, shrinkage was minimized, and good contact between the Plexiglas wall and the curing cement was ensured to avoid sound attenuation and signal loss. Ultrasonic measurements were conducted with a commercial ultrasonic pulse generator (GE USM 23) driving two identical transducers of nominal frequency (500 kHz). The waveforms were recorded with an A/D converter using LabView home-built software. All experiments were performed at room temperature for a minimum time period of 80 days and conducted four times. The values of the ultrasonic velocity measurements provided here are the mean values of the four experiments. The estimated uncertainty for the ultrasonic velocity measurements was ± 30 m/s (calculated as $\frac{1}{2}$ ($V_{max} - V_{min}$), where V_{max} and V_{min} are the maximum and minimum velocity values as measured from the experiments). For the NMR experiments, samples were taken from the same mixtures used in the ultrasonic measurements to ensure identical preparation conditions. Immediately after mixing with water, the samples were sealed into NMR glass tubes (9 mm in diameter and 30 mm in length) using a Parafilm® membrane to minimize the evaporation of water. ^1H-T_1 experiments were conducted using a home-built circular Halbach array magnet, suitable for low-field NMR measurements [55]. The field at the magnet center was 0.29 T, corresponding to a proton resonance frequency. The magnet was coupled to a broadband spectrometer operating in the frequency range of 5–800 MHz. T_1 was measured using a standard saturation recovery technique (($\pi/2$) – t – ($\pi/2$) – τ – (π)) with the interpulse delay, t, ranging between 100 μs and 6 s on a logarithmic scale. The signal was detected by the common Hahn echo pulse sequence with a τ value of 60 μs. All experiments were performed at room temperature and the hydration process for each sample was monitored for 28 days. Time intervals between successive experiments ranged from minutes and several hours in the initial hydration stage up to full days at the later hydration stages. At the early stage of hydration, T_1 was characterized by a single exponential function. However, with progressive hydration, a multiexponential behavior 12.1718 developed, which was resolved by means of an inverse Laplace transform [55]. The numerical Laplace inversion of the ^1H- of 12.1718 MHz NMR saturation recovery curves was obtained using a modified CONTIN algorithm [56], which was constrained to a positive output for 30 logarithmically distributed points between T_{1min} = 0.01 ms and T_{1max} = 10.000 ms.

Water diffusion experiments were also conducted in the same Halbach magnet with a magnetic field gradient of 1.03 T m^{-1}. Measurements were carried out using the Hahn echo pulse sequence (($\pi/2$) – τ – (π) – τ – echo) with the interpulse delay, τ, ranging from 20 to 8 ms. All experiments were performed at room temperature and the hydration process for each sample was monitored for 300 h with time intervals between successive experiments ranging from minutes to several hours. The uncertainty in the T_1 values was estimated at ±0.01 ms.

From the same batch mixtures that were used for ultrasonic and NMR analyses, an additional set of cement-TiO_2 specimens were prepared to measure microstructural properties during hydration. Each specimen was set in prismatic molds (20 × 20 × 80 mm) and left to hydrate at a relative humidity of 98 ± 2% and a temperature of 21 ± 1 °C inside a curing chamber. The molds were membrane-sealed and covered with glass sheets to avoid water evaporation. After two days, the specimens were removed from the molds. These prismatic specimens were used to measure density at progressing hydration ages using the standard Archimedes method with water. All specimens were examined during the first 28 days of hydration with scanning electron microscopy (SEM) using a FEI Quanta Inspect coupled with an energy-dispersive spectroscopy (EDS) unit. Specimens were vacuum dried using ethanol and diethyl ether prior to analysis and gold-coated. Standard Vicat measurements were performed and isothermal calorimetry measurements were carried out in an I-CAL 2000 HPC, Calmetrix at 20.0 ± 0.5 °C. The hydration of the mixtures was monitored by simultaneous differential thermal analysis and thermogravimetry (DTA/TG) in a Perkin Elmer Pyris 2000 thermal analyzer [57].

3. Results and Discussion

3.1. Vicat, Isothermal Calorimetry and DTA/TG Measurements

Figure 1 exhibits the normalized rate of hydration (per gram of cement + TiO_2) for all mixtures, according to isothermal calorimetry measurements performed for the first seven days. Note that the w/c ratio is the same for all mixtures.

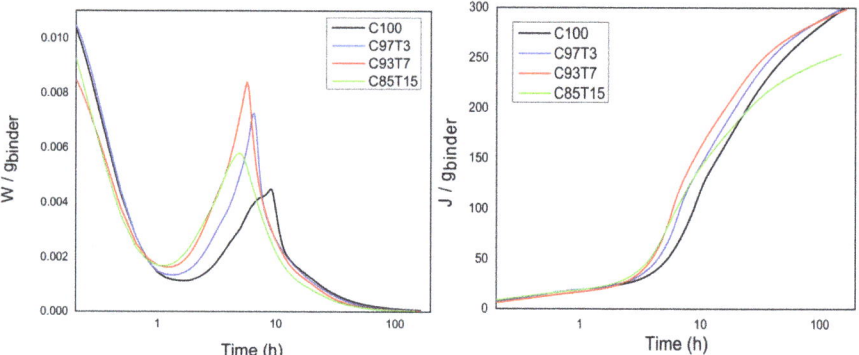

Figure 1. Isothermal calorimetry measurement. Power versus hydration time and heat release versus hydration time divided by the total solid mass.

These calorimetry data show that the peak heights were increased with the addition of the inert TiO_2 nanoparticles, thus the addition of TiO_2 greatly affects the early-age hydration. The TiO_2-doped samples' heat release curves around 5 h are shifted to the left, indicating the early-age acceleration effect of TiO_2. Although Titania is inert, the rate of reaction of the clinker component is enhanced. The main peak is higher for the samples with 7% titania, while for all doped samples, it is higher than the reference sample, and the acceleration slope is steeper for all the doped samples compared to the reference sample. The cumulative heat of the C85T15 sample falls lower than the reference sample after approximately 24 h. Considering that J in Figure 1 is referred to as the extent of the hydration process, the differentiation of mixture C85T15 is attributed to the sorter retardation/deceleration period and the beginning of the long-term reactions at earlier stages. Overall, the full curve of C85T15 is shifted to the left.

Standard Vicat measurements were performed in all mixtures for determining the initial and final setting time (see Table 1 in Section 3.5), using the same amount of water.

Table 1. Percolation parameters (p_c) obtained from fitting NMR and ultrasonic data to Equations (8) and (9) and weighted standard deviation. Vicat measurements with uncertainty values.

Sample	Percolation Threshold, p_c		Vicat Setting Times (h)	
	Ultrasonic/h	NMR/h	Initial	Final
C100	4.7 ± 0.08	3.6 ± 0.1	4.18 ± 0.02	4.88 ± 0.02
C97T3	4.3 ± 0.1	3.3 ± 0.1	3.66 ± 0.02	4.26 ± 0.02
C93T7	3.7 ± 0.2	2.8 ± 0.2	3.35 ± 0.02	3.80 ± 0.02
C85T15	3.2 ± 0.4	2.2 ± 0.3	2.78 ± 0.02	3.15 ± 0.02

The setting time determined by different methods and the hydration process are presented and discussed in Table 2 (Section 3.5).

Table 2. Periods of the hydration process for all four samples.

	Samples	C100	C97T3	C93T7	C85T15
NMR	Start of acceleration (h)	1.63	1.57	1.20	1.08
	Formation of "shoulder" (h)	4.88	4.71	-	-
	Start of diffusion (h)	89.0	43.0	29.0	23.0
UltraSounic	Initial detection of signal (h)	2.75	2.73	2.22	2.20
	End of acceleration (h)	8.8	6.0	4.9	3.9
	End of deceleration (h)	692	570	490	306
Calorimetry	Start of acceleration (h)	1.67	1.37	1.26	1.06
	End of acceleration/start of deceleration (h)	9.04	6.48	5.66	4.87

Based on the peaks appeared in the DTA curves, the total amount of chemically bound water (attributed to the hydrated C–S–H and C–A–H phases) was calculated gravimetrically (TG%) at different setting periods in the range of 75–320 °C [58,59]. Samples were heated in an air atmosphere in three steps: from 25 to 60 °C, using a heating rate of 5 °C/min; a hold-step at 60 °C for 1 h (aiming to remove any physically adsorbed water; and heating up to 400 °C with a rate of 10 °C/min. The amount of chemically bound water was compared to that calculated in a reference cement specimen (C100R) after setting for two years at the same curing conditions.

The hydration rate (R_H) of each different cement-TiO$_2$ mixture was calculated (Figure 2) by dividing the amount of bound water at different setting periods by the amount of bound water in the fully hydrated cement (C100R) [60].

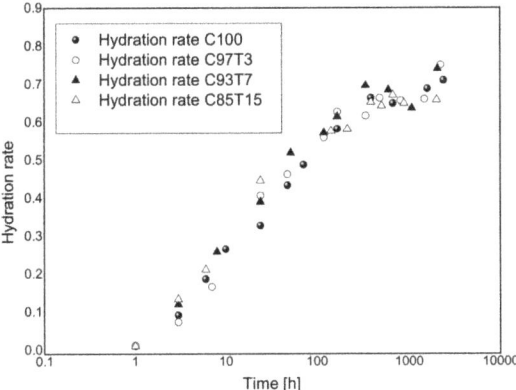

Figure 2. Normalized hydration rates of the four samples with respect to the fully hydrated C100R sample.

3.2. Ultrasonic Experiments

The hydration process for all samples was monitored by means of continuously measuring the longitudinal ultrasonic wave velocity as a function of hydration time. Figure 3 demonstrates the longitudinal ultrasonic velocity V_L as a function of hydration time. As seen from Figure 3, the hydration periods of cement [27,61,62] are distinguished in the time evolution of V_L except for the "dormant period", which occurs at the early time (below 2 h) of hydration, where cement paste behaves as a liquid suspension of particles. In our measurements, no longitudinal velocity signals were observed at the dormant period (left region of the dashed line A, Figure 3) as entrapped air led to strong attenuation of the longitudinal waves [25,26,62]. Air bubbles within the binder blocked the higher frequencies at the early-age, allowing them to transmit at a later-age [26,45,63–65]. The evolution of the frequency follows the evolution of the ultrasonic velocity and asymptotically reaches the carrier frequency at later times. Between ~2 and 3 h of hydration (dashed lines A and B), ultrasonic waves become detectable and V_L increases rapidly with hydration time.

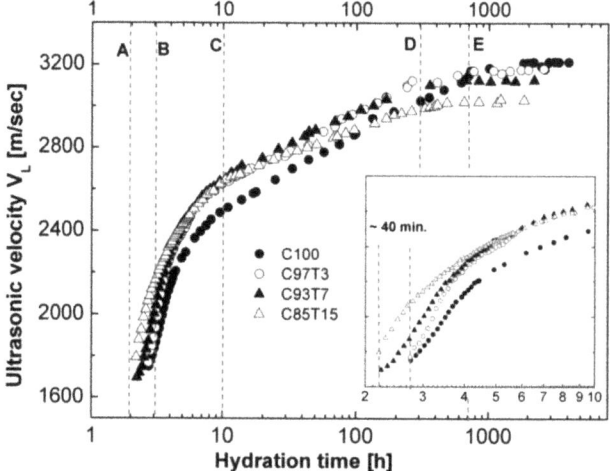

Figure 3. Ultrasonic longitudinal velocity versus hydration time for the samples used in this study. The inset presents the data during the first ten hours of hydration time. The vertical dashed lines are presented for clarification and to indicate the different hydration periods as discussed within the text.

Samples with higher TiO_2 content exhibited higher V_L values. The V_L value of the sample with the highest TiO_2 content, C85T15, was recorded approximately 40 min earlier compared to the reference sample, C100 (notice the signal between the two blue lines in the inset of the figure). This indicates that substituting cement with TiO_2 causes a strong acceleration of the early hydration. In the "acceleration period" (~3–0 h, lines B and C), an increased hydration rate is observed for all the samples with the lowest V_L values for the reference sample. The rapid growth of cement gel in this period facilitates solid pathways, which enables the transmission and detection of longitudinal waves [28,45]. After approximately 10 h of hydration, the "deceleration period" began with a slower increase in V_L for all samples compared to the acceleration period. This period lasted for up to ~300 h of hydration for the TiO_2-containing samples (between lines C and D) and up to ~700 h for the reference sample (between lines C and E). The data show a nonlinear behavior for the TiO_2-containing specimens and a linear behavior for the reference sample, a characteristic of adding fine aggregates to cement [66–70]. The addition of fine aggregates accelerates the evolution of the microstructure [51]. Therefore, additives such as TiO_2 caused a reduction of the distances among solid particles, and smaller amounts of hydration products are needed to form the first connection path for the ultrasonic signal to transmit. This "filler effect" is also observed in the SEM images of the samples and presented later in this section (Figure 8). A further observation of the data at the end of this period indicates that the V_L value of the reference sample overpassed the V_L values of the other samples, indicating its solid paths became fully developed. This is opposite to what was observed at the early stage of hydration (between the two blue lines).

Furthermore, all samples reach a saturation value in their ultrasound velocity after 1000 h. The saturation values from Figure 3 are calculated as 3018 m/s for sample C85T15, 3123 m/s for sample C93T7, 3169 m/s for sample C97T3 and 3207 m/s for sample C100.

These values are inversely proportional to the amount of TiO_2 in the sample, indicating the reduction of the constrained modulus with increasing TiO_2 content in cement. In this connection, the constrained modulus ($M = \rho V_L^2$) of the samples is calculated, and the results are presented in Figure 4.

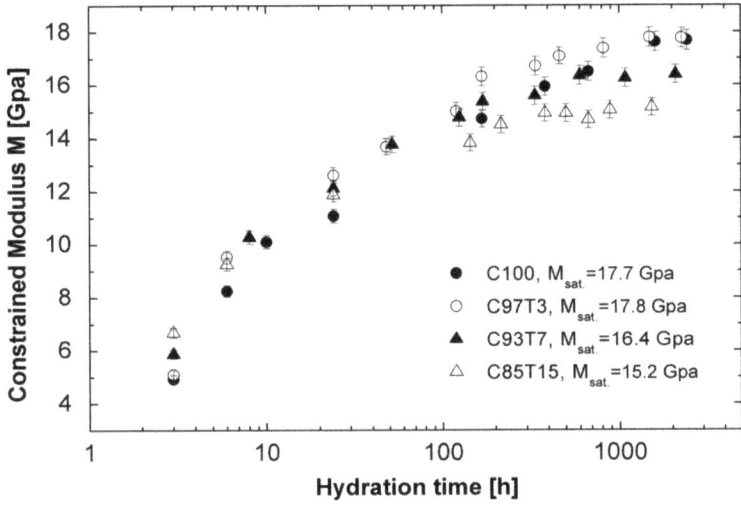

Figure 4. Evolution of constrained modulus M for the samples versus hydration time.

The saturation values of M_{sat} are 15.2, 16.4, 17.7 and 17.8 Gpa for C85T15, C93T7, C100 and C97T3, respectively. It is known that TiO_2 is added to construction materials (cement, concrete, tiles and windows) for its sterilizing, deodorizing and antifouling properties [3–8]. The results of Figure 4 show that if too much TiO_2 (15%, as in C85T15) is used, the mechanical properties of the material will be affected.

3.3. ^{1}H-Spin–Lattice Relaxation Time (T_1) Measurements

The hydration process for the samples was monitored by measuring ^1H-T_1, which is less liable to artifacts caused by water molecule diffusion in magnetic field in-homogeneities. Figure 5 shows T_1 distribution profiles, obtained by the inverse Laplace transform [55], as a function of hydration time. For all samples and at early hours of hydration, the magnetization of water protons relaxes uniformly due to the fast exchange between water spins in the various environments, represented by a single T_1 component [20,71,72]. T_1 for this peak is ~ 100 ms. With proceeding hydration, a growing number of hydration products develop and the paste becomes rigid. The surface area of the pore network increases, which causes a reduction of T_1, as predicted by Equation (3). Monitoring the change in T_1 allows the observation of the evolution of hydration and the growth of the pore structure of cement. At approximately 12 h of hydration, T_1 distribution splits into two components. The short and long T_1 peaks are attributed to water in gel pores and large capillary pores, respectively [55,73]. The observation of these two components indicates the formation of both gel and capillary pores, with different pore geometries reflected by their T_1 values. At longer hydration times, the cement paste hardens and a further reduction of T_1 is observed, as a result of increasing pore surface areas according to Equation (3). Beyond 12 h of hydration, the peak assigned to the gel pore (short T_1) grew larger in area while the capillary peak (long T_1) decreased slightly, providing a measure of the change of the two pore populations. The two T_1 peaks are visible over different time intervals depending on the percentage of the TiO_2 in each sample (with the reference sample lasting longer). The two T_1 peaks are visible from 12 h for all samples and retain until 18 h, 1 day, 3 days and 14 days for C85T15, C93T7, C97T3 and C100, respectively. Beyond these times, both peaks merge into one. This behavior can be related to the deceleration period of the samples, as deduced from ultrasonic results (Figure 3), where the deceleration period lasted longer for the reference sample. Another observation on the two peaks of T_1 in Figure 5 is that the gel pore peak (short T_1) attains relatively higher initial values in the TiO_2-containing samples compared to the reference sample, while it subsequently shifts to lower T_1

values with progressing hydration. This might be due to the creation of an initial "more open" gel pore structure.

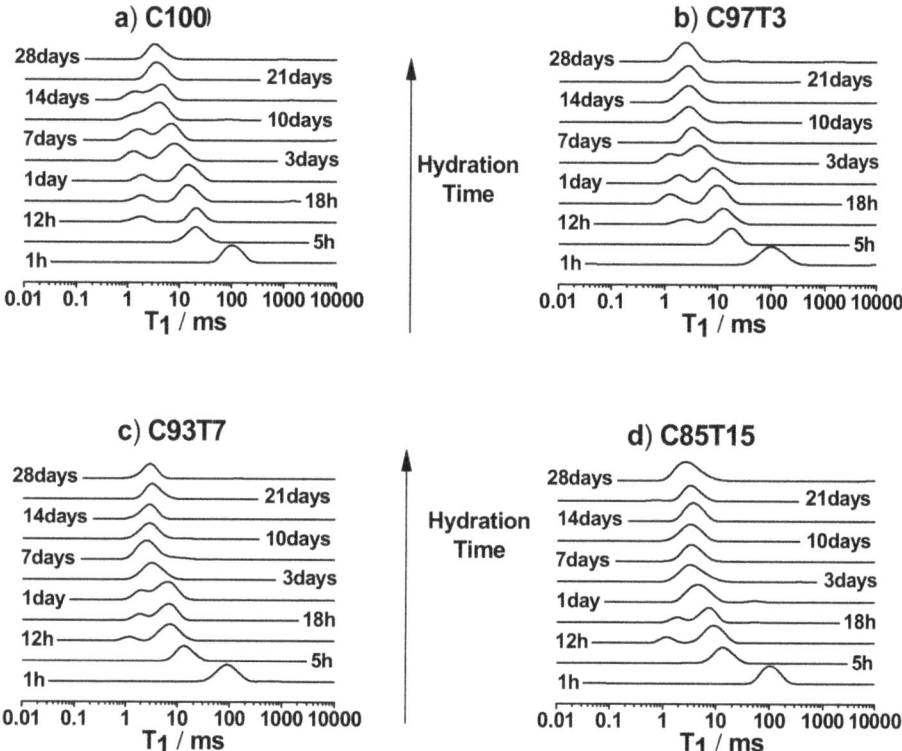

Figure 5. NMR spin–lattice relaxation T_1 distribution profile versus hydration time obtained from the inverse Laplace transform of the NMR data for (**a**) C100, (**b**) C97T3, (**c**) C93T7 and (**d**) C85T15, samples. Note the change of the T_1 profile from one, two, to one component at short, intermediate and long hydration times, respectively.

Figure 6 shows the evolution of the average T_1 as a function of hydration time for the four cement samples. A formation of a "shoulder" in the T1 versus hydration time curve is observed at about 5 h for both C97T3 and C100 samples but is more visible in the C100 sample. The formation of this shoulder is known to relate with the *w/c* ratio and is more pronounced at a higher *w/c* ratio [12], related to a second water release originating from the melting of solid substructures, mainly ettringite crystals into monosulfate [41,42]. Although the samples used in this study have the same *w/c* ratios, C93T7 and C85T15 do not exhibit such a shoulder contrary to C100 and C97T3 samples. This behavior can only be attributed to the role of titania in withholding water, making the samples with a larger titian content not exhibit such a shoulder curve.

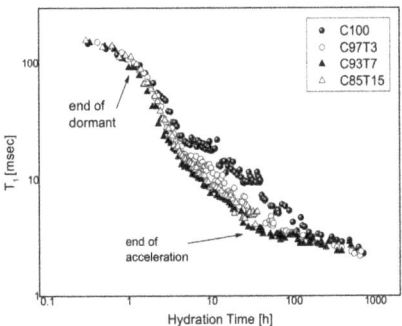

Figure 6. ^1H-T_1 NMR measurements for all four samples versus hydration time.

The ratios of the mean relaxation rates $(1/T_1)_{norm} = \dfrac{\frac{1}{\langle T_1 \rangle}}{\frac{1}{\langle T_1 \rangle_{max}}}$ versus hydration time are obtained using the procedure described in [55]. This quantity is the average relaxation rate and is a measure of the development of the fine porous system by the increase of the total surface area of the porous system. The increase of the total surface area is related to the formation of hydration products and follows the hydration rate of the samples [17,39,40], regarded as a well-defined parameter for a heterogeneous multiphase system such as cement (gel and capillary pores) [59]. Figure 7 shows the connection between the hydration rates derived from DTA/TG measurements and the normalized NMR $1/T_1$ rates. For the samples C100 and C97T3, there is a significant deviation between DGA/TA and NMR results. This deviation is related to the formation of the shoulder in the T_1 measurements in those samples (Figure 6). This behavior could be explained by the hydrophilicity of TiO_2 nanoadditives and the consequent stronger van der Waals forces between water and titania. This has a consequent effect on the water molecules' mobility (considerably reduced by increasing the amount of TiO_2) and the resulting ^1H T_1 NMR measurements. The TG measurements are affected, since the fast heating rate (10 °C/min) cannot overcome the strong hydrophilic forces between water molecules, resulting in a decrease of the hydration related water and a better fit to NMR results. On the other hand, T_1 NMR monitors all mobile water molecules both in gel and capillary pores. For the samples C93T7 and C85T15, which do not manifest a shoulder in the T_1 results, a very good coherence between these two techniques is shown.

An increase in relaxation rates is observed for the higher-doped samples. Increasing relaxation rates correspond to an increase in the pore surface areas, as expected by Equation (3). The enhancement is believed to be due to two reasons: (1) TiO_2 grains promotes C–S–H gel production and the grains act as nucleation centers for cement gel growth, causing an increase in the short T_1 peak area versus hydration time (Figure 5); (2) TiO_2 speeds up the rupture of the gelatinous coating created around cement grains [31,74], therefore the dormant period is shorter for the doped samples, as seen at the beginning of the hydration time of Figure 6.

For the SEM images, the specimens were cast on freshly cleaved flat mica surfaces and left to cure for two days at 21 °C and 95% RH. Samples were then placed on SEM aluminum stubs and the mica surface was removed. Images were collected from the surface of each specimen exposed to mica. Figure 8 shows the SEM images of the reference sample, C100, and the 7% TiO_2 sample, C93T7. The latter appears to have a denser and more packed structure compared to C100 due to the high production of early hydration products. This can also be related to the NMR results (shown in Figure 5; Figure 6), which revealed (from $1/T_1$) an increased production of cement gel in doped samples compared to those in the reference sample.

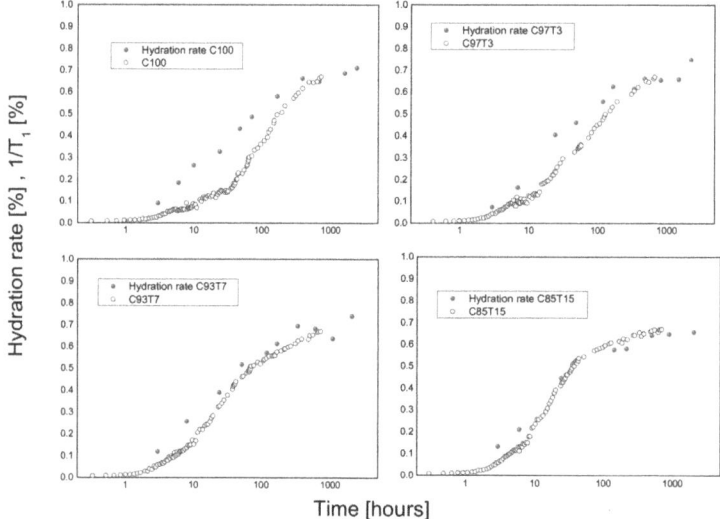

Figure 7. Hydration rates and normalized mean relaxation rates $\left[\frac{1}{T_1}\right]_{norm}$ versus hydration time for all four samples calculated by DTA/TG measurements.

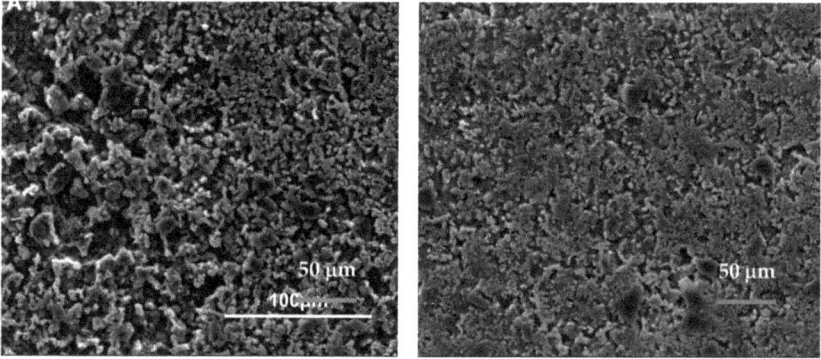

Figure 8. Scanning electron microscopy (SEM) images for specimens C100 (left) and C93T7 (right) after two days of hydration. The 7% TiO$_2$ (right image) sample exhibits a significantly denser internal structure compared to the reference sample (left image).

3.4. ^{1}H-NMR Diffusion Measurements

Water mobility inside the porous system of hardening cement was monitored through the evaluation of self-diffusion coefficient, D_{eff}, as determined from the NMR diffusion data using Equations (5)–(7). The results are shown in Figure 9. At the beginning of the dormant period, D_{eff} remains practically unchanged, with the TiO$_2$-containing specimens exhibiting higher values. Going more into this period, a sharp decrease is seen in the coefficient value for all specimens. This indicates a restriction of water mobility as the pore system develops due to the increased production of cement gel. The decrease in D_{eff} directly corresponds to an increase in the pore size area. It is interesting to note that the aforementioned decrease in D_{eff} takes place after an amount of time proportional to TiO$_2$ content, as depicted by the dotted lines in Figure 9 (at approximately 8, 9 and 14 h for C100, C93T7 and C85T15, respectively).

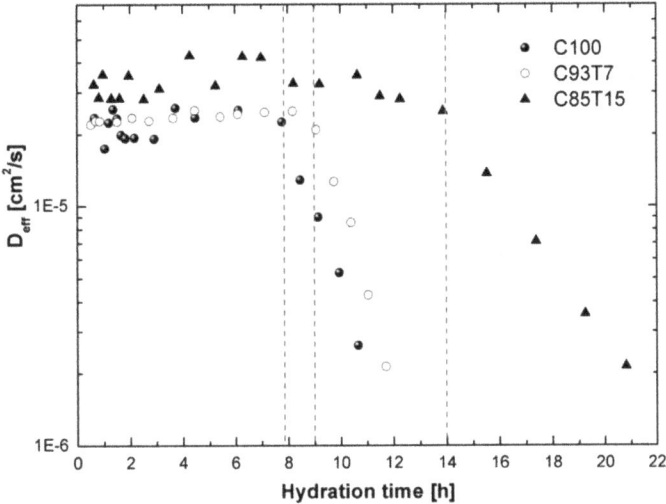

Figure 9. Self-diffusion coefficient, D_{eff} versus hydration time obtained from NMR data by using Equations (5)–(7). The time at which D_{eff} has decreased depends on the TiO$_2$ concentration. For the 15% TiO$_2$ sample, this started at a longer time compared to the other samples.

3.5. Correlation between NMR and Ultrasonic Results

Ultrasonic longitudinal velocity V_L evolves with hydration time and is closely related to the formation of the pore network. On the other hand, NMR relaxation rate $1/T_1$ is proportional to the pore sizes through Equation (3) and provides a measure of the development of the fine pore system. Therefore, a correlation between these two properties can be made. Figure 9 presents the normalized V_L values versus the normalized $1/T_1$ values, considering hydration time as an implicit parameter. The $1/T_1$ values were normalized with respect to the $1/T_1$ value measured for the reference sample after a setting period of two years. V_L values were normalized with respect to the saturation values obtained from Figure 3. In Figure 9, two distinct regions of linear correlations between V_L and $1/T_1$ can be observed. The first linear correlation appears immediately after the initial setting and continues for 7 to 10 h of hydration time. The second linear region appears directly after this period and continues for the entire time of the analysis (28 days). The steeper slope observed in the first region indicates that the ultrasonic technique is more sensitive than NMR during early hydration. V_L increases rapidly when the first solid pathways are formed, but water is still unhindered and pore size distributions (measured through $1/T_1$) are not yet distinguishable on the NMR time scale. With progressing hydration, water either reacts with cement or becomes confined within the porous system, resulting in an observable change in the NMR signal. In Figure 10, the slope of the second linear region is shifted toward the NMR $1/T_1$ axis, indicating that the NMR technique is more sensitive at later hydration times in detecting the development of the porous network. Accordingly, both techniques monitor the formation of hydration products through different yet complementary mechanisms.

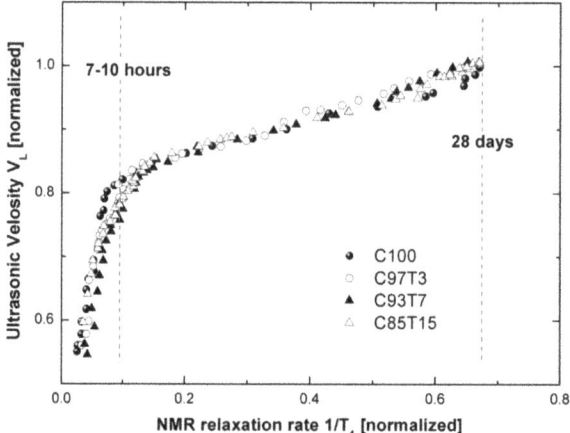

Figure 10. Correlation between ultrasonic longitudinal velocity and NMR relaxation rates. Both are normalized and the hydration times are taken as implicit parameters. Two distinct regions of linear correlations between V_L and $1/T_1$ are observed from the starting time until ~7–10 h and ~7–10 h until 28 days.

As discussed previously, ^{1}H-NMR experiments define the evolution of cement microstructure and the formation of a pore network over time by probing the water molecules and their interaction with the solid surface of the pores. ^{1}H $1/T_1$ rates are a measure of hydration by probing the development and evolution of the pore network of the cement paste during the hydration process, as described by Equation (3). Ultrasonic measurements probe the development of the solid matrix by monitoring changes in the elastic modulus and density of the system through the transmitted ultrasonic waves. Percolation is a critical element in defining the performance of cement-based materials. To quantitatively characterize phase percolation, a complete understanding of microstructural changes in three dimensions is necessary. Experimentally, this can be challenging because commonly used techniques such as scanning electron microscopy or mercury intrusion porosimetry provide partial information or demand a significant amount of time to perform or to analyze [75]. On the other hand, the two techniques used in this study can be used to define percolation during cement hydration. In this context, each method defines the percolation of cement hydration in a different manner. As per ultrasonic methods, the percolation threshold can be defined as the point in time upon when the connectivity of the solid hydration products in the cement paste reaches a critical content, allowing for enhanced propagation of ultrasonic waves from that point onwards (acoustic percolation) [61,76]. From the NMR T_1 relaxation perspective, the percolation threshold can be defined as the point in time when the interconnectivity of the pore network reaches a low critical value, following which T_1 relaxation is controlled mostly by gel porosity (NMR relaxation percolation) [21,55,77]. By extending the power law equation used previously by Scherer et al. [76] (in ultrasonic, similar to Equation (8)) and NMR (Equation (9)), the percolation thresholds for V_L and $1/T_1$ can be calculated as:

$$\left(V^2 - V_0^2\right) \propto (p - p_c)^\gamma \qquad (8)$$

$$\left(\frac{1}{T_1}\right) - \left(\frac{1}{T_{10}}\right) \propto (p - p_c)^\gamma \qquad (9)$$

where V_0 and T_{10} are the initial values of V_L and T_1, respectively. In order to determine the critical exponent γ and the percolation threshold, p_c, ultrasonic and NMR data were fitted to the above equations, and a sample of the results is shown in Figure 11. From the fitted curves, the percolation parameters were obtained for all samples and presented in Table 1. The initial and final setting times are also presented in Table 1. These are obtained by Vicat measurements.

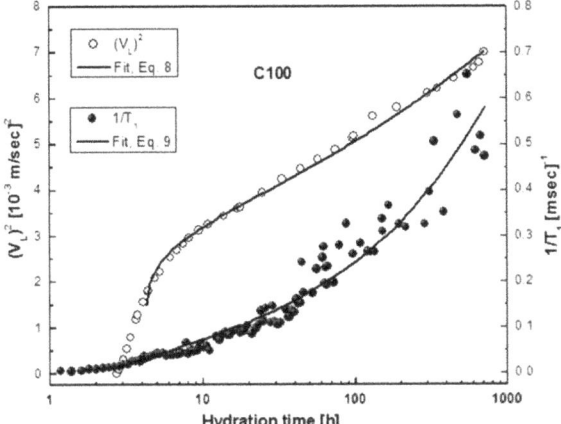

Figure 11. Fit of Equations (8) and (9) to the NMR data (black circles) and ultrasonic data (white circles) for the reference specimen C100 to obtain the percolation parameters p_c as described in the text.

The threshold values obtained with ultrasonic, NMR and Vicat techniques (Table 1) are inversely proportional to the concentration of TiO_2 in the samples. The ultrasonic parameters, p_c, indicate a formation of solid pathways after ~5 h of hydration for the reference sample C100 and after ~3.5 h for the samples containing TiO_2. Similarly, from the NMR part, the percolation threshold, p_c, for pore critical low interconnectivity is reached after ~4 h for the reference sample and after ~3 h for the TiO_2 samples. The initial setting times from the Vicat measurements correlate with the NMR percolation times (Columns 2 and 3 of Table 1), while the final setting times correlate with the ultrasonic percolation times (Columns 1 and 4 of Table 1).

Furthermore, in the following table, all the important hydration periods are presented, as detected from all the methods used. The start of the acceleration derived from the NMR measurements coincides excellently with the start of acceleration derived from the isothermal calorimetry measurements (Rows 2 and 8 of Table 2, bold). The end of the acceleration, as detected from the ultrasonic measurements, is verified from the main peak position of the isothermal calorimetry measurements (Rows 6 and 9 of Table 2, shaded).

4. Conclusions

The effect of titanium dioxide TiO_2 in cement hydration has been demonstrated, using two noninvasive techniques (NMR and ultrasonic). Spin–lattice relaxation times T_1 and diffusion measurements were used to monitor the dynamics of water molecules confined in cement pores and their interaction with the pore surface. Critical information on the role of TiO_2 on the pore developments during hydration of cement paste was obtained. Ultrasonic wave velocity measurements provided information on bulk mechanical properties of cement in the presence of TiO_2 and the evolution of the cement-solid matrix.

An acceleration of early hydration kinetics in TiO_2-containing samples was observed with NMR, specifically as an enhancement in early hydration and C–S–H gel production. The existence of two pore reservoirs was also revealed: a small gel pore reservoir (short T_1 component) and a large capillary pore reservoir (long T_1 component), evolving separately from each other with the progress of hydration. The presence of TiO_2 appeared to favor the formation of gel pores more than capillary pores, providing further proof of its role as a nanofiller.

Hydration rates derived from DTA/TG measurements and the normalized NMR $1/T_1$ rate were well correlated for the highly doped samples. Ultrasonic results showed that TiO_2 enhances the mechanical properties of cement paste during early hydration. The ultrasonic signal was detected in earlier hydration time for samples with TiO_2, an indication that the additive improved cement paste consistency. On the other hand, opposite behavior was observed at later hydration times, as the elastic properties (defined by the saturation values of the ultrasonic velocities) and the constrained modulus M of cement pastes (after 28 days of hydration) were inversely proportional to TiO_2 content. At later stages, TiO_2 appears to hamper hydration, presumably by hindering the transfer of water molecules to access unhydrated cement grains. Isothermal calorimetry measurements further support our conclusions.

The percolation threshold parameters were calculated using data of both techniques. These parameters are inversely proportional to the concentration of TiO_2 in the samples. Using the ultrasonic methodology, the acoustic percolation threshold can be identified as the moment when solid pathways are first formed and sound propagation is enhanced. NMR detected the percolation threshold as the point when the interconnectivity of the pore network reaches a low critical value, after which T_1 relaxation is controlled mostly by gel porosity. These values were well correlated with initial and final setting times determined by the Vicat method.

Author Contributions: G.D. & M.K. (Marios Katsiotis) designed the initial study and analyzed the experimental data. G.D. & J.H. did the uncertainty calculations for the results and wrote the main manuscript. J.H. provided funding. M.F. & G.P. helped in the NMR measurements and the conclusions. I.K. & S.A. contributed in the SEM images and analysis. M.K. (Marina Karagianni) did the calculation for the thermal analysis. All authors have read and agreed to the published version of the manuscript.

Funding: This work is partly based upon work supported by the Khalifa University of Science and Technology under Award No. CIRA-2020-051.

Conflicts of Interest: The authors declare no conflict of interest.

Competing Financial Interests: The author(s) declare no competing financial interests.

References

1. Park, J.H.; Kim, S.; Bard, A.J. Novel carbon-doped TiO_2 nanotube arrays with high aspect ratios for efficient solar water splitting. *Nano Lett.* **2006**, *6*, 24–28. [CrossRef] [PubMed]
2. Nazeeruddin, M.K.; Kay, A.; Rodicio, I.; Humphry-Baker, R.; Mueller, E.; Liska, P.; Vlachopoulos, N.; Graetzel, M. Conversion of light to electricity by cis-X2bis(2,2'-bipyridyl-4,4'-dicarboxylate)ruthenium(II) charge-transfer sensitizers (X = Cl-, Br-, I-, CN-, and SCN-) on nanocrystalline titanium dioxide electrodes. *J. Am. Chem. Soc.* **1993**, *115*, 6382–6390. [CrossRef]
3. Parkin, I.P.; Palgrave, R.G. Self-cleaning coatings. *J. Mater. Chem.* **2005**, *17*, 1689–1695. [CrossRef]
4. Lackhoff, M.; Prieto, X.; Nestle, N.; Dehn, F.; Niessner, R. Photocatalytic activity of semiconductor-modified cement—Influence of semiconductor type and cement ageing. *Appl. Catal. B-Environ.* **2003**, *43*, 205–216. [CrossRef]
5. Hüsken, G.; Hunger, M.; Ballari, M.M.; Brouwers, H.J.H. The effect of various process conditions on the photocatalytic degradation of NO. In *NICOM3*; Bittnar, Z., Bartos, P.J.M., Nemecek, J., Smilauer, V., Zeman, J., Eds.; Springer: Prague, Czech Republic, 2009; pp. 223–229.
6. Fujishima, A.; Hashimoto, K.; Watanabe, T. *TiO_2 Photocatalysis, Fundamentals and Applications*; BKC Inc.: Tokyo, Japan, 1999.
7. VVallée, F.; Ruot, B.; Bonafous, L.; Guillot, L.; Pimpinelli, N.; Cassar, L.; Strini, A.; Mapelli, E.; Schiavi, L.; Gobin, C.; et al. Innovative self-cleaning and de-polluting facade surfaces. In Proceedings of the CIB World Building Congress, Toronto, Canada, 2–7 May 2004; pp. 1–7.
8. Ballari, M.M.; Hunger, M.; Hüsken, G.; Brouwers, H.J.H. Heterogeneous Photocatalysis Applied to Concrete Pavement for Air Remediation. In *NICOM3*; Bittnar, Z., Bartos, P.J.M., Nemecek, J., Smilauer, V., Zeman, J., Eds.; Springer: Prague, Czech Republic, 2009; pp. 409–414.
9. *Guidebook on Non-Destructive Testing of Concrete Structures*; International Atomic Energy Agency: Vienna, Austria, 2002.

10. Jennings, H.M.; Bullard, J.W.; Thomas, J.J.; Andrade, J.E.; Chen, J.J.; Scherer, G.W. Characterization and Modeling of Pores and Surfaces in Cement Paste: Correlations to Processing and Properties. *J. Adv. Concr. Technol.* **2008**, *6*, 5–29. [CrossRef]
11. Jehng, J.-Y.; Sprague, D.; Halperin, W. Pore structure of hydrating cement paste by magnetic resonance relaxation analysis and freezing. *Magn. Reson. Imaging* **1996**, *14*, 785–791. [CrossRef]
12. Boguszyńska, J.; Tritt-Goc, J. 1H NMR Cryoporometry Study of the Melting Behavior of Water in White Cement. *Z. Nat. A* **2004**, *59*, 550–558. [CrossRef]
13. Papavassiliou, G.; Milia, F.; Fardis, M.; Rumm, R.; Laganas, E. ^{1}H Nuclear Magnetic Resonance Imaging of Water Diffusion in Hardened Cement Pastes. *J. Am. Ceram. Soc.* **1993**, *76*, 2109–2111. [CrossRef]
14. Prado, P.J.; Balcom, B.J.; Beyea, S.D.; Bremner, T.W.; Armstrong, R.L.; Pishe, R.; Gratten-Bellew, P.E. Spatially resolved relaxometry and pore size distribution by single-point MRI methods: Porous media calorimetry. *J. Phys. D Appl. Phys.* **1998**, *31*, 2040. [CrossRef]
15. Jaffer, S.; Lemaire, C.; Hansson, C.; Peemoeller, H. MRI: A complementary tool for imaging cement pastes. *Cem. Concr. Res.* **2007**, *37*, 369–377. [CrossRef]
16. Friedemann, K.; Stallmach, F.; Kärger, J. NMR diffusion and relaxation studies during cement hydration—A non-destructive approach for clarification of the mechanism of internal post curing of cementitious materials. *Cem. Concr. Res.* **2006**, *36*, 817–826. [CrossRef]
17. Nestle, N.; Galvosas, P.; Kärger, J. Liquid-phase self-diffusion in hydrating cement pastes—Results from NMR studies and perspectives for further research. *Cem. Concr. Res.* **2007**, *37*, 398–413. [CrossRef]
18. Neuman, C. Spin echo of spins diffusing in a bounded medium. *J. Chem. Phys.* **1974**, *60*, 4508–4511. [CrossRef]
19. De Swiet, T.M.; Sen, P.N. Decay of nuclear magnetization by bounded diffusion in a constant field gradient. *J. Chem. Phys.* **1994**, *100*, 5597–5604. [CrossRef]
20. Nestle, N. A Simple Semiempiric Model for NMR Relaxometry Data of Hydrating Cement Pastes. *Cem. Concr. Res.* **2004**, *34*, 447–454. [CrossRef]
21. Papavassiliou, G.; Fardis, M.; Laganas, E.; Leventis, A.; Hassanien, A.; Milia, F.; Papageorgiou, A.; Chaniotakis, E. Role of the surface morphology in cement gel growth dynamics: A combined nuclear magnetic resonance and atomic force microscopy study. *J. Appl. Phys.* **1997**, *82*, 449–452. [CrossRef]
22. Blinc, R.; Dolinsek, J.; Lahajnar, G.; Sepe, A.; Zupančič, I.; Zumer, S.; Milia, F.; Pintar, M.M. Spin-Lattice Relaxation of Water in Cement Gels. *Z. Nat. A* **1988**, *43*, 1026–1038. [CrossRef]
23. Laganas, E.; Papavassiliou, G.; Fardis, M.; Leventis, A.; Milia, F.; Chaniotakis, E.; Meletiou, C. ^{1}H Nuclear Magnetic Resonance Relaxation Measurements in Developing Porous Structures: A Study in Hydrating Cement. *J. Appl. Phys.* **1995**, *77*, 3343–3348. [CrossRef]
24. D'Angelo, R.; Plona, T.J.; Schwartz, L.M.; Coveney, P. Ultrasonic measurements on hydrating cement slurries. *Adv. Cem. Based Mater.* **1995**, *2*, 8–14. [CrossRef]
25. Boumiz, A.; Vernet, C.; Tenoudji, F.C. Mechanical properties of cement pastes and mortars at early ages: Evolution with time and degree of hydration. *Adv. Cem. Based Mater.* **1996**, *3*, 94–106. [CrossRef]
26. Sayers, C.; Dahlin, A. Propagation of ultrasound through hydrating cement pastes at early times. *Adv. Cem. Based Mater.* **1993**, *1*, 12–21. [CrossRef]
27. Keating, J.; Hannant, D.J.; Hibbert, A.P. Comparison of shear modulus and UPV techniques to measure the build-up of structure in fresh cement pastes used in oil well cementing. *Cem. Concr. Res.* **1989**, *19*, 554–566. [CrossRef]
28. Ye, G.; Van Breugel, K.; Fraaij, A.L.A. Experimental study on ultrasonic pulse velocity evaluation of the microstructure of cementitious material at early age. *Heron* **2001**, *46*, 161–167.
29. Jalal, M. Durability enhancement of concrete by incorporating titanium dioxide nanopowder into binder. *J. Am. Sci.* **2012**, *8*, 289–294.
30. Jalal, M.; Fathi, M.; Farza, M. Effects of fly ash and TiO_2 nanoparticles on rheological, mechanical, microstructural and thermal properties of high strength self compacting concrete. *Mech. Mater.* **2013**, *61*, 11–27. [CrossRef]
31. Lawrence, P.; Cyr, M.; Ringot, E. Mineral admixtures in mortars-Effect of inert materials on short-term hydration. *Cem. Concr. Res.* **2003**, *33*, 1939–1947. [CrossRef]
32. Jayapalan, A.; Lee, B.; Kurtis, K. Effect of Nano-sized Titanium Dioxide on Early Age Hydration of Portland Cement. In *NICOM3*; Bittnar, Z., Bartos, P.J.M., Nemecek, J., Smilauer, V., Zeman, J., Eds.; Springer: Prague, Czech Republic, 2009; pp. 267–273.

33. Folli, A.; Pade, C.; Hansen, T.B.; De Marco, T.; Macphee, D.E. TiO$_2$ photocatalysis in cementitious systems: Insights into self-cleaning and depollution chemistry. *Cem. Concr. Res.* **2012**, *42*, 539–548. [CrossRef]
34. Chen, J.; Kou, S.-C.; Poon, C.-S. Hydration and properties of nano-TiO$_2$ blended cement composites. *Cem. Concr. Compos.* **2012**, *34*, 642–649. [CrossRef]
35. Nazari, A. The effects of curing medium on flexural strength and water permeability of concrete incorporating TiO2 nanoparticles. *Mater. Struct.* **2011**, *44*, 773–786. [CrossRef]
36. Noorvand, H.; Ali, A.A.A.; Demirboga, R.; Farzadnia, N.; Noorvand, H. Incorporation of nano TiO$_2$ in black rice husk ash mortars. *Constr. Build. Mater.* **2013**, *47*, 1350–1361. [CrossRef]
37. Li, H.; Zhang, M.-H.; Ou, J.-P. Abrasion resistance of concrete containing nano-particles for pavement. *Wear* **2006**, *260*, 1262–1266. [CrossRef]
38. Zhang, M.-H.; Li, H. Pore structure and chloride permeability of concrete containing nano-particles for pavement. *Constr. Build. Mater.* **2011**, *25*, 608–616. [CrossRef]
39. Soleymani, F. The filler effects TiO$_2$ nanoparticles on increasing compressive strength of limestone aggregate-based concrete. *J. Am. Sci.* **2012**, *8*, 3.
40. Rashad, A.M. A Synopsis about the Effect of Nano-Titanium Dioxide on Some Properties of Cementitious Materials-a Short Guide for Civil Engineer. *Rev. Adv. Mater. Sci.* **2015**, *40*, 72–88.
41. Meng, T.; Yu, Y.; Qian, X.; Zhan, S.; Qian, K. Effect of nano-TiO$_2$ on the mechanical properties of cement mortar. *Constr. Build. Mater.* **2012**, *29*, 241–245. [CrossRef]
42. Lee, B.Y. Effect of Titanium Dioxide Nanoparticles on Early Age and Long Term Properties of Cementitious Materials. Ph.D. Thesis, Georgia Institute of Technology, Atlanta, GA, USA, 2012.
43. Behfarnia, K.; Keivan, A.; Keivan, A. The effects of TiO$_2$ and ZnO nanoparticles on physical and mechanical properties of normal concrete. *Asian J. Civ. Eng.* **2013**, *14*, 517–531.
44. Leslie, J.; Cheesman, W. An ultrasonic method of studying deterioration and cracking in concrete structures. *J. Am. Concr. Inst.* **1949**, *21*, 17–36.
45. Lee, H.K.; Lee, K.M.; Kim, Y.H.; Yim, H.; Bae, D.B. Ultrasonic in-situ monitoring of setting process of high-performance concrete. *Cem. Concr. Res.* **2004**, *34*, 631–640. [CrossRef]
46. Robeyst, N.; Gruyaert, E.; Grosse, C.U.; De Belie, N. Monitoring the setting of concrete containing blast-furnace slag by measuring the ultrasonic p-wave velocity. *Cem. Concr. Res.* **2008**, *38*, 1169–1176. [CrossRef]
47. Gallegos, D.P.; Munn, K.; Smith, D.S.; Stermer, D.L. A NMR Technique for the Analysis of Pore Structure: Application to Materials with Well-Defined Pore Structure. *J. Colloid Interface Sci.* **1987**, *119*, 127–140. [CrossRef]
48. Schreiner, L.J.; MacTavish, J.C.; Miljković, L.; Pintar, M.; Blinc, R.; Lahajnar, G.; Lasic, D.; Reeves, L.W. NMR Line Shape-Spin-Lattice Relaxation Correlation Study of Portland Cement Hydration. *J. Am. Ceram. Soc.* **1985**, *68*, 10–16. [CrossRef]
49. McDonald, P.J.; Korb, J.P.; Mitchell, J.; Monteilhet, L. Surface relaxation and chemical exchange in hydrating cement pastes: A two-dimensional NMR relaxation study. *Phys. Rev. E* **2005**, *72*, 011409. [CrossRef] [PubMed]
50. Hürlimann, M.D. Effective gradients in porous media due to susceptibility differences. *J. Magn. Reson.* **1998**, *131*, 232–240. [CrossRef] [PubMed]
51. Valckenborg, R.M.E. *NMR on Technological Porous Materials*; Eindhoven University of Technology: Eindhoven, The Netherlands, 2001.
52. Taylor, H.F. *Cement Chemistry*; Thomas Telford: London, UK, 1997.
53. Allen, A.J.; Thomas, J.J.; Jennings, H.M. Composition and density of nanoscale calcium-silicate-hydrate in cement. *Nat. Mater.* **2007**, *6*, 311–316. [CrossRef] [PubMed]
54. EUROPEAN STANDARD NORME196 Part 3, Methods of testing cement. In *Determination of Setting Times and Soundness*; EU: Brussel, Belgium, 2005.
55. Karakosta, E.; Diamantopoulos, G.; Katsiotis, M.S.; Fardis, M.; Papavassiliou, G.; Pipilikaki, P.; Protopapas, M.; Panagiotaras, D. In Situ Monitoring of Cement Gel Growth Dynamics. Use of a Miniaturized Permanent Halbach Magnet for Precise 1H NMR Studies. *Ind. Eng. Chem. Res.* **2010**, *49*, 613–622. [CrossRef]
56. Provencher, S.W. A Constrained Regularization Method for Inverting Data Represented by Linear Algebraic or Integral Equations. *Comput. Phys. Commun.* **1982**, *27*, 213–227. [CrossRef]
57. Parrott, L.J.; Geiker, M.; Gutteridge, W.A.; Killoh, D. Monitoring Portland cement hydration: Comparison of methods. *Cem. Concr. Res.* **1990**, *20*, 919–926. [CrossRef]

58. Tsivilis, S.; Kakali, G.; Chaniotakis, E.; Souvaridou, A. A study on the hydration of Portland limestone cement by means of TG. *J. Therm. Anal. Calorim.* **1998**, *52*, 863–870. [CrossRef]
59. Cioffi, R.; Marroccoli, M.; Santoro, L.; Valenti, G. DTA study of the hydration of systems of interest in the field of building materials manufacture. *J. Therm. Anal. Calorim.* **1992**, *38*, 761–770. [CrossRef]
60. Seki, S.; Kasahara, K.; Kuriyama, T.; Kawasumi, M. Effects of Hydration of cement on compressive strength, modulus of elasticity and creep of concrete. In Proceedings of the Fifth International Symposium on the Chemistry of Cement, Tokyo, Japan, 7–11 October 1968; pp. 175–185.
61. Ye, G.; Lura, P.; van Breugel, K.; Fraaij, A.L.A. Study on the development of the microstructure in cement-based materials by means of numerical simulation and ultrasonic pulse velocity measurement. *Cem. Concr. Res.* **2004**, *26*, 491–497. [CrossRef]
62. Keating, J.; Hannant, D.J.; Hibbert, A.P. Correlation between cube strength, ultrasonic pulse velocity and volume change for oil well cement slurries. *Cem. Concr. Res.* **1989**, *19*, 715–726. [CrossRef]
63. Gaunaurd, G.; Überall, H. Resonance theory of bubbly liquids. *J. Acoust. Soc. Am.* **1981**, *69*, 362–370. [CrossRef]
64. Feylessoufi, A.; Tenoudji, F.C.; Morin, V.; Richard, P. Early ages shrinkage mechanisms of ultra-high-performance cement-based materials. *Cem. Concr. Res.* **2001**, *31*, 1573–1579. [CrossRef]
65. Bentz, D.P.; Coveney, P.V.; Garboczi, E.J.; Kleyn, M.F.; Stutzman, P.E. Cellular automaton simulations of cement hydration and microstructure development. *Model. Simul. Mater. Sci. Eng.* **1994**, *2*, 783. [CrossRef]
66. Trtnik, G.; Turk, G. Influence of superplasticizers on the evolution of ultrasonic P-wave velocity through cement pastes at early age. *Cem. Concr. Res.* **2013**, *51*, 22–31. [CrossRef]
67. Liu, Z.; Zhang, Y.; Jiang, Q.; Sun, G.; Zhang, W. In situ continuously monitoring the early age microstructure evolution of cementitious materials using ultrasonic measurement. *Constr. Build. Mater.* **2011**, *25*, 3998–4005. [CrossRef]
68. Trtnik, G.; Valič, M.I.; Kavčič, F.; Turk, G. Comparison between two ultrasonic methods in their ability to monitor the setting process of cement pastes. *Cem. Concr. Res.* **2009**, *39*, 876–882. [CrossRef]
69. Reinhardt, H.; Grosse, C. Continuous monitoring of setting and hardening of mortar and concrete. *Constr. Build. Mater.* **2004**, *18*, 145–154. [CrossRef]
70. Buchwald, A.; Tatarin, R.; Stephan, D. Reaction progress of alkaline-activated metakaolin-ground granulated blast furnace slag blends. *J. Mater. Sci.* **2009**, *44*, 5609–5617. [CrossRef]
71. Overloop, K.; Gerven, L.V. NMR relaxation in adsorbed water. *J. Magn. Reson.* **1992**, *100*, 303–315. [CrossRef]
72. Blinc, R.; Lahajnar, G.; Zumer, S.; Pintar, M.M. NMR Study of the Time Evolution of the Fractal Geometry of Cement Gels. *Phys. Rev. B* **1988**, *38*, 2873–2875. [CrossRef] [PubMed]
73. Karakosta, E.; Lagkaditi, L.; ElHardalo, S.; Biotaki, A.; Kelessidis, V.C.; Fardis, M.; Papavassiliou, G. Pore structure evolution and strength development of G-type elastic oil well cement. A combined 1 H NMR and ultrasonic study. *Cem. Concr. Res.* **2015**, *72*, 90–97. [CrossRef]
74. Lee, B.Y.; Kurtis, K.E. Influence of TiO2 nanoparticles on early C3S hydration. *J. Am. Ceram. Soc.* **2010**, *93*, 3399–3405. [CrossRef]
75. Bentz, D.P.; Garboczi, E.J. Percolation of phases in a three-dimensional cement paste microstructural model. *Cem. Concr. Res.* **1991**, *21*, 325–344. [CrossRef]
76. Scherer, G.W.; Zhang, J.; Quintanilla, J.A.; Torquato, S. Hydration and percolation at the setting point. *Cem. Concr. Res.* **2012**, *42*, 665–672. [CrossRef]
77. Gussoni, M.; Greco, F.; Bonazzi, F.; Vezzoli, A.; Botta, D.; Dotelli, G.; Sora, I.N.; Pelosato, R.; Zetta, L. 1H NMR spin-spin relaxation and imaging in porous systems: An application to the morphological study of white portland cement during hydration in the presence of organics. *Magn. Reson. Imaging* **2004**, *22*, 877–889. [CrossRef] [PubMed]

Publisher's Note: MDPI stays neutral with regard to jurisdictional claims in published maps and institutional affiliations.

© 2020 by the authors. Licensee MDPI, Basel, Switzerland. This article is an open access article distributed under the terms and conditions of the Creative Commons Attribution (CC BY) license (http://creativecommons.org/licenses/by/4.0/).

Article

Monitoring the Effect of Calcium Nitrate on the Induction Period of Cement Hydration via Low-Field NMR Relaxometry

Mihai M. Rusu [1], David Faux [2] and Ioan Ardelean [1,*]

[1] Department of Physics and Chemistry, Technical University of Cluj-Napoca, 400114 Cluj-Napoca, Romania
[2] Department of Physics, University of Surrey, Guildford GU2 7XH, UK
* Correspondence: ioan.ardelean@phys.utcluj.ro

Abstract: The hydration process of Portland cement is still not completely understood. For instance, it is not clear what produces the induction period, which follows the initial period of fast reaction, and is characterized by a reduced reactivity. To contribute to such understanding, we compare here the hydration process of two cement samples, the simple cement paste and the cement paste containing calcium nitrate as an accelerator. The hydration of these samples is monitored during the induction period using two different low-field nuclear magnetic resonance (NMR) relaxometry techniques. The transverse relaxation measurements of the ^{1}H nuclei at 20 MHz resonance frequency show that the capillary pore water is not consumed during the induction period and that this stage is shortened in the presence of calcium nitrate. The longitudinal relaxation measurements, performed at variable Larmor frequency of the ^{1}H nuclei, reveal a continuous increase in the surface-to-volume ratio of the capillary pores, even during the induction period, and this increase is faster in the presence of calcium nitrate. The desorption time of water molecules from the surface was also evaluated, and it increases in the presence of calcium nitrate.

Keywords: cement hydration; induction period; accelerator; NMR relaxometry; Fast Field Cycling; 3-Tau model

Citation: Rusu, M.M.; Faux, D.; Ardelean, I. Monitoring the Effect of Calcium Nitrate on the Induction Period of Cement Hydration via Low-Field NMR Relaxometry. *Molecules* **2023**, *28*, 476. https://doi.org/10.3390/molecules28020476

Academic Editor: Igor Serša

Received: 26 November 2022
Revised: 22 December 2022
Accepted: 29 December 2022
Published: 4 January 2023

Copyright: © 2023 by the authors. Licensee MDPI, Basel, Switzerland. This article is an open access article distributed under the terms and conditions of the Creative Commons Attribution (CC BY) license (https://creativecommons.org/licenses/by/4.0/).

1. Introduction

There are various applications of cement-based materials where the process of cement hydration requires acceleration. Some examples include urgent repair works, an increase in the productivity of precast concrete elements, or 3D printing technology [1]. An accelerated hydration process can be obtained by raising the curing temperature [2] using a cement powder of higher fineness [3] by introducing calcium silicate hydrate seeds [4] or by adding different types of accelerators [5,6]. One such accelerator is calcium nitrate (Ca(NO$_3$)$_2$), which is used for its corrosion inhibitor properties as a substitute for calcium chloride in steel-reinforced concrete structures [7].

The hydration of Portland cement is a complicated chemical reaction that starts immediately after mixing the cement grains with water molecules and is characterized by five evolution intervals: initial period, induction period, acceleration period, deceleration period, and continuous slow hydration period [8]. A comprehensive description of cement hydration, containing also the cement chemistry terminology and abbreviations, can be found in Refs. [8,9]. The duration of these evolution intervals is strongly influenced by the composition and the size of the cement grains, the amount of water in the mixture, and by the curing temperature [2,3]. Although it has been investigated for a long time, the hydration process is still not fully understood [10,11]. For instance, it is not clear what produces the induction period, which is characterized by a reduced reactivity [12]. There are two main hypotheses used to explain the low reactivity of cement grains during the induction period, the so called protective membrane theory [13,14] and the dissolution theory [11,15], but none of them are fully accepted by the cement research community.

The protective membrane theory [10] assumes that a layer of monosulfoaluminoferrite hydrates (AFm) [8,16] and metastable calcium silicate hydrate (C-S-H) [13] forms on the surface of cement grains during the initial period (first minutes of hydration) and this layer covers the reacting surface, thus preventing the fast dissolution and the appearance of the induction period. At the end of the induction period, a stable but permeable C-S-H forms from the precipitation of the solution, and the metastable layer disappears; thus, water access to the surface is improved, and the hydration process is accelerated. The existence of such a layer is questioned by some authors, especially because it has not yet been directly visualized despite the progress of microscopic techniques [10].

Dissolution theory assumes that the dissolution rate of a crystalline material is influenced by its interfacial properties and the surrounding solution [15]. The interfacial properties of the crystalline material depend on chemical composition, type of bond, impurities, and lattice defects. The solution properties depend on the nature of the solvent, ionic composition, or temperature and are characterized by the so called undersaturation, which reflects deviation from equilibrium [15]. In the case of the cement clinker, the dissolution rate of alite (C_3S), the most abundant component of the clinker, decreases as undersaturation of the system decreases by moving toward equilibrium [11,15]. Thus, at very high undersaturation levels, immediately after placing water molecules in contact with C_3S, vacancy islands nucleate on the surface of C_3S. As undersaturation decreases it is not possible to create etch pits on plain surfaces but only on defects such as dislocations. At lower levels of undersaturation, etch pits can be created only at a lower rate, and the slow dissolution occurs, thus the system is in the induction period. According to this model, the onset of the acceleration period occurs when a high enough concentration of Ca^{2+} ions is present in solution, and portlandite (CH) can precipitate. Portlandite precipitation increases the degree of undersaturation and thus determines an accelerated C-S-H precipitation [15].

The presence of an accelerator complicates the hydration process but can also help us understand it better [6]. If the accelerator is calcium nitrate, its presence affects both the hydration of tricalcium aluminate (C_3A) and tricalcium silicate (C_3S). Thus, the presence of calcium nitrate may lead to an increased formation of mono-sulfoaluminoferrite hydrates (AFm) and trisulfoaluminoferrite hydrates (AFt), in which the most important phase is ettringite [8,17], and accelerates C_3S reaction due to the fast dissolution of $Ca(NO_3)_2$, which produces a faster supersaturation with Ca^{2+} ions of the solution, which in turn produces earlier crystallization of CH [11,15].

The influence of accelerators, or other admixtures, on the hydration process can be studied by a combination of different techniques, such as calorimetric studies [18], X-ray diffraction [19], mercury porosimetry [20], scanning electron microscopy [21], or ultrasonic tests [22]. Some of these techniques require special sample preparation or stopping of the hydration process when utilized. That is why, when studying water evolution and pore structure development inside cement-based materials, it is favorable to use ^1H low-field nuclear magnetic resonance (NMR) relaxometry techniques [3,22–26]. The advantage of NMR relaxometry over other techniques is that it allows nonperturbative investigations even during the hydration process without the requirement to stop the hydration process and without any pretreatment of the sample.

The most applied variant of NMR relaxometry is based on transverse relaxation measurements using the well-known Carr-Purcell-Meiboom-Gill (CPMG) [27] pulse sequence. This technique provides the relaxation time distribution of water molecules and, thus, information about molecule–surface interaction, saturation degree, and the pore size distribution. Another technique that can be successfully applied to cement-based materials is the Fast Field Cycling (FFC) relaxometry [2,28–30]. The technique allows the determination of longitudinal relaxation time as a function of the proton Larmor frequency (10 kH–10 MHz in the present work), thus becomes sensitive to a wider range of molecular mobilities compared to the measurements performed at a single frequency [28].

In the present work, both CPMG and FFC NMR relaxometry will be applied to study the influence introduced by an accelerator on the hydration dynamics of cement paste during the induction period. In the case of the FFC NMR relaxometry technique, the relaxation dispersion curves recorded at different hydration times for the simple cement paste and the cement paste containing the accelerator will be evaluated using the 3-Tau relaxation model [31–35]. By comparing the outcomes of the two techniques, information about the interaction of water molecules with the surface of cement grains and about surface evolution during the induction period will be extracted.

2. Results and Discussions

To monitor water evolution during cement hydration, CPMG experiments were performed first (see Section 3). The two cement-based materials under investigation were the simple cement paste (CP) and the cement paste containing the accelerator (CP+2% $Ca(NO_3)_2$). The relaxation decay curves were formed of 1000 echoes recorded for short echo time intervals of 0.08 ms, both for reducing the internal gradient effects and detecting short relaxation time components present in the sample. The recorded CPMG echo trains were then used to extract the relaxation time distributions using a numerical Laplace transformation [36,37].

Figure 1 shows the relaxation time distributions of the simple cement paste (Figure 1a) and the cement paste containing the accelerator (Figure 1b). The first CPMG data were recorded at 10 min from mixing start and the last at 300 min. One can observe a distribution of relaxation times with two peaks corresponding to two distinct water environments in the mixture. The first peak, arising at about 1 ms, can be associated to the embedded water inside the flocculation of the cement grains. A second peak of higher area and longer T_2 (between 10–30 ms) corresponds to the capillary pore water. As can be noticed, both peaks evolve during hydration with a faster evolution in the case of the sample containing 2% of $Ca(NO_3)_2$, by cement mass. The decrease in the area of the capillary water peak during hydration can be associated with the transformation of the capillary water into solid components such as C-S-H, ettringite, and CH (not visible in a low-field NMR experiment due to the short relaxation times). Note that, the porous structure of C-S-H also contains water molecules of lower mobility, the so-called gel water which manifests itself by an increase in the area of the first peak after a certain hydration time. However, here we will concentrate only on the capillary water peak because it is also studied in the FFC NMR relaxation measurements.

To quantify the consumption of capillary water and the pore size evolution during hydration, we have evaluated in Figure 2 the capillary peak area (Figure 2a) and the relaxation rate $1/T_2^{max}$ corresponding to the position of the peak maximum (Figure 2b). We restricted our evaluation to hydration times up to 240 min, when one can clearly separate the capillary water component from the gel water (intra and inter C-S-H) [38]. As can be observed in Figure 2a, the peak area decreases in the first 20 min (initial period) for both samples, indicating a fast water consumption and the rapid transformation into hydration products [6,10]. However, after 40 min of hydration the reaction seems to stop, and no capillary water consumption is observed after this time until 160 min in the case of simple cement paste (CP) and 80 min in the case of cement paste containing the accelerator (CP + 2% $Ca(NO_3)_2$). This behavior demonstrates the effect of shortening the induction period of cement hydration in the presence of calcium nitrate. It also demonstrated an acceleration of the hydration kinetics in the presence of calcium nitrate.

If we compare the relaxation times corresponding to the capillary peak maximum (Figure 2b), we notice a continuous variation of the relaxation time for both samples, and this variation is again accelerated in the presence of $Ca(NO_3)_2$. However, if we compare the relaxation rates (Figure 2b) with the capillary water evolution (Figure 2a), we notice that a variation of the relaxation time may exist without water consumption. According to Equation (2) (see Section 3), there are two reasons for such an evolution: increase in the surface-to-volume ratio of the pores and increase in surface relaxivity. However, separation of these contributions is not possible in a CPMG experiment. That is why we complete here the investigations on the hydration of the two samples with Fast Field Cycling NMR relaxation measurements, which better separate the two contributions, provided that a suitable relaxation model is used [31,32].

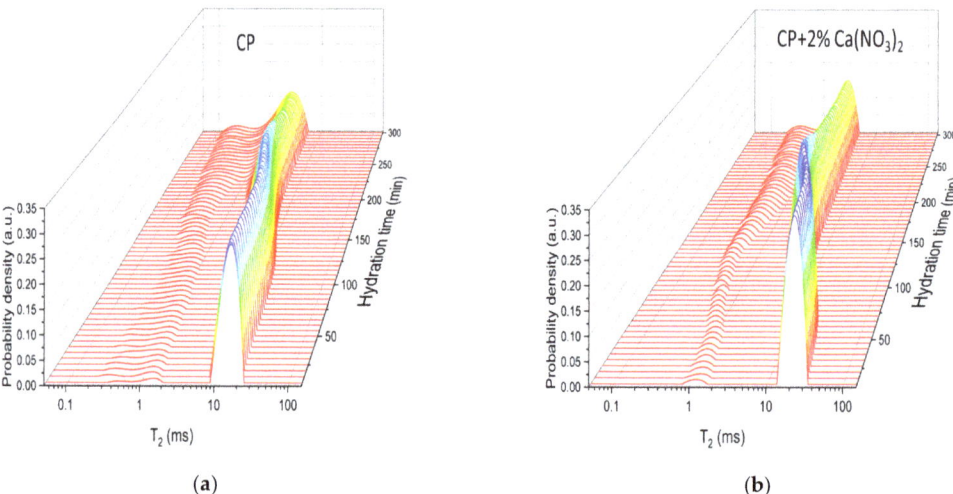

Figure 1. Relaxation time distributions of the two cement pastes during the early hydration: (**a**) the simple cement paste CP; (**b**) the cement paste prepared with 2% accelerator (CP + 2% Ca(NO$_3$)$_2$), by cement weight. The smaller peak corresponds to the embedded water while the larger peak corresponds to the capillary water.

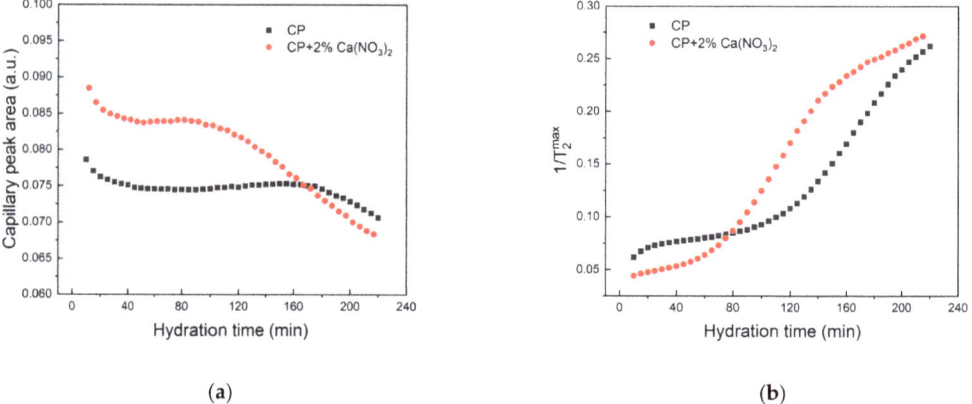

Figure 2. (**a**) Capillary peak area versus hydration time for the two cement pastes. (**b**) Relaxation rates corresponding to the maximum of the capillary versus hydration time.

Longitudinal relaxation measurements were performed on both samples using the FFC techniques described in Section 3. The switching time of 2 ms was set to reduce the contribution of the embedded water component to the detected signal and to ensure that only the capillary water is detected. Due to the required time for the adjustment of the instrument parameters, the first relaxation dispersion curve could be detected at 20 min from mixing start. The subsequent measurements were performed every 30 min, up to 170 min. The last measurement was recorded at 170 min due to the fact that, after this hydration time, it is difficult to separate the embedded water from capillary water contribution, as one can directly observe in Figure 1.

The relaxation dispersion curves recorded for the two samples at different hydration times are shown in Figure 3. One can observe a faster evolution in the case of the sample containing 2% of Ca(NO$_3$)$_2$, by cement mass (Figure 3b) as compared to the simple cement paste (Figure 3a) and this evolution takes place at all frequencies. To extract information on the dynamics of confined molecules, the relaxation dispersion curves were fitted with the 3-Tau model [31–35], briefly described in Section 3.3. The fitting approach follows the comprehensive description in Ref. [39] and uses the software package provided by Kogon and Faux [40]. The continuous lines represent the best fits of the experimental data with the 3-Tau model. On this basis the following fitting parameters could be extracted (see Section 3.3): τ_l-surface diffusion time; τ_b-bulk diffusion time; τ_d-desorption time; N_{par}-number density of paramagnetic ions; $S\delta/V$-surface layer volume to pore volume ratio.

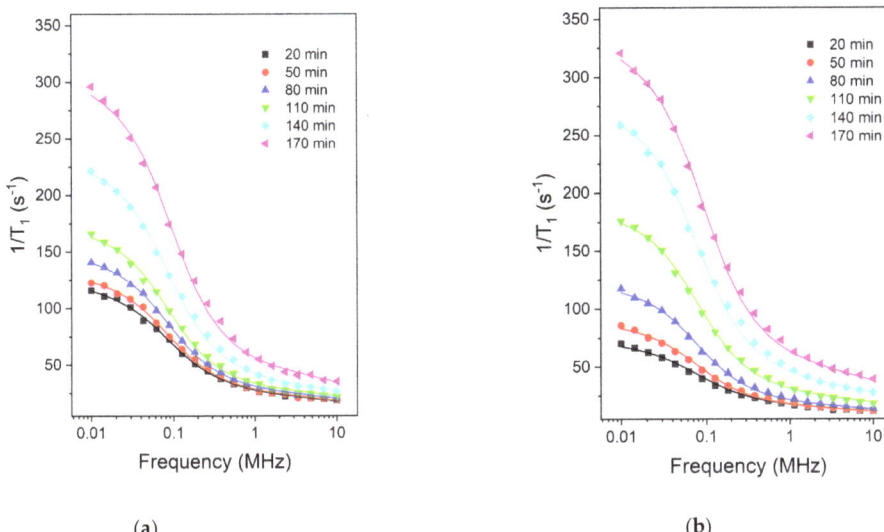

(a) (b)

Figure 3. Relaxation dispersion curves recorded for the two samples during the early hydration: (a) the simple cement paste CP; (b) the cement paste prepared with 2% accelerator (CP + 2% Ca(NO$_3$)$_2$). The hydration times are indicated in the legend. The lines represent the best fits obtained with the 3-Tau model.

To fit the relaxation dispersion curves with the 3-Tau model, we started the fitting approach by setting all five parameters free. However, after several fitting rounds, we noticed that two of them, τ_l and τ_d, do not vary and can be kept constant during the first 170 min of hydration. Thus, the surface diffusion time of $\tau_l = 0.24$ µs and the bulk diffusion time of $\tau_b = 17.8$ ps could be considered fixed for both samples and all hydration times. The desorption time parameter τ_d is, however, different in the case of CP sample as compared with the CP + 2% Ca(NO$_3$)$_2$ sample. Thus, for hydration times of up to 140 min the desorption time value $\tau_d = 3.65$ µs was extracted in the case of cement paste and $\tau_d = 4.22$ µs for the sample containing the accelerator. The longer desorption time in the case of the sample containing the accelerator shows a stronger interaction of water molecules with the surface. Note, however, that, for the hydration time of 170 min, both samples revealed the same value of desorption time $\tau_d = 3.16$ µs. This indicates identical surface properties of the two samples at 170 min of hydration. The surface diffusion time τ_l for the two samples during the interval 20–140 min allows the extraction of a surface diffusion coefficient $D_l = 5.06 \times 10^{-14}$ m^2/s that is three orders of magnitude smaller than that extracted on the basis of Korb's relaxation model [29]. The surface water diffusion coefficient found here is typical of "hard" solid porous material such as pastes (plaster/silica), clay, and rocks independent of the model [33]. The diffusion coefficient is low because the surface layer of water is both chemi- and physiosorbed to the surface. The surface water can diffuse by breaking away from the surface before moving over the potential energy barrier to a neighboring vacant site. However, most surface sites are filled unless a water molecule has desorbed into the bulk to create a vacancy. Desorption itself is hindered by the blocking effect of the mobile water in the second hydration layer.

The number density of paramagnetic ions N_{par} of the two samples was also determined from the fitting approach. The values are represented in Figure 4a as a function of hydration time. One can observe a dependence of this parameter on hydration time for both samples. However, even if the cement pastes are made of identical cement powders, the smaller number densities are observed for the sample containing the accelerator at initial hydration times. That behavior could be explained by the presence of a thicker layer of hydration products, which isolates the capillary water from the surface [6], which is equivalent with a smaller effective ion density. Another observation is that a rapid increase in N_{par} occurs at hydration times of about 100 min (CP) and 50 min, respectively (CP + 2% Ca(NO$_3$)$_2$). This increase in N_{par} can be associated with an increase in the permeability for water molecules to reach the grain surface and the beginning of the acceleration process. We notice that the hydration times for which the increase in N_{par} is observed in Figure 4a corresponds to the hydration times when changes in the relaxation rate behavior (Figure 2b) are observed, that is the end of induction stage.

Another parameter that can be extracted from the fitting approach is $S\delta/V$, which represents the ratio between the volume of the surface layer and the pore volume. Figure 4b shows an increase of the surface volume to pore volume ratio during the induction period. The increase of $S\delta/V$ during induction period can be associated with a continuous increase of the pore surface due to continuous formation of etch pits on the surface of cement grains [15]. The formation of these pits does not consume the capillary water as can be observed from the constant capillary peak area (Figure 2a). A similar conclusion about the $S\delta/V$ increase was drawn from a fractal analysis of the pore surface based on transverse NMR relaxation measurements [24]. Figure 4b shows a faster saturation of the increase in the $S\delta/V$ in the presence of the accelerator (CP + 2% Ca(NO$_3$)$_2$, by cement mass), as compared with the simple cement paste (CP). The dependence of $S\delta/V$ parameter on hydration time allows, in the approximation of plane pores, extraction of an effective pore size h. The dependence of the pore size on hydration time is shown in Figure 4c. Note, however, that the extracted values are affected by the roughness of the pore surface and must be considered with caution. Nevertheless, the extracted values are in the same range of dimensions as those obtained on similar samples using a different approach, based on diffusion in internal gradients [41].

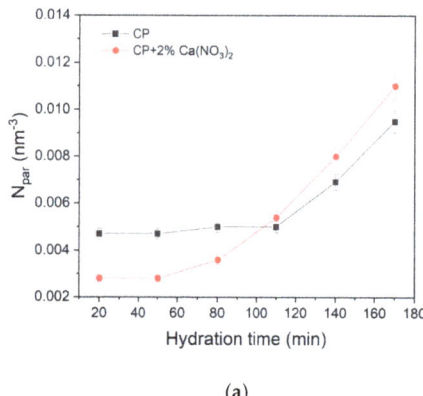

(a)

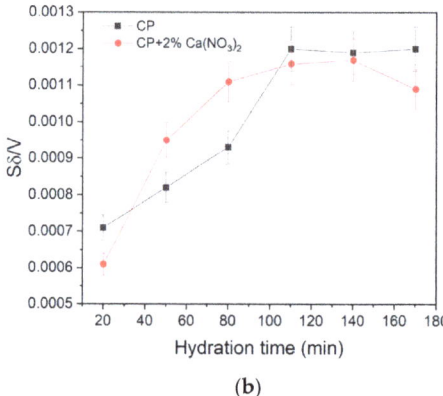

(b)

Figure 4. Cont.

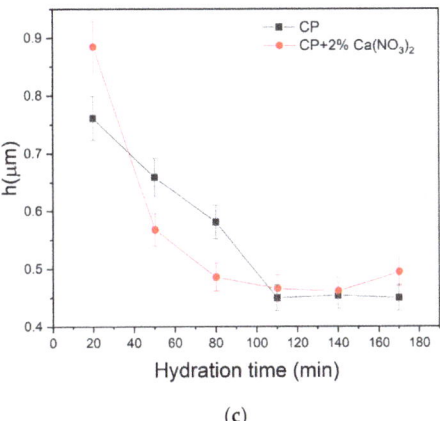

(c)

Figure 4. The fitting parameters, extracted from the data in Figure 3 by using the 3-Tau model. (**a**) Number density of paramagnetic ions versus hydration time. (**b**) The ratio the ratio between the volume of the surface layer and the pore volume versus hydration time. (**c**) The pore size versus hydration time in the approximation of plane pores.

3. Materials and Methods

3.1. Sample Preparation

Two cement-based samples were prepared using white Portland cement (CEM I 52.5R Holcim, Romania), for a water-to-cement ratio of 0.4, by weight. The chemical composition of the cement powder was determined in Ref. [23] and is specific to a white Portland cement with low iron content: CaO (70.7%), SiO_2 (16.3%), and Al_2O_3 (4.7%), with gypsum derived sulphate (5.8%) and other impurities based on MgO (0.9%), Na_2O (0.7%), KO (0.6%), and Fe_2O_3 (0.3%). The grain size distribution indicates an average size and a maximum size of 5 μm and 25 μm, respectively. The white cement, with a low content of iron oxide, was chosen here with the aim of reducing the internal gradient effects on echo attenuation in the CPMG experiment [42]. The first sample was a simple cement paste (CP) obtained by mixing the cement grains with distilled water. The second sample further contains the accelerator $(Ca(NO_3)_2)$, acquired from Nordic Chemicals SRL, Romania). The amount of accelerator is 2%, by cement mass, and was introduced into the mixture by first dissolving it in water. The ingredients were mixed at room temperature for 5 min using a mechanical mixer. The first NMR experiments started at 10 min from the beginning of mixing in the case of CPMG measurements and at 20 min in the case of FFC relaxometry measurements. They were performed under identical temperature conditions at 35 °C.

3.2. The NMR Methodology

NMR relaxation measurements of the molecules confined inside porous media provide information about the porous structure, the molecule-surface interaction, and the liquid distribution inside pores. To extract such information, two types of experiments are usually performed: (i) transverse relaxation time distribution determinations for confined molecules and (ii) longitudinal relaxation measurements of ^1H nuclei as a function of Larmor frequency. Of course, these experiments must be supplemented with theoretical models describing relaxation phenomena under specific experimental conditions. Here, we will shortly describe the two experiments performed in our investigations and the specific relaxation model.

Transverse relaxation time distributions can be obtained from the echo trains recorded using the well-known CPMG pulse sequence shown in Figure 5a (top). The pulse sequence consists of a series of hard 180-degree radiofrequency (RF) pulses following a first 90-degree pulse. An echo train is composed of the echoes recorded at evolution intervals $t = 2n\tau$. The amplitude of the n-th echo in the echo train attenuates according to the formula [38]:

$$A_n = A_0 \int_0^\infty P(T_2) e^{-\frac{2n\tau}{T_2}} dT_2 \qquad (1)$$

where A_0 is a constant affected by the sample magnetization, temperature, and hardware characteristics of the NMR instrument. $P(T_2)$ represents the relaxation time probability density, and T_2 is the relaxation time of molecules confined inside a given pore. As the above formula suggests, a numerical Laplace inversion of the recorded echo train amplitudes provides the relaxation time distribution [36,37].

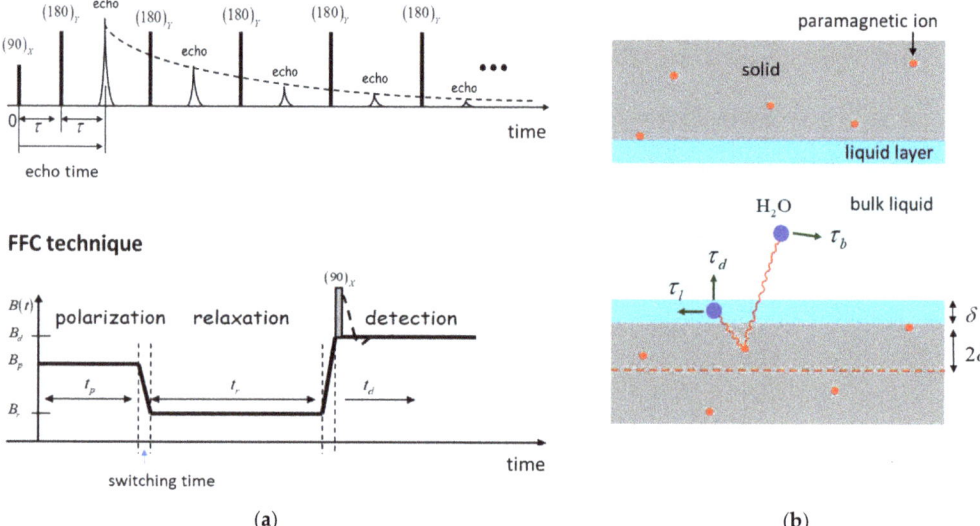

Figure 5. (a) Top: CPMG pulse sequence generating the echo trains for transverse relaxation measurements. Bottom: The variation of the main magnetic field along a Fast Field Cycling relaxation experiment. (b) The schematic representation of the interactions accounted in the frame of the 3-Tau model [32].

The transverse relaxation time depends on the surface-to-volume ratio S/V of the pore by the formula [38]

$$\frac{1}{T_2} = \frac{1}{T_2^{bulk}} + \rho \frac{S}{V} \qquad (2)$$

where $1/T_2^{bulk}$ represents the relaxation rate of the bulk-like liquid inside the pores, ρ represents the relaxivity, a constant depending on the paramagnetic impurity content of the pore surface; wettability of the filling liquid; temperature; and the strength of the main magnetic field of the instrument. Considering the above equation, the relaxation time distribution extracted via Equation (1) allows the determination of the pore size distribution, provided that the relaxivity is known from an independent experiment. Even if the relaxivity is not known, the relative distribution of pore sizes can be obtained. Note that, the above equation applies only for saturated pores and by neglecting diffusion effects on echo train attenuation [38,42].

Frequency dependent longitudinal relaxation measurements can be performed with the so-called Fast Field Cycling technique [28]. This technique allows the polarization and detection of the nuclear spins at higher fields while relaxation can take place at lower fields. With this approach, a significant increase in the detection sensitivity can be obtained as compared with the case in which polarization and detection would be performed at lower fields that are identical to the relaxation field. Thus, a main advantage of the FFC technique is the possibility of making longitudinal relaxation measurements at different Larmor frequencies (10 kHz–10 MHz, in the present work).

The schematic representation of an FFC experiment is shown in Figure 5a (bottom) and consists of three evolution intervals: polarization, relaxation, and detection [28]. Thus, during the polarization interval, the sample polarizes in a higher field B_p for a duration t_p until it reaches saturation. Then the magnetic field is rapidly switched to a value B_r and the nuclear spins relax inside this field for a relaxation time t_r. Immediately after, the field is switched up to a detection field B_d and the remaining magnetization is measured by applying a 90-degree RF pulse. Note that the switching time from one field to another cannot be made arbitrarily short and that limits the detection of very short relaxation components. On the other side, the switching time can be used as a filter for short relaxation times, as was done in our experiments.

The CPMG relaxation measurements were performed using a low-field NMR instrument (Minispec MQ20, Bruker BioSpin GmbH, Rheinstetten, Germany), operating at a resonance frequency of 20 MHz. This low frequency value, together with the short echo time interval, provides the conditions for reduced diffusion effects on echo train attenuation and thus justifies the implementation of Equation (2) for the relaxation rate [38,42]. A commercially available Fast Field Cycling NMR relaxometer (SMARtracer, Stelar S.R.L, Mede, Italy) was used for recording the relaxation dispersion curves in a range of proton resonance frequencies between 10 kHz and 10 MHz. All the measurements were performed at 35 °C.

3.3. The Relaxation Model

In the case of CPMG relaxation measurements, the frequency dependence of the transverse relaxation rate is not important. This is because the measurements are performed at a single proton resonance frequency (here, 20 MHz) and a simple relation, Equation (2), connects the surface-to-volume ratio with the relaxation time. However, in the case of the FFC relaxation measurements, which provide the dependence of the longitudinal relaxation rate on the Larmor frequency (the so-called relaxation dispersion curves), a relaxation model, adapted to the system studied, must be used to extract the dynamical information. A valuable model describing relaxation of water molecules confined inside the pores of cement-based materials is the 3-Tau model developed by Faux and collaborators [31–35]. This model was comprehensively discussed in the previous publications of Faux and collaborators [31–35] and here we will only summarize its main features and outcomes. Moreover, to facilitate the applications of the 3-Tau model to real samples a fitting package was provided and tested on different samples by Faux and collaborators [39,40].

In the frame of the 3-Tau model, the relaxation rate of water molecules confined inside porous materials containing paramagnetic impurities is dominated by the thermally modulated dipolar interactions of the proton spins with the paramagnetic ions (here Fe^{3+}) inside the solid matrix. According to the 3-Tau model shown in Figure 5b, water molecules can be found in two distinct environments: in the bulk-like state and in a surface monolayer of thickness $\delta = 0.27$ nm (the size of one water molecule). These molecules encounter displacements along the surface and in bulk (indicated by arrows on Figure 5b), which modulate the dipolar interaction between the nuclear spins and the electronic spin of the paramagnetic ions inside the solid structure. As a result of these modulations the longitudinal relaxation rate is a complicated function of three-time constants (3-Tau), the density of paramagnetic ions and the surface-to-volume ratio:

$$\frac{1}{T_1(\omega)} = f\left(\tau_l, \tau_b, \tau_d, N_{para}, \frac{S\delta}{V}\right) \quad (3)$$

In the above expression, τ_l is a characteristic time that describes the displacement of water molecules on the surface and is related to the surface water diffusion coefficient D_l by the formula $\tau_l = \delta^2/6D_l$. The bulk-like water molecules are characterized by a diffusion time constant τ_b which describes the diffusional displacement of bulk molecules over a distance δ. It is related to the bulk diffusion coefficient D_b by the formula $\tau_b = \delta^2/6D_b$. Note that, in the case of pure water, at room temperature, $\tau_b = 5.3$ ps but it would be longer for water confined inside the pores of cement paste due to the presence of dissolved ions, which interact with water and hinder diffusion, but also due to their slower mobility inside the second hydration layer at the surface, which dominates the bulk relaxation [39]. Assuming that the surface spins desorb as $\exp(-t/\tau_d)$, the third constant τ_d characterizes the time water molecule spend at the surface before desorption. According with molecular dynamic simulations τ_d and τ_l should be in the same order of magnitude [31].

The number of paramagnetic ions per unit volume N_{para} is considered here as an effective parameter corresponding to the position of an effective layer (dashed line in Figure 5b located 2δ under the surface of the pores. Note that if a layer of hydration products precipitates on the surface of cement grains, N_{para} should only indirectly depend on the paramagnetic impurity content of the cement grains. $S\delta/V$ represents the ratio between the volume of the surface layer and the pore volume. Assuming planar pores, this ratio allows finding of another key quantity, the pore size h as the distance between the two planes. However, in the case of an evolving porous structure, as is the case of cement-based materials, the change in the surface to volume ratio can be also determined by the change in fractal dimension [24], and the pore size parameter extracted from the fitting approach must be used here with caution when characterizing the pore size evolution during hydration.

4. Conclusions

Two low-field NMR relaxometry techniques were implemented to extract information about the influence of calcium nitrate on early hydration of a white cement paste. Transverse relaxation measurements, performed with CPMG technique, have demonstrated that water contained inside the capillary pores is not consumed during the induction period. However, the duration of the induction stage is shortened in the presence of calcium nitrate. Longitudinal relaxation dispersion curves, recorded with the FFC NMR relaxometry technique, could be well-fitted with the 3-Tau relaxation model for all hydration times. The results of the fitting revealed a continuous increase in the surface-to-volume ratio of the capillary pores, even during the induction period, and this increase is faster in the presence of calcium nitrate. It was also observed that the desorption time of water molecules from the capillary pore surface increases in the presence of calcium nitrate but is constant during the induction period.

Author Contributions: Conceptualization, I.A. and M.M.R.; methodology, I.A.; software, D.F.; validation, M.M.R., D.F. and I.A.; formal analysis, I.A.; investigation, M.M.R. and I.A.; resources, I.A.; writing—original draft preparation, I.A.; writing—review and editing, M.M.R., D.F. and I.A.; funding acquisition, I.A. All authors have read and agreed to the published version of the manuscript.

Funding: This research was funded by a grant from the Romanian Ministry of Education and Research, CNCS-UEFISCDI, project number PN-III-P4-ID-PCE-2020-0533, within PNCDI III.

Institutional Review Board Statement: Not applicable.

Informed Consent Statement: Not applicable.

Data Availability Statement: Data may be provided on request from corresponding author.

Conflicts of Interest: The authors declare no conflict of interest.

Sample Availability: Not applicable.

References

1. Bos, F.; Wolfs, R.; Ahmed, Z.; Salet, T. Additive Manufacturing of Concrete in Construction: Potentials and Challenges of 3D Concrete Printing. *Virtual Phys. Prototyp.* **2016**, *11*, 209–225. [CrossRef]
2. Badea, C.; Pop, A.; Mattea, C.; Stapf, S.; Ardelean, I. The Effect of Curing Temperature on Early Hydration of Gray Cement via Fast Field Cycling-NMR Relaxometry. *Appl. Magn. Reson.* **2014**, *45*, 1299–1309. [CrossRef]
3. Alesiani, M.; Pirazzoli, I.; Maraviglia, B. Factors Affecting Early-Age Hydration of Ordinary Portland Cement Studied by NMR: Fineness, Water-to-Cement Ratio and Curing Temperature. *Appl. Magn. Reson.* **2007**, *32*, 385–394. [CrossRef]
4. Wang, B.; Yao, W.; Stephan, D. Preparation of Calcium Silicate Hydrate Seeds by Means of Mechanochemical Method and Its Effect on the Early Hydration of Cement. *Adv. Mech. Eng.* **2019**, *11*, 1–7. [CrossRef]
5. Aïtcin, P.C. Accelerators. In *Science and Technology of Concrete Admixtures*; Woodhead Publishing: Sawston, UK, 2016; pp. 405–413. [CrossRef]
6. Dorn, T.; Blask, O.; Stephan, D. Acceleration of Cement Hydration—A Review of the Working Mechanisms, Effects on Setting Time, and Compressive Strength Development of Accelerating Admixtures. *Constr. Build. Mater.* **2022**, *323*, 126554. [CrossRef]
7. Gaidis, J.M. Chemistry of Corrosion Inhibitors. *Cem. Concr. Compos.* **2004**, *26*, 181–189. [CrossRef]
8. Marchon, D.; Flatt, R.J. *Science and Technology of Concrete Admixtures*; Elsevier: Amsterdam, The Netherlands, 2016; ISBN 9780081006931.
9. Bullard, J.W.; Jennings, H.M.; Livingston, R.A.; Nonat, A.; Scherer, G.W.; Schweitzer, J.S.; Scrivener, K.L.; Thomas, J.J. Mechanisms of Cement Hydration. *Cem. Concr. Res.* **2011**, *41*, 1208–1223. [CrossRef]
10. Scrivener, K.; Ouzia, A.; Juilland, P.; Kunhi Mohamed, A. Advances in Understanding Cement Hydration Mechanisms. *Cem. Concr. Res.* **2019**, *124*, 105823. [CrossRef]
11. Juilland, P.; Nicoleau, L.; Arvidson, R.S.; Gallucci, E. Advances in Dissolution Understanding and Their Implications for Cement Hydration. *RILEM Tech. Lett.* **2017**, *2*, 90–98. [CrossRef]
12. Scrivener, K.L.; Juilland, P.; Monteiro, P.J.M. Advances in Understanding Hydration of Portland Cement. *Cem. Concr. Res.* **2015**, *78*, 38–56. [CrossRef]
13. Gartner, E.M.; Jennings, H.M. Thermodynamics of Calcium Silicate Hydrates and Their Solutions. *J. Am. Ceram. Soc.* **1987**, *70*, 743–749. [CrossRef]
14. Jennings, H.M.; Pratt, P.L. An Experimental Argument for the Existence of a Protective Membrane Surrounding Portland Cement during the Induction Period. *Cem. Concr. Res.* **1979**, *9*, 501–506. [CrossRef]
15. Juilland, P.; Gallucci, E.; Flatt, R.; Scrivener, K. Dissolution Theory Applied to the Induction Period in Alite Hydration. *Cem. Concr. Res.* **2010**, *40*, 831–844. [CrossRef]
16. Matschei, T.; Lothenbach, B.; Glasser, F.P. The AFm Phase in Portland Cement. *Cem. Concr. Res.* **2007**, *37*, 118–130. [CrossRef]
17. Balonis, M.; Mędala, M.; Glasser, F.P. Influence of Calcium Nitrate and Nitrite on the Constitution of AFm and AFt Cement Hydrates. *Adv. Cem. Res.* **2011**, *23*, 129–143. [CrossRef]
18. Gawlicki, M.; Nocuń-Wczelik, W.; Bąk, Ł. Calorimetry in the Studies of Cement Hydration: Setting and Hardening of Portland Cement-Calcium Aluminate Cement Mixtures. *J. Therm. Anal. Calorim.* **2010**, *100*, 571–576. [CrossRef]
19. Scrivener, K.L.; Füllmann, T.; Gallucci, E.; Walenta, G.; Bermejo, E. Quantitative Study of Portland Cement Hydration by X-Ray Diffraction/Rietveld Analysis and Independent Methods. *Cem. Concr. Res.* **2004**, *34*, 1541–1547. [CrossRef]
20. Cook, R.A.; Hover, K.C. Mercury Porosimetry of Cement-Based Materials and Associated Correction Factors. *Constr. Build. Mater.* **1993**, *7*, 231–240. [CrossRef]
21. Stutzman, P. Scanning Electron Microscopy Imaging of Hydraulic Cement Microstructure. *Cem. Concr. Compos.* **2004**, *26*, 957–966. [CrossRef]
22. Diamantopoulos, G.; Katsiotis, M.; Fardis, M.; Karatasios, I.; Alhassan, S.; Karagianni, M.; Papavassiliou, G.; Hassan, J. The Role of Titanium Dioxide on the Hydration of Portland Cement: A Combined NMR and Ultrasonic Study. *Molecules* **2020**, *25*, 5364. [CrossRef]
23. Rusu, M.M.; Vilau, C.; Dudescu, C.; Pascuta, P.; Popa, F.; Ardelean, I. Characterization of the Influence of an Accelerator upon the Porosity and Strength of Cement Paste by Nuclear Magnetic Resonance (NMR) Relaxometry. *Anal. Lett.* **2022**, *56*, 303–311. [CrossRef]
24. Ardelean, I. The Effect of an Accelerator on Cement Paste Capillary Pores: NMR Relaxometry Investigations. *Molecules* **2021**, *26*, 5328. [CrossRef] [PubMed]
25. Wang, B.; Faure, P.; Thiéry, M.; Baroghel-Bouny, V. 1H NMR Relaxometry as an Indicator of Setting and Water Depletion during Cement Hydration. *Cem. Concr. Res.* **2013**, *45*, 1–14. [CrossRef]
26. Pop, A.; Badea, C.; Ardelean, I. The Effects of Different Superplasticizers and Water-to-Cement Ratios on the Hydration of Gray Cement Using T2-NMR. *Appl. Magn. Reson.* **2013**, *44*, 1223–1234. [CrossRef]
27. Meiboom, S.; Gill, D. Modified Spin-Echo Method for Measuring Nuclear Relaxation Times. *Rev. Sci. Instrum.* **1958**, *29*, 688–691. [CrossRef]

28. Kimmich, R.; Anoardo, E. Field-Cycling NMR Relaxometry. *Prog. Nucl. Magn. Reson. Spectrosc.* **2004**, *44*, 257–320. [CrossRef]
29. Korb, J.-P.; Monteilhet, L.; McDonald, P.J.; Mitchell, J. Microstructure and Texture of Hydrated Cement-Based Materials: A Proton Field Cycling Relaxometry Approach. *Cem. Concr. Res.* **2007**, *37*, 295–302. [CrossRef]
30. Ardelean, I. Applications of Field-Cycling NMR Relaxometry to Cement Materials. In *Field-Cycling NMR Relaxometry: Instrumentation, Model Theories and Applications*; Kimmich, R., Ed.; Royal Society of Chemistry: Cambridge, UK, 2019; pp. 462–488.
31. Faux, D.A.; Mcdonald, P.J.; Howlett, N.C. Nuclear Magnetic Resonance Relaxation Due to the Translational Diffusion of Fluid Confined to Quasi-Two-Dimensional Pores. *Phys. Rev. E* **2017**, *95*, 033116. [CrossRef]
32. Faux, D.A.; McDonald, P.J. A Model for the Interpretation of Nuclear Magnetic Resonance Spin-Lattice Dispersion Measurements on Mortar, Plaster Paste, Synthetic Clay and Oil-Bearing Shale. *Microporous Mesoporous Mater.* **2018**, *269*, 39–42. [CrossRef]
33. Faux, D.A.; McDonald, P.J. Explicit Calculation of Nuclear-Magnetic-Resonance Relaxation Rates in Small Pores to Elucidate Molecular-Scale Fluid Dynamics. *Phys. Rev. E* **2017**, *95*, 033117. [CrossRef]
34. Faux, D.A.; McDonald, P.J. Nuclear-Magnetic-Resonance Relaxation Rates for Fluid Confined to Closed, Channel, or Planar Pores. *Phys. Rev. E* **2018**, *98*, 063110. [CrossRef]
35. Faux, D.; Kogon, R.; Bortolotti, V.; McDonald, P. Advances in the Interpretation of Frequency-Dependent Nuclear Magnetic Resonance Measurements from Porous Material. *Molecules* **2019**, *24*, 3688. [CrossRef] [PubMed]
36. Venkataramanan, L.; Song, Y.-Q.; Hurlimann, M.D. Solving Fredholm Integrals of the First Kind with Tensor Product Structure in 2 and 2.5 Dimensions. *IEEE Trans. Signal Process.* **2002**, *50*, 1017–1026. [CrossRef]
37. Provencher, S.W. CONTIN: A General Purpose Constrained Regularization Program for Inverting Noisy Linear Algebraic and Integral Equations. *Comput. Phys. Commun.* **1982**, *27*, 229–242. [CrossRef]
38. Bede, A.; Scurtu, A.; Ardelean, I. NMR Relaxation of Molecules Confined inside the Cement Paste Pores under Partially Saturated Conditions. *Cem. Concr. Res.* **2016**, *89*, 56–62. [CrossRef]
39. Kogon, R.; Faux, D. 3TM: Software for the 3-Tau Model. *SoftwareX* **2022**, *17*, 100979. [CrossRef]
40. Kogon, R.; Faux, D. 3TM: Fitting Software for Fast Field Cycling NMR Dispersion Date. Available online: https://zenodo.org/record/5774107#.Y7UXsBVByUk (accessed on 14 May 2022). [CrossRef]
41. Stepišnik, J.; Ardelean, I. Usage of Internal Magnetic Fields to Study the Early Hydration Process of Cement Paste by MGSE Method. *J. Magn. Reson.* **2016**, *272*, 100–107. [CrossRef]
42. Zielinski, L.J. Effect of Internal Gradients in the Nuclear Magnetic Resonance Measurement of the Surface-to-Volume Ratio. *J. Chem. Phys.* **2004**, *121*, 352–361. [CrossRef]

Disclaimer/Publisher's Note: The statements, opinions and data contained in all publications are solely those of the individual author(s) and contributor(s) and not of MDPI and/or the editor(s). MDPI and/or the editor(s) disclaim responsibility for any injury to people or property resulting from any ideas, methods, instructions or products referred to in the content.

Article

MR Study of Water Distribution in a Beech (*Fagus sylvatica*) Branch Using Relaxometry Methods

Urša Mikac [1], Maks Merela [2], Primož Oven [2], Ana Sepe [1] and Igor Serša [1,*]

[1] Department of Condensed Matter Physics, Jožef Stefan Institute, 1000 Ljubljana, Slovenia; urska.mikac@ijs.si (U.M.); ana.sepe@ijs.si (A.S.)
[2] Department of Wood Science and Technology, Biotechnical Faculty, University of Ljubljana, 1000 Ljubljana, Slovenia; maks.merela@bf.uni-lj.si (M.M.); primoz.oven@bf.uni-lj.si (P.O.)
* Correspondence: igor.sersa@ijs.si

Citation: Mikac, U.; Merela, M.; Oven, P.; Sepe, A.; Serša, I. MR Study of Water Distribution in a Beech (*Fagus sylvatica*) Branch Using Relaxometry Methods. *Molecules* **2021**, *26*, 4305. https://doi.org/10.3390/molecules26144305

Academic Editor: José A. González-Pérez

Received: 24 June 2021
Accepted: 13 July 2021
Published: 16 July 2021

Publisher's Note: MDPI stays neutral with regard to jurisdictional claims in published maps and institutional affiliations.

Copyright: © 2021 by the authors. Licensee MDPI, Basel, Switzerland. This article is an open access article distributed under the terms and conditions of the Creative Commons Attribution (CC BY) license (https://creativecommons.org/licenses/by/4.0/).

Abstract: Wood is a widely used material because it is environmentally sustainable, renewable and relatively inexpensive. Due to the hygroscopic nature of wood, its physical and mechanical properties as well as the susceptibility to fungal decay are strongly influenced by its moisture content, constantly changing in the course of everyday use. Therefore, the understanding of the water state (free or bound) and its distribution at different moisture contents is of great importance. In this study, changes of the water state and its distribution in a beech sample while drying from the green (fresh cut) to the absolutely dry state were monitored by 1D and 2D ^{1}H NMR relaxometry as well as by spatial mapping of the relaxation times T_1 and T_2. The relaxometry results are consistent with the model of homogeneously emptying pores in the bioporous system with connected pores. This was also confirmed by the relaxation time mapping results which revealed the moisture transport in the course of drying from an axially oriented early- and latewood system to radial rays through which it evaporates from the branch. The results of this study confirmed that MRI is an efficient tool to study the pathways of water transport in wood in the course of drying and is capable of determining the state of water and its distribution in wood.

Keywords: magnetic resonance imaging (MRI); relaxation times; beech (*Fagus sylvatica*); wood; moisture content (MC)

1. Introduction

Wood is a hygroscopic porous and permeable material that is widely used in everyday life. It interacts with water from humid air causing a constantly changing moisture content (MC), especially in the outdoor use where it is exposed to dynamic moisture cycles. The MC changes affect the wood properties and are responsible for shrinkage and swelling of wood, moisture-induced stresses and mechanosorptive effects, which may lead to cracking or loss of loadbearing capacity. Wood contains macromolecules that link water by hydrogen bonding [1,2]. Thus, water in wood exists as bound and free water. Free water is in the form of liquid or vapor in cell lumina and bound water is hydrogen bonded in the cell wall material. Changed in the MC in the range between the absolutely dry wood (MC = 0%) and the wood at the fiber saturation point (FSP) (approximately 30%) where all water is bound in the cell walls cause alterations in physical and mechanical properties of wood. At higher MCs, water also exists as free water with almost no effect on the physical and mechanical properties. It is established that the optimal fungal growth is achieved at MC = 35–50% on the basis of dry weight. Therefore, the knowledge about the state of water and moisture transport in wood is of utmost importance for understanding its utilization, durability and wood product quality [3].

Different methods such as traditional gravimetrical determination, methods based on the electrical properties of moist wood and titration, for instance, are used to measure the MC of wood [4]. Among other methods, nuclear magnetic resonance (NMR) and magnetic resonance imaging (MRI) have been successfully employed for studying the MC in wood [5–9] as well as its spatial distribution in wood samples [10–20]. The spin–lattice (T_1) and spin–spin (T_2) relaxation times of the protons in wood change with the MC. This is because the NMR relaxation times depend on the local environment of protons as they determine the mobility of molecules and thus influence the T_1/T_2 ratio. This ratio is higher in the environment with molecules of higher mobility [21]. The T_2 of protons in solid wood is in tens of microseconds, the T_2 of bound water with hindered local motion is in the range from hundreds of microseconds to several milliseconds, while the T_2 of free water in the cell lumina is in the range from tens to hundreds of milliseconds [6,22,23]. In addition, the T_2 values of free water depend on cell dimensions i.e., the T_2 of free water is longer in cells with larger lumina [24]. Therefore, four peaks are observed in the T_2 distribution of wood. The first two peaks are associated with free water and are therefore at higher T_2 values. Their amplitudes decrease with a decreasing MC and they vanish at MCs below the FSP. The third peak at shorter T_2 is associated with bound water. Its amplitude is constant with MCs above the FSP but it starts to decrease with MCs lower than the FSP. The fourth peak is associated with solid wood and it is at the shortest T_2 values. The amplitude of this peak is constant with any MC [6,13,22,25,26]. Different relaxation time values thus enable determination of the water state in the wood. The simplest are one-dimensional (1D) T_1 and T_2 spectra which enable distinction between bound and free water. More complex are two-dimensional (2D) T_1–T_2 and T_2–T_2 correlation spectra with which improvement of the resolution and information on water states in the wood is significant. T_1–T_2 correlation spectra enable distinction between the two types of bound water in cell walls, while T_2–T_2 correlation spectra can identify the water exchange between cell walls and the free water in the lumina, enabling measurement of the corresponding exchange rates [27]. These methods have also been used to study the adsorption mechanisms in earlywood and latewood [28], determine the structural changes of wood after thermal modification [29] and the effect of wood aging at the molecular level [9] and to characterize the decay process of wood due to fungal decomposition [30].

Proton density-weighted MRI produces a signal that is proportional to free water in wood, but it cannot detect bound states of water and solid wood. This is because the NMR signal of bound water and solid wood decays before detection with standard imaging methods. More precise information on the state of water in wood can be obtained from T_1, T_2 and apparent diffusion coefficient (ADC) maps [31]. T_2 maps are, in particular, important to get better contrast between free water in different wood structures [17,30].

The goal of this study is to demonstrate that NMR relaxometry is a powerful technique that allows studying the distribution and movement of water, free or bound, in different anatomical structures of wood in the course of its drying. Specifically, 1D T_1 and T_2 distributions, 2D T_1–T_2 correlation spectra and T_1 and T_2 maps of a beechwood sample at different MCs in the range from 90% (green state) to an absolutely dry sample were measured in this study to follow changes of the water state and distribution in the course of wood drying.

2. Results and Discussion

2.1. 1D T_1 and T_2 Distributions at Different MCs

A multiexponential analysis of T_1 and T_2 relaxation decay curves was used to determine the relaxation time distributions. Figure 1 shows the T_1 and T_2 distributions for different MCs. T_1 distributions consisted of two peaks: an intense peak in the range of hundreds of milliseconds and a small peak at few milliseconds. With the decreasing MC (wood drying), the position of the intense peak first decreased, reached a minimum value of 210 ms at MC = 25% and then increased with the decreasing MC (Figure 1a). The values of the shorter T_1 components were in the range of 10 ms. This peak was almost constant with drying until MC = 20% and then increased with a decreasing MC, up to 50 ms at MC = 9%. In the course of drying, the integrated intensities of both peaks slightly decreased until MC = 42%. Then, the integral of the longer T_1 component decreased and the integral of the shorter T_1 component increased in the MC range between 42% and 20%, whereas at MCs below 20%, the integral of the longer component increased and the integral of the shorter component decreased and was no longer observed at MC = 0% (Figure 1a).

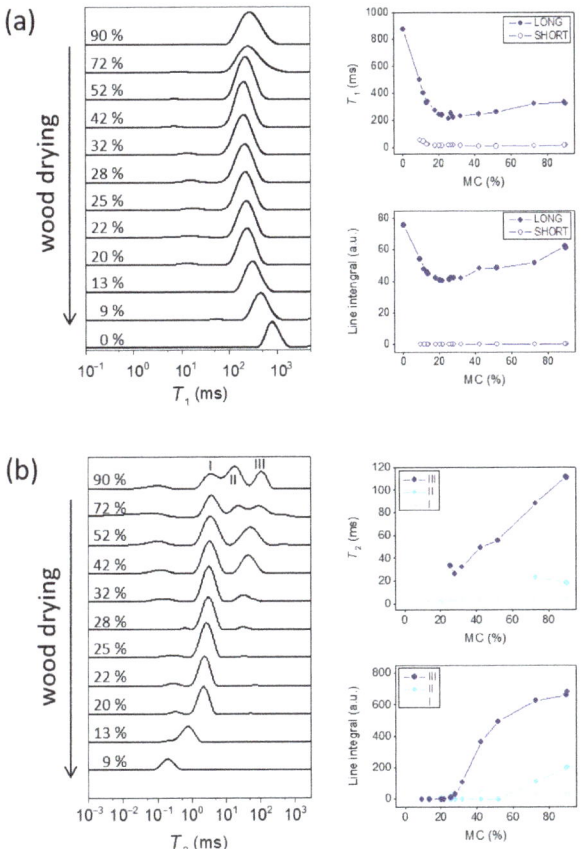

Figure 1. NMR relaxation time distributions together with the central relaxation time and the integrated intensities of the peaks at different MCs for the beech branchwood sample: (**a**) T_1 and (**b**) T_2. The labels LONG and SHORT in the graphs in panel (**a**) correspond to the T_1 values of an intense peak in the range of hundreds of milliseconds (long) and to a small peak at few milliseconds (short), while the labels I, II and III in the graphs in panel (**b**) correspond to short, medium and long T_2 values of three distinct peaks in T_2 distributions.

The T_2 distributions are, however, different (Figure 1b). A small and broad peak was observed in the T_2 distribution at 0.1 ms that remained almost constant throughout the sample drying. In addition, three peaks I, II and III were observed at higher MCs. With the decreasing MC (wood drying), peak I remained at the same position until MC = 20% and shifted to lower values at lower MCs. The T_2 of peak II slightly increased when MC decreased from 90% to 72% and then overlapped with peak III or I at lower MCs. Peak III shifted to lower values with the decreasing MC. The integrated intensity of peak I increased with the decreasing MC until 52%, remained constant until MC = 25% and decreased with MC further decreasing, while the integral of peak III decreased with the decreasing MC and was no longer observed at MCs lower than 25%. As in the previous studies [25,26,29,30,32–35], the peaks I, II and III were assigned to bound water, free water in cells with smaller lumina and free water in cells with larger lumina, respectively.

2.2. Two-Dimensional T_1–T_2 Correlation Spectra at Different MCs

To further evaluate the T_1 and T_2 results, 2D T_1–T_2 correlation spectra were measured for three different MCs (Figure 2). At MC = 90%, five peaks (labeled A1, A2, B, C and D, see Figure 2) were observed, with two different T_1 and four different T_2 values. The peaks A1, A2 and C were just below the diagonal $T_1 = T_2$, while the peaks B and C had similar T_2 but different T_1. The intensities and positions of the peaks kept changing with MC. At MC = 35%, intensities of the peaks A1 and A2 decreased significantly and could not be distinguished, and the intensity of peak C increases compared to its intensity at MC = 90%. At MC = 6%, the peaks A1 and A2 were no longer observed, and peak C had a very low intensity. The T_1 values of all the peaks decreased when the MC decreased from 90% to 35% and increased again when the MC decreased to 6%. Peaks B and C had similar T_2 values at MC = 90% and 35%; however, their T_2 values decreased at MC = 6%. The peaks in the T_1–T_2 correlation spectra could be identified on the basis of previous analyses [27]. The peaks with longer T_1 and the longest T_2 (A1 and A2) arose from water with the highest molecular mobility, i.e., free water in lumina with different diameters. Peak B with shorter T_2 corresponded to bound water, peak D with the shortest T_2—to solid-like protons. Peak C with shorter T_1 and the same T_2 as peak B was assigned to the water absorbed in wood polymers.

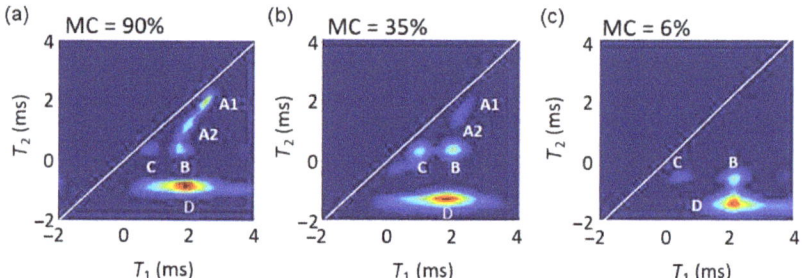

Figure 2. Two-dimensional T_1–T_2 correlation spectra of the beech branchwood sample at: (a) MC = 90%, (b) MC = 35% and (c) MC = 6%. The five peaks are attributed to free water in cell lumina (A1 and A2), protons of bound water (B and C) and solid wood protons (D) as discussed in the text.

2.3. MR Imaging: Proton Density Images and T_1 and T_2 Maps

The spatial distributions of relaxation times in the wood sample were measured by T_1 and T_2 mapping. For the proton density imaging, the first image with the shortest echo time of a sequence of echo images for T_2 map determination was used. It should be noted that the shortest echo time was still too long to allow detection of a signal from protons in solid wood as their T_2 values are in the range of tens of microseconds. The imaging method which was used allows detection of signals with T_2 values over a millisecond. For the same reason, the signal of bound water with T_2 of hundreds of microseconds produces a low signal that is, therefore, not completely detected. Thus, the signal of proton density images consists mainly of free water. Relaxation time maps were calculated by the complete set of echo images using the best fit to the monoexponential decaying function. The resolution of the images is lower than the size of a wood cell and therefore each pixel of the image consists of several cells with the cell lumina and cell walls. This implies that the multi-component decaying exponential function would yield a more accurate fit to the data and determine the relaxation times of all the states of water and solid protons in each pixel. However, due to the insufficient signal-to–noise ratio (SNR), the monoexponential fit was used. In addition, T_2 values measured using the spin-echo imaging pulse sequence at various echo times are underestimated due to diffusional loss of the signal during read gradients [31,36]. Therefore, the T_2 values cannot be directly compared to the spectroscopically determined

T_2 values, especially for protons with longer T_2 values. Nevertheless, the T_1 and T_2 maps can still give valuable information on the water in different wood structures.

Proton density images, T_1 and T_2 maps are shown in Figure 3. The brightness of these images is proportional to proton density, T_1 and T_2 relaxation times, respectively. The proton density image at MC = 90% shows different anatomical structures: annual rings with earlywood and latewood and rays. The annual rings and rays are also clearly shown on the T_1 and T_2 maps. It can be seen from the maps that both relaxation times were longer in the earlywood compared to the latewood and the shortest in the rays (Table 1). As the MC decreased, the contrast between different wood tissues increased. Signal intensity in the rays increased due to an increased amount of free water with longer T_2 relaxation time. In contrast, the signal of the annual rings decreased due to a decrease of free water amount as well as T_2 reduction in partially filled lumina. At MC = 32%, the rays were either filled with free water or already empty, which can be seen in the corresponding MR image and maps as indicated by high or no signal intensity.

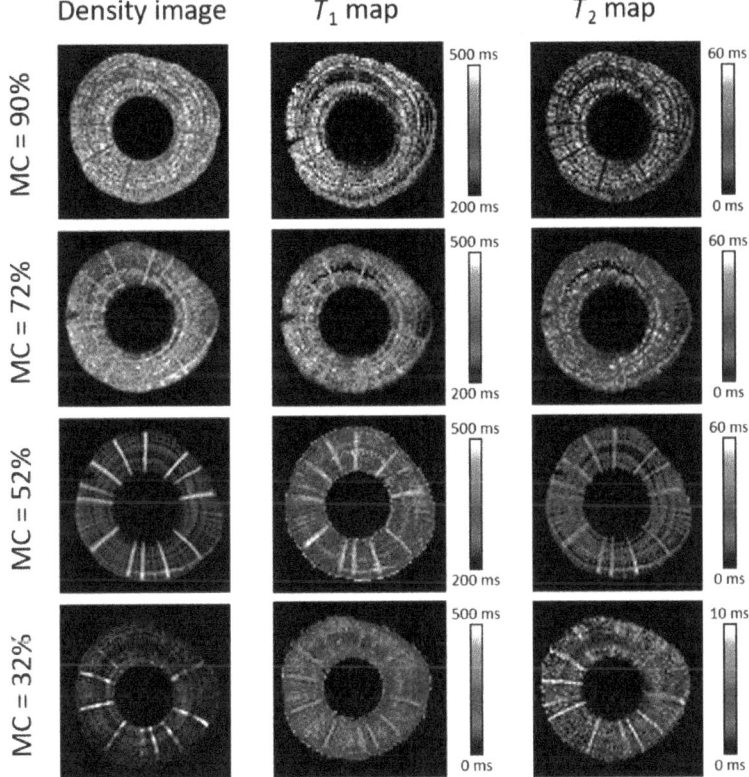

Figure 3. Density images, T_1 and T_2 maps of beech branchwood at different MCs. Note that the scales for the T_1 and T_2 maps are different for MC = 32% than for the higher MCs.

In some MR maps, a dark region with shorter T_1 and T_2 values or even with non-defined relaxation times values is observed due to too low SNR. It is interesting to note that the relaxation times of the rays in this region remained the same as for the rays elsewhere in the sample. This region is not observed in images at all the MCs because the sample was removed from the magnet after measurement at each MC, and the slices of the images at different MCs might be slightly different.

Table 1. T_1 and T_2 values of earlywood, latewood and ray regions obtained from the T_1 and T_2 maps at different MCs.

MC		T_1 (ms)	T_2 (ms)
90%	Earlywood	350 ± 30	35 ± 5
	Latewood	290 ± 20	17 ± 2
	Ray	290 ± 20	12 ± 2
72%	Earlywood	330 ± 30	30 ± 5
	Latewood	300 ± 20	17 ± 2
	Ray	400 ± 20	27 ± 2
52%	Earlywood	320 ± 30	24 ± 3
	Latewood	260 ± 20	13 ± 2
	Ray	400 ± 20	35 ± 2
32%	Earlywood	230 ± 30	5 ± 1
	Latewood	180 ± 20	3 ± 1
	Ray	280 ± 20	8 ± 1

2.4. Discussion

Wood contains two main proton compartments: solid wood material (cellulose, hemicellulose and lignin) and water that can be observed in cell cavities as lumen water (free water) or bound in cell walls (bound water). It should be noted that the relaxation times of lumen water depend on the cell size [1]. Since wood generally contains a continuous distribution of cell sizes, the analysis of relaxation time distributions using inverse Laplace transformation (LT) is more appropriate than a multiexponential analysis using a model function equal to the sum of a predefined number of exponentially decaying functions. In the study, 1D inverse Laplace transformation was applied to the experimental data obtained by the inversion recovery (IR) and Carr–Purcell–Meiboom–Gill (CPMG) pulse sequences to calculate 1D distributions (spectra) of the T_1 and T_2 relaxations times, respectively. The drawback of the 1D LT relaxation time distribution analysis is that it cannot always resolve all different proton compartments in wood, particularly in cases when different proton compartments have similar T_1 or T_2 values and the spectral peaks overlap. However, if these protons have similar T_2 but different T_1 values or vice versa, then it is possible to resolve these different proton compartments by 2D T_1–T_2 correlation spectroscopy. This was performed using 2D LT of the data acquired by the IR–CPMG sequence. Two-dimensional T_1–T_2 correlation spectra were measured at three different MCs in order to differentiate the overlapping peaks in the 1D relaxation time spectra. To obtain differences in relaxation times for different wood structures, 2D T_1 and T_2 maps were measured as well.

The T_1 distributions had two peaks (Figure 1a). The two peaks in the T_1 distributions were attributed to different T_1 values of earlywood and latewood in red cedar and hemlock [37] or the fast exchange between free and bound water [25,27]. Results of the T_1 maps (Figure 3, Table 1) yielded values in the earlywood, latewood and ray regions in the range of the longer T_1 component, i.e., 100–700 ms. These results are therefore more consistent with the fast exchange scenario.

Differences in the T_2 distributions (Figure 1b) in the course of sample drying show that T_2 and integrated intensity of peak III decreased with the decreasing MC and the peak vanished at MC = 25%. This value is close to that of the FSP where all free water evaporates and only bound water remains. Peak III can therefore be assigned to free water in cell lumina. Peak II could not be distinguished from peaks I and III at MCs below 72%. It is interesting to note that the T_2 value of peak II increased when MC decreased from 90% to 72%. In the previous studies, these two peaks were associated with free water in cell lumina of different sizes [25,29,32–35] as the T_2 value is directly proportional to the pore size [24]. Peak III was assigned to free water in earlywood vessels, peak II—to free water in smaller latewood vessels and ray cells. Another study suggested that peak III corresponds to free water in tracheid (fiber) cells, peak II—to free water in ray cell lumina, pits and tracheid lumen ends [32]. The T_2 value of peak I was constant down to MC = 20% (just below the FSP) and then decreased with the decreasing MC. The dependence of the integrated line intensity of peak I on MC is interesting. The integrated intensity first increased, then it was constant and finally decreased again below MC = 20%. This can be explained by the model of a bioporous system with connected pores [38]. The T_2 value and the integrated intensity of peak III decreased in the course of drying indicating the homogeneous decrease of water in large pores. The increase in the integrated intensity of peak I shows that the larger and smaller pores were connected, and emptying of the large pores left some liquid films along the walls. The water of the liquid film has a much shorter T_2 that could overlap with the T_2 values of the smaller pores or even with the T_2 values of bound water. This result was also supported by the T_1–T_2 correlation spectra (Figure 2) where two peaks with different T_2 values and an identical T_1 value were observed, i.e., peaks B and C. The intensity of peak C, i.e., the bound water with higher mobility (higher T_1/T_2) increased as the MC decreased from 90% to 35% than the bound water assigned to peak B (lower T_1/T_2), At MC = 35%, almost no free water was in the cell lumina (low intensity of the peaks A1 and A2). This result can be explained by an increasing proportion of liquid film on cell walls with decreasing MCs (wood drying). The signal of the liquid film can be assigned to the peak of bound water with higher mobility (peak C). With further drying of the sample below the FSP, the intensity of peak C decreased while the intensity of peak B was almost the same, i.e., highly mobile bound water evaporates first, causing the decrease of peak II at MCs below 20%.

In addition to three peaks (I, II and III) in the T_2 distribution, there was also a peak at much shorter T_2 values of around 100 µs corresponding to peak D in the T_1–T_2 correlation spectra. This peak remained constant throughout drying of the sample and was assigned to solid wood. However, the T_2 values of the solid wood are in the range of several tens of microseconds. This is too short for signal detection with the CPMG sequence at the parameters used in this study. Thus, most probably only the part of the spectrum with the longest T_2 values of solid wood was successfully measured while the actual T_2 of this peak was below our detection limit.

The spatial distribution of the T_2 value at various MCs is shown in T_2 maps (Figure 3 and Table 1). Shorter T_2 value for latewood than for earlywood at all MCs was observed, which is in agreement with a previous study [17] and is the consequence of larger lumina of earlywood cells compared to latewood cells. The T_2 value of the rays first increased with the MC decrease down to 52%. At this MC, the T_2 value of the ray tissue was even higher than the T_2 value of earlywood. As the MC decreased to 32%, the T_2 value of rays decreased as well but was still higher than the T_2 value of larger earlywood vessels at this MC. The multiseriate rays were larger than the earlywood vessels. Therefore, an additional reason for the longer T_2 value was a higher amount of water in ray cells; namely, the T_2 value increased with the water concentration in pores [38]. These results indicate that in the course of drying of a wood sample, water is diffused from the annual rings to the rays before evaporating from the sample. The spatial distribution of the T_2 value at different MCs can also explain vanishing of peak II below MC = 72%, i.e., high above the FSP. The T_2 value of the ray cells at high MC contributed to peak II. As the ray cells were filled with more water at lower MCs, the T_2 of free water in the cells increased and began to overlap with the T_2 value of peak III. Free water in latewood cells also contributed to peak II at high MCs. However, as the amount of water decreased in the course of drying, the T_2 value of the latewood decreased to several milliseconds such that the T_2 value of free water in partially empty latewood cells could overlap with peak I.

This study was performed on small samples due to the sample size limitations of the MRI scanner that was used in the study. The scanner was optimized for spatial resolution (for MR microscopy) and therefore had very sensitive but small RF probes. The largest RF probe had a diameter of only 27 mm and this was also the largest sample size that could be scanned. However, the identical methodology used in this study can be used on a much larger scale, e.g., with clinical scanners, where the samples can be up to ten times larger than in this study.

3. Materials and Methods

3.1. Plant Material

Five 15-mm-long samples of a young forest beech tree (*Fagus sylvatica* L.) were cut from fresh branches with a diameter of approximately 8 mm and the annual growth ring width of 0.2 mm. Pith and bark were removed from the samples to avoid large variations of MCs in the samples. The samples were then dried in a desiccator until the MC of the samples decreased from the initial 88% (in the green state) to below 20%. This was needed in order to reach the state of wood below the fiber saturation point (FSP) with only bound water. To moisten the samples to different well-defined MCs, they were equilibrated in a desiccator over different salt solutions ensuring different relative air humidities (RH): $MgCl_2$ (RH = 33%), K_2CO_3 (RH = 44%), $NaNO_2$ (RH = 65%), NaCl (RH = 75%), and $ZnSO_4$ (RH = 85%). After all the MR experiments were finished, the samples were completely dried in the oven at 103 °C for several hours until their masses were equilibrated. The MCs were determined gravimetrically using the Equation (1).

$$MC = \frac{m - m_0}{m_0} \times 100\% \quad (1)$$

where m is the mass of a moist sample and m_0 is the mass of an absolutely dry sample. Wood density in the absolutely dry state was 580 kg/m^3.

3.2. NMR and MRI Experiments

The NMR and MRI experiments were performed on a system consisting of a superconducting 2.35-T (^{1}H NMR frequency of 100 MHz) horizontal bore magnet (Oxford Instruments, Abingdon, UK) equipped with gradients and RF coils for MR microimaging (Bruker, Ettlingen, Germany) using a Tecmag Apollo (Tecmag, Houston, TX, USA) NMR/MRI spectrometer. For the MR experiments, the wood sample was taken out of the desiccator at appropriate time intervals, weighted and inserted into a glass tube that was sealed with a Teflon cap to prevent sample drying during the scanning. The sample was reoriented in the magnet in such a way that it allowed the imaging of an axial slice (parallel to the radial–tangential plane) in 2D MRI experiments. Each sample was weighted before and after the MR measurements. The maximal change of weight during MR experiments was less than 2% and observed only for the samples with high MC, while the mass differences were negligible for the samples with MCs less than 30%.

The spin–spin relaxation times T_2 were measured using the Carr–Purcell–Meiboom–Gill (CPMG) sequence $90°–\tau–[180°–\tau–AQ–\tau]^N$ with the echo time τ of 150 µs and loop repetitions N of 3000 in order to enable measurement of a wide range of T_2 values for the sample with different MCs. To measure the spin–lattice relaxation time T_1, the inversion recovery (IR) pulse sequence $180°–\tau_1–90°–AQ$ was used, with the logarithmically increasing IR delay τ_1 (from 20 µs to 10 s; 36 different τ_1 values). To further validate the relaxation results, 2D $T_1–T_2$ relaxation correlations were measured at three different MCs, 90%, 35% and 6%, using the IR-CPMG sequence, where the IR part was followed by the CPMG loop [39]. The IR delays were the same as for 1D T_1 measurements. The echo delays in the CPMG loop were equal to 350 µs, 50 µs and 25 µs, with the number of loops of 2048, 1024 and 512 for the samples with the MC of 90%, 35% and 6%, respectively.

The experimental data of T_1, T_2 and $T_1–T_2$ measurements were processed via a multiexponential analysis using the Prospa software that was provided by Prof. P. Callaghan [36,39]. The analysis based on multidimensional inverse Laplace transformation allows the resolution and quantification of various components in the relaxation distribution to some extent.

Two-dimensional T_1 and T_2 relaxation time maps were measured using a modified spin-echo imaging pulse sequence. Specifically, the inversion recovery spin-echo (IR-SE) imaging sequence was used for T_1 mapping, i.e., a hard 180° pulse followed by the time interval τ_1 added before the standard 2D spin-echo imaging sequence. T_1 maps were determined from the IR-SE images measured with the time interval τ_1 ranging from 40 µs to 10 s (nine different τ_1 values); the echo time was equal to TE = 3.6 ms and the repetition time was TR = 10 s. T_2 maps were determined from the standard 2D spin-echo images measured with the echo time (TE) varying between 3.6 ms (the shortest possible TE) and 300 ms (nine different values). The other imaging parameters for 2D images were as follows: field of view (FOV) = 13 mm, matrix size of 128 × 128 and slice thickness = 1 mm with the in-plane resolution of 0.1 mm. Proton density-weighted images were selected as the images with the shortest echo time (TE = 3.6 ms) of the sequence used for T_2 map calculation.

4. Conclusions

The present study demonstrates that a combination of 1D T_1 and T_2 spectra, 2D T_1–T_2 correlation spectra and their spatial distributions given by the T_1 and T_2 maps provides valuable information about changes in wood in the course of drying. The obtained results enabled precise analysis of moisture redistribution in the course of drying between different anatomic regions of wood. It also enabled determination of the ratio between the amounts of bound and free water as well as the amount of water in wood cells of different lumina. The advantage of the proposed method is also that it is non-destructive, non-invasive and non-contact and therefore enables MC analysis of the same sample during different stages of its drying.

Author Contributions: U.M., M.M., P.O., A.S. and I.S. conceived and designed the experiments; M.M. provided the samples; U.M., A.S. and I.S. performed the experiments; U.M. analyzed the data; U.M. wrote the original draft; M.M., P.O. and I.S. reviewed and edited the manuscript. All authors discussed the results and commented on the manuscript. All authors have read and agreed to the published version of the manuscript.

Funding: This research was funded by the Slovenian Research Agency through the Research Core Funding (Nos. P1-0060, P4-0015 and J1-7042).

Institutional Review Board Statement: Not applicable.

Informed Consent Statement: Not applicable.

Data Availability Statement: The data presented in this study are available on request from the corresponding author.

Acknowledgments: The authors thank Kanza Awais for proofreading an earlier version of the manuscript.

Conflicts of Interest: The authors declare no conflict of interest.

Sample Availability: Not available.

References

1. Bowyer, J.L.; Shmulsky, R.; Haygreen, J.G. *Forest Products and Wood Science*; Blackwell Publishing Professional: Ames, IA, USA, 2007.
2. Rostom, L.; Care, S.; Courtier-Murias, D. Analysis of water content in wood material through 1D and 2D H-1 NMR relax-ometry: Application to the determination of the dry mass of wood. *Magn. Reson. Chem.* **2021**, *59*, 614–627. [CrossRef]
3. Skaar, C. *Wood-Water Relations*; Springer Science and Business Media LLC: Berlin, Germany, 1988.
4. Dietsch, P.; Franke, S.; Franke, B.; Gamper, A.; Winter, S. Methods to determine wood moisture content and their applicability in monitoring concepts. *J. Civ. Struct. Health* **2015**, *5*, 115–127. [CrossRef]
5. Merela, M.; Oven, P.; Serša, I.; Mikac, U. A single point NMR method for an instantaneous determination of the moisture content of wood. *Holzforschung* **2009**, *63*, 348–351. [CrossRef]
6. Menon, R.; Mackay, A.L.; Hailey, J.R.T.; Bloom, M.; Burgess, A.E.; Swanson, J.S. An NMR determination of the physiological water distribution in wood during drying. *J. Appl. Polym. Sci.* **1987**, *33*, 1141–1155. [CrossRef]
7. Hartley, I.; Kamke, F.A.; Peemoeller, H. Absolute Moisture Content Determination of Aspen Wood Below the Fiber Saturation Point using Pulsed NMR. *Holzforschung* **1994**, *48*, 474–479. [CrossRef]
8. Thygesen, L.G. PLS calibration of pulse NMR free induction decay for determining moisture content and basic density of softwood above fiber saturation. *Holzforschung* **1996**, *50*, 434–436.
9. Rostom, L.; Courtier-Murias, D.; Rodts, S.; Care, S. Investigation of the effect of aging on wood hygroscopicity by 2D 1H NMR relaxometry. *Holzforschung* **2019**, *74*, 400–411. [CrossRef]
10. Bucur, V. *Nondestructive Characterization and Imaging of Wood*; Springer Science and Business Media LLC: Berlin, Germany, 2003.
11. Merela, M.; Oven, P.; Sepe, A.; Serša, I. Three-dimensional in vivo magnetic resonance microscopy of beech (*Fagus sylvatica* L.) wood. *Magma Magn. Reson. Mater. Phys. Biol. Med.* **2005**, *18*, 171–174. [CrossRef]

12. Brownstein, K. Diffusion as an explanation of observed NMR behavior of water absorbed on wood. *J. Magn. Reson. (1969)* **1980**, *40*, 505–510. [CrossRef]
13. Araujo, C.D.; Mackay, A.L.; Hailey, J.R.T.; Whittall, K.P.; Le, H. Proton Magnetic-Resonance Techniques for Character-ization of Water in Wood—Application to White Spruce. *Wood Sci. Technol.* **1992**, *26*, 101–113. [CrossRef]
14. Araujo, C.D.; Mackay, A.L.; Whittall, K.P.; Hailey, J.R.T. A Diffusion-Model for Spin-Spin Relaxation of Compart-mentalized Water in Wood. *J. Magn. Reson. Ser. B* **1993**, *101*, 248–261. [CrossRef]
15. Oven, P.; Merela, M.; Mikac, U.; Serša, I. 3D magnetic resonance microscopy of a wounded beech branch. *Holzforschung* **2008**, *62*, 322–328. [CrossRef]
16. Oven, P.; Merela, M.; Mikac, U.; Serša, I. Application of 3D magnetic resonance microscopy to the anatomy of woody tissues. *IAWA J.* **2011**, *32*, 401–414. [CrossRef]
17. Javed, M.A.; Kekkonen, P.M.; Ahola, S.; Telkki, V.-V. Magnetic resonance imaging study of water absorption in thermally modified pine wood. *Holzforschung* **2015**, *69*, 899–907. [CrossRef]
18. Žlahtič, M.; Mikac, U.; Serša, I.; Merela, M.; Humar, M. Distribution and penetration of tung oil in wood studied by magnetic resonance microscopy. *Ind. Crops Prod.* **2017**, *96*, 149–157. [CrossRef]
19. Zupanc, M.Ž.; Mikac, U.; Serša, I.; Merela, M.; Humar, M. Water distribution in wood after short term wetting. *Cellulose* **2018**, *26*, 703–721. [CrossRef]
20. Almeida, G.; Leclerc, S.; Perre, P. NMR imaging of fluid pathways during drainage of softwood in a pressure membrane chamber. *Int. J. Multiph. Flow* **2008**, *34*, 312–321. [CrossRef]
21. Abragam, A. *The Principles of Nuclear Magnetism*; Clarendon Press: Oxford, UK, 1961.
22. Almeida, G.; Gagné, S.; Hernández, R.E. A NMR study of water distribution in hardwoods at several equilibrium moisture contents. *Wood Sci. Technol.* **2006**, *41*, 293–307. [CrossRef]
23. Dvinskikh, S.; Henriksson, M.; Berglund, L.; Furó, I. A multinuclear magnetic resonance imaging (MRI) study of wood with adsorbed water: Estimating bound water concentration and local wood density. *Holzforschung* **2011**, *65*, 103–107. [CrossRef]
24. Brownstein, K.R.; Tarr, C.E. Importance of classical diffusion in NMR studies of water in biological cells. *Phys. Rev. A* **1979**, *19*, 2446–2453. [CrossRef]
25. Telkki, V.-V.; Yliniemi, M.; Jokisaari, J. Moisture in softwoods: Fiber saturation point, hydroxyl site content, and the amount of micropores as determined from NMR relaxation time distributions. *Holzforschung* **2013**, *67*, 291–300. [CrossRef]
26. Elder, T.; Houtman, C. Time-domain NMR study of the drying of hemicellulose extracted aspen (*Populus tremuloides* Michx.). *Holzforschung* **2013**, *67*, 405–411. [CrossRef]
27. Cox, J.; McDonald, P.J.; Gardiner, B. A study of water exchange in wood by means of 2D NMR relaxation correlation and exchange. *Holzforschung* **2010**, *64*, 259–266. [CrossRef]
28. Bonnet, M.; Courtier-Murias, D.; Faure, P.; Rodts, S.; Care, S. NMR determination of sorption isotherms in earlywood and latewood of Douglas fir. Identification of bound water components related to their local environment. *Holzforschung* **2017**, *71*, 481–490. [CrossRef]
29. Cai, C.; Javed, M.A.; Komulainen, S.; Telkki, V.-V.; Haapala, A.; Heräjärvi, H. Effect of natural weathering on water absorption and pore size distribution in thermally modified wood determined by nuclear magnetic resonance. *Cellulose* **2020**, *27*, 4235–4247. [CrossRef]
30. Hiltunen, S.; Mankinen, A.; Javed, M.A.; Ahola, S.; Venäläinen, M.; Telkki, V.-V. Characterization of the decay process of Scots pine caused by *Coniophora puteana* using NMR and MRI. *Holzforschung* **2020**, *74*, 1021–1032. [CrossRef]
31. Callaghan, P. *Principles of Nuclear Magnetic Resonance Microscopy*; Oxford University Press: New York, NY, USA, 1991.
32. Fredriksson, M.; Thygesen, L.G. The states of water in Norway spruce (*Picea abies* (L.) Karst.) studied by low-field nuclear magnetic resonance (LFNMR) relaxometry: Assignment of free-water populations based on quantitative wood anatomy. *Holzforschung* **2017**, *71*, 77–90. [CrossRef]
33. Xu, K.; Lu, J.; Gao, Y.; Wu, Y.; Li, X. Determination of moisture content and moisture content profiles in wood during drying by low-field nuclear magnetic resonance. *Dry. Technol.* **2017**, *35*, 1909–1918. [CrossRef]
34. Gezici-Koç, Ö.; Erich, S.J.F.; Huinink, H.P.; Van Der Ven, L.G.J.; Adan, O.C.G. Bound and free water distribution in wood during water uptake and drying as measured by 1D magnetic resonance imaging. *Cellulose* **2017**, *24*, 535–553. [CrossRef]
35. Xu, K.; Yuan, S.; Gao, Y.; Wu, Y.; Zhang, J.; Li, X.; Lu, J. Characterization of moisture states and transport in MUF resin-impregnated poplar wood using low field nuclear magnetic resonance. *Dry. Technol.* **2020**, *39*, 791–802. [CrossRef]
36. Godefroy, S.; Callaghan, P. 2D relaxation/diffusion correlations in porous media. *Magn. Reson. Imaging* **2003**, *21*, 381–383. [CrossRef]
37. Xu, Y.; Araujo, C.; Mackay, A.; Whittall, K. Proton Spin–Lattice Relaxation in Wood—T_1 Related to Local Specific Gravity Using a Fast-Exchange Model. *J. Magn. Reson. Ser. B* **1996**, *110*, 55–64. [CrossRef]
38. Lerouge, T.; Maillet, B.; Coutier-Murias, D.; Grande, D.; Le Droumaguet, B.; Pitois, O.; Coussot, P. Drying of a Compressible Biporous Material. *Phys. Rev. Appl.* **2020**, *13*, 044061. [CrossRef]
39. Song, Y.-Q.; Venkataramanan, L.; Hurlimann, M.; Flaum, M.; Frulla, P.; Straley, C. T_1–T_2 Correlation Spectra Obtained Using a Fast Two-Dimensional Laplace Inversion. *J. Magn. Reson.* **2002**, *154*, 261–268. [CrossRef] [PubMed]

Article

Reactivity of Waterlogged Archeological Elm Wood with Organosilicon Compounds Applied as Wood Consolidants: 2D ^{1}H–^{13}C Solution-State NMR Studies

Magdalena Broda [1,*] and Daniel J. Yelle [2]

1. Department of Wood Science and Thermal Techniques, Faculty of Forestry and Wood Technology, Poznan University of Life Sciences, ul. Wojska Polskiego 38/42, 60-637 Poznan, Poland
2. Forest Biopolymers Science and Engineering, Forest Products Laboratory, USDA Forest Service, One Gifford Pinchot Drive, Madison, WI 53726, USA; daniel.j.yelle@usda.gov
* Correspondence: magdalena.broda@mail.up.poznan.pl

Abstract: Some organosilicon compounds, including alkoxysilanes and siloxanes, proved effective in stabilizing the dimensions of waterlogged archaeological wood during drying, which is essential in the conservation process of ancient artifacts. However, it was difficult to determine a strong correlation between the wood stabilizing effect and the properties of organosilicon compounds, such as molecular weight and size, weight percent gain, and the presence of other potentially reactive groups. Therefore, to better understand the mechanism behind the stabilization effectiveness, the reactivity of organosilicons with wood polymers was studied using a 2D ^{1}H–^{13}C solution-state NMR technique. The results showed an extensive modification of lignin through its demethoxylation and decarbonylation and also the absence of the native cellulose anomeric peak in siloxane-treated wood. The most substantial reactivity between wood polymers and organosilicon was observed with the (3-mercaptopropyl)trimethoxysilane treatment, showing complete removal of lignin side chains, the lowest syringyl/guaiacyl ratio, depolymerization of cellulose and xylan, and reactivity with the C6 primary hydroxyls in cellulose. This may explain the outstanding stabilizing effectiveness of this silane and supports the conclusion that extensive chemical interactions are essential in this process. It also indicates the vital role of a mercapto group in wood stabilization by organosilicons. This 2D NMR technique sheds new light on the chemical mechanisms involved in organosilicon consolidation of wood and reveals what chemical characteristics are essential in developing future conservation treatments.

Keywords: archaeological wood; silane; siloxane; wood consolidation; 2D NMR; chemical reactivity; solution-state NMR; wood conservation; waterlogged wood

Citation: Broda, M.; Yelle, D.J. Reactivity of Waterlogged Archeological Elm Wood with Organosilicon Compounds Applied as Wood Consolidants: 2D ^{1}H–^{13}C Solution-State NMR Studies. *Molecules* 2022, 27, 3407. https://doi.org/10.3390/molecules27113407

Academic Editor: Igor Serša

Received: 29 April 2022
Accepted: 23 May 2022
Published: 25 May 2022

Publisher's Note: MDPI stays neutral with regard to jurisdictional claims in published maps and institutional affiliations.

Copyright: © 2022 by the authors. Licensee MDPI, Basel, Switzerland. This article is an open access article distributed under the terms and conditions of the Creative Commons Attribution (CC BY) license (https://creativecommons.org/licenses/by/4.0/).

1. Introduction

The oldest known method for the conservation of waterlogged wooden artifacts dates back to the mid-1800s, when hot solutions of alum salts (KAl(SO$_4$)$_2$·12H$_2$O) were used for this purpose for the first time [1,2]. More recent standard conservation procedures employ mainly polyethylene glycols of different molecular weights and various sugars [3–9]. However, since none of these methods has been entirely satisfying and some of them, such as alum and PEG treatment, turned out to be even detrimental to wooden artifacts in the long term [1,10–14], the search for more reliable solutions continues. One of the newly tested methods for waterlogged wood conservation is the application of organosilicon compounds, among which some (e.g., alkoxysilanes) can polymerize inside the wood structure, forming a stabilizing 3D network [15–17].

Our foregoing research on organosilicons allowed us to identify several compounds effective in stabilizing waterlogged wood dimensions during drying, including methyltrimethoxysilane, (3-mercaptopropyl)trimethoxysilane, (3-aminopropyl)triethoxysilane,

1,3-bis(3-aminopropyl)-1,1,3,3-tetramethyldisiloxane, or 1,3-bis-[(diethylamino)-3- (propoxy) propan-2-ol]-1,1,3,3-tetramethyldisiloxane [15,16]. They differ significantly in molecular weight, size, and chemical structure, which suggests different stabilizing mechanisms. Surprisingly, among the tested organosilicons that turned out ineffective in wood stabilization were some with a similar structure to those effective ones, differing only in the presence/absence of a particular side group or the length of the side chain. That indicates that not only the formation of a spatial network inside the wood tissue by polymerized organosilicon compounds but also that their chemical reactivity with wood polymers may contribute to their stabilizing efficiency.

There are several potential reactive sites in both wood polymers and organosilicon compounds that enable the formation of covalent or hydrogen bonds between their molecules. In wood polymers, they include primary (at **C6**) and secondary (at **C2** and **C3**) hydroxyls in cellulose [18,19], free hydroxyls present on all sugar units in hemicelluloses [20,21], and phenolic α-O-4 and β-O-4 linkages, as well as aliphatic and phenolic hydroxyl groups in lignin [22,23]. In turn, alkoxysilanes have highly reactive alkoxy groups that enable their polymerization by reacting with other silane molecules or different chemicals [24]. Additionally, organosilicons can contain several different functional groups, such as mercapto, thiocyanate, amine, vinyl, epoxy, etc., that may facilitate interactions with wood polymers [25,26].

Although it has already been confirmed that various silanes can react with cellulose [27–33], lignin [33–38], and wood [39–43], and the results of our previous FT-IR analyses on waterlogged wood treated with organosilicons confirmed the formation of new chemical bonds between them [15,43], the details of the interactions and potential preferences of silanes to react with individual wood polymers remain not fully understood. Therefore, to unveil the mechanism of waterlogged wood stabilization by organosilicon compounds, further research is necessary that will help to better understand the wood–silane interactions, especially in highly decayed wood where the usual cellulose/lignin ratio and the regular chemical composition and structure of wood polymers are altered by degradation processes.

One of the methods that provide insights into changes in wood chemistry caused by its modification is two-dimensional solution-state nuclear magnetic resonance (NMR) spectroscopy. The technique has been successfully employed to study the reactivity of wood polymers with various chemicals and modification agents, including phenol-formaldehyde adhesive [44], functionalized benzoic acids [45], polymeric methylene diphenyl diisocyanate [46], or N-methylimidazole (NMI) and acetic anhydride [47]. It is also helpful in qualitative and quantitative analyses of cell wall polymers in plant tissues [48], allows us to study of interactions between them [49], and facilitates the identification of structural changes in lignin, cellulose, and hemicelluloses caused by wood-decaying fungi [50,51] or hydrothermal pretreatment and enzymatic hydrolysis [52].

In the present study, the two-dimensional solution-state nuclear magnetic resonance (NMR) method was used to address four crucial questions: (1) whether any new chemical bonds are formed between organosilicons applied as wood consolidants and the cell wall polymers that remained in the degraded waterlogged wood; (2) which active sites in wood polymers interact with particular groups present in organosilicon molecules; (3) whether silanes have any preference to individual wood polymers; and, finally, (4) what makes an organosilicon an effective stabilizer (from the perspective of the chemical structure and reactivity). Understanding the interactions of organosilicons with wood polymers and the resulting wood stabilization mechanism will enable the design of more effective consolidants for waterlogged wood. It will also help develop new functional lignocellulosic materials modified with organosilicon compounds for different industrial purposes.

2. Results

2.1. Effectiveness of Organosilicon Compounds in Waterlogged Wood Stabilization

Keeping the original dimensions of waterlogged wooden artifacts during drying is a primary goal of successful conservation. Therefore, the effectiveness of conservation agents applied as waterlogged wood consolidants is usually evaluated based on parameters that measure dimensional wood stability, including shrinkage and anti-shrink efficiency [15,53,54].

Table 1 presents the efficacy of selected organosilicons used to stabilize highly degraded waterlogged archeological elm. Alkoxysilanes and siloxanes are labeled with the A and S letters, respectively, followed by the consecutive numbers (the full names of the chemicals are given in Section 4.1 Materials). The most effective wood-stabilizing treatment (with anti-shrink efficiency (ASE) over 80%) was that with (3-mercaptopropyl)trimethoxysilane (**A3**), 3-bis(3-aminopropyl)-1,1,3,3-tetramethyldisiloxane (**S2**), 1,3-bis-[(diethylamino)-3-(propoxy)propan-2-ol]-1,1,3,3-tetramethyldisiloxane (**S3**), 3-[3-(hydroxy)(polyethoxypropyl)] 1,1,1,3,5,5,5-heptamethyltrisiloxane (**S7**), and methyltrimethoxysilane (**A1**). Pretty good stabilization with ASE over 70% was achieved using 1,3,5,7-tetrakis(1-(diethylamino)-3-(propoxy)propane-2-ol)-1,3,5,7-tetramethylcyclotetrasiloxane (**S5**) and (3-thiocyanatopropyl)trimethoxysilane $NCS(CH_2)_3Si(OCH_3)_3$ (**A4**). The other organosilicons used in the study were less effective, with ASE values of about 50%, which is insufficient from the conservation perspective.

Table 1. The parameters measured/calculated for selected organosilicons or wood samples treated with them: S_v, volumetric wood shrinkage; ASE_v, anti-shrink efficiency of the individual organosilicon compound; WPG, weight percent gain; MW, molecular weight of an organosilicon monomer; C, untreated waterlogged wood used as a control; the full names of organosilicon compounds are given in Section 4.1 Materials.

Organosilicon Applied	S_v [%]	ASE_v [%]	MW [g/mol]	WPG [%]
C	55.1 ± 4.9	-	-	-
A1	9.7 ± 1.3	82.4	136.22	231.9 ± 6.8
A2	29.2 ± 7.3	47.0	277.82	328.8 ± 1.3
A3	0.7 ± 0.5	98.7	196.34	136.9 ± 9.4
A4	15.9 ± 3.5	71.1	221.35	212.5 ± 1.9
S1	24.9 ± 1.7	54.8	362.61	227.3 ± 0.5
S2	4.5 ± 1.4	91.8	248.51	236.2 ± 1.9
S3	4.5 ± 1.4	91.8	508.88	219.8 ± 5.9
S4	29.7 ± 2.5	46.1	482.80	234.1 ± 2.2
S5	15.0 ± 0.9	72.8	989.62	231.8 ± 1.3
S6	26.3 ± 1.1	52.3	1762.40	270.8 ± 2.1
S7	5.3 ± 2.5	90.4	588.95	227.3 ± 1.4

Considering the general molecular structure (alkoxysilanes and siloxanes), the molecular weight, and the weight percent gain of the organosilicons applied (Table 1), it was difficult to determine any simple correlation between these parameters and the wood dimensional stabilization achieved with the treatment. Amongst the alkoxysilanes and siloxanes, the most effective were agents contained molecules as small as 136 g mol^{-1} and 196 g mol^{-1} (**A1** and **A3**), medium size of 248 g mol^{-1} (**S2**), or as large as 508 g mol^{-1} (**S3**) and 588 g mol^{-1} (**S7**). Weight percent gain in the range of 212% to 236% was obtained for both the most effective (**S2**, **S3**, **S7**, **A1**, **S5**, and **A4**) and less effective (**S1** and **S4**) chemicals. Interestingly, the best stabilizer (**A3**) was characterized by one of the smallest molecular weights (196 g mol^{-1}) and the lowest WPG (only 137% g mol^{-1}), too.

Additionally, there was no direct correlation between the chemical structure and the stabilizing effectiveness of the organosilicon compounds used in this study. All alkoxysilanes (Figure 1) had a similar structure with the presence of three methoxyl groups. They differed only in the fourth group, which varied from a simple methyl group in A1 through a longer alkyl chain (propyl) terminated with a pyridinium chloride (**A2**), a thiol group (**A3**), or a thiocyanate group (**A4**). However, their waterlogged wood-stabilizing efficiency differed significantly (Table 1), which suggested the critical role of the fourth additional chemical group bound to the silicon atom.

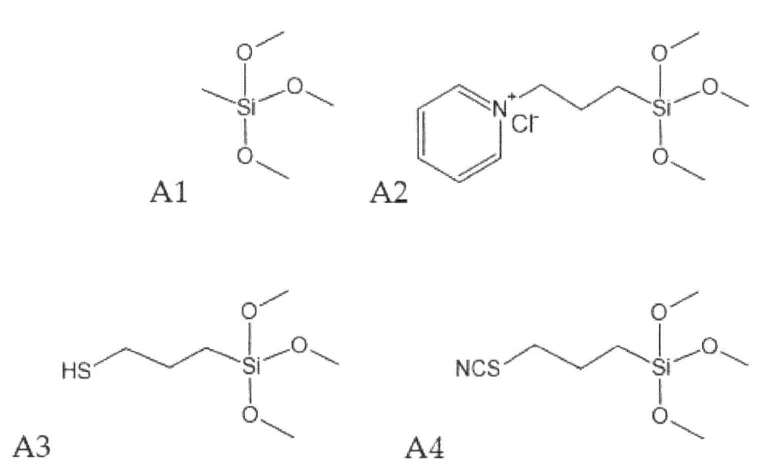

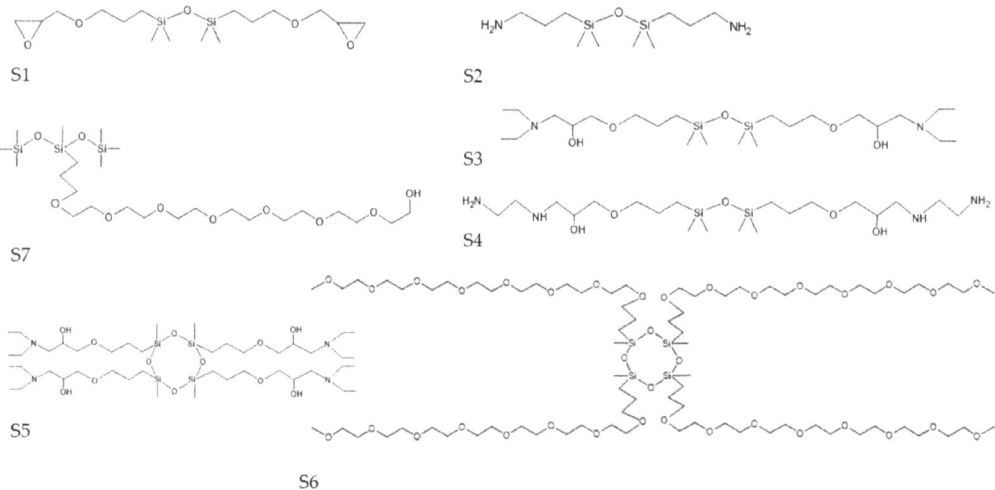

Figure 1. Chemical structures of alkoxysilanes used for waterlogged wood conservation.

The structures of siloxanes used in the research were more varied (Figure 2). They included disiloxanes with shorter (**S1** and **S2**) and longer alkyl chains (**S3**, **S4**), with additional amino (**S2**, **S3**, **S4**) or epoxy (**S1**) groups, a trisiloxane with a long polyethoxypropyl chain (**S7**), as well as more complex cyclic tetrasiloxanes (**S5** and **S6**). Similarly to alkoxysilanes, it was difficult to find a correlation between the structure and stabilizing effectiveness of these chemicals because the best-performing ones (**S2**, **S3**, **S7**, and **S5**) differed in the length of a side chain and the presence of reactive groups, while some of the less effective ones had side chains of similar length and also contained a reactive group that could interact with wood polymers (e.g., **S1** vs. **S3** or **S4** vs. **S3**).

Figure 2. Chemical structures of siloxanes used for waterlogged wood conservation.

The results presented above suggest that more than one mechanism must be involved in the stabilizing effect of the organosilicon compounds on waterlogged wood. Hence the idea to use the two-dimensional (2D) ^{1}H–^{13}C single-bond correlation NMR technique to investigate the reactivity between organosilicons and wood polymers.

2.2. NMR Spectra

Two-dimensional (2D) ^{1}H–^{13}C single-bond correlation NMR spectra were acquired on the whole cell walls of alkoxysilane- and siloxane-treated archaeological elm wood. Through this analysis, the native wood cell wall polymers were semi-quantifiable, and the detailed chemistry between a treatment and the wood could be visualized, thus revealing clues as to the mechanisms involved in how each treatment stabilizes the wood. Figures 3–5 display partial 2D NMR spectra for all samples studied here. Figure 3 includes a chemical structure key to the color-coded contours that are referenced in each spectrum. Figures 6 and 7 are bar charts showing the NMR integration values for the major wood polymers present in each of the spectra relative to the lignin methoxyl group (for alkoxysilane- and siloxane-treated wood, respectively). Not all of the spectra displayed the presence of the major wood polymers due to the overwhelming intensities of the organosilicon contours. For example, the 2D NMR spectra of archaeological elm wood treated with **A2**, **S1**, **S6**, and **S7** showed intense organosilicon peaks that overlapped with the wood cell wall polymer peaks, making the signals from wood visually obscured. The organosilicon treatments may be grouped into two types: alkoxysilanes and siloxanes. The following will describe the chemical characteristics found in the wood cell walls after each treatment.

2.2.1. Alkoxysilane-Treated Wood

Archaeological wood treated with alkoxysilanes included methyltrimethoxysilane (**A1**), 1-[3-(trimethoxysilyl)propyl]pyridinium chloride (**A2**), (3-mercaptopropyl)trimethoxysilane (**A3**), and (3-thiocyanatopropyl)trimethoxysilane (**A4**). The chemical structures for these treatments are shown in Figure 3.

Treatment A1 was the simplest structure of all the organosilicons. From Figure 4, the NMR spectrum showed quite similar characteristics to the control (**C**) degraded wood (Figure 3). For example, all the main lignin linkages (β-O-4, β-5, and β-β), the syringyl (S) and guaiacyl (G) units, and major polysaccharides cellulose (Glc) and xylan (Xyl) were present. The oxidized aromatic units were evidenced by the presence of α-carbonyl versions of syringyl and guaiacyl units, as depicted by S′ and G′. From the integration data shown in Figure 6, the S/G ratio of A1 decreased by 31% compared to the control.

Treatment A2 was the only organosilicon based on a salt. In the spectrum shown in Supplementary Figure S1, the major wood polymers were not able to be detected. The only functional group detectable from wood was that of the lignin methoxyl, and even this group was considered a weak signal. The high intensity of the treatment contours seemed to overwhelm the weaker wood polymer signals.

Treatment A3 contained a mercaptopropyl group. The NMR spectrum shown in Figure 4 displayed a dramatic degradation of the wood polymers. For example, the spectrum was devoid of all the major lignin linkages, as well as the predominant lignin aromatic units; the only aromatic units present were the α-carbonyl versions of syringyl and guaiacyl units (S′ and G′). The contour peak for p-hydroxyphenyl units (H) was shown to be enhanced, while the S and G contour peaks were depleted, showing evidence of methoxyl removal. Similarly, the major polysaccharides were also heavily cleaved; the α- and β-reducing end groups (α_{red} and β_{red}) of cellulose (Glc) and xylan (Xyl) showed intense signals. Therefore, this treatment resulted in a high amount of wood degradation. However, we also detected partial reactivity between the treatment and the Glc_6 position, showing new contours labeled $R-Glc_6$ (yellow). These new contours showed evidence that this treatment does react with cellulose. In the integration data, shown in Figure 6, the S/G ratio of A3 decreased by 54% as compared to the control.

Treatment A4 contained a thiocyanatopropyl group. The NMR spectrum in Figure 4 displayed similar characteristics to spectra of the control degraded wood (Figure 3) and treatment A1. All of the major lignin linkages were present, as well as the aromatic lignin units (S and G). The α-carbonyl versions syringyl and guaiacyl units (S′ and G′) were also present. Cellulose and xylan were also evident. From Figure 6, the S/G ratio of A4 decreased by 27% compared to the control.

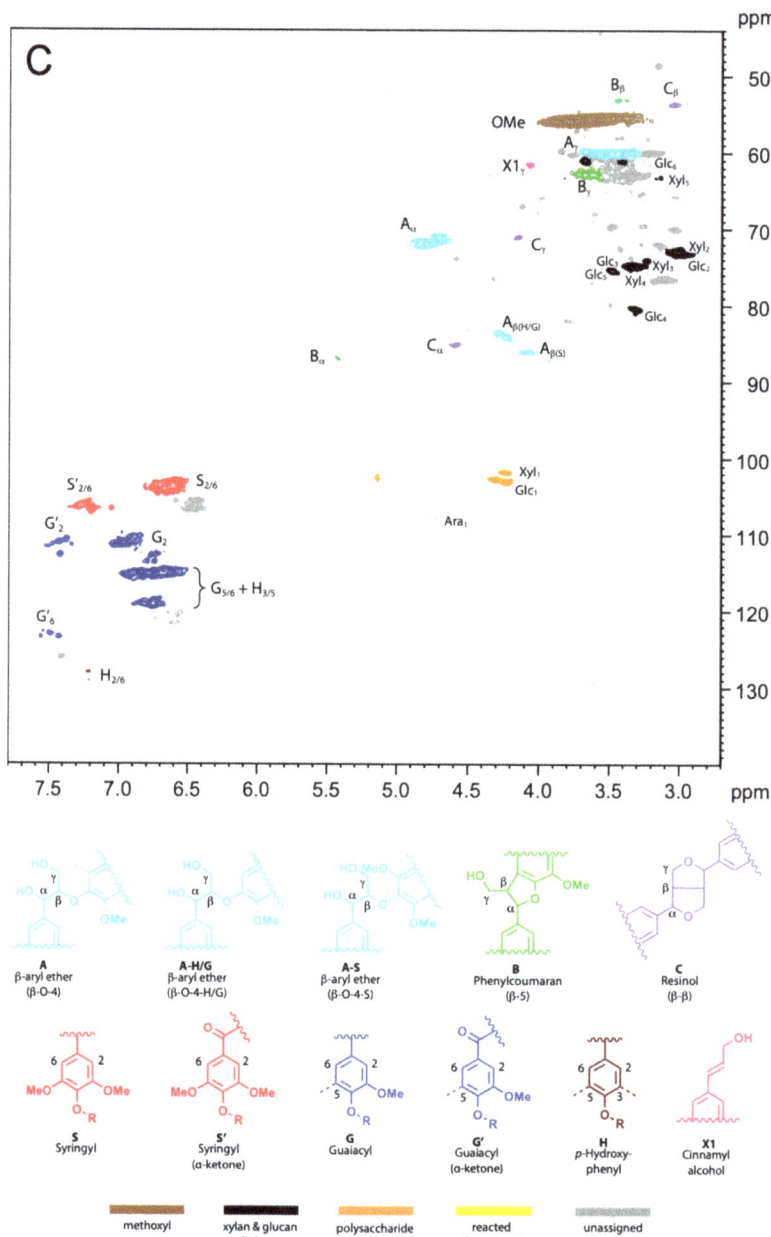

Figure 3. The partial 2D NMR spectrum of the control degraded archaeological elm (**C**) (**top**) and the chemical structures (**bottom**) of the main wood polymers present in the spectrum and the following figures.

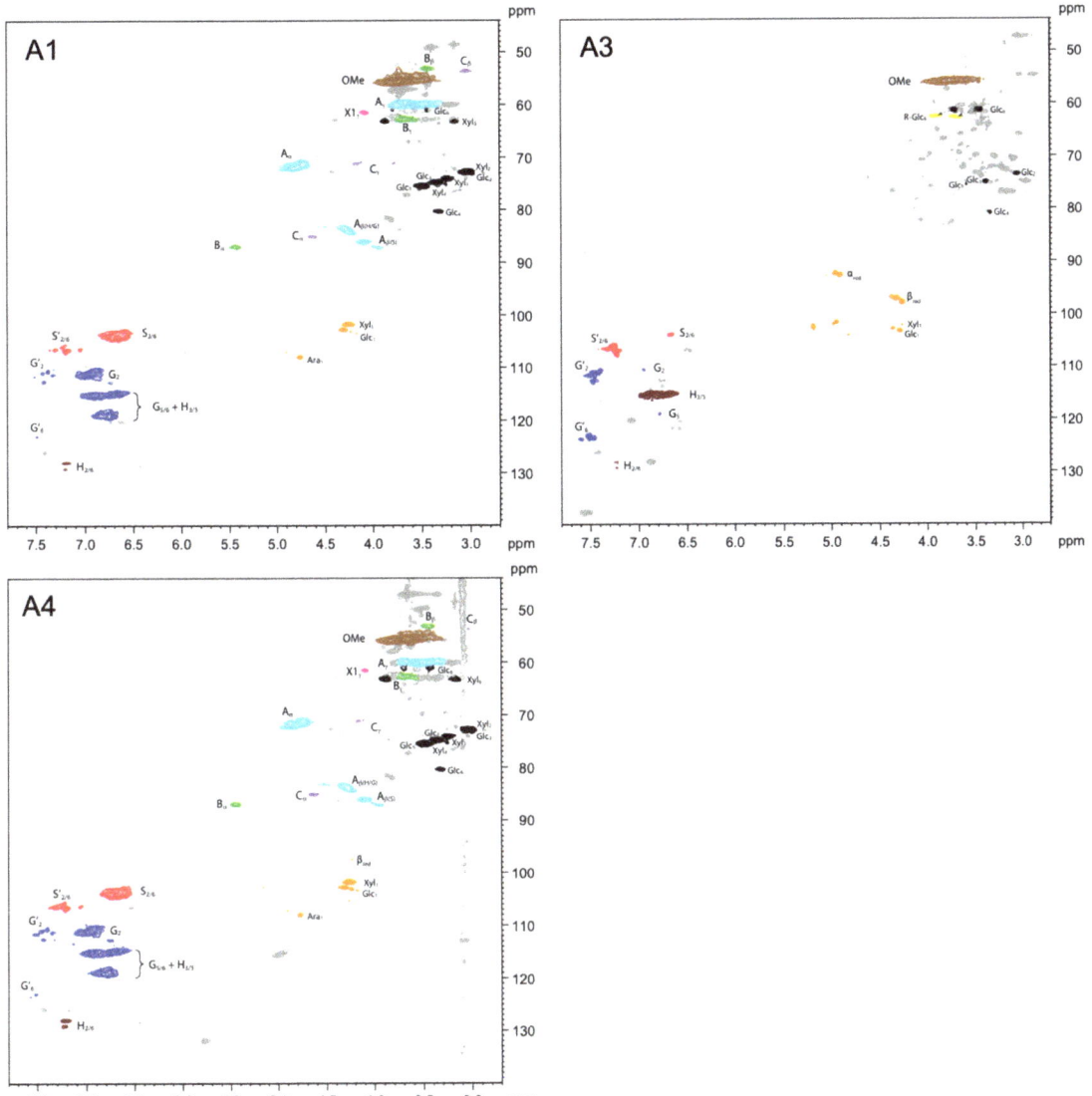

Figure 4. Partial 2D NMR spectra of alkoxysilane-treated wood showing the effects of the treatment with methyltrimethoxysilane (**A1**), (3-mercaptopropyl)trimethoxysilane (**A3**), and (3-thiocyanatopropyl)trimethoxysilane (**A4**). The colored contours and labels correspond to the chemical structures shown in Figure 3.

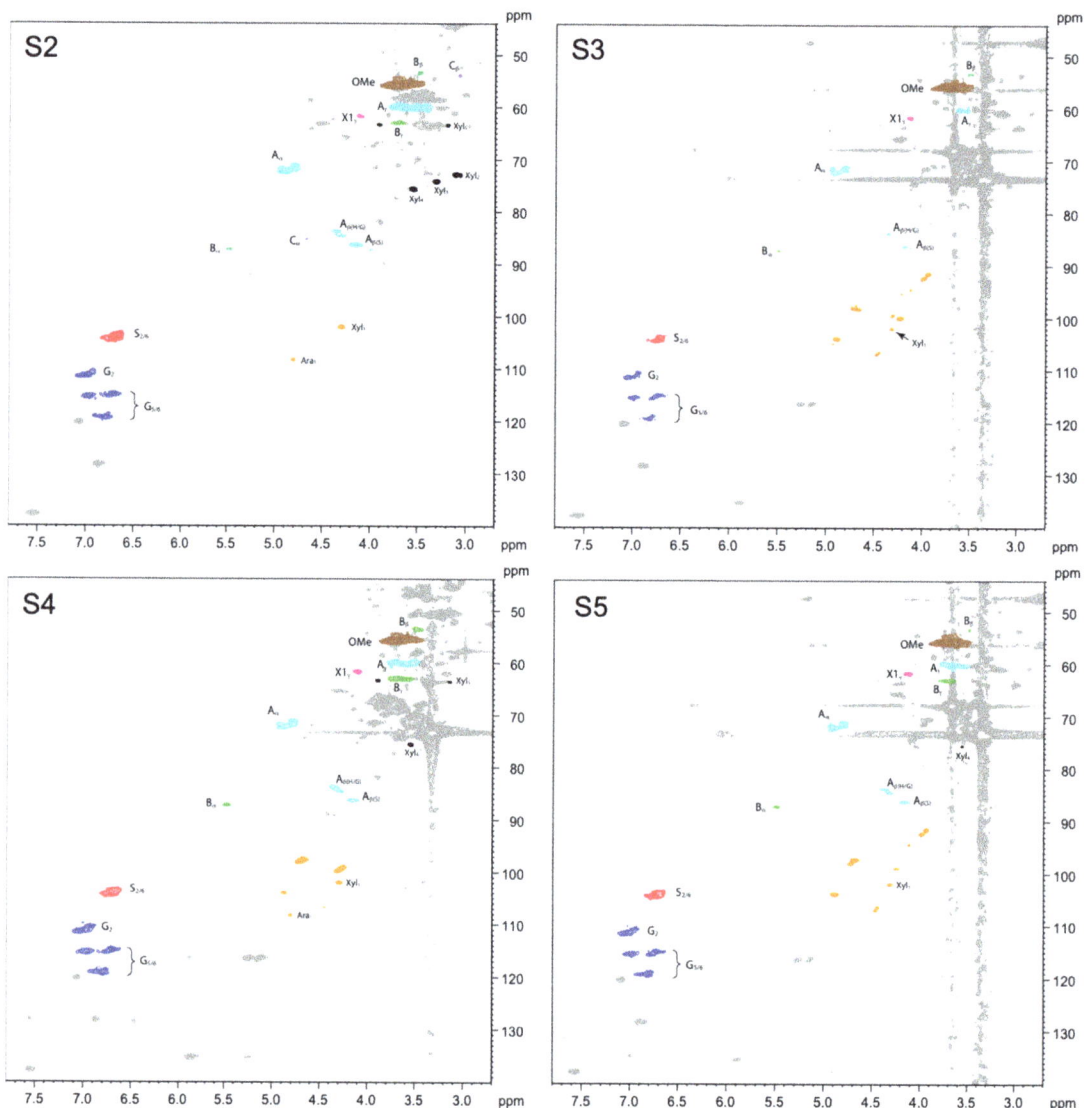

Figure 5. Partial 2D NMR spectra of siloxane-treated wood showing the effects of the treatment with (1,3-bis(3-aminopropyl)-1,1,3,3-tetramethyldisiloxane (**S2**), 1,3-bis-[(diethylamino)-3-(propoxy)propan-2-ol]-1,1,3,3-tetramethyldisiloxane (**S3**), 1,3-bis-[(ethylenodiamino)-3-(propoxy)propan-2-ol]-1,1,3,3-tetramethyldisiloxane (**S4**), 1,3,5,7-tetrakis(1-(diethylamino)-3-(propoxy)propan-2-ol)-1,3,5,7-tetramethylcyclotetrasiloxane (**S5**). The colored contours and labels correspond to the chemical structures shown in Figure 3.

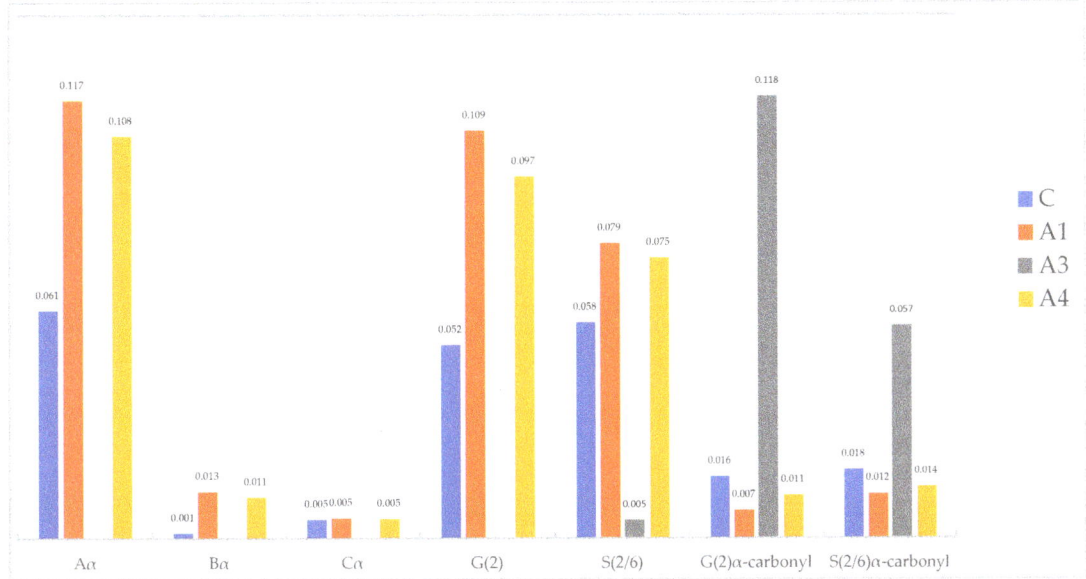

Figure 6. Bar chart summarizing the 2D NMR integrations of peaks from lignin subunits present in each spectrum of alkoxysilane-treated wood (**A1**, **A3**, and **A4**) with control degraded wood (**C**). The numbers above the bars indicate the actual value of the integral. All integrations are relative to the lignin methoxyl peak.

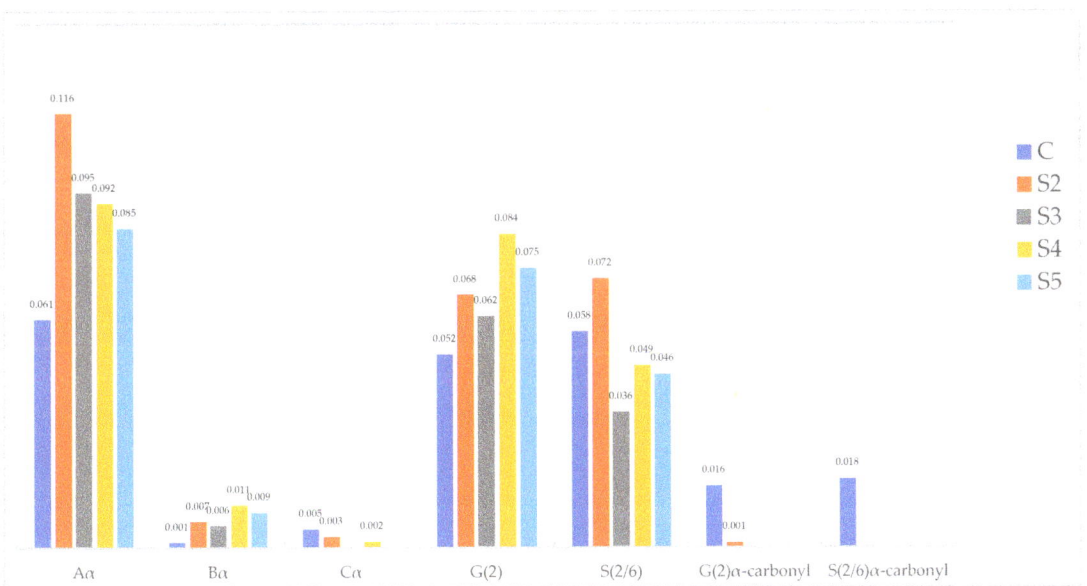

Figure 7. Bar chart summarizing the 2D NMR integrations of peaks from lignin subunits present in each spectrum of siloxane-treated wood (**S2**, **S3**, **S4**, and **S5**) with control degraded wood (**C**). The numbers above the bars indicate the actual value of the integral. All integrations are relative to the lignin methoxyl peak.

2.2.2. Siloxane-Treated Wood

Archaeological wood treated with siloxanes included 1,3-bis(3-glycidyloxypropyl)-1,1,3,3-tetramethyldisiloxane (**S1**), 1,3-Bis(3-aminopropyl)-1,1,3,3-tetramethyldisiloxane (**S2**), 1,3-bis-[(diethylamino)-3-(propoxy)propan-2-ol]-1,1,3,3-tetramethyldisiloxane (**S3**), 1,3-bis-[(ethylenodiamino)-3-(propoxy)propan-2-ol]-1,1,3,3-tetramethyldisiloxane (**S4**), 1,3,5,7-tetrakis(1-(diethylamino)-3-(propoxy)propan-2-ol)-1,3,5,7-tetramethylcyclotetrasiloxane (**S5**), 1,3,5,7-tetrakis(3-polyethoxypropyl)-1,3,5,7-tetramethyltetracyclosiloxane, methoxy terminated (**S6**), and 3-[3-(hydroxy)(polyethoxypropyl)]1,1,1,3,5,5,5-heptamethyltrisiloxane (**S7**). The chemical structures for these treatments are shown in Figure 2.

In all NMR spectra obtained for wood treated with siloxanes of the molecular weight over 300 g mol^{-1} (**S1**, **S3–7**), the high-intensity peaks coming from the treatments that overwhelmed the wood polymers signals were seen.

Treatments **S1**, **S2**, **S3**, and **S4** all contained a tetramethyldisiloxane group. Treatments **S2**, **S3**, and **S4** showed similar characteristics in their NMR spectra (Figure 5). For example, the major lignin linkages β-O-4 and β-5 were present as well as the S and G units. However, the S′ and G′ units were absent, which suggests the reduction or removal of the α-carbonyl functionality. From the integration data, shown in Figure 7, the S/G ratios for treatments **S2**, **S3**, and **S4** all showed a decrease of 7%, 47%, and 48%, respectively, compared to the control. The polysaccharides also showed several different contour peaks as compared to the control, suggesting reactivity between the treatment and the polysaccharides, especially for treatments **S3** and **S4**. Treatment **S1** was devoid of major wood polymer peaks, most likely due to the overwhelming peaks from the organosilicon treatment, but did show the presence of polysaccharide peaks in the anomeric region and S units in the aromatic region (Figure S1).

Treatments **S5** and **S6** both contained a tetramethyltetracyclosiloxane group. Interestingly, treatment **S5** displayed similar characteristics to treatments **S3** and **S4** in the NMR spectra in that the major lignin linkages β-O-4 and β-5 were present, as well as the S and G units, and the polysaccharide peaks looked similar in the anomeric region (Figure 5). On the other hand, treatment **S6** was devoid of major lignin polymer peaks, most likely due to the overwhelming peaks from the organosilicon treatment, but did show polysaccharide peaks in the anomeric region (Figure S1). From Figure 7, the S/G ratio for treatment **S5** showed a decrease of 45% compared to the control

Treatment **S7** contained a heptamethyltrisiloxane group. Similar to treatment **S6** characteristics, this treatment was also mostly devoid of wood polymer peaks with the exception of weak aromatic units (S and G) and polysaccharide peaks in the anomeric region (Figure S1). The S/G ratio was not measurable, given the weak intensity of the aromatic unit peaks.

3. Discussion

The NMR results obtained shed new light on the interactions between the wood cell wall polymers and organosilicon compounds applied as consolidants to stabilize waterlogged wood dimensions.

Our previous FT-IR studies [15,43] showed that hydrolysis and condensation of alkoxysilane monomers occurred in the treated wood, leading to the formation of a stabilizing polymer network inside the wood structure. Moreover, it seemed that also new chemical bonds between wood hydroxyls and alkoxysilanes were formed due to the treatment, in particular when methyltrimethoxysilane (**A1**) and (3-mercaptopropyl)trimethoxysilane (**A3**) were applied. The reduction of available hydroxyls on the cell walls was additionally confirmed by dynamic water sorption experiments that showed the decrease in equilibrium moisture content and the sorption hysteresis of treated archaeological wood compared to untreated wood [55,56]. However, the new NMR data only showed the chemical modification of C6 primary hydroxyls in cellulose in the wood sample treated with **A3**; no other evidence of wood hydroxyls' modification is visible.

On the other hand, we observed demethoxylation of lignin S and G units, which was also seen in our previous FT-IR spectra [43] and other studies on wood modification with alkoxysilanes [57,58]. From the integration data in Figure 6, it was evident that the major lignin linkages in all samples treated with alkoxysilanes were much—in some cases two times—higher than that found in the control degraded wood samples, with the exception of the β-β linkage. This result, in conjunction with the overwhelmingly consistent decrease in the S/G ratios in the treated woods compared to the control, suggests that the lignin was undergoing demethoxylation during alkoxysilane treatment; thus, alkoxysilanes have the ability to oxidize these methoxyl groups on the aromatic ring of lignin.

The NMR results were, then, surprising, especially since several researchers observed the reactivity of alkoxysilanes (via alkoxy groups) with wood hydroxyls [24,39,41,58]. However, some catalysts are usually applied to promote the reactivity of alkoxysilanes with wood, which were not used in the case of our waterlogged wood treatments. This fact may explain why we did not observe the modification of the wood hydroxyls in our NMR spectra of **A1**, **A2**, and **A4**. It is worth mentioning here that the NMR spectra obtained contained some unidentified peaks that arose from the alkoxysilanes applied. For example, in the spectra shown in Figure 4 we do see peaks from the trimethoxy groups around 3.0–3.5/45–50 ppm; most of the other peaks from alkoxysilanes did not interfere with wood polymer peaks. However, we cannot currently assign any other peaks related to new plausible chemical bonds between the silanes and lignin/polysaccharides units in **A1**, **A2**, and **A4**. Further research on this phenomenon is planned.

For alkoxysilane treatment **A3** ((3-mercaptopropyl)trimethoxysilane), we were able to tentatively assign the new peaks of reacted cellulose **C6** hydroxyls, labeled as R-Glc$_6$, at 3.73/63.1 ppm and 3.93/63.1 ppm (Figure 4). Treatment **A3** also showed complete removal of lignin side chains, the lowest S/G ratio (0.52), and depolymerization of cellulose and xylan as evidenced by the presence of their intensified α- and β-reducing end groups at 4.97/92.8, 4.93/93.0 and 4.34/97.5, 4.28/98.3 ppm, respectively. This strong reactivity of the most effective wood stabilizer with all wood polymers indicates that chemical interactions are essential in stabilizing waterlogged wood dimensions during drying. It also points to the conclusion about a vital role the mercapto group plays in these interactions and wood stabilization. Even though the observed reactivity may hinder the reversibility of the treatment with this silane, which is required by conservation ethics, the SEM images of the treated wood showed that the silane locates in/on the cell wall [15], leaving the lumina empty for further re-treatment, which potentially does not exclude the chemical from the conservation practice.

Interestingly, in the case of siloxane treatment, we could mainly observe lignin modification employing demethoxylation of S and G units and removal (decarboxylation) of the α-carbonyl versions of syringyl and guaiacyl units (S' and G'). From the integration data in Figure 7, as with what we observed for the alkoxysilane treatments, the major lignin linkages in all of the samples treated with siloxanes were much higher than those found in control degraded wood samples, with the exception of the β-β linkage. This result suggests that the lignin can undergo demethoxylation during organosilicon treatment, regardless of treatment type. From our previous FT-IR research [43], we learned that the methoxyl groups in lignin might contribute to the interaction with 1,3-bis-[(diethylamino)-3-(propoxy)propan-2-ol]-1,1,3,3-tetramethyldisiloxane (**S3**). Perhaps all siloxanes applied in this study react with lignin similarly. However, considering their diverse effectiveness as wood consolidants, this reactivity seems not to play a crucial role in the stabilizing mechanism.

When it comes to decarbonylation, the literature data indicate that siloxanes, including tetramethyldisiloxane, can reduce α,β-unsaturated carbonyl derivatives [59,60]. A tetramethyldisiloxane unit is present in all siloxanes used in our research. However, it contains methyl groups attached to the silicon atom instead of hydrogen, which is necessary for the reductive activity. Therefore, it is difficult to say if the treatment conditions or the solvent used for NMR analysis (DMSO) could cause demethylation of the silicon atom and foster the reductive properties of siloxanes; this question requires further study to be answered.

Another aspect that was evident from the NMR analysis of siloxane-treated wood was the absence of the native cellulose anomeric peak normally observed at 4.25/103.7 ppm. Cellulose was clearly present in all of the alkoxysilane-treated wood samples; so, this indicates that the siloxane treatments are able to modify cellulose heavily. Several new anomeric peaks were observed in the siloxane spectra of S3, S4, and S5 (Figure 5) and **S1**, **S6**, and **S7** (Figure S1), but were not currently assigned here. Further research is needed to assign these new anomeric peaks. Since the effectiveness of the applied siloxanes varied, it is difficult to explicitly state if polysaccharide modification is essential in stabilizing waterlogged wood dimensions during drying.

Considering the results of the presented research and the data on organosilicon compounds used for waterlogged wood conservation from the previous studies [15,16,43], it can be concluded that the mechanism of waterlogged archaeological wood dimensional stabilization by alkoxysilanes is based on (1) bulking the cell wall, (2) forming a stabilizing polymer network, and (3) chemically interacting with wood polymers (at least in the case of alkoxysilanes containing mercapto groups). On the other hand, in the case of siloxanes, wood stabilization seems to be mainly based on filling the cell lumina. However, the absence of the native cellulose anomeric peak and the absence of the α-carbonyls in aromatic lignin units in all the siloxane treatments demonstrate that cellulose and lignin modifications are also intimately involved in stabilizing siloxane-treated wood.

4. Materials and Methods

4.1. Materials

The research material was waterlogged elm (*Ulmus* spp.) heartwood: the remnants of a medieval bridge excavated from the sediments of the Lednica Lake in the Wielkopolska Region, Poland. The wood was highly degraded, with reduced cellulose and hemicelluloses content and the loss of wood substance estimated at about 70–80% [15,43].

Organosilicon compounds for waterlogged wood treatment were obtained by hydrosilylation of relevant olefins with Si–H-containing compounds in the presence of platinum catalysts [61] at the Adam Mickiewicz University Foundation, Poznań Science and Technology Park, Poznań, Poland [15]:

- methyltrimethoxysilane $CH_3Si(OCH_3)_3$ (**A1**);
- 1-[3-(trimethoxysilyl)propyl]pyridinium chloride $(C_5H_5NCl)C_3H_6Si(OCH_3)_3$ (**A2**);
- (3-mercaptopropyl)trimethoxysilane $HS(CH_2)_3Si(OCH_3)_3$ (**A3**);
- (3-thiocyanatopropyl)trimethoxysilane $NCS(CH_2)_3Si(OCH_3)_3$ (**A4**);
- 1,3-bis(3-glycidyloxypropyl)-1,1,3,3-tetramethyldisiloxane $[CH_2(O)CHCH_2O(CH_2)_3Si(CH_3)_2]_2O$ (**S1**);
- 1,3-bis(3-aminopropyl)-1,1,3,3-tetramethyldisiloxane $[H_2N(CH_2)_3Si(CH_3)_2]_2O$ (**S2**);
- 1,3-bis-[(diethylamino)-3-(propoxy)propan-2-ol]-1,1,3,3-tetramethyldisiloxane $[(C_2H_5)_2NCH_2CH(OH)CH_2O(CH_2)_3Si(CH_3)_2]_2O$ (**S3**);
- 1,3-bis-[(ethylenodiamino)-3-(propoxy)propan-2-ol]-1,1,3,3-tetramethyldisiloxane $[(NH_2(CH2)_2HNCH_2CH(OH)CH_2O(CH_2)_3Si(CH_3)_2]_2O$ (**S4**);
- 1,3,5,7-tetrakis(1-(diethylamino)-3-(propoxy)propan-2-ol)-1,3,5,7-tetramethylcyclotrasiloxane $[(C_2H_5)_2NCH_2CH(OH)CH_2O(CH_2)_3Si(CH_3)O]_4$ (**S5**);
- 1,3,5,7-tetrakis(3-polyethoxypropyl)-1,3,5,7-tetramethyltetracyclosiloxane methoxy terminated $[CH_3O(CH_2CH_2O)_7(CH_2)_3Si(CH_3)O]_4$ (**S6**);
- 3-[3-(hydroxy)(polyethoxypropyl)]1,1,1,3,5,5,5-heptamethyltrisiloxane $HO(CH_2CH_2O)_7(CH_2)_3Si(CH_3)[OSi(CH_3)_3]_2$ (**S7**).

For simplicity, the numbers of consecutive alkoxysilanes (**A**) and siloxanes (**S**) listed above, instead of their full chemical names, are used throughout the manuscript.

Dimethylsulfoxide-d_6 (DMSO-d_6, 99.5% D) and 1-methylimidazole-d_6 (NMI-d_6) for NMR analysis were supplied by Aldrich Chemical Company (Milwaukee, WI, USA).

4.2. Methods

4.2.1. Waterlogged Wood Treatment

Waterlogged elm log was cut into small samples with the dimensions of 20 mm × 20 mm × 10 mm (radial × tangential × longitudinal direction). To ensure the greatest possible homogeneity of the wood degradation degree, thus reproducibility of the results, the specimens were sampled from a selected part of the log and a similar distance from the pit, since the number of suitable wooden pieces was limited.

The specimens were dehydrated by soaking them in 96% ethanol for 4 weeks and then treated with 50% ethanol solutions of selected organosilicon compounds. An oscillating-pressure method was used for the wood treatment, applying a −0.9 bar vacuum for 0.5 h and then 10 bars of pressure for 6 h. The cycle was repeated six times every 24 h. Between the cycles, the wood was left submerged in the organosilicon solution under atmospheric pressure to ensure continuous treatment. After the treatment, the samples were removed from the conservation solution and air-dried at room temperature (21 ± 1 °C) for 4 weeks. As a result, five replicates of each treatment were obtained, and five more untreated specimens were air-dried from the waterlogged state and used as a standard control for this type of wood.

To evaluate the effectiveness of the treatment, weight percent gain (WPG) for each organosilicon compound was calculated according to the standard Equation (1):

$$WPG = \frac{W_1 - W_0}{W_0} \times 100 \qquad (1)$$

where W_0 is the estimated dry mass of the specimen before treatment, and W_1 is the dry mass of the sample treated with a selected organosilicon compound [15].

The evaluation of the stabilizing effect of particular conservation agents was based on the values of volumetric shrinkage (S_v) and volumetric anti-shrink efficiency coefficient (ASE_v) calculated according to the standard Equations (2) and (3):

$$S_v = \frac{V_0 - V_1}{V_0} \times 100 \qquad (2)$$

where V_0 is the initial volume of a waterlogged specimen, and V_1 is the final volume of the specimen (untreated or treated, respectively) after air-drying, and

$$ASE_v = \frac{S_{vu} - S_{vt}}{S_{vu}} \times 100 \qquad (3)$$

where S_{vu} is the volumetric shrinkage of the untreated specimen, and S_{vt} is the volumetric shrinkage of the treated specimen.

4.2.2. Wood Preparation for NMR

Air-dried archaeological wood samples were sliced with a knife in the radial direction to obtain sections approximately 1 mm thick. Each sliced specimen was placed into a 50 mL ZrO_2 jar followed by three 20 mm ZrO_2 balls and loaded into a Retsch PM-400 planetary ball mill (Newtown, PA, USA). The wood was milled for 24 h (300 rpm, 20 min milling, 10 min pause; these conditions allowed us to keep the wood temperature below 50 °C, which prevented its thermal degradation and changes in the chemical composition). Then, the 20 mm balls were removed and shaken in the copper sieve to recover the wooden material that remained on their surface. Afterward, ten 10 mm ZrO_2 were added to pre-milled wood, and the milling was continued for another 24 h under the same conditions. Following ball-milling, the 10 mm balls were removed and shaken in the copper sieve, as performed previously, and the milled wood was scraped from the jar and weighed.

4.2.3. Wood Cell Wall Dissolution

About 30 mg of each sample was placed in the NMR tube (5 mm in diameter, 17.8 cm in length) and dissolved using 400 µL DMSO-d_6. To expedite the wood dissolution, the tubes were sonicated at 35 °C for 1 h [62]. For the samples that dissolved successfully, giving homogeneous and transparent solutions, an additional 100 µL DMSO-d_6 was added to reach the final solvent volume of 500 µL. For the samples that did not fully dissolve, 50 µL of 1-methylimidazole-d_6 (NMI-d_6), a non-degradative co-solvent with DMSO-d_6, was added to facilitate the disruption of hydrogen bonds. NMI-d_6 was added to samples **A2**, **A3**, **S1**, **S2**, **S3**, **S4**, **S5**, **S6**, and **S7**; it was omitted in samples **C**, **A1**, and **A4** because dissolution proceeded in DMSO-d_6 without the need for NMI-d_6. Adding NMI-d_6 as a co-solvent did not affect the NMR chemical shifts of the wood cell wall polymers in this study. Then, the samples were sonicated until homogeneous and clear solutions evolved. In the end, 50 µL DMSO-d_6 was added to the tubes with NMI-d_6 to reach the final solvent volume of 500 µL.

4.2.4. NMR Analysis

NMR spectra for untreated and treated archaeological wood were acquired using a Bruker-Biospin (Rheinstetten, Germany) AVANCE III HD^{TM} 500 MHz spectrometer fitted with a nitrogen-cooled 5 mm Prodigy^{TM} TCI gradient cryoprobe with inverse geometry. The one-bond ^{1}H–^{13}C correlation (HSQC) spectra were obtained using the adiabatic Bruker pulse program hsqcetgpsisp2.2 and processed as previously described [63]. For semi-quantitative analysis of the wood polymer structures present in the spectra, specific chemical shifts of native structural units, such as the β-aryl ether, phenylcoumaran, and resinol subunits in lignin or arabinoxylan and glucomannan units in hemicellulose were integrated and referenced to the lignin methoxyl group (since it is known as the most stable functional group) using Bruker TopSpin 3.6.2 software.

5. Conclusions

The conservation of ancient wooden artifacts is critical in preserving history and retelling stories that would otherwise be lost. Developing unique and effective methods in wood conservation requires an understanding of the mechanisms involved in stabilizing the wood. Organosilicons have been proven to be highly effective as wood stabilizers. Here, we explored and characterized the detailed chemistry occurring between organosilicon treatments and the wood cell wall polymers using 2D ^{1}H–^{13}C solution-state NMR. The results of this study on the reactivity of organosilicon compounds applied as consolidants for waterlogged archaeological wood with wood cell wall polymers revealed an extensive modification of lignin and polysaccharides due to the treatment. In the case of alkoxysilanes, mainly lignin demethoxylation was observed. However, in the case of (3-mercaptopropyl)trimethoxysilane treatment, which was the most effective in stabilizing wood dimensions, more comprehensive interactions with wood polymers were observed, including depolymerization of cellulose and xylan, reactivity with the C6 primary hydroxyls in cellulose, complete removal of lignin side chains, and the lowest syringyl/guaiacyl unit ratio. In turn, siloxane treatments caused severe modification of lignin aromatics, including its α-decarboxylation and demethoxylation, as well as cellulose modification.

In answering the questions presented in our research objectives, we can state that:

- New chemical bonds were formed between (3-mercaptopropyl)trimethoxysilane and cellulose in waterlogged wood. In the case of other organosilicons, it was difficult to assign unidentified peaks in NMR spectra to potential new bonds formed between them and wood polymers. This problem is planned to be solved in future research.
- The active sites in wood polymers that interacted with organosilicons were **C6** primary hydroxyls in cellulose (in the case of (3-mercaptopropyl)trimethoxysilane treatment), as well as methoxyl (in both types of organosilicon treatments) and α-carbonyl groups in aromatic lignin units (in the case of siloxane treatment).

- In general, alkoxysilanes appear to preferentially react with lignin, while siloxanes can modify lignin and polysaccharides; only (3-mercaptopropyl)trimethoxysilane was confirmed to react also with cellulose.
- Since a similar modification of wood polymers was observed for both groups of organosilicons used in this study, but their effectiveness as wood stabilizers was different, we cannot state if lignin demethoxylation or modification of lignin aromatics by organosilicons plays a crucial role in the stabilizing mechanism; on the other hand, we can clearly state that the extensive chemical modification of 3-mercaptopropyl) trimethoxysilane (containing a reactive mercapto group) with wood polymers is crucial for the excellent stabilization of waterlogged wood dimensions during drying.

Supplementary Materials: The following supporting information can be downloaded at https://www.mdpi.com/article/10.3390/molecules27113407/s1. Figure S1: Partial 2D NMR spectra of organosilicon-treated wood showing 1-[3-(trimethoxysilyl)propyl]pyridinium chloride (**A2**), 1,3-bis(3-glycidyloxypropyl)-1,1,3,3-tetramethyldisiloxane (**S1**), 1,3,5,7-tetrakis(3-polyethoxypropyl)-1,3,5,7-tetramethyltetracyclosiloxane methoxy terminated (**S6**), 3-[3-(hydroxy)(polyethoxypropyl)]1,1,1,3,5,5,5-heptamethyltrisiloxane (**S7**). The colored contours and labels correspond to the chemical structures shown in Figure 3. Table S1: Chemical shifts of functional groups present in selected organosilicon compounds used in this study. References [64–73] are cited in Supplementary Materials.

Author Contributions: Conceptualization, M.B. and D.J.Y.; methodology, D.J.Y. and M.B.; investigation, D.J.Y. and M.B.; data curation, D.J.Y. and M.B.; writing—original draft preparation, D.J.Y. and M.B.; writing—review and editing, M.B. and D.J.Y.; visualization, M.B. and D.J.Y. All authors have read and agreed to the published version of the manuscript.

Funding: Waterlogged wood treatment was funded by the Polish Ministry of Science and Higher Education as part of the "Cultural heritage—research into innovative solutions and methods for historic wood conservation" project within the National Program for the Development of Humanities in 2015–2018 (project No. 2bH 15 0037 83). NMR studies were supported by the Polish–U.S. Fulbright Commission through a Fulbright Senior Award 2020/21 granted to Magdalena Broda and funded by the USDA Forest Products Laboratory, Madison, WI, USA.

Institutional Review Board Statement: Not applicable.

Informed Consent Statement: Not applicable.

Data Availability Statement: The data presented in this study are available on request from the corresponding author.

Conflicts of Interest: The authors declare no conflict of interest. The funders had no role in the design of the study, in the collection, analyses, or interpretation of data, in the writing of the manuscript, or in the decision to publish the results.

References

1. Braovac, S.; Kutzke, H. The Presence of Sulfuric Acid in Alum-Conserved Wood–Origin and Consequences. *J. Cult. Herit.* 2012, *13*, S203–S208. [CrossRef]
2. Braovac, S.; Tamburini, D.; Lucejko, J.J.; McQueen, C.; Kutzke, H.; Colombini, M.P. Chemical Analyses of Extremely Degraded Wood Using Analytical Pyrolysis and Inductively Coupled Plasma Atomic Emission Spectroscopy. *Microchem. J.* 2016, *124*, 368–379. [CrossRef]
3. Hoffmann, P. On the Stabilization of Waterlogged Oakwood with PEG. II. Designing a Two-Step Treatment for Multi-Quality Timbers. *Stud. Conserv.* 1986, *31*, 103–113. [CrossRef]
4. Hoffmann, P. On the Stabilization of Waterlogged Softwoods with Polyethylene Glycol (PEG). Four Species from China and Korea. *Holzforschung* 1990, *44*, 87–93. [CrossRef]
5. Hocker, E.; Almkvist, G.; Sahlstedt, M. The Vasa Experience with Polyethylene Glycol: A Conservator's Perspective. *J. Cult. Herit.* 2012, *13*, S175–S182. [CrossRef]
6. Collett, H.; Bouville, F.; Giuliani, F.; Schofield, E. Structural Monitoring of a Large Archaeological Wooden Structure in Real Time, Post PEG Treatment. *Forests* 2021, *12*, 1788. [CrossRef]
7. Tahira, A.; Howard, W.; Pennington, E.R.; Kennedy, A. Mechanical Strength Studies on Degraded Waterlogged Wood Treated with Sugars. *Stud. Conserv.* 2017, *62*, 223–228. [CrossRef]

8. Morgós, A.; Imazu, S.; Ito, K. Sugar Conservation of Waterlogged Archaeological Finds in the Last 30 Years. In Proceedings of the 2015 Conservation and Digitalization Conference, London, UK, 19–22 May 2015; pp. 15–20.
9. Hoffmann, P. To Be and to Continue Being a Cog: The Conservation of the Bremen Cog of 1380. *Int. J. Naut. Archaeol.* **2001**, *30*, 129–140. [CrossRef]
10. Broda, M.; Hill, C.A.S. Conservation of Waterlogged Wood–Past, Present and Future Perspectives. *Forests* **2021**, *12*, 1193. [CrossRef]
11. McQueen, C.M.; Tamburini, D.; Lucejko, J.J.; Braovac, S.; Gambineri, F.; Modugno, F.; Colombini, M.P.; Kutzke, H. New Insights into the Degradation Processes and Influence of the Conservation Treatment in Alum-Treated Wood from the Oseberg Collection. *Microchem. J.* **2017**, *132*, 119–129. [CrossRef]
12. McQueen, C.M.A.; Mortensen, M.N.; Caruso, F.; Mantellato, S.; Braovac, S. Oxidative Degradation of Archaeological Wood and the Effect of Alum, Iron and Calcium Salts. *Herit. Sci.* **2020**, *8*, 32. [CrossRef]
13. Mortensen, M.N.; Egsgaard, H.; Hvilsted, S.; Shashoua, Y.; Glastrup, J. Characterisation of the Polyethylene Glycol Impregnation of the Swedish Warship Vasa and One of the Danish Skuldelev Viking Ships. *J. Archaeol. Sci.* **2007**, *34*, 1211–1218. [CrossRef]
14. Mortensen, M.N.; Egsgaard, H.; Hvilsted, S.; Shashoua, Y.; Glastrup, J. Tetraethylene Glycol Thermooxidation and the Influence of Certain Compounds Relevant to Conserved Archaeological Wood. *J. Archaeol. Sci.* **2012**, *39*, 3341–3348. [CrossRef]
15. Broda, M.; Dąbek, I.; Dutkiewicz, A.; Dutkiewicz, M.; Popescu, C.-M.; Mazela, B.; Maciejewski, H. Organosilicons of Different Molecular Size and Chemical Structure as Consolidants for Waterlogged Archaeological Wood–a New Reversible and Retreatable Method. *Sci. Rep.* **2020**, *10*, 2188. [CrossRef]
16. Broda, M.; Mazela, B.; Dutkiewicz, A. Organosilicon Compounds with Various Active Groups as Consolidants for the Preservation of Waterlogged Archaeological Wood. *J. Cult. Herit.* **2019**, *35*, 123–128. [CrossRef]
17. Kavvouras, P.K.; Kostarelou, C.; Zisi, A.; Petrou, M.; Moraitou, G. Use of Silanol-Terminated Polydimethylsiloxane in the Conservation of Waterlogged Archaeological Wood. *Stud. Conserv.* **2009**, *54*, 65–76. [CrossRef]
18. Hidayat, B.J.; Felby, C.; Johansen, K.S.; Thygesen, L.G. Cellulose Is Not Just Cellulose: A Review of Dislocations as Reactive Sites in the Enzymatic Hydrolysis of Cellulose Microfibrils. *Cellulose* **2012**, *19*, 1481–1493. [CrossRef]
19. Heinze, T. Cellulose: Structure and Properties. In *Cellulose Chemistry and Properties: Fibers, Nanocelluloses and Advanced Materials*; Rojas, O.J., Ed.; Advances in Polymer Science; Springer International Publishing: Cham, Switzerland, 2016; pp. 1–52. ISBN 978-3-319-26015-0.
20. Peng, F.; Ren, J.L.; Xu, F.; Sun, R.-C. Chemicals from Hemicelluloses: A Review. In *Sustainable Production of Fuels, Chemicals, and Fibers from Forest Biomass*; ACS Symposium Series; American Chemical Society: Washington, DC, USA, 2011; Volume 1067, pp. 219–259. ISBN 978-0-8412-2643-2.
21. Sun, R.; Sun, X.F.; Tomkinson, J. Hemicelluloses and Their Derivatives. *ACS Symp. Ser.* **2004**, *864*, 2–22. [CrossRef]
22. Varila, T.; Romar, H.; Luukkonen, T.; Hilli, T.; Lassi, U. Characterization of Lignin Enforced Tannin/Furanic Foams. *Heliyon* **2020**, *6*, e03228. [CrossRef]
23. Antonino, L.D.; Gouveia, J.R.; de Sousa Júnior, R.R.; Garcia, G.E.S.; Gobbo, L.C.; Tavares, L.B.; dos Santos, D.J. Reactivity of Aliphatic and Phenolic Hydroxyl Groups in Kraft Lignin towards 4,4′ MDI. *Molecules* **2021**, *26*, 2131. [CrossRef]
24. Donath, S.; Militz, H.; Mai, C. Wood Modification with Alkoxysilanes. *Wood Sci. Technol.* **2004**, *38*, 555–566. [CrossRef]
25. Tesoro, G. Yulong Wu Silane Coupling Agents: The Role of the Organofunctional Group. *J. Adhes. Sci. Technol.* **1991**, *5*, 771–784. [CrossRef]
26. Tingaut, P.; Weigenand, O.; Mai, C.; Militz, H.; Sèbe, G. Chemical Reaction of Alkoxysilane Molecules in Wood Modified with Silanol Groups. *Holzforschung* **2006**, *60*, 271–277. [CrossRef]
27. Brochier Salon, M.-C.; Abdelmouleh, M.; Boufi, S.; Belgacem, M.N.; Gandini, A. Silane Adsorption onto Cellulose Fibers: Hydrolysis and Condensation Reactions. *J. Colloid Interface Sci.* **2005**, *289*, 249–261. [CrossRef] [PubMed]
28. de Oliveira Taipina, M.; Ferrarezi, M.M.F.; Yoshida, I.V.P.; Gonçalves, M.D.C. Surface Modification of Cotton Nanocrystals with a Silane Agent. *Cellulose* **2013**, *20*, 217–226. [CrossRef]
29. Nakatani, H.; Iwakura, K.; Miyazaki, K.; Okazaki, N.; Terano, M. Effect of Chemical Structure of Silane Coupling Agent on Interface Adhesion Properties of Syndiotactic Polypropylene/Cellulose Composite. *J. Appl. Polym. Sci.* **2011**, *119*, 1732–1741. [CrossRef]
30. Salon, M.-C.B.; Gerbaud, G.; Abdelmouleh, M.; Bruzzese, C.; Boufi, S.; Belgacem, M.N. Studies of Interactions between Silane Coupling Agents and Cellulose Fibers with Liquid and Solid-State NMR. *Magn. Reson. Chem.* **2007**, *45*, 473–483. [CrossRef]
31. Robles, E.; Csóka, L.; Labidi, J. Effect of Reaction Conditions on the Surface Modification of Cellulose Nanofibrils with Aminopropyl Triethoxysilane. *Coatings* **2018**, *8*, 139. [CrossRef]
32. Neves, R.M.; Ornaghi, H.L.; Zattera, A.J.; Amico, S.C. The Influence of Silane Surface Modification on Microcrystalline Cellulose Characteristics. *Carbohydr. Polym.* **2020**, *230*, 115595. [CrossRef]
33. Siuda, J.; Perdoch, W.; Mazela, B.; Zborowska, M. Catalyzed Reaction of Cellulose and Lignin with Methyltrimethoxysilane—FT-IR, 13C NMR and 29Si NMR Studies. *Materials* **2019**, *12*, 2006. [CrossRef]
34. Prasetyo, E.N.; Kudanga, T.; Fischer, R.; Eichinger, R.; Nyanhongo, G.S.; Guebitz, G.M. Enzymatic Synthesis of Lignin–Siloxane Hybrid Functional Polymers. *Biotechnol. J.* **2012**, *7*, 284–292. [CrossRef] [PubMed]
35. Kabir, M.M.; Wang, H.; Lau, K.T.; Cardona, F. Effects of Chemical Treatments on Hemp Fibre Structure. *Appl. Surf. Sci.* **2013**, *276*, 13–23. [CrossRef]

36. Li, H.; Bunrit, A.; Li, N.; Wang, F. Heteroatom-Participated Lignin Cleavage to Functionalized Aromatics. *Chem. Soc. Rev.* **2020**, *49*, 3748–3763. [CrossRef] [PubMed]
37. Zhang, J.; Chen, Y.; Brook, M.A. Reductive Degradation of Lignin and Model Compounds by Hydrosilanes. *ACS Sustain. Chem. Eng.* **2014**, *2*, 1983–1991. [CrossRef]
38. Zhu, J.; Xue, L.; Wei, W.; Mu, C.; Jiang, M.; Zhou, Z. Modification of Lignin with Silane Coupling Agent to Improve the Interface of Poly(L-Lactic) Acid/Lignin Composites. *BioResources* **2015**, *10*, 4315–4325. [CrossRef]
39. Baur, S.I.; Easteal, A.J. Improved Photoprotection of Wood by Chemical Modification with Silanes: NMR and ESR Studies. *Polym. Adv. Technol.* **2013**, *24*, 97–103. [CrossRef]
40. Wang, Q.; Xiao, Z.; Wang, W.; Xie, Y. Coupling Pattern and Efficacy of Organofunctional Silanes in Wood Flour-Filled Polypropylene or Polyethylene Composites. *J. Compos. Mater.* **2015**, *49*, 677–684. [CrossRef]
41. Sèbe, G.; Tingaut, P.; Safou-Tchiama, R.; Pétraud, M.; Grelier, S.; Jéso, B.D. Chemical Reaction of Maritime Pine Sapwood (Pinus Pinaster Soland) with Alkoxysilane Molecules: A Study of Chemical Pathways. *Holzforschung* **2004**, *58*, 511–518. [CrossRef]
42. Grubbström, G.; Holmgren, A.; Oksman, K. Silane-Crosslinking of Recycled Low-Density Polyethylene/Wood Composites. *Compos. Part A Appl. Sci.* **2010**, *41*, 678–683. [CrossRef]
43. Popescu, C.-M.; Broda, M. Interactions between Different Organosilicons and Archaeological Waterlogged Wood Evaluated by Infrared Spectroscopy. *Forests* **2021**, *12*, 268. [CrossRef]
44. Yelle, D.J.; Ralph, J. Characterizing Phenol–Formaldehyde Adhesive Cure Chemistry within the Wood Cell Wall. *Int. J. Adhes. Adhes.* **2016**, *70*, 26–36. [CrossRef]
45. Namyslo, J.C.; Drafz, M.H.H.; Kaufmann, D.E. Durable Modification of Wood by Benzoylation—Proof of Covalent Bonding by Solution State NMR and DOSY NMR Quick-Test. *Polymers* **2021**, *13*, 2164. [CrossRef] [PubMed]
46. Yelle, D.J.; Ralph, J.; Frihart, C.R. Delineating PMDI Model Reactions with Loblolly Pine via Solution-State NMR Spectroscopy. Part 2. Non-Catalyzed Reactions with the Wood Cell Wall. *Holzforschung* **2011**, *65*, 145–154. [CrossRef]
47. Lu, F.; Ralph, J. Non-Degradative Dissolution and Acetylation of Ball-Milled Plant Cell Walls: High-Resolution Solution-State NMR. *Plant J.* **2003**, *35*, 535–544. [CrossRef]
48. Yelle, D.J. *Multifaceted Approach for Determining the Absolute Values for Lignin Subunits in Lignocellulosic Materials*; Res. Note FPL-RN-0384; US Department of Agriculture, Forest Service, Forest Products Laboratory: Madison, WI, USA, 2020; p. 384.
49. Miyagawa, Y.; Tobimatsu, Y.; Lam, P.Y.; Mizukami, T.; Sakurai, S.; Kamitakahara, H.; Takano, T. Possible Mechanisms for the Generation of Phenyl Glycoside-Type Lignin–Carbohydrate Linkages in Lignification with Monolignol Glucosides. *Plant J.* **2020**, *104*, 156–170. [CrossRef]
50. Yelle, D.J.; Ralph, J.; Lu, F.; Hammel, K.E. Evidence for Cleavage of Lignin by a Brown Rot Basidiomycete. *Environ. Microbiol.* **2008**, *10*, 1844–1849. [CrossRef]
51. Yelle, D.J.; Kapich, A.N.; Houtman, C.J.; Lu, F.; Timokhin, V.I.; Fort Jr, R.C.; Ralph, J.; Hammel, K.E. A Highly Diastereoselective Oxidant Contributes to Ligninolysis by the White Rot Basidiomycete Ceriporiopsis Subvermispora. *Appl. Environ. Microbiol.* **2014**, *80*, 7536–7544. [CrossRef]
52. Yelle, D.J.; Kaparaju, P.; Hunt, C.G.; Hirth, K.; Kim, H.; Ralph, J.; Felby, C. Two-Dimensional NMR Evidence for Cleavage of Lignin and Xylan Substituents in Wheat Straw Through Hydrothermal Pretreatment and Enzymatic Hydrolysis. *Bioenerg. Res.* **2013**, *6*, 211–221. [CrossRef]
53. Giachi, G.; Capretti, C.; Macchioni, N.; Pizzo, B.; Donato, I.D. A Methodological Approach in the Evaluation of the Efficacy of Treatments for the Dimensional Stabilisation of Waterlogged Archaeological Wood. *J. Cult. Herit.* **2010**, *11*, 91–101. [CrossRef]
54. Grattan, D.W. 3-Waterlogged Wood. In *Conservation of Marine Archaeological Objects*; Pearson, C., Ed.; Butterworth-Heinemann: Oxford, UK, 1987; pp. 55–67. ISBN 978-0-408-10668-9.
55. Broda, M.; Majka, J.; Olek, W.; Mazela, B. Dimensional Stability and Hygroscopic Properties of Waterlogged Archaeological Wood Treated with Alkoxysilanes. *Int. Biodeter. Biodegr.* **2018**, *133*, 34–41. [CrossRef]
56. Broda, M.; Curling, S.F.; Spear, M.J.; Hill, C.A. Effect of Methyltrimethoxysilane Impregnation on the Cell Wall Porosity and Water Vapour Sorption of Archaeological Waterlogged Oak. *Wood Sci. Technol.* **2019**, *53*, 703–726. [CrossRef]
57. Van Opdenbosch, D.; Dörrstein, J.; Klaithong, S.; Kornprobst, T.; Plank, J.; Hietala, S.; Zollfrank, C. Chemistry and Water-Repelling Properties of Phenyl-Incorporating Wood Composites. *Holzforschung* **2013**, *67*, 931–940. [CrossRef]
58. Hill, C.A.S.; Farahani, M.R.M.; Hale, M.D.C. The Use of Organo Alkoxysilane Coupling Agents for Wood Preservation. *Holzforschung* **2004**, *58*, 316–325. [CrossRef]
59. Pesti, J.; Larson, G.L. Tetramethyldisiloxane: A Practical Organosilane Reducing Agent. *Org. Process Res. Dev.* **2016**, *20*, 1164–1181. [CrossRef]
60. Larson, G.L.; Fry, J.L. Ionic and Organometallic-Catalyzed Organosilane Reductions. In *Organic Reactions*; John Wiley & Sons, Ltd.: Hoboken, NJ, USA, 2008; pp. 1–737. ISBN 978-0-471-26418-7.
61. Marciniec, B. Hydrosilylation: A Comprehensive Review on Recent Advances. In *Advances in Silicon Science*; Springer: Dordrecht, Germany, 2009.
62. Hossain, M.A.; Rahaman, M.S.; Yelle, D.; Shang, H.; Sun, Z.; Renneckar, S.; Dong, J.; Tulaphol, S.; Sathitsuksanoh, N. Effects of Polyol-Based Deep Eutectic Solvents on the Efficiency of Rice Straw Enzymatic Hydrolysis. *Ind. Crops. Prod.* **2021**, *167*, 113480. [CrossRef]

63. Yelle, D.J.; Ralph, J.; Frihart, C.R. Characterization of Nonderivatized Plant Cell Walls Using High-Resolution Solution-State NMR Spectroscopy. *Magn. Reson. Chem.* **2008**, *46*, 508–517. [CrossRef]
64. Methyltrimethoxysilane(1185-55-3) 13C NMR. Available online: https://www.chemicalbook.com/SpectrumEN_1185-55-3_13 CNMR.htm (accessed on 29 April 2022).
65. Safaei-Ghomi, J.; Nazemzadeh, S.H. Ionic Liquid-Attached Colloidal Silica Nanoparticles as a New Class of Silica Nanoparticles for the Preparation of Propargylamines. *Catal. Lett.* **2017**, *147*, 1696–1703. [CrossRef]
66. 1-(Aminoformylmethyl)Pyridinium Chloride(41220-29-5) 13C NMR. Available online: https://www.chemicalbook.com/SpectrumEN_41220-29-5_13CNMR.htm (accessed on 29 April 2022).
67. Wieszczycka, K.; Filipowiak, K.; Wojciechowska, I.; Buchwald, T.; Siwińska-Ciesielczyk, K.; Strzemiecka, B.; Jesionowski, T.; Voelkel, A. Novel Highly Efficient Ionic Liquid-Functionalized Silica for Toxic Metals Removal. *Sep. Purif. Technol.* **2021**, *265*, 118483. [CrossRef]
68. Scott, A.F.; Gray-Munro, J.E.; Shepherd, J.L. Influence of Coating Bath Chemistry on the Deposition of 3-Mercaptopropyl Trimethoxysilane Films Deposited on Magnesium Alloy. *J. Colloid Interface Sci.* **2010**, *343*, 474–483. [CrossRef]
69. The NMR of (3-Mercaptopropyl)Trimethoxysilane. Available online: http://www.hanhonggroup.com/nmr/nmr_en/MR040137.html (accessed on 29 April 2022).
70. Wipfelder, E.; Höhn, K. Epoxysiloxane Resins by the Condensation of 3-Glycidyloxypropyltrimethoxysilane with Diphenylsilanediol. *Angew. Makromolek. Chem.* **1994**, *218*, 111–126. [CrossRef]
71. 3-Glycidoxypropyltrimethoxysilane(2530-83-8) 13C NMR. Available online: https://www.chemicalbook.com/SpectrumEN_2530-83-8_13CNMR.htm (accessed on 29 April 2022).
72. Li, S.; Kong, X.; Feng, S. Preparation of Uniform Poly(Urea–Siloxane) Microspheres through Precipitation Polymerization. *RSC Adv.* **2015**, *5*, 90313–90320. [CrossRef]
73. Karasiewicz, J.; Krawczyk, J. Thermodynamic Analysis of Trisiloxane Surfactant Adsorption and Aggregation Processes. *Molecules* **2020**, *25*, 5669. [CrossRef] [PubMed]

Article

Diffusion Spectrum of Polymer Melt Measured by Varying Magnetic Field Gradient Pulse Width in PGSE NMR

Aleš Mohorič [1,2,*], Gojmir Lahajnar [2] and Janez Stepišnik [1]

[1] Department of Physics, Faculty of Mathematics and Physics, University of Ljubljana, 1000 Ljubljana, Slovenia; janez.stepisnik@fmf.uni-lj.si
[2] Institute Josef Stefan, 1000 Ljubljana, Slovenia; gojmir.lahajnar@fmf.uni-lj.si
* Correspondence: ales.mohoric@fmf.uni-lj.si; Tel.: +386-1-4766500

Academic Editor: Elena G. Bagryanskaya
Received: 3 November 2020; Accepted: 7 December 2020; Published: 9 December 2020

Abstract: The translational motion of polymers is a complex process and has a big impact on polymer structure and chemical reactivity. The process can be described by the segment velocity autocorrelation function or its diffusion spectrum, which exhibit several characteristic features depending on the observational time scale—from the Brownian delta function on a large time scale, to complex details in a very short range. Several stepwise, more-complex models of translational dynamics thus exist—from the Rouse regime over reptation motion to a combination of reptation and tube-Rouse motion. Accordingly, different methods of measurement are applicable, from neutron scattering for very short times to optical methods for very long times. In the intermediate regime, nuclear magnetic resonance (NMR) is applicable—for microseconds, relaxometry, and for milliseconds, diffusometry. We used a variation of the established diffusometric method of pulsed gradient spin-echo NMR to measure the diffusion spectrum of a linear polyethylene melt by varying the gradient pulse width. We were able to determine the characteristic relaxation time of the first mode of the tube-Rouse motion. This result is a deviation from a Rouse model of polymer chain displacement at the crossover from a square-root to linear time dependence, indicating a new long-term diffusion regime in which the dynamics of the tube are also described by the Rouse model.

Keywords: diffusion; PGSE; NMR; Rouse; reptation

1. Introduction

Molten polymers are macromolecular systems with complex translational dynamics of entangled chains and their segments being characterized by a large span of spatial and temporal scales. These dynamics are an important factor in the functionality and reactivity of the molecules. High-density entanglements, chain-bonds and cross-links prevent the formulation of an explicit theory of translational dynamics, even on larger intra-molecular length scales. In general, translation is described in a $6N$ phase space (N is the number of Kuhn segments of the chain) and studied by computer simulations [1,2]. Simpler models with fewer parameters can be set up. The simplest one-parameter model describing the chain motion is center-of-mass Brownian self-diffusion. In this model, the self-diffusion coefficient relates the chain center-of-mass mean square displacement (MSD) to the diffusion time: $\langle r^2 \rangle = 6 D_c t$. Compared to the coefficients of simple liquids, this coefficient is several orders of magnitude smaller. This approach considers a polymer melt as a simple liquid and is suitable only on a long-time scale. Measurements of diffusion on a shorter time scale show anomalous diffusion [3–5], indicating translation more complex than that described by the Brownian model. In this case, the self-diffusion coefficient is not a constant but depends on the diffusion time and can suitably be described by its diffusion spectrum. The diffusion spectrum is the Fourier transformation of the chain segment velocity autocorrelation function.

To describe anomalous diffusion on a shorter time scale, the Rouse model [6] is used. In the Rouse model, the polymer chain is approximated by a string of Kuhn segments, connected by bonds modeled as springs, diffusing in viscous surroundings described by its effective friction drag ζ. No topological effects of the surrounding chains are considered. The MSD of a segment along a chosen axis is expressed as a sum of modes [7]:

$$\langle \Delta z_R^2(t) \rangle = 2D_c t + \frac{4}{3} \sum_{p=1}^{N} \langle X_p^2 \rangle \left(1 - e^{-\frac{p^2 t}{\tau_R}} \right) \quad (1)$$

Here, N is the number of Kuhn segments of length a, $D_c = \frac{k_B T}{N\zeta}$, k_B is the Boltzmann constant, $\langle X_p^2 \rangle = \frac{Na^2}{2\pi^2 p^2}$ is the squared amplitude of displacement of the p-th mode, and $\tau_R = \frac{2\langle X_1^2 \rangle}{3D_c}$ is the Rouse relaxation time of the chain. This model can be further simplified for the case of a long chain [7] to

$$\langle \Delta z_R^2(t) \rangle \approx \frac{4\sqrt{\pi}}{3} \langle X_1^2 \rangle \sqrt{\frac{t}{\tau_R}} \quad (2)$$

Intermolecular entanglements in a dense polymer prevent the lateral motion of a chain and localize it inside a curved tube. The Rouse model can be used to model a shorter part of the chain, with N_e Kuhn segments, between the adjacent entanglements in the short time limit. In the intermediate time regime, segments reach the tube walls, and their motion is restrained. The polymer chain can only move along the tube in a reptation process [8,9]. As the polymer chain is released from the tube, the correlation with the initial conformation is lost. The described progression in translational mechanisms causes successively different time dependencies of the MSD: proportional to $t^{1/2}$ at short times, to $t^{1/4}$ and back to $t^{1/2}$ in the intermediate reptation regime, and to t for the chain disengagement [10]. Considering the collective motion of chains as a single chain moving in a fixed tube is a simplification. In real polymers, adjacent chains also move, causing constraint release [11,12], and the tube itself behaves as a coarser-grained Rouse chain [8,13], with the relaxation time proportional to the lifetime of the obstacles. The combined model of chain reptation inside the tube, exhibiting Rouse motion, predicts the segmental MSD starting as $t^{1/2}$ and evolving to a t time dependency. According to [11], the longest relaxation time for the tube-Rouse motion τ in a mono-dispersed polymer melt is equal to the terminal time of the chain's reptation in the tube, which amounts to almost equal contributions of both processes to the MSD $t^{1/2}$ dependence.

An alternative to the description of the MSD in the time domain is the transformation of the translation dynamics to the frequency domain, where it is described by the power spectrum $D(\omega) = -\omega 2/2 \, FT[\langle \Delta z2(t) \rangle]$, where FT is the Fourier transformation. As the MSD exhibits different translational modes expressed through changes in the power of the MSD time progression, so does $D(\omega)$. It starts as a constant D_c at low frequencies and increases as $\omega^{1/2}$ and $\omega^{3/4}$ in the tube/reptation regime. It passes into the Rouse regime at the frequency $2\pi/\tau_R$ and exhibits $\omega^{1/2}$ dependence again until it levels off at high frequencies [14,15]. The tube/reptation model replaces the many-chain problem by a single chain moving in a tube of topological constraints exerted by the surrounding chains. This model oversimplifies the actual dynamics, because the surrounding chains are not static but are moving as well. This motion is responsible for the constraint release and relaxation of the tube [11,12]. The tube relaxation time for constraint release, or tube reorganization, is determined by obstacle lifetime [8]. Various models account for the impermanence of entanglements [16–18]. The theory of constraint release involves tube dilation and tube-Rouse motion [13]. In this model, the constraint release is considered as a Rouse motion of the tube with coarser segments and slower relaxation than the Rouse motion of the chain. The chain relaxation in polymer melts results from two independent and concurrent processes: reptation inside the tube and tube-Rouse motion as the tube reorganization. The diffusion spectrum at low frequencies starts from a constant for both processes and changes to the

$\omega^{1/2}$ at inverse values of the longest relaxation time of each process. In the case of a mono-dispersed polymer melt, the tube reorganization is slower than reptation and must have a small effect on the diffusion properties [16,17]; nevertheless, tube reorganization significantly affects the viscoelastic properties of the polymer melt, presumably because of the difference between the spectrum of the tube-Rouse modes and the spectrum of reptation [11].

The structural and dynamical properties of polymers predicted by the models are in qualitative agreement with experimental data resulting from different methods, some of which are not limited to a macroscopic, rheological scale and offer insight into the chain dynamics on the segmental scale [12]. In a polymer melt, the chains exhibit a complex hierarchy of dynamic processes. Very fast and local conformational rearrangements on the picosecond scale can be measured by neutron scattering [19]. Slow, diffusive and cooperative motion extending into the range of seconds can be observed by methods of nuclear magnetic resonance (NMR), optical methods or viscosity measurements in rheology [20]. NMR is sensitive to polymer dynamics on a wide range of time scales; for example, the diffusion coefficient can be measured in the interval from milliseconds to seconds [3], either indirectly with NMR relaxometry [10,21–23] or directly by measuring the effect of spin-bearing-molecule displacement on the gradient spin-echo (GSE) attenuation in the applied magnetic field gradient [3,10].

The chain translation dynamics influence NMR relaxation because the dipolar coupling between adjacent spins depends on mutual orientation. The orientational fluctuations mirror the segmental dynamics through the magnetic dipole–dipole correlation function [24,25]. The correlation function includes intramolecular and intermolecular contributions. Intramolecular interactions fluctuate due to molecular rotation. Intermolecular couplings also depend on the relative translational motion of the chains. The presence of internal field gradients (conditioned by voids in polymer melts) in high-molecular-mass polymers has been suggested in [10] based on an accelerated transversal relaxation rate obtained from free induction decay, which is effectively reduced by the application of a Carr–Purcell–Meeboom–Gill pulse sequence. Other phenomena can lead to accelerated relaxation, e.g., dipole–dipole interactions not averaged by molecular motion arising due to the anisotropy of the motion of the chain segments, which is typical for entangled polymer chains. Reorientational and translational dynamics must be discerned in order to study polymer dynamics by NMR relaxometry. This is achieved by different techniques, e.g., by the isotope dilution technique in field-cycling and transverse NMR relaxometry [10,26], by combining NMR relaxometry and dielectric spectroscopy [27,28], or by double-quantum NMR experiments [22,23,29]. Different models and approximations of polymer dynamics have been discussed in the context of different spin relaxation studies, failing to provide an exact form of the correlation function; however, these experiments generally confirm the scaling laws of the reptation model [25].

GSE methods can be roughly divided into two classes, modulated and pulsed GSE. Modulated GSE (MGSE) uses an applied magnetic field gradient modulated in a way to harmonically change spin dephasing, thus measuring the diffusion spectrum at the modulation frequency [30]. Pulsed GSE (PGSE) employs (two) short gradient pulses separated by a defined time interval. In the limit of short pulses, this time interval can be considered as the diffusion time. The first applied gradient pulse defocuses spins, encoding their position in their phase; the second, decoding pulse refocuses all stationary spins in the spin echo. The moving spins do not refocus completely, causing the attenuation of the echo, which thus becomes sensitive to translational motion. If the time between the pulses, the diffusion time, is longer than the terminal relaxation time, defined as the asymptotic viscous decay of the polymer in rheology, the PGSE method can provide the polymer center-of mass diffusion coefficient in the polymer melt [4]. PGSE can also measure anomalous diffusion. The shortest diffusion time interval is limited by the strongest applicable gradients, and the longest diffusion time interval is limited by the decoherence of spins (transversal relaxation). This puts the limits of the segment displacement that can be detected by GSE NMR somewhere in the range of several hundred nanometers, assuming the self-diffusion coefficient of the high-molecular-weight polymer melt is on the order of 10^{-15}–10^{-12} m$^2 \cdot$s^{-1}. Polymer chain reptation displacements are smaller than 100 nm and are not detectable with a conventional PGSE experiment. Conflicting reports on the self-diffusion N scaling power follow from poorly determining the center-of-mass diffusion coefficient without considering the crossover to the anomalous diffusion regime at the same time [3].

Internal gradients, caused by susceptibility mismatches or paramagnetic centers, can cause artifacts, leading to the overestimation of the self-diffusion coefficient. There are common NMR diffusion techniques that can be used to reduce artifacts by internal gradients, such as bipolar gradients [31]. However, when the background gradients are spatially non-uniform, molecular diffusion introduces a temporal modulation of the background gradients, defeating the simple bipolar gradient suppression of background gradients in diffusion-related measurements. Several other methods have thus been proposed to minimize the effect of the internal gradient [32–35], among which is also the method presented in the paper [14], where the data from the PGSE measurements are explained with a crossover to the anomalous diffusion regime in polymer melts with the addition of the internal gradient effect. In certain cases, the effect of internal gradients can provide valuable information on the dynamics, topology or composition of the material studied [36]. Measurements of molten polydisperse polymers provide a diffusion coefficient that scales as N^2 for polymers with numbers of Kuhn segments larger than the entanglement number and as power N for those below [4]. However, the subsequent PGSE measurements of very mono-dispersed molten polymer [37,38] do not confirm this result but provide a scaling power larger than 2 for the total range of polymer lengths without any crossover to the power 1 for short chains. These conflicting data could result from a mis-defined crossover to the anomalous diffusion regime as shown in [30]. There are also reports that crossover is mis-defined because a strong internal susceptibility magnetic field at the interstices of voids in a polymer melt spoils the measurement [12,15]. Internal gradients are, aside from paramagnetic centers, caused by voids in the melt. Voids in polymer melts are statistically varying formations, which can be characterized by their sizes and mean lifetimes. For example, in polybutadiene, these voids are adjacent to the reptating chain segments and characterized by a diameter ~0.5 nm. However, such voids can affect the diffusion NMR experiment only if the diameters of the voids and mean lifetimes are at least of the order of magnitude of the covered diffusion paths and the diffusion times, respectively [39]. A method that also avoids the effects of the internal gradient is the MGSE method. Its results for self-diffusion measurements of mono-dispersed molten polymers [37,38] provide scales for the total range of polymer lengths and the transition into the regime of entanglement at Kuhn steps, which are below theoretical predictions. A test of the tube/reptation model by measuring the diffusion of nanoscopic strands of linear, mono-disperse poly (ethylene oxide) embedded in artificial cross-linked methacrylate matrices is described in [40]. PGSE studies of polymer dynamics are well described by the Rouse model in the case of dilute and semi-dilute polymers [41–44]. However, the PGSE measurements of diffusion in dense polymers do not clearly support the tube/reptation model [5,10,37,45]. The MGSE method, which measures the velocity autocorrelation spectrum, shows that in a polymer melt, the tube-Rouse motion has a prevailing role at long diffusion times, and this indicates faster tube reorganization than expected [30].

NMR measurements in a magnetic gradient field are sensitive to the MSD, $\langle \Delta z^2 \rangle$, in the direction of the applied magnetic field gradient $\boldsymbol{G} = \nabla |\boldsymbol{B}|$, here taken to be along the z axis. The attenuation of the spin echo is given by:

$$\ln \frac{S_0}{S} = \frac{1}{\pi} \int_0^\infty |q(\omega)|^2 D(\omega) d\omega \qquad (3)$$

where S is the spin-echo amplitude, S_0 is the amplitude of the echo without the applied gradient (in the limit $G \to 0$) and $|q(\omega)|^2$ is the sampling function tailored by the gradient–radiofrequency pulse sequence. The gradient sampling function for the Hahn-echo PGSE sequence is given by [34]:

$$|q(\omega)|^2 = 16 \gamma^2 G^2 \frac{\sin^2 \frac{\omega \delta}{2} \sin^2 \frac{\omega \Delta}{2}}{\omega^4} \qquad (4)$$

Here, G is the strength of the gradient, γ is the gyro-magnetic ratio, δ is the width of the gradient pulse and Δ is the time interval between the leading edges of the two gradient pulses. The gradient sampling function Equation (4) is shown superimposed on the diffusion spectrum given by Equation (5) in Figure 1.

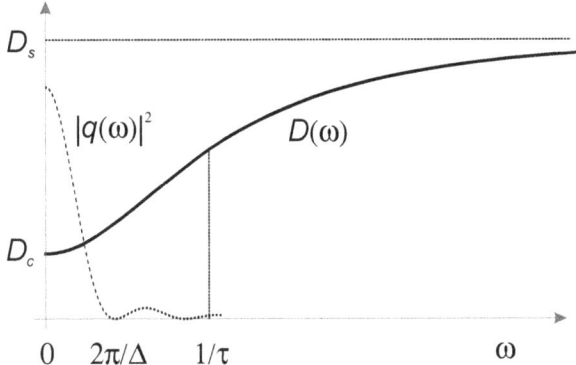

Figure 1. The diffusion spectrum $D(\omega)$ of a polymer melt and the sampling function $|q(\omega)|^2$ of the pulsed gradient spin-echo (PGSE) sequence.

This paper presents a study of anomalous self-diffusion in a linear polyethylene melt by the PGSE method. A special short diffusion time sensitivity is achieved by the variation of the gradient pulse width δ, contrary to usual measurements of anomalous diffusion with variable inter-pulse separation Δ. By changing only δ, artifacts induced by internal gradients can also be reduced as described by Equation (A2) in [14]. Measurements with PGSE are more effective for long diffusion times or, conversely, the low-frequency part of the diffusion spectrum. A problem arises if we want to measure the diffusion spectrum at high frequencies, as a short Δ together with strong magnetic gradients must be used to achieve the desired attenuation of the spin echo. This is experimentally hard to implement. Additional attenuation caused by the background or internal magnetic field gradient and the effect of transverse relaxation must also be accounted for when the inter-pulse separation is changed. Here, we set out to verify the results of the measurements of a polymer melt diffusion spectrum with the MGSE method reported in [15] by an alternative method of PGSE. The results in [15] show that the observed dynamics in the low-frequency range belong to tube-Rouse motion [13] and can be described by the formula

$$D(\omega) = D_c + D_s \frac{\omega^2 \tau^2}{1 + \omega^2 \tau^2} \qquad (5)$$

where D_c is the center-of-mass diffusion coefficient, D_s is the diffusion rate of the tube segments and τ is the tube-Rouse time, corresponding to the characteristic time of the crossover. This spectrum is shown in Figure 1 and overlaid with the gradient sampling function of the PGSE sequence.

For the diffusion spectrum model given in Equation (5), the spin-echo attenuation Equation (3) becomes:

$$\ln \frac{S_0}{S} = \gamma^2 G^2 \delta^2 \left(\Delta - \frac{\delta}{3} \right) D_c + \gamma^2 G^2 \tau^3 D_s \left[2 e^{\frac{\Delta}{\tau}} \left\{ 1 + \left(\frac{\delta}{\tau} - 1 \right) e^{\frac{\delta}{\tau}} \right\} - \left(e^{\frac{\delta}{\tau}} - 1 \right)^2 \right] e^{-\frac{\delta + \Delta}{\tau}}. \qquad (6)$$

The standard evaluation of the diffusion data measured with the PGSE is calculating the effective diffusion coefficient D_e, defined with:

$$\ln \frac{S_0}{S} = -b D_e, \qquad (7)$$

where the b factor is given by $b = \gamma^2 G^2 \delta^2 \left(\Delta - \frac{\delta}{3}\right)$. The effective diffusion coefficient is a constant for all possible parameters in the case of Brownian diffusion, but in the case of anomalous diffusion, it is interpreted as a time-dependent diffusion coefficient, in our case, given by:

$$D_e = D_c + \frac{\tau^3}{\delta^2\left(\Delta - \frac{\delta}{3}\right)} D_s \left[2e^{\frac{\Delta}{\tau}}\left\{1 + \left(\frac{\delta}{\tau} - 1\right)e^{\frac{\delta}{\tau}}\right\} - \left(e^{\frac{\delta}{\tau}} - 1\right)^2\right] e^{-\frac{\delta + \Delta}{\tau}}. \tag{8}$$

2. Results and Discussion

Polyethylene melt diffusion was measured by the PGSE method. In the experiment, the spin-echo amplitude was recorded by varying the gradient pulse width δ at several different strengths of the applied gradient pulse. Figure 2a shows the spin-echo amplitude and (b) the derived effective self-diffusion coefficient as defined in Equation (7), both as a function of the gradient pulse width δ and for all the applied gradient strengths. The effective diffusion coefficient in Figure 2b clearly shows signs of anomalous diffusion.

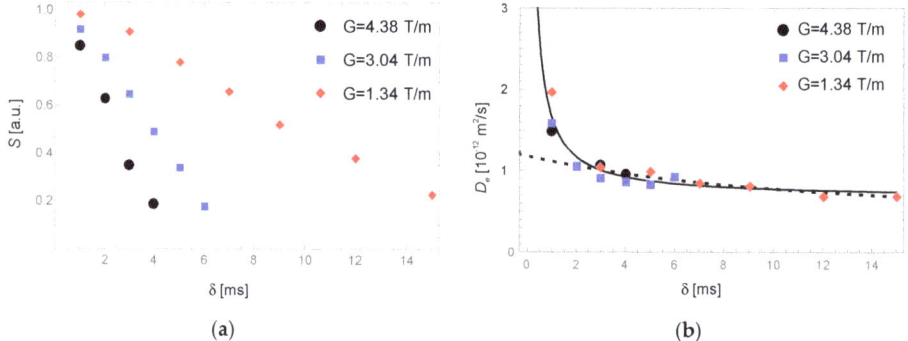

Figure 2. Measurements of polyethylene melt self-diffusion with the PGSE and a fixed diffusion time $\Delta = 80$ ms at temperature 426 K: (**a**) The echo amplitude as a function of the applied gradient pulse width for different applied gradient strengths; (**b**) The effective diffusion coefficient as a function of the applied gradient pulse width. Included are the least-square fits of the model in Equation (9) with the best-fit parameters for all the data (solid line) and excluding the shortest δ points (dashed line).

The model describing the data is given by Equation (8), and a few simplifications can be made, since it is reasonable to assume from previous measurements [30] that $\Delta \gg \tau$, to obtain a simpler model:

$$D_e = D_c + \frac{2\tau^3}{\delta^2\left(\Delta - \frac{\delta}{3}\right)} D_s \left[e^{-\frac{\delta}{\tau}} + \left(\frac{\delta}{\tau} - 1\right)\right]. \tag{9}$$

Both models return the same fitting parameters with the measured data. A least-square non-linear fit of the model to the data gives the parameters presented in Table 1.

Table 1. Parameters of the effective diffusion model of Equation (9) of the polyethylene melt SRM 1482 at 426 K.

Parameter	Estimate	Standard Error	t-Statistic	p-Value
D_c	7.0×10^{-13} m^2/s	8.0×10^{-14} m^2/s	9.3	2.2×10^{-7}
D_s	5.8×10^{-11} m^2/s	7.0×10^{-10} m^2/s	0.51	0.62
τ	0.75 ms	0.60 ms	1.25	0.23

The only value estimated with high certainty is the chain diffusion coefficient D_c. Both the tube segment diffusion coefficient and the relaxation time appear in the model (Equation (9)) together as $\tau^2 D_s$ to the first order of τ/δ, and any change in one can be compensated by an according change in the other and does not significantly alter the fit. Thus, the δ used in the measurements should be accordingly short, or at least one of the parameters should be determined separately. The results for the chain diffusion coefficient match within the error with the results in [14] and [46]. The result for the relaxation time matches the tube displacement per obstacle lifetime $L_{eq}/\sqrt{\tau_{ob}}$, if the number of Kuhn segments between the entanglements N_e is 25 (compared to the 120 total number of segments per chain) since $L_{eq} = \frac{N}{N_e} a$ and $\tau = \frac{N^2}{N_e^2} \tau_{ob}$.

The dashed line in Figure 2b represents the best fit of the D_e model to the data without the points measured at $\delta = 1$ ms. The fitting parameters in this case differ significantly: $D_c = 3.3 \times 10^{-13}$ m^2/s, $D_s = 1.8 \times 10^{-11}$ m^2/s and $\tau = 3.8$ s. This demonstrates the sensitivity of the model to the input data without measurements at a short-enough δ, which should be short enough to include the increase in D_e at a short δ. It also demonstrates that caution using the approximation $\Delta \gg \tau$ for the model in Equation (9) is warranted, since τ was determined to be longer than Δ in the case of short δ data points being excluded.

We have shown here that the PGSE method enables the measurement of the segmental translation of polymeric chains by the variation of the gradient pulse width. This approach can also effectively take into account the effect caused by the internal gradient, which commonly affects PGSE measurements but requires knowledge of the interplay between the molecular motion and the buildup of spin phase structure during the magnetic field gradient action. By combining the PGSE sampling function and the segmental diffusion spectrum rendered from the model of tube-Rouse motion [13], we obtain the dependence of the PGSE signal attenuation on the gradient pulse width. The data obtained by the measurements of molten polyethylene well fit the predictions and provide evidence of the tube-Rouse motion model proposed in [13]. The model, which was already confirmed for other polymer samples by the MGSE method [30], reveals the tube segmental motion in the range of milliseconds. The sample polymer data M_w and M_n indicate a sharp distribution of fragment sizes; thus, the effects of polydispersity, which may cause a deviation from the model, can be neglected in our case.

This study presented here is a reevaluation of the study in [14]. The study of polymer diffusion by the MGSE method [30] shows slow dynamics that can be attributed to the reorganization of the polymer tube with temporal and spectral resemblance to Rouse motion. This description matches the theory of tube-Rouse motion put forward in [13]. In [14], PGSE measurements were used to trace the crossover of the chain Rouse dynamics from $\sqrt{t}$ to t dependence because of the constraint release. The constraint release was originally termed tube reorganization by Pierre-Gilles de Gennes, where the obstacle lifetime determines the tube relaxation times [8]. Various models account for the impermanence of entanglements [16–18], among which is also the theory of constraint release involving the tube dilation and tube-Rouse motion [13]. In this theory, the constraint release is considered as the tube-Rouse motion, and the relaxation time is proportional to the lifetime of the obstacles [13]. In the previous paper [14], we followed a quite common approach to considering polymer chain dynamics described by the Rouse model in the range where the dynamics cross from square-root to linear time dependence in the MSD [47], to explain the anomalous effective diffusion obtained from the PGSE measurements in polymers. In the original experiment, the internal gradient artifacts were not suppressed by any of the numerous methods, because the system was considered homogenous enough and the measurements fitted well to the model used for the larger part of the measured interval. However, the results deviate from the model in the limit of the short δ. In [14], it was proposed that the deviation was a result of internal gradients (caused by a susceptibility mismatch) adding to the effect of the external applied gradient. An extra term based on internal gradients was added to the attenuation factor, resulting in a better fit. According to [39], this would require unrealistic conditions, and a search for a better explanation was fruitful, since the results are here satisfactorily described with the new tube-Rouse model and without recourse to the effect of the internal gradient. This is also in accordance with subsequent measurements with the MGSE method [30], which indicate that the polymers in the millisecond time range exhibit some new dynamics that are not related to the motion of the polymer chain inside the tube, but can be explained by the theory of polymer tube reorganization, where the tube behaves in a similar way to a chain, and therefore, this motion can be called tube-Rouse motion [13]. In this paper, we show that the new interpretation fits better to the results of our PGSE measurements. We show that the data can be well fitted to this model of tube dynamics, and this is a deviation from the previous results based on long-chain approximation (Equations (4) and (6) in [14]).

3. Materials and Methods

We studied a sample of linear polyethylene Standard Reference Material 1482 with a narrow molecular weight distribution ($M_n = 11,400$ g mol^{-1}, $M_w = 13,600$ g mol^{-1}) prepared by NIST, Washington, DC, USA. Measurements were performed on a melted polyethylene sample at 426 K.

The measurements were performed on a home-made pulsed NMR spectrometer (Ljubljana, Slovenia) at a 60 MHz proton NMR frequency and equipped with a magnetic field gradient coil system described in [48]. The PGSE sequence is shown in Figure 3. The widths of the $\pi/2$ radiofrequency (RF) pulses used were 1.2 microseconds. The π RF pulse was applied symmetrically between the gradient pulses. The gradient pulse followed the RF pulse with a delay short enough to be neglected in the signal analysis. The same is true for the echo following the second gradient pulse; however, the echo followed the second gradient pulse with a delay large enough that no artifacts were introduced because of the finite gradient fall time. The PGSE attenuation dependence on the duration of the gradient pulses was measured by changing the pulse width δ from 1 to 15 ms, with the diffusion time (the interval between the gradient pulses) fixed at $\Delta = 80$ ms. The measurements were performed with the gradient fields 4.38, 3.04 and 1.34 T/m.

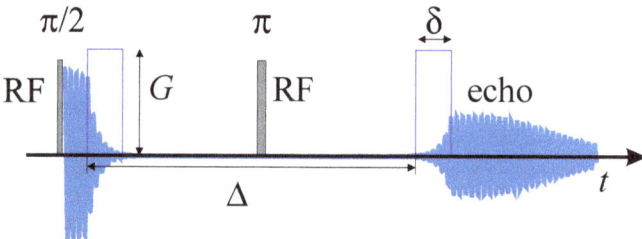

Figure 3. The PGSE sequence used to measure self-diffusion—shown are the RF and gradient pulses and the NMR signal: free induction decay at short times and echo after time Δ (blue line).

Author Contributions: Conceptualization, A.M. and J.S.; methodology, A.M. and J.S.; validation, A.M., J.S. and G.L.; formal analysis, A.M.; investigation, G.L.; resources, G.L.; data curation, A.M. and G.L.; writing—original draft preparation, A.M.; writing—review and editing, A.M., J.S. and G.L.; visualization, A.M.; supervision, J.S.; project administration, J.S.; funding acquisition, A.M. All authors have read and agreed to the published version of the manuscript.

Funding: This research was funded by Javna Agencija za Raziskovalno Dejavnost RS (P1-0060).

Conflicts of Interest: The authors declare no conflict of interest. The funders had no role in the design of the study; in the collection, analyses, or interpretation of data; in the writing of the manuscript; or in the decision to publish the results.

References

1. Theodorou, D.N. *Diffusion in Polymers*; Neogi, P., Ed.; Marcel Dekker: New York, NY, USA, 1996; ISBN 0-8247-9530-X.
2. Theodorou, D.N. Hierarchical modelling of polymeric materials. *Chem. Eng. Sci.* **2007**, *62*, 5697–5714. [CrossRef]

3. Komlosh, M.E.; Callaghan, P.T. Segmental motion of entangled random coil polymers studied by pulsed gradient spin echo nuclear magnetic resonance. *J. Chem. Phys.* **1998**, *109*, 10053–10067. [CrossRef]
4. Fleischer, G. The chain length dependence of self-diffusion in melts of polyethylene and polystyrene. *Coll. Polym. Sci.* **1987**, *265*, 89–95. [CrossRef]
5. Appel, M.; Fleischer, G.; Kaerger, J.; Fujara, F.; Chang, I. Anomalous segment diffusion in polymer melts. *Macromolecules* **1994**, *27*, 4274–4277. [CrossRef]
6. Rouse, P. A theory of the linear viscoelastic properties of dilute solutions of coiling polymer. *J. Chem. Phys.* **1953**, *21*, 1272–1279. [CrossRef]
7. Perez-Aparicio, R.; Colmenero, J.; Alvarez, F.; Padding, J.T.; Briels, W.J. Chain dynamics of poly(ethylene-altpropylene) melts by means of coarse-grained simulations based on atomistic molecular dynamics. *J. Chem. Phys.* **2010**, *132*, 024904. [CrossRef]
8. De Gennes, P. Reptation of a polymer chain in the presence of fixed obstacles. *J. Chem. Phys.* **1971**, *55*, 572–579. [CrossRef]
9. Doi, M.; Edwards, S.F. Dynamics of concentrated polymer system. *J. Chem. Soc. Faraday Trans.* **1978**, *74*, 1789–1801. [CrossRef]
10. Kimmich, R.; Fatkullin, N. Polymer chain dynamics and NMR. *Adv. Polym. Sci.* **2004**, *170*, 1–113. [CrossRef]
11. Graessley, W. Entangled linear branched and network polymer systems–molecular theories. *Adv. Polym. Sci.* **1982**, *47*, 67–117. [CrossRef]
12. McLeish, T.C.B. Tube theory of entangled polymer dynamics. *Adv. Phys.* **2002**, *51*, 1379–1527. [CrossRef]
13. Viovy, J.; Rubinstein, M.; Colby, R. Constraint release in polymer melts: Tube reorganization versus tube dilation. *Macromolecules* **1991**, *24*, 3587–3596. [CrossRef]
14. Stepišnik, J.; Lahajnar, G.; Zupančič, I.; Mohorič, A. Study of translational dynamics in molten polymer by variation of gradient pulse-width of PGSE. *J. Magn. Reson.* **2013**, *236*, 41–46. [CrossRef] [PubMed]
15. Stepišnik, J.; Mohorič, A.; Serša, I.; Lahajnar, G. Analysis of polymer dynamics by NMR modulated gradient spin echo. *Macromol. Symp.* **2011**, *305*, 55–62. [CrossRef]
16. Klein, J. The Onset of Entangled Behavior in Semidilute and Concentrated Polymer Solutions. *J. Macromol.* **1978**, *11*, 852–858. [CrossRef]
17. Daoud, M.; de Gennes, P.G. Some remarks on the dynamics of polymer melts. *J. Polym. Sci. Polym. Phys. Ed.* **1979**, *17*, 1971–1981. [CrossRef]
18. Leygue, A.; Bailly, C.; Keunings, R. A differential formulation of thermal constraint release for entangled linear polymers. *J. Non-Newton. Fluid Mech.* **2005**, *128*, 23–28. [CrossRef]
19. Ewen, B.; Richter, D. Neutron Spin Echo Investigations on the Segmental Dynamics of Polymers in Melts, Networks and Solutions. *Adv. Polym. Sci.* **1997**, *134*, 1–129. [CrossRef]
20. Schowalter, W.R. *Mechanics of Non-Newtonian Fluids*; Pergamon Press: Oxford, UK, 1978; ISBN 0-08-021778-8.
21. Stapf, S.; Kimmich, R. Field-cycling nuclear magnetic resonance relaxometry and field-gradient nuclear magnetic resonance diffusometry of polymers confined in porous glasses: Evidence for a restricted-geometry effect. *Macromolecules* **1996**, *29*, 1638–1641. [CrossRef]
22. Graf, R.; Heuer, A.; Spiess, H.W. Chain-order effects in polymer melts probed by hydrogen double-quantum NMR spectroscopy. *Phys. Rev. Lett.* **1998**, *26*, 5738–5741. [CrossRef]
23. Chávez, F.V.; Saalwächter, K. NMR observation of entangled polymer dynamics: Tube model predictions and constraint release. *Phys. Rev. Lett.* **2010**, *104*, 198305. [CrossRef]
24. Cohen-Addad, J.P. Single chain diffusion process and proton magnetic relaxation in polymer melts. *J. Phys.* **1982**, *43*, 1509–1528. [CrossRef]
25. Fatkullin, N.; Mattea, C.; Stapf, S. On the theory of double quantum NMR in polymer systems: The second cumulant approximation for many spin I = 1/2 systems. *J. Chem. Phys.* **2013**, *139*, 194905. [CrossRef] [PubMed]
26. Herrmann, A.; Kresse, B.; Wohlfahrt, M.; Bauer, I.; Privalov, A.F.; Kruk, D.; Fatkullin, N.; Fujar, F.; Rössler, E.A. Mean Square Displacement and Reorientational Correlation Function in Entangled Polymer Melts Revealed by Field Cycling 1H and 2H NMR Relaxometry. *Macromolecules* **2012**, *45*, 6516–6526. [CrossRef]
27. Kruk, D.; Herrmann, A.; Rössler, E. Field-cycling NMR relaxometry of viscous liquids and polymers. *Prog. Nucl. Magn. Reson. Spectrosc.* **2012**, *63*, 33–64. [CrossRef] [PubMed]
28. Meier, R.; Herrmann, A.; Hofmann, M.; Schmidtke, B.; Kresse, B.; Privalov, A.F.; Kruk, D.; Fujara, F.; Rössler, E.A. Iso-Frictional Mass Dependence of Diffusion of Polymer Melts Revealed by 1H NMR Relaxometry. *Macromolecules* **2013**, *46*, 5538–5548. [CrossRef]

29. Saalwächter, K. Proton multiple-quantum NMR for the study of chain dynamics and structural constraints in polymeric soft materials. *Prog. Nucl. Magn. Reson. Spectrosc.* **2007**, *51*, 1–35. [CrossRef]
30. Stepišnik, J.; Mohorič, A.; Mattea, C.; Stapf, S.; Serša, I. Velocity autocorrelation spectra in molten polymers measured by NMR modulated gradient spin-echo. *Eur. Phys. Lett.* **2014**, *106*, 27007. [CrossRef]
31. Cotts, R.M.; Hoch, M.J.R.; Sun, T.; Markert, J.T. Pulsed field gradient stimulated echo methods for improved NMR diffusion measurements in heterogeneous systems. *J. Magn. Reson.* **1989**, *83*, 252–266. [CrossRef]
32. Bar-Shir, A.; Avram, L.; Ozarslan, E.; Basser, P.; Cohen, Y. The effect of the diffusion time and pulse gradient duration ratio on the diffraction pattern and the structural information estimated from q-space diffusion MR: Experiments and simulations. *J. Magn. Reson.* **2008**, *194*, 230–236. [CrossRef]
33. Price, W.S. Pulsed-field gradient nuclear magnetic resonance as a tool for studying translational diffusion: Part II. Experimental aspects. *Concepts Magn. Reson.* **1998**, *10*, 197–237. [CrossRef]
34. Mohorič, A. A modified PGSE for measuring diffusion in the presence of static magnetic field gradients. *J. Magn. Reson.* **2005**, *174*, 223–228. [CrossRef] [PubMed]
35. Sun, P.Z.; Seland, J.G.; Cory, D. Background gradient suppression in pulsed gradient stimulated echo measurements. *J. Magn. Reson.* **2003**, *161*, 168–173. [CrossRef]
36. Stepišnik, J.; Ardelean, I. Usage of internal magnetic fields to study the early hydration process of cement paste by MGSE method. *J. Magn. Reson.* **2016**, *272*, 100–107. [CrossRef]
37. Von Meerwall, E.; Ozisik, R.; Mattice, W.L.; Pfister, P.M. Self-diffusion of linear and cyclic alkanes measured with pulsed-gradient spin-echo nuclear magnetic resonance. *J. Chem. Phys.* **2003**, *118*, 3867–3873. [CrossRef]
38. Von Meerwall, E.D.; Dirama, N.; Mattice, W.L. Diffusion in Polyethylene Blends: Constraint Release and Entanglement Dilution. *Macromolecules* **2007**, *40*, 3970–3976. [CrossRef]
39. Kärger, J.; Pfeifer, H.; Heink, W. Principles and Applications of Self-Diffusion Measurements by Nuclear Magnetic Resonance. *Adv. Magn. Reson.* **1988**, *12*, 1–89. [CrossRef]
40. Fischer, E.; Beginn, U.; Fatkullin, N.; Kimmich, R. Nanoscopic poly(ethylene oxide) strands embedded in semi-interpenetrating methacrylate networks. Preparation method and quantitative characterization by field-gradient NMR diffusometry. *Macromolecules* **2004**, *37*, 327786. [CrossRef]
41. Callaghan, P.T.; Coy, A. Evidence for reptation motion and the entanglement tube in semidilute polymer solutions. *Phys. Rev. Lett.* **1992**, *68*, 3176–3179. [CrossRef]
42. Griffiths, P.C.; Stilbs, P.; Yu, G.E.; Booth, C. Role of molecular architecture in polymer diffusion: A PGSE-NMR study of linear and cyclic poly(ethylene oxide). *J. Chem. Phys.* **1995**, *55*, 16752–16756. [CrossRef]
43. Callaghan, P.; Komlosh, M.; Nyden, M. High magnetic field gradient PGSE NMR in the presence of a large polarizing field. *J. Magn. Reson.* **1998**, *133*, 177–182. [CrossRef] [PubMed]
44. Roding, M.; Bernin, D.; Jonasson, J.; Sarkka, A.; Topgaard, D.; Rudemo, M.; Nyden, M. The gamma distribution model for pulsed-field gradient NMR studies of molecular-weight distributions of polymers. *J. Magn. Reson.* **2012**, *222*, 105–111. [CrossRef] [PubMed]
45. Fatkullin, N.; Kimmich, R. Theory of field-gradient nmr diffusometry of polymer segment displacements in the tube-reptation motion. *Phys. Rev. Lett.* **1995**, *802*, 5738416. [CrossRef]
46. Pearson, D.S.; Ver Strate, G.; von Meerwall, E.; Schilling, F.C. Viscosity and Self-Diffusion Coefficient of Linear Polyethylene. *Macromolecules* **1987**, *20*, 1133–1141. [CrossRef]
47. Callaghan, P.T.; Pinder, D.N. A Pulsed Field Gradient NMR Study of Self-Diffusion in a Polydisperse Polymer System: Dextran in Water. *Macromolecules* **1983**, *16*, 968–973. [CrossRef]
48. Zupancic, I.; Lahajnar, G.; Blinc, R.; Reneker, D.; Vanderhart, D. Nmr self-diffusion study of polyethylene and paraffin melts. *J. Polym. Sci.* **1985**, *23*, 387–404. [CrossRef]

Sample Availability: Samples of the compounds are not available.

Publisher's Note: MDPI stays neutral with regard to jurisdictional claims in published maps and institutional affiliations.

© 2020 by the authors. Licensee MDPI, Basel, Switzerland. This article is an open access article distributed under the terms and conditions of the Creative Commons Attribution (CC BY) license (http://creativecommons.org/licenses/by/4.0/).

Article

Non-Exponential ^{1}H and ^{2}H NMR Relaxation and Self-Diffusion in Asphaltene-Maltene Solutions

Kevin Lindt *, Bulat Gizatullin, Carlos Mattea and Siegfried Stapf *

Department of Technical Physics II/Polymer Physics, Institute of Physics, Faculty of Mathematics and Natural Science, Ilmenau University of Technology, P.O. Box 100565, D-98684 Ilmenau, Germany; bulat.gizatullin@tu-ilmenau.de (B.G.); carlos.mattea@tu-ilmenau.de (C.M.)
* Correspondence: kevin.lindt@tu-ilmenau.de (K.L.); siegfried.stapf@tu-ilmenau.de (S.S.)

Abstract: The distribution of NMR relaxation times and diffusion coefficients in crude oils results from the vast number of different chemical species. In addition, the presence of asphaltenes provides different relaxation environments for the maltenes, generated by steric hindrance in the asphaltene aggregates and possibly by the spatial distribution of radicals. Since the dynamics of the maltenes is further modified by the interactions between maltenes and asphaltenes, these interactions—either through steric hindrances or promoted by aromatic-aromatic interactions—are of particular interest. Here, we aim at investigating the interaction between individual protonic and deuterated maltene species of different molecular size and aromaticity and the asphaltene macroaggregates by comparing the maltenes' NMR relaxation (T_1 and T_2) and translational diffusion (D) properties in the absence and presence of the asphaltene in model solutions. The ratio of the average transverse and longitudinal relaxation rates, describing the non-exponential relaxation of the maltenes in the presence of the asphaltene, and its variation with respect to the asphaltene-free solutions are discussed. The relaxation experiments reveal an apparent slowing down of the maltenes' dynamics in the presence of asphaltenes, which differs between the individual maltenes. While for single-chained alkylbenzenes, a plateau of the relaxation rate ratio was found for long aliphatic chains, no impact of the maltenes' aromaticity on the maltene–asphaltene interaction was unambiguously found. In contrast, the reduced diffusion coefficients of the maltenes in presence of the asphaltenes differ little and are attributed to the overall increased viscosity.

Keywords: asphaltenes; maltenes; NMR; relaxation; diffusion

Citation: Lindt, K.; Gizatullin, B.; Mattea, C.; Stapf, S. Non-Exponential ^{1}H and ^{2}H NMR Relaxation and Self-Diffusion in Asphaltene-Maltene Solutions. *Molecules* **2021**, *26*, 5218. https://doi.org/10.3390/molecules26175218

Academic Editor: Igor Serša

Received: 2 August 2021
Accepted: 23 August 2021
Published: 28 August 2021

Publisher's Note: MDPI stays neutral with regard to jurisdictional claims in published maps and institutional affiliations.

Copyright: © 2020 by the authors. Licensee MDPI, Basel, Switzerland. This article is an open access article distributed under the terms and conditions of the Creative Commons Attribution (CC BY) license (https://creativecommons.org/licenses/by/4.0/).

1. Introduction

Crude oils are complex fluids consisting of thousands of chemical species of different structures and sizes, which can be fractionated into saturates, aromatics, resins, and asphaltenes (SARA) based on their solubility in different solvents. Saturates, aromatics and resins are defined as maltenes and are soluble in n-alkane solvents, for example, n-pentane or n-heptane, whereas the molecule fraction insoluble in n-alkane solvents is defined as asphaltenes. The presence of asphaltenes in crude oil can be the origin of severe issues related to production, refinery, and transportation, affecting the economic potential of the crude oil. The precipitation of asphaltenes, caused by changes in temperature, pressure, or the composition of the crude oil, can lead to fouled and clogged pipes in the extraction process or a blocked catalytic network in the refinery process, putting the process to a halt and requiring extensive cleaning. The asphaltene molecules themselves consist of polycyclic aromatic hydrocarbons (PAH) with alkyl side chains and contain heteroatoms like O, N, and S. Their tendency to self-aggregate distinguishes them from the other oil components and results, after the formation of nanoaggregates and clusters, in asphaltene macroaggregates, which are colloidally dispersed in the crude oil or a solvent until they precipitate as black, friable solids up to dense solid deposits [1].

To assess the risk of asphaltene precipitation and to choose the extraction parameters, one needs to gain insight into the conditions in the oil reservoir such as oil saturation or capillary pressures as well as the properties of the oil, i.e., its qualitative and quantitative composition, which ideally are obtained in situ downhole. Nowadays, nuclear magnetic resonance (NMR) techniques are well established among the standard downhole measurement techniques. This technique determines longitudinal (T_1) and transverse (T_2) relaxation times of the maltene nuclei, as well as the self-diffusion coefficients (D) of these species. The parameters are further characterized by a broad distribution due to the complexity of the crude oil. The relaxation times are dominated by molecular rotations, eventually modulated by diffusion, whereas the diffusion coefficient measured by NMR reflects the translational motion on typical scales of µm. Both quantities are related to each other and depend on different ways on the viscosity, which is a result of the oil composition. This was investigated in detail for linear alkanes [2,3]. For the increased complexity in crude oils due to the vast amount of components, two-dimensional techniques that correlate two of the three parameters were applied to assess the oil composition [4]. Most references agree that one average relaxation time be considered for a particular maltene, and empiric correlations attempt to define the maltene distribution in crude oil, e.g., by the average T_1/T_2 ratio in a T_1-T_2 map.

The situation is further complicated by the presence of asphaltenes, which provide an additional relaxation mechanism due to geometrical hindrance or unpaired electrons from ions such as VO^{2+}, which are present in the form of vanadyl porphyrins in asphaltene aggregates [5–7], or persistent free radicals [8–10] and therefore affect the relaxation of the maltenes significantly, while the diffusion properties are hardly affected [4]. As mentioned above, the possibility of non- or multiexponential relaxation for maltenes is frequently not considered in asphaltene-containing crude oils, and similar correlations are being established while neglecting the influence of the presence of asphaltenes on the maltenes' behavior.

Our previous studies [11,12] showed a significant (factor 20) difference in the T_1/T_2 ratios of fluorinated aromatic ring molecules (higher ratio) (benzene-f_6 and toluene-f_8) and aliphatic chain molecules (lower ratio) (octane-f_{18} and pentadecane-f_{32}) in an oil containing 13 wt% asphaltenes, which is much smaller in an asphaltene-free resinic oil and vanishes in an asphaltene- and resin-free oil. This difference in the T_1/T_2 ratio persists in a solution of chloroform-d and 3 vol% of the tracer molecule (benzene-f_6 and octane-f_{18}) containing 10 wt% asphaltenes [13], extracted from the previous investigated asphaltenic oil. Furthermore, a strong dispersion of T_1 at Larmor frequencies between 10^6 Hz and 10^8 Hz was observed for the aromatic molecules in the asphaltenic oil, which is in accordance with the T_1 dispersion of the maltene mixtures present in crude oils containing asphaltenes [14–16]. In contrast, a significantly weaker dispersion is observed for octane-f_{18} in the asphaltenic oil as well as for the aromatic molecules in the asphaltene-free oil [11]. Additionally, it was found [17] that in comparison with saturated ^1H fraction, an aromatic ^1H fraction of the crude oil exhibits stronger interaction with asphaltene molecules. Furthermore, molecular dynamics and NMR relaxation of maltenes in the proximity of asphaltenes cannot be characterized by a simple model as in the case of only high molecular weight compounds solution even containing paramagnetic impurities [18].

The aim of this work is to study the maltene–asphaltene interaction for individual maltenes rather than a mixture of maltenes. NMR relaxometry and diffusometry methods are used to access the microscopic and macroscopic maltene motion. Since the non-monoexponential relaxation of maltenes in crude oils persists in the absence of asphaltenes, its origin is located in the large number of different maltenes present in crude oils [19]. On the other hand, the broadening of the T_1 distribution for lower frequencies in asphaltenic crude oils indicates maltene populations in different relaxation environments as an additional cause for their non-monoexponential relaxation in the presence of asphaltenes. The investigation of individual maltenes eliminates the non-monoexponential decay due to different maltene species and thus allows us to study the influence of the asphaltenes on the maltene's relaxation. The focus on individual maltenes rather than a maltene mixture further opens the possibility to study the influence of the maltene's size, shape, and aromaticity on the maltene–asphaltene interaction.

2. Theory

For protons, the most dominant NMR relaxation mechanism is the dipolar coupling between spin-containing nuclei, which consists of an intra- and an intermolecular contribution. However, in the presence of unpaired electrons, e.g., due to free radicals and paramagnetic ions in asphaltenes, the relaxation may be dominated by the dipolar coupling between the nucleus and the free electron. The relaxation, i.e., the return of the total magnetization towards its equilibrium, is enabled by the modulation of the local magnetic field environment of the nucleus. This field modulation occurs due to the molecular rotational and translational motion and is therefore influenced by the viscosity of the solution and the molecule's size.

The intermolecular part of the dipole–dipole interaction is predominantly influenced by the translational motion of the molecule since the interacting nuclei are located at different molecules, while the intramolecular dipolar coupling occurs between nuclei on the same molecule. For small molecules, the intramolecular part is therefore dominated by the rotational motion of the molecule, while for larger molecules the internal motions can gain in importance. The motion relevant for the dipolar coupling between a nucleus and a free electron depends on whether the unpaired electron is located on a different molecule or within the molecule hosting the interacting nucleus. Due to the presence of an electric quadrupole moment, the dominant relaxation mechanism for deuterons is the quadrupolar coupling, which is the interaction of the quadrupolar moment with the electric field gradient at the site of the nucleus. The quadrupolar coupling is a very strong relaxation mechanism and the dipolar coupling becomes negligible for relaxation so that the relaxation of deuterons solely reflects the rotational motion of the molecule. The relaxation rates $R_{1,2}$ of the molecule result additively from the different contributions of the relaxation mechanisms:

$$R_{1,2}^{^1H} = \underbrace{R_{1,2}^{D,intra} + R_{1,2}^{D,inter}}_{\substack{\text{intra- and intermolecular}\\\text{contribution to the dipolar}\\\text{coupling between nuclei}}} + \underbrace{R_{1,2}^{D,elec}}_{\substack{\text{dipolar coupling}\\\text{between nuclei and}\\\text{unpaired electrons}}} \quad (1)$$

$$R_{1,2}^{^2H} = \underbrace{R_{1,2}^{Q}}_{\substack{\text{quadrupolar coupling}\\\text{between nuclei}}} + \underbrace{R_{1,2}^{D,elec}}_{\substack{\text{dipolar coupling}\\\text{between nuclei and}\\\text{unpaired electrons}}} \quad (2)$$

In case of the protonic maltenes, the reference solution consists of only 5 vol% of proton bearing molecules, so that the intermolecular contribution to the dipolar coupling can be neglected. This no longer applies for the asphaltene-containing solution since asphaltenes are mainly composed of carbons and protons ($n(^1H)/n(C) \approx 1...1.3$ [20,21]) and the asphaltene concentration (≈ 17 wt%) is quite high (asphaltene concentration in crude oils ranges from 0 wt% up to approximately 20 wt% [21–23]).

As mentioned above, the relaxation requires a fluctuation of the local magnetic field in the environment of the nucleus, e.g., caused by the molecular motion. The fluctuation of the local magnetic field is described by an autocorrelation function, and its normalized Fourier transform, the reduced spectral density $\mathcal{I}(\omega)$, contains the Larmor frequency dependence of the relaxation rates R [24,25]:

$$R_1^{D,II} = \frac{1}{5}C_D[\mathcal{I}(\omega_I) + 4\mathcal{I}(2\omega_I)]$$
$$R_2^{D,II} = \frac{1}{10}C_D[3\mathcal{I}(0) + 5\mathcal{I}(\omega_I) + 2\mathcal{I}(2\omega_I)]$$
(3)

$$R_1^{D,IS} = \frac{1}{15}C_D[\mathcal{I}(\omega_I - \omega_S) + 3\mathcal{I}(\omega_I) + 6\mathcal{I}(\omega_I + \omega_S)]$$
$$R_2^{D,IS} = \frac{1}{30}C_D[4\mathcal{I}(0) + \mathcal{I}(\omega_I - \omega_S) + 3\mathcal{I}(\omega_I) + 6\mathcal{I}(\omega_S) + 6\mathcal{I}(\omega_I + \omega_S)]$$
(4)

with $C_D = \left(\frac{\mu_0}{4\pi}\right)^2 \gamma_I^2 \gamma_S^2 \hbar^2 S(S+1) C_{int}$, $C_{int} = \begin{cases} \frac{1}{r^6} & \text{for intramolecular coupling} \\ 72\frac{N_I}{d^3} & \text{for intermolecular coupling} \end{cases}$

and the Larmor frequency $\omega = \gamma B_0$. The prefactor C_D incorporates the vacuum permeability μ_0, the gyromagnetic ratio γ of the considered nucleus (γ_I) and the nucleus coupled to it (γ_S), as well as the spin number of the coupled nucleus S. In the homonuclear case (Equation (3)), where two nuclei of the same type are coupled together, $S = I$ applies. The relaxation rate for the nucleus–electron interaction can be obtained from the equations for heteronuclear coupling (Equation (4)). The difference between intra- and intermolecular contribution lies in the prefactor C_{int}, as well as in the shape of the reduced spectral density (see Equations (6) and (7)). The prefactor C_{int} contains the intramolecular nuclei–nuclei distance r or, in case of intermolecular coupling, the distance of closest approach d between the interacting nuclei and N, the number of spins I per volume.

The relaxation rates for the quadrupolar relaxation [25] in case of, e.g., deuterons differs only by the prefactors from the ones describing the intramolecular, homonuclear dipolar coupling (Equations (3) and (6)):

$$R_1^Q = \frac{3}{80}C_Q[\mathcal{I}(\omega_I) + 4\mathcal{I}(2\omega_I)]$$
$$R_2^Q = \frac{3}{160}C_Q[3\mathcal{I}(0) + 5\mathcal{I}(\omega_I) + 2\mathcal{I}(2\omega_I)]$$
(5)

with $C_Q = \left(\frac{e^2 qQ}{\hbar}\right)^2 \left(1 + \frac{\eta^2}{3}\right)$

The first term of the quadrupolar prefactor C_Q is known as quadrupole coupling constant, consisting of the electric field gradient q and the quadrupole moment Q, while the second term reflects the asymmetry of the electric field by the asymmetry parameter η.

The reduced spectral density contains, via the frequency dependence of the relaxation rates, the dependence on the molecular motion, which can be represented by a correlation time τ. The properties of molecular motion are not only reflected in different correlation times, but also in the functional form of the reduced spectral density. For isotropic rotation, this function is described by a Lorentzian [25]

$$\mathcal{I}_{intra}(\omega) = \mathcal{I}_{rot}(\omega) = \frac{2\tau_{rot}}{1 + \omega^2 \tau_{rot}^2}.$$
(6)

The Lorentzian shape of the reduced spectral density function results from an exponential correlation function, which derives from the assumption of random motion. Assuming the force-free-hard-sphere model, the reduced spectral density of translational motion, which dominates the intermolecular relaxation, can be expressed as [26]

$$\mathcal{I}_{inter}(\omega) = \mathcal{I}_{trans}(\omega) = \int_0^\infty \frac{u^2}{81 + 9u^2 - 2u^4 + u^6} \frac{u^2 \tau_{trans}}{u^4 + \omega^2 \tau_{trans}^2} du$$
(7)

$$\text{with} \quad \tau_{trans} = \frac{d^2}{D_{IS}},$$

where d denotes the distance of closest approach of the interacting species, D the relative translational diffusion coefficient as a sum of the diffusion coefficients of the spin-bearing molecules $D_{IS} = D_I + D_S$, and u a dimensionless integration variable.

In the BPP model, i.e., when dipolar relaxation is dominated by isotropic rotation, the frequency dependence of T_1 is proportional to ω^2 until T_1 becomes independent of the Larmor frequency in the extreme narrowing limit, when $\omega\tau \ll 0.707$, and $T_1 = T_2$, e.g., in low viscosity liquids. In the region $\omega\tau \gg 0.707$, which can be found, for example, for high viscosity liquids, the ratio of T_1 and T_2 grows with increasing correlation times. The extreme narrowing region reflects therefore motions with correlation times faster than $1/\omega$, i.e., fast dynamics, while the region of large T_1/T_2 ratios reflects motions with correlation times comparable to or slower than $1/\omega$, i.e., slow dynamics. This is true for protons and deuterons in rotation dominated molecules since the shape of Equation (3) and Equation (5) do not differ.

The free electrons present in asphaltenes provide a highly effective ($\omega_e \approx 658\, \omega_{1_H}$) variant of the dipolar relaxation to the maltenes, as the maltenes' nuclear dipole interacts with a dipole of the free electron. Since the interacting dipoles are located on different molecules, the intermolecular interaction between maltenes and asphaltenes becomes important for the maltenes' relaxation. In addition, the self-aggregated structures of asphaltene molecules in crude oil or asphaltene-solvent solutions interfere with the motions of the maltenes, slowing them down so that the reorientations of the maltenes no longer fulfill the extreme narrowing limit, and T_1/T_2 ratios larger than one and a frequency dependence of T_1 are observed [27]. Investigations of maltenes' relaxation in the presence of asphaltenes show faster relaxation with higher asphaltene content and a stronger increase in the transverse than the longitudinal relaxation rate [13,19]. The latter observation indicates significant importance of the decreased maltene mobility due to contact and entanglement with the asphaltene structures, as the proximity to free electrons influences the longitudinal and transverse relaxation equally. The BPP assumption of random motion is therefore not valid for these systems, and models taking a correlated motion into account were developed to explain the relaxation behavior of the maltenes.

The current understanding of maltene–asphaltene interaction is based on the work of Zielinski et al. and Korb et al., who both found a strong T_1 dispersion for the maltenes in asphaltenic-resinic, but not in asphaltene-free crude oils [14,15]. The reorientations of the maltene molecules in these asphaltene-free oils are hence fast enough to fulfill the extreme narrowing limit, i.e., T_1 is frequency independent, while the presence of asphaltenes results in slower maltene reorientations, which do not fulfill the extreme narrowing limit anymore. Since the viscosities of the studied asphaltene-free and asphaltenic oils did not differ substantially, the relaxation dispersion is an effect of the maltene–asphaltene interaction, which couples the normally fast motion of the maltenes to the slow motion of the asphaltene macrostructures. Although the asphaltene content of the oil is the dominating factor for the relaxation dispersion, the influence of the oil's total composition is not negligible. In case of a low resin content (<15 wt%) and a lack of asphaltenes, often no dispersion is found (for example sample 2 in [15], oil A-D in [14], but not oil 10 in [28]), while in crude oils with a similar asphaltene content, a stronger dispersion is found for the more resinic oil (compare oils E and J, as well as G and I in [14]). The influence of the resin on the dispersion may be seen in [14] (although the composition of the used oil, besides its asphaltene content, is unknown), where the removal of the asphaltene content of an oil results in a partial disappearance of the dispersion below 1 MHz, which is consistent with asphaltenes forming the largest structures and hence are the slowest components in crude oils. Furthermore, in the asphaltenic-resinic oils, the shape of the T_1 distribution becomes narrower with increasing frequency, while this is not observed in the asphaltene-free oils. Korb et al. were able to fit the T_1 distribution with a bimodal log-normal distribution at low frequencies and a single log-normal distribution at high frequencies (15 MHz) according to a model based on the translational motion of maltenes on a locally flat surface containing unpaired electrons. While the principal distribution of T_1 (and also T_2) is attributed to the vast amount of different maltenes present in the crude oil, Korb and coworkers concluded from their bimodal log-normal distribution fit different relaxation environments of the maltenes, which depend on their proximity to the asphaltene macrostructures. Although the models of Zielinski and Korb differ slightly regarding the detailed interaction between the maltenes and the asphaltenes, the overall concept is similar: The relaxation behavior of the maltenes is caused by their motion through the slowly rotating porous asphaltene macrostructures. In contact with the asphaltenes, the maltenes diffusion is correlated, while the diffusion between the asphaltene structures is only influenced by the global composition of the solution, i.e., its viscosity. Furthermore, fast exchange between the surface region and the bulk region is assumed.

Since the model is developed from the investigations of crude oils, i.e., a mixture of various different maltenes, descriptions of the interaction of individual maltene species with the asphaltene structures are limited in the literature. However, it is reasonable to assume that the size of the maltene is an important parameter affecting the contact time and the maltene-asphaltene distance, i.e., the distance between the maltenes' nuclei and corresponding spins on the asphaltenes, as it influences possible entanglements and the trapping of maltene molecules in the asphaltene structures. For example, the increase in relaxation rates was found to be larger for longer and less mobile hydrocarbon chains with increasing asphaltene concentration [19], though the contact time between the maltenes and asphaltenes is still short, as the overall diffusion of the maltenes is not severely hindered [4,16,19].

In addition to a purely steric interaction between the maltenes and the asphaltenes, interactions between aromatic parts of the maltenes and the asphaltenes are conceivable and may contribute to the contact time and the minimum maltene–asphaltene distance. These aromatic interactions, often referred to as π-stacking, describe attractive forces between aromatic ring molecules, due to interactions between their aromatic π electron clouds. The result can be a rather stable structure, as it is for example reported for the benzene dimer (dissociation energy (moderate hydrogen bonds: $D_0 = (4-15)\,\text{kcal/mol}$ [29], van der Waals bonds: $D_0 \lesssim 1\,\text{kcal/mol}$) $D_0 = (2.0-2.7)\,\text{kcal/mol}$) [30,31], though the importance of π-stacking, that means the face-centered stacking arrangement of aromatic molecules, for aromatic–aromatic interactions is still under discussion [32]. Nevertheless, π-π interactions are considered as an important interaction among others responsible for asphaltene aggregation and precipitation [33–36]. Recently, an unusual type of parallel π-stacking, the so-called pancake bonding, which occurs between radicals with highly delocalized π electrons, as they occur in asphaltenes, gained attention with regard to the aggregation of asphaltenes [37]. The pancake bonding differs from the π-stacking by a stronger interaction, closer contact distances, and a preferred orientation for direct atom-to-atom overlapping due to an energy lowering overlapping of the singly occupied molecular orbitals of the radicals [38].

Some studies have addressed the interactions of maltenes and asphaltenes by density functional theory [39–41] and molecular dynamics simulations [42,43], and experimental studies have investigated the aggregation ability of asphaltenes in different solvents [44–46]. Although some of these studies found evidence of a stronger interaction between asphaltenes and aromatic molecules, the aromatic–aromatic interactions do not appear to be exclusively responsible for the asphaltene–maltene interactions, as strong anionic, polar and acid-base interactions were also found.

While the direct comparison of the shape of the relaxation decays is a straightforward method to identify monoexponential and non-monoexponential relaxations, the T_1/T_2 ratio allows an estimation of the strength of restriction the asphaltene structures impose on the maltene's motion. Since the T_1/T_2 ratio is close to unity in the extreme narrowing limit, but larger for motions slower than the inverse of the Larmor frequency, a higher ratio indicates a slower motion. Furthermore, the T_1/T_2 ratio is an often used parameter in borehole analysis to classify the extracted oil. It is obtained from a line in the T_1-T_2 plot parallel to $T_1 = T_2$ and allows summarizing properties of a mixture of substances by a common parameter.

3. Materials and Methods

Maltene molecules and solvents were purchased from different vendors (see Appendix A) and were used without further purification or degassing. Benzene and naphthalene were chosen to represent typical aromatic molecules. To investigate the influence of the aromaticity of the maltenes on the interaction with the asphaltene, the set is extended by n-decane, a saturated hydrocarbon, and the cycloalkanes cyclohexane and decalin. In addition, several alkylbenzenes (toluene, ethylbenzene, propylbenzene, butylbenzene, and decylbenzene) were selected to explore a possible impact of a side chain of increasing length attached to an aromatic core on the maltene–asphaltene interactions. Furthermore, fully deuterated (benzene-d_6, naphthalene-d_8, decalin-d_{18}, cyclohexane-d_{12}, nonane-d_{20}, toluene-d_8, ethylbenzene-d_{10}) variants were included to study directly the impact of the asphaltene on the rotational motion of the maltenes since the relaxation of the deuterated maltenes is expected to be dominated only by intramolecular contributions. In addition, due to the lack of deuterons in asphaltenes, there is no signal overlap of the maltenes and asphaltenes and therefore the sole maltene dynamic is observed.

3.1. Sample Preparation

The asphaltene-free reference solutions consist of 95 vol% of a solvent, benzene-d_6 in case of the protonic maltenes and benzene for the deuterated variants, and 5 vol% of the maltene. As the impact of the presence of asphaltenes on the relaxation times of the maltene molecules increases with the asphaltene concentration [13,19], a high asphaltene concentration is desirable. Hence, an asphaltene concentration of 150 g per liter of reference solution ($\approx$17 wt%) is used for the asphaltene solutions. This concentration is well above the concentration where asphaltene cluster formation starts [47–49].

The asphaltene was provided by Schlumberger Doll Research and was extracted from oil A13 used in references [11,12]. It is the same as in reference [13], where its radical content was determined by EPR to $(145 \pm 7) \times 10^{14}$ spins per mg.

The lack of a sufficient amount of the asphaltene and the large set of samples required a sequential sample preparation and the reuse of the asphaltene. Therefore, the asphaltene solutions with the protonic maltenes were prepared directly in the 5 mm NMR tube, which was flame sealed afterward. After the measurement, the samples were reopened and kept in an oven at 100 °C until the liquid components had evaporated (checked by weight) and only the solid asphaltene remained. The asphaltene is unchanged by this process and due to the in-tube preparation the amount of asphaltene in the tube is known so that a new sample can be prepared in the reopened tube. The repetition of this process is limited by the minimum required length of the tube to fit into the spectrometer. In this work, the initial amount of asphaltene in a tube was used for two samples. Attention was paid that the subsequential samples consisted of chemically similar maltenes and solvents, e.g., the sample consisting of benzene-d_6, toluene and asphaltene is followed by the sample consisting of benzene, toluene-d_8 and asphaltene.

The in-tube preparation of the asphaltene samples was carried out in the following way. After the asphaltene (30 mg) was placed in the tube and filled up with the reference solution (200 µL), the tube was flame sealed. To assist the permeation process of the reference solution in the air-filled pores between the asphaltene flakes, especially in the case of the reused asphaltene, the sample was manually centrifuged for one minute. Afterward, the sample was held for two minutes on a laboratory shaker with 1000 rpm before being placed in an ultrasonic bath at room temperature for 30 min. The samples were stored for at least 12 h at 5 °C in the refrigerator before measurement.

3.2. Measurement and Evaluation

The experiments were carried out at room temperature at a magnetic field strength of 7.05 T with a Bruker Avance III 300. The spectral resolution obtained allows a distinction between aromatic and aliphatic nuclei (see Appendix B) so that their relaxation is evaluated separately. The inversion recovery (IR) and the Carr–Purcell–Meiboom–Gill (CPMG) pulse sequence were used to obtain the longitudinal (T_1) and transverse (T_2) recovery functions. The relaxation curves were either fitted with a single exponential function or a sum of two exponential functions of the form:

$$S_j(t) = S_\infty + \sum_{i=1}^{n} A_i e^{-\frac{t}{T_{k_i}}} \quad \text{with} \quad n = 1,2 \tag{8}$$

$j = z$, $k = 1$ for longitudinal relaxation(IR)
$j = xy$, $k = 2$, $S_\infty = 0$ for transverse relaxation (CPMG)

The fit covers a range between t_0 and the loss of the signal below noise level, but with a maximum $t_{end}^{CPMG} = 10\,\text{s}$ in case of the transverse relaxation decays. This results from a compromise regarding the echo time in the CPMG sequence, which was chosen as $\tau_e = 1\,\text{ms}$ in all T_2 experiments to ensure the comparability of the experiments with the asphaltene-free and the asphaltenic samples. The selected echo time is short enough to enable a good resolution of the signal decay in the asphaltenic samples, as well as long enough to capture at least the signal decay below $1/e$ in the asphaltene-free samples. The 90° and 180° pulse durations were 5.75 µs and 11.5 µs for the proton and 56 µs and 112 µs for the deuteron measurements, respectively.

Additionally, the diffusion coefficients of the protonic maltenes in the absence and presence of the asphaltene were determined with the Pulse Gradient Stimulated Echo (PGSTE) sequence using half-sine shaped gradient pulses, with a duration of $\delta = 2\,\text{ms}$ and a separation between the gradient pulses of $\Delta = 20\,\text{ms}$. The gradient strength g was varied up to 1 T/m to determine the self-diffusion coefficients of the protonic maltenes. The time between the first two radio frequency pulses is about 4.2 ms. In the absence of asphaltene, the data were fitted with a single exponential function, while in the presence of asphaltene, a deviation from a monoexponential decay was observed at larger gradients g, which is attributed to the significantly slower diffusion of the asphaltene. The signal decay is therefore fitted in the following way:

$$S_D(g^2) = \sum_{i=1}^{n} A_i e^{-D_i 4\pi^2 \gamma_{1H}^2 \delta^2 (\Delta - \frac{\delta}{3}) g^2} \quad \text{with} \quad n = \begin{cases} 1 & \text{asphaltene-free} \\ 2 & \text{with asphaltene} \end{cases} \tag{9}$$

Since the measurements aimed at the determination of the diffusion coefficient of the maltenes, the obtained data is not sufficient to reliably determine the diffusion coefficient of the asphaltene in the different maltene solutions. Including D_2 as a free parameter in the fit results in an order of $10^{-10}\,\text{m}^2/\text{s}$ for the reasonable fits, while some decays could not be fitted. Therefore, the parameter D_2 was fixed to the fastest diffusion coefficient obtained for the aliphatic asphaltene protons in a solution consisting of 150 g/L asphaltene in benzene-d_6 (see Section 4.4). The obtained fits describe the data well and provide reliable diffusion coefficients for the maltenes.

4. Results

The results of the experiments are evaluated by a direct comparison of the signal decay, as well as the relaxation time constants T_1 and T_2 and their respective relative weights p_1 and p_2. Furthermore, the mean relaxation rate constants R_1 and R_2 are computed to obtain a single variable classifying the longitudinal and transverse relaxation of the different samples.

$$\overline{R}_k = \sum_{i=1}^{n} p_i \cdot \frac{1}{T_{k_i}} \quad \text{with} \ k = 1, 2, \tag{10}$$

where p_i denotes the relative weight of the $T_{1_i,2_i}$ component and is calculated from the amplitudes A_i of the different exponential functions in the fit:

$$p_i = \frac{A_i}{\sum_{i=1}^{n} A_i} \quad \text{with} \ n = 1, 2 \ \text{and} \ \sum_{i=1}^{n} p_i = 1. \tag{11}$$

In addition, the mean relaxation rate constants enable the construction of the R_2/R_1 ratio, equivalent to the widely used T_1/T_2 ratio, known from borehole analysis to evaluate the crude oil regardless of the detailed form of the relaxation time constant distributions. The mean relaxation rate constant is chosen, since the relaxation rate constant is, unlike the relaxation time constant, directly proportional to the spectral density function (see Equations (3)–(5)) and therefore the mean relaxation rate constant is proportional to the mean spectral density function.

The mean relaxation rate constants are also used to enable the qualitative comparison of the relaxation decays of the different samples regarding their monoexponentiality. The time axis of the relaxation decay of each sample is rescaled with the corresponding mean relaxation rate constant

$$t^* = \overline{R}_{1,2} \cdot t, \tag{12}$$

so that t^* reflects multiples of $\overline{R}_{1,2}^{-1}$. Furthermore, the signal is normalized according to:

$$S_z^* = \frac{S_z - S_\infty}{\sum_{i=1}^{n} A_i} \tag{13}$$

$$S_{xy}^* = \frac{S_{xy}}{\sum_{i=1}^{n} A_i} \tag{14}$$

As a result, the signal S_j^* is normalized to unity at $t^* = 0$ and all monoexponential decays collapse to $S_j^* = e^{-t^*}$ (see Figure 1a). Non-monoexponential decays differ from this relation, and the visualization of $\log S^*$ over t^* allows the qualitative comparison of the monoexponentiality of the relaxation.

As mentioned in Section 3.2 the relaxation decays were fitted with a single exponential or a sum of two exponential functions to obtain the parameters T_{k_i} and A_i. The decision between a monoexponential and biexponential fit is based on the obtained data, regarding which fit describes the observed relaxation decay sufficiently. In all cases of non-monoexponential decays, we identified a biexponential fit as sufficient and therefore did not attempt to fit a sum of more than two exponential functions. This does not mean that the relaxation consists of two components, but that it could be described sufficiently with the specified time constants.

4.1. Relaxation Decays

The relaxation decays of the protonic and deuterated maltenes in the asphaltene-free and the asphaltenic solution are displayed in Figures 1 and 2 and Figures 3 and 4, respectively. The different maltenes are distinguished by different colors, while the symbols indicate the chosen fit to obtain the relaxation time constants. A monoexponential decay is indicated by a red line.

(**a**) Longitudinal relaxation (0 g/L).

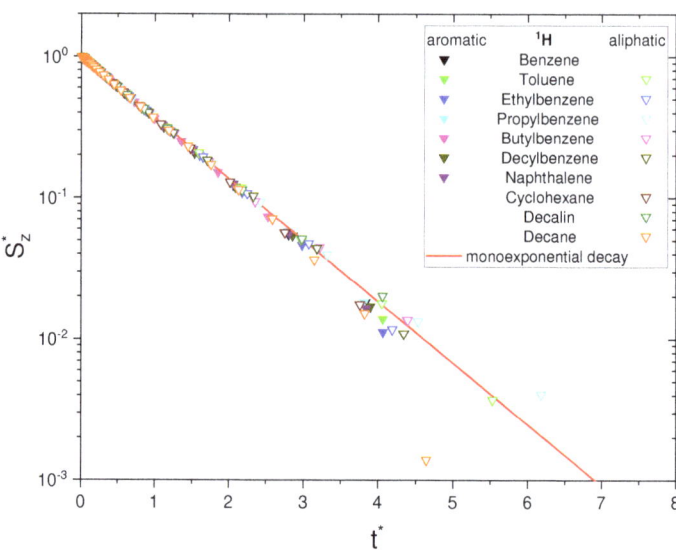

(**b**) Transverse relaxation (0 g/L).

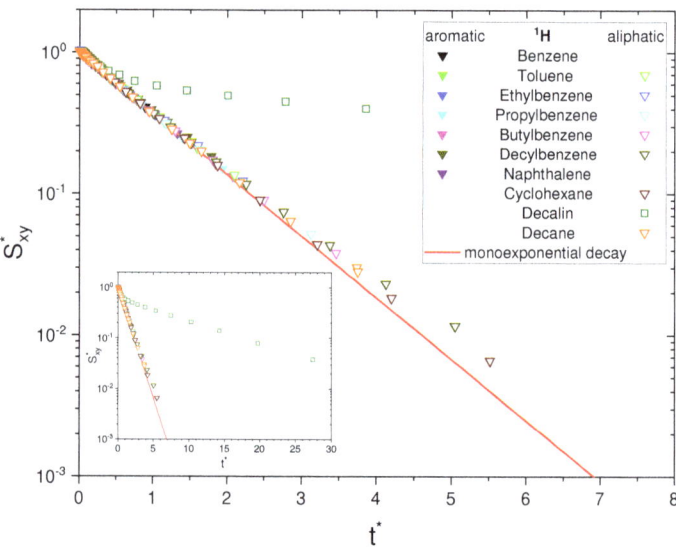

Figure 1. Longitudinal (**a**) and transverse (**b**) relaxation decays of the maltene protons in the asphaltene-free solution, consisting of 95 vol% benzene-d_6 and 5 vol% of the investigated maltene. The inset in (**b**) shows the whole measured range of the relaxation decay of decalin. For a qualitative comparison, the time axis is rescaled (see Equation (12)) to reflect multiples of $\overline{R}_{1,2}^{-1}$ and the signal is normalized to unity at $t^* = 0$ (see Equations (13) and (14)). The needed parameters were obtained from a monoexponential ($\triangledown$) or biexponential ($\square$) fit.

In the absence of asphaltene (Figure 1) the relaxation decays of the maltene protons can be described sufficiently well by a monoexponential fit, except for the transverse relaxation of decalin, which can be very well fitted by a sum of two exponential functions using Equation (8). The biexponential relaxation of decalin results from the coexistence of a mixture of cis- and trans-isomers of decalin (see Appendix D). In presence of the asphaltene (Figure 2), most of the relaxation decays become non-monoexponential and are therefore fitted with a sum of two exponential functions. In fact, only the longitudinal relaxation of benzene, toluene, and naphthalene can still be described sufficiently well by a monoexponential decay. Furthermore, the deviation from a monoexponential relaxation is more pronounced for the transverse relaxation than the longitudinal relaxation.

Since asphaltenes are a proton-rich species consisting of many different molecular structures, the signal of the protons located on the asphaltenes and the maltenes overlap, as can be seen, for example, in Figure 5. Since the time constants characterizing the relaxation of the asphaltene protons are of a comparable order to the relaxation time constants of the maltene protons (Section 4.2), the observed relaxation decays reflect the dynamic of both proton species. Therefore, relaxation experiments with a different nucleus, not abundant in asphaltenes, are necessary to unequivocally link the non-monoexponential relaxation to the maltenes. As one can see in Figure 4, the non-monoexponential character of the relaxation persists for the deuterated maltenes with the similarity that the deviations from a monoexponential decay are more pronounced for the transverse relaxation. The relaxation of the deuterated maltenes in the absence of the asphaltene is well described by a monoexponential fit for all investigated maltenes (see Figure 3).

4.2. Relaxation Time Constants

As mentioned before, the relaxation decays of the maltene protons consist of the signals of both the protons located on the maltenes and on the asphaltenes. To access the relaxation of the asphaltene, a sample consisting of 150 g/L asphaltene in benzene-d_6 is prepared. For this sample, the longitudinal (a) and transverse (b) relaxation curves of the aromatic and aliphatic protons are displayed in Figure 6. The fit parameter and the relative weight derived according to Equation (11) can be found in Table 1.

Table 1. Fit parameters used to fit the in Figure 6 shown relaxation curves of protons in a solution consisting of 150 g/L asphaltene in benzene-d_6, using Equations (13) and (14).

	Longitudinal Relaxation			Transverse Relaxation		
Parameter	Aromatic	Aliphatic	Parameter	Aromatic	Aliphatic
S_0	1.001(2)	0.997(2)	S_0	0	0
A_1	−1.11(2)	−1.890(4)	A_1	0.34(8)	1.80(6)
A_2	−0.82(2)		A_2	0.79(8)	0.48(4)
T_{1_1} [s]	0.562(7)	0.529(3)	T_{2_1} [s]	0.09(2)	0.018(1)
T_{1_2} [s]	2.15(4)		T_{2_2} [s]	0.25(2)	0.098(6)
p_1 [%]	57.4(9)		p_1 [%]	30(7)	79(2)
p_2 [%]	42.6(9)		p_2 [%]	70(7)	21(2)

(**a**) Longitudinal relaxation (150 g/L).

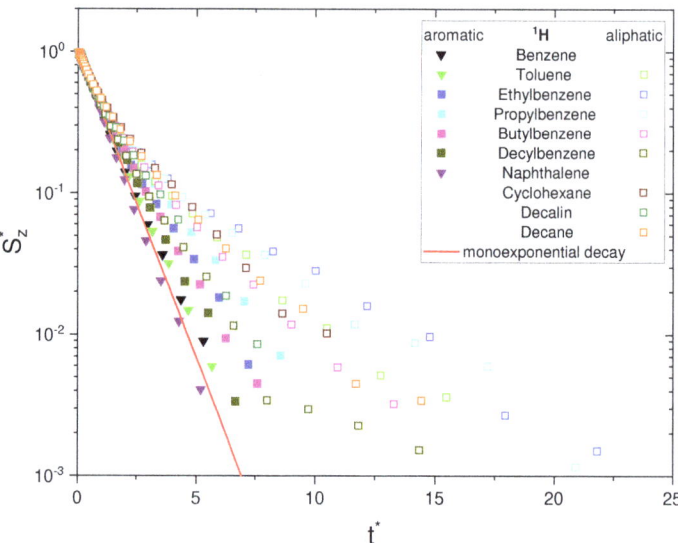

(**b**) Transverse relaxation (150 g/L).

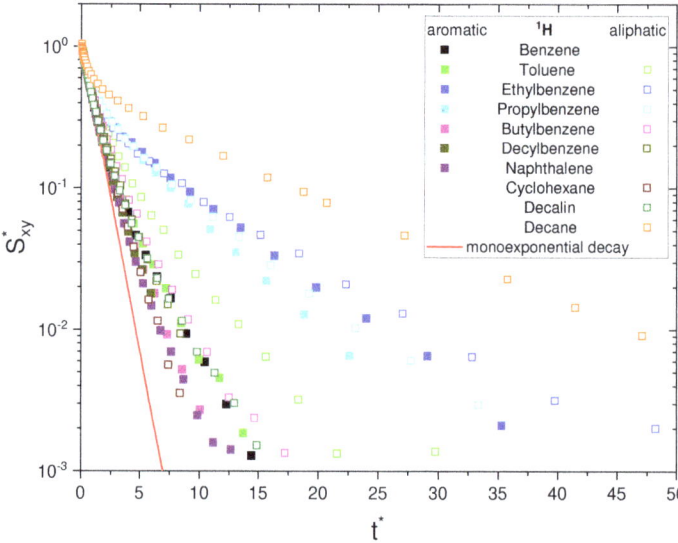

Figure 2. Longitudinal (**a**) and transverse (**b**) relaxation decays of the maltene protons in the asphaltenic solution, consisting of 95 vol% benzene-d_6 and 5 vol% of the investigated maltene with additional 150 g/L asphaltene. For a qualitative comparison, the time axis is rescaled (see Equation (12)) to reflect multiples of $\overline{R}_{1,2}^{-1}$, and the signal is normalized to unity at $t^* = 0$ (see Equations (13) and (14)). The needed parameters were obtained from a monoexponential ($\triangledown$) or biexponential ($\square$) fit.

(**a**) Longitudinal relaxation (0 g/L).

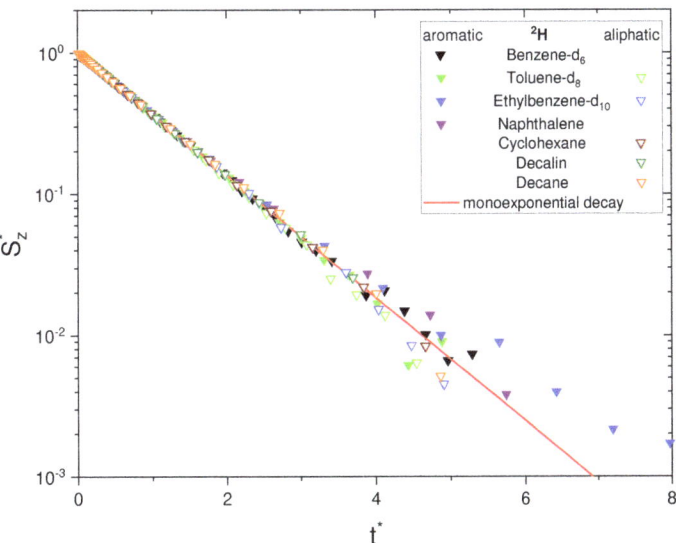

(**b**) Transverse relaxation (0 g/L).

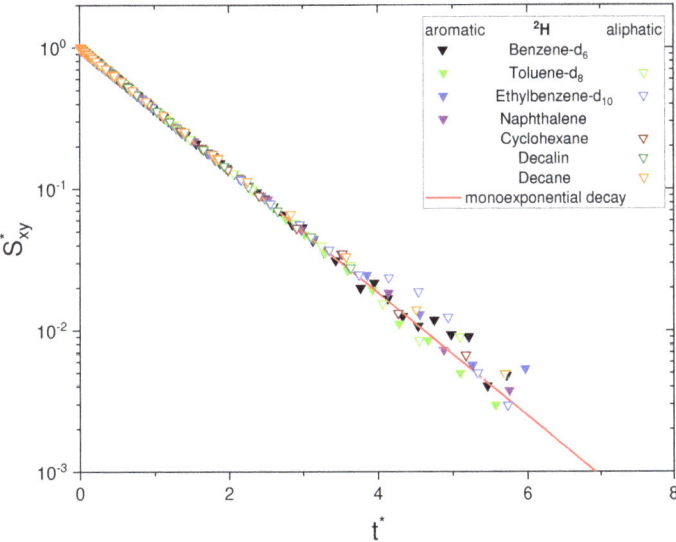

Figure 3. Longitudinal (**a**) and transverse (**b**) relaxation decays of the maltene deuterons in the asphaltene-free solution, consisting of 95 vol% benzene and 5 vol% of the investigated deuterated maltene. For a qualitative comparison, the time axis is rescaled (see Equation (12)) to reflect multiples of $\overline{R}_{1,2}^{-1}$, and the signal is normalized to unity at $t^* = 0$ (see Equations (13) and (14)). The needed parameters were obtained from a monoexponential ($\triangledown$) fit.

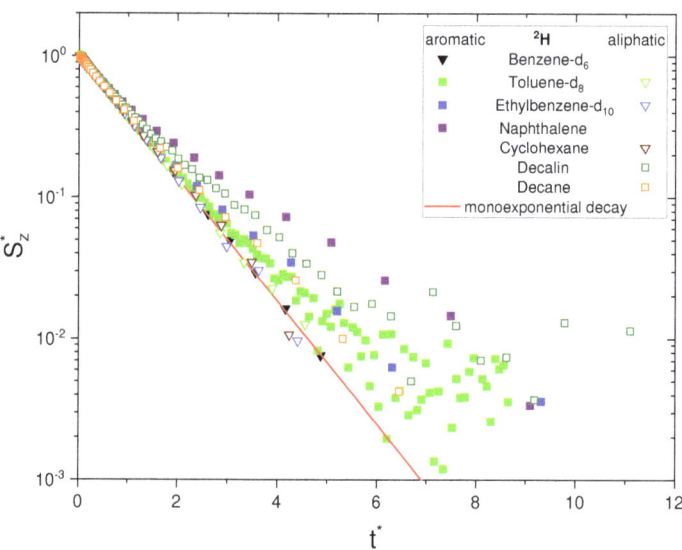

(**a**) Longitudinal relaxation (150 g/L).

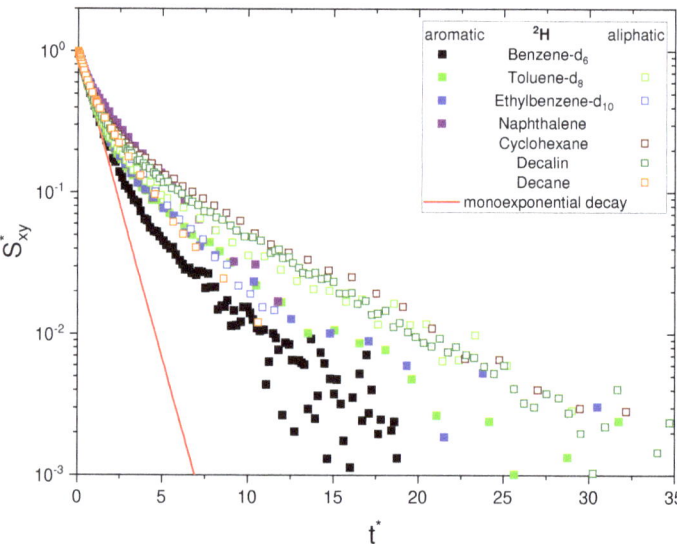

(**b**) Transverse relaxation (150 g/L).

Figure 4. Longitudinal (**a**) and transverse (**b**) relaxation decays of the maltene deuterons in the asphaltenic solution, consisting of 95 vol% benzene and 5 vol% of the investigated deuterated maltene with additional 150 g/L asphaltene. For a qualitative comparison, the time axis is rescaled (see Equation (12)) to reflect multiples of $\overline{R}_{1,2}^{-1}$ and the signal is normalized to unity at $t^* = 0$ (see Equations (13) and (14)). The needed parameters were obtained from a monoexponential (▽) or biexponential (□) fit.

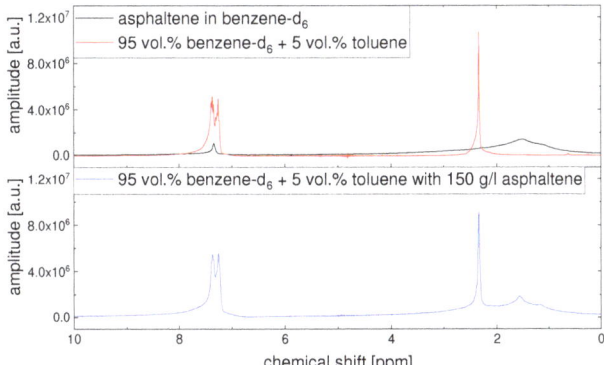

Figure 5. Example for the signal overlap of maltene and asphaltene protons. Top: ^{1}H-NMR spectra of the asphaltene in benzene-d$_6$ and the asphaltene-free toluene solution. Bottom: ^{1}H-NMR spectrum of the asphaltenic toluene solution. The ppm scale of the red and blue spectra is calibrated to the toluene CH$_3$ signal at 2.34 ppm. Since the aromatic signal in the black spectrum results not only from the asphaltene, but also from non-deuterated benzene molecules (see Section 4.2), the ppm value of this signal is set to 7.34 ppm. The amplitudes were normalized by the number of scans used to acquire the spectrum.

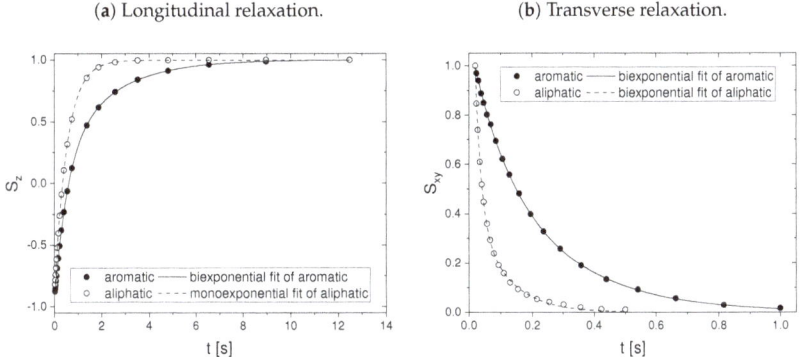

Figure 6. Longitudinal and transverse relaxation curves of the aromatic and aliphatic protons in a solution consisting of 150 g/L asphaltene in benzene-d$_6$.

Except for the longitudinal relaxation of the aliphatic protons, the proton relaxation is characterized by two time constants. The signal of the aromatic asphaltene protons overlaps with the signal of the remaining non-deuterated benzene molecules. Regarding the deuteration degree of 99.6 atom%, the non-deuterated benzene-d$_6$ protons correspond approximately to about 15% of the total number of aromatic protons in the asphaltene-benzene-d$_6$ solution. The longer relaxation time constant of the aromatic protons can therefore not be fully attributed to the smaller, more mobile non-deuterated benzene molecules. The relaxation time constants of the aromatic protons reflect rather mainly the dynamic of the aromatic asphaltene protons and to a minor but not negligible extent the dynamic of the non-deuterated benzene molecules. In contrast, the relaxation of the aliphatic protons reflects purely the relaxation of aliphatic protons in the asphaltene molecule.

Note the similar time constant components ($T_1 \approx 550$ ms and $T_2 \approx 95$ ms) of the relaxation of the aromatic and aliphatic protons.

4.2.1. Protonic Maltenes

In Figures 7–9 the relaxation time constants of the maltenes' proton relaxation are displayed, as well as the relative weight of the two exponential functions used for the biexponential fit of the maltenes' relaxation in presence of the asphaltene. The relaxation time constants of the protons' relaxation in the asphaltene-benzene-d_6 solution are indicated by a black line, and a gray area indicates the error margins. The maltenes are divided into two subsets differentiating between maltenes with an aliphatic chain and purely cyclic maltenes. Benzene is also included in the first subset as aromatic side-chain free reference maltene.

In absence of the asphaltene, T_1 and T_2 are similar for all maltenes except for cyclohexane and decalin. The remarkably short T_2 of cyclohexane is, as well as the short T_2 component of decalin, a consequence of an interconversion process present in the cis isomers (see Appendix D). Furthermore, the shorter time constants are found for the larger, less mobile maltenes.

In presence of the asphaltene, two time constants are required to describe the proton relaxation sufficiently, except for the longitudinal relaxation of benzene, toluene, and naphthalene.

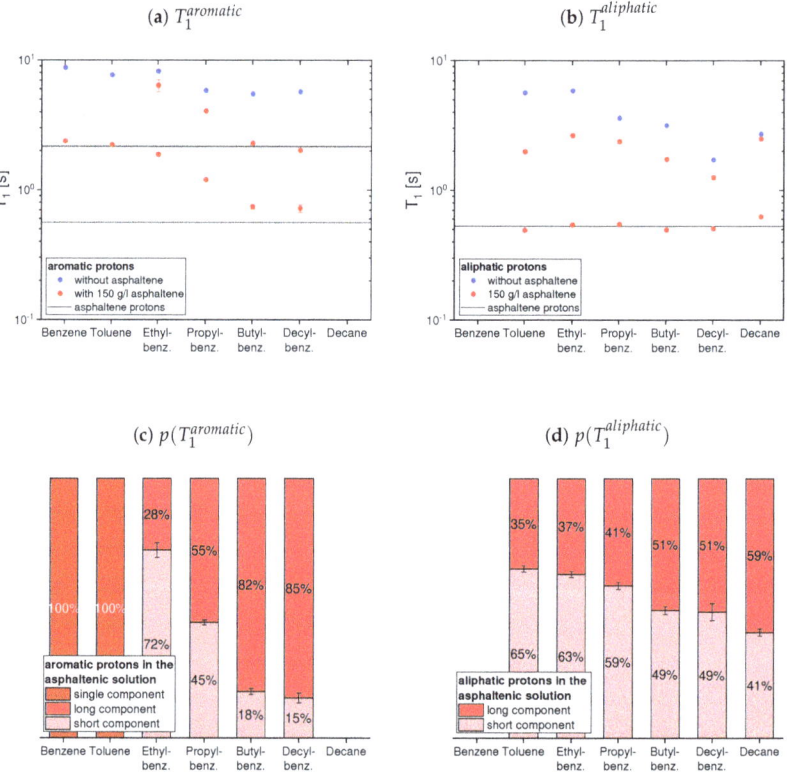

Figure 7. Top: Longitudinal relaxation time constants T_1 of the aromatic (**a**) and aliphatic (**b**) protons in the asphaltene-free and the asphaltenic solution. The black lines indicate the T_1 constants of the asphaltene protons in a benzene-d_6 solution. Bottom: Relative weight of the T_1 components of the aromatic (**c**) and aliphatic (**d**) protons in the asphaltenic solution. The x-axis distinguishes the different solutions, consisting of 5 vol% of the specified maltene and 95 vol% benzene-d_6 without or with 150 g/L asphaltene.

Regarding the relaxation of the first subset of maltenes in presence of the asphaltenes, the short T_1 component of the aliphatic protons, as well as the short T_2 component of both, aromatic and aliphatic protons, are quite constant to slightly increasing with the length of the side chain. The values of these short components are close to a relaxation time constant found for the asphaltene in benzene-d_6. For the short T_1 component of the aromatic protons, a decrease with the increase of the length of the side chain, from the longer T_1 to the shorter T_1 of the asphaltene protons in benzene-d_6, is found. The single T_1 constants of benzene and toluene fit in the trend of the short T_1 components of the remaining alkylbenzenes' aromatic protons and differ little from the long asphaltene T_1.

The long T_1 component of the aromatic protons decreases similar to the short component, approaching the long T_1 of the asphaltenes for butyl- and decylbenzene. After the long T_1 component of the aliphatic protons increases from toluene to ethylbenzene, it decreases with the increase of the length of the side chain.

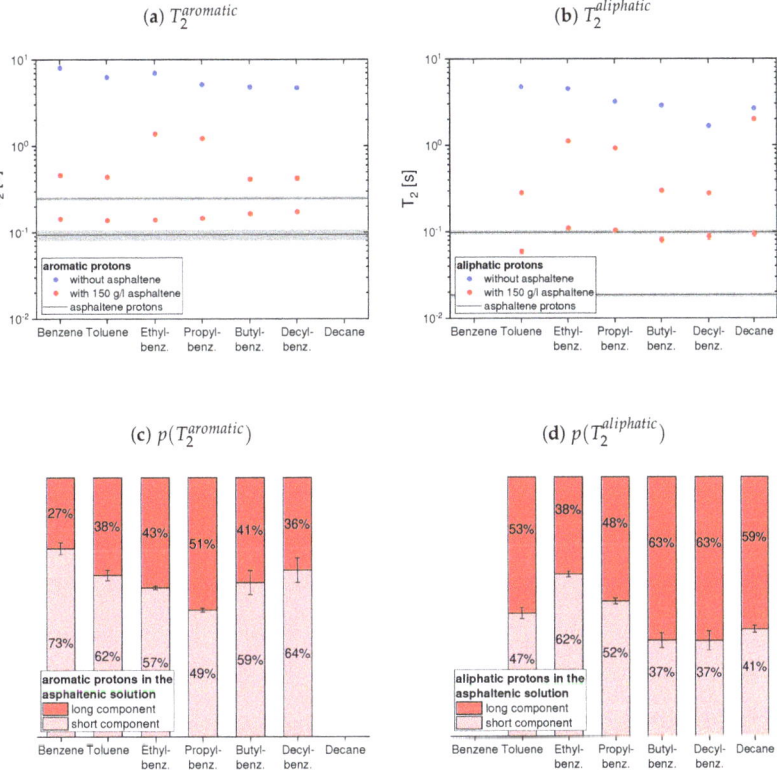

Figure 8. Top: Transverse relaxation time constants T_2 of the aromatic (**a**) and aliphatic (**b**) protons in the asphaltene-free and the asphaltenic solution. The black lines indicate the T_2 constants of the asphaltene protons in a benzene-d_6 solution. Bottom: Relative weight of the T_2 components of the aromatic (**c**) and aliphatic (**d**) protons in the asphaltenic solution. The x-axis distinguishes the different solutions, consisting of 5 vol% of the specified maltene and 95 vol% benzene-d_6 without or with 150 g/L asphaltene.

Similar to the aromatic and aliphatic short T_2 components, the long T_2 components are constant for the "short-chained" maltenes benzene and toluene, as well as for the "long-chained" maltenes butyl- and decylbenzene. The long T_2 components of ethyl- and propylbenzene are also similar, but about three times longer than the long T_2 components of the other alkylbenzenes' protons. The long T_1 and T_2 component of decane is only slightly shorter than its relaxation time constant in the asphaltene-free solution and significantly larger than the corresponding components of decylbenzene, while the short components are only slightly larger.

The relative weight of the long T_1 component increases with the increase of the length of the side chain. This is more pronounced for the aromatic protons than for the aliphatic ones, while the relative weight of the T_2 components is roughly the same.

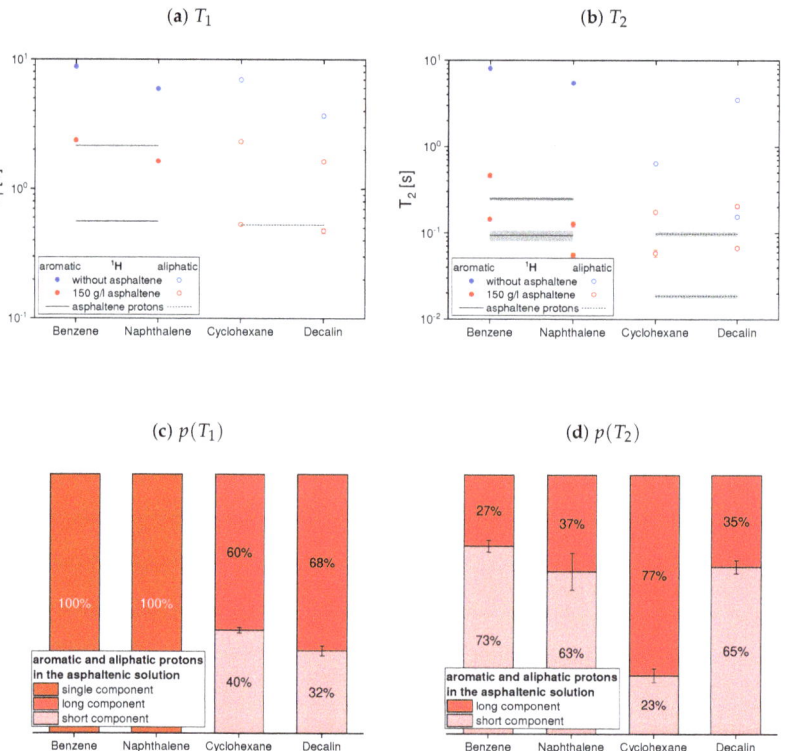

Figure 9. Top: Longitudinal (**a**) and transverse (**b**) relaxation time constants of the aromatic and aliphatic protons of cyclic maltenes in the asphaltene-free and the asphaltenic solution. The black lines indicate the T_1 and T_2 constants of the asphaltene protons in a benzene-d_6 solution. Bottom: Relative weight of the T_1 (**c**) and T_2 (**d**) components of the maltenes' protons in the asphaltenic solution. The x-axis distinguishes the different solutions, consisting of 5 vol% of the specified maltene and 95 vol% benzene-d_6 without or with 150 g/L asphaltene.

The relaxation time constants of the second maltene subset, representing aromatic and aliphatic cyclic maltenes, are presented with their relative weights in Figure 9. As mentioned above, only the longitudinal proton relaxation of the aromatic cyclic maltenes is described by a single time constant. Like in the asphaltene-free solution, the shorter time constants are found for the bicyclic maltene, except for the transverse relaxation of the aliphatic cyclic maltenes, where the T_2 components are similar.

The long T_1 component of the aliphatic cyclic maltenes is similar to the single T_1 of the aromatic ones, which are close to the long T_1 of the aromatic asphaltene protons in benzene-d_6, while for the aliphatic cyclic maltenes the short T_1 component is close to the T_1 of the aliphatic asphaltene protons. The relative weight of the T_1 components are similar, with a relative weight of about 65% for the long T_1 component.

The T_2 components of the cyclic maltenes are all close to time constants found for the transverse relaxation of aromatic and aliphatic asphaltene protons in benzene-d_6. The short T_2 component is the major component with a relative weight higher than 60% except for cyclohexane, where the long component is the major component.

4.2.2. Deuterated Maltenes

The deuteron relaxation time constants of the deuterated maltenes and their relative weights are displayed in Figures 10–12. Similar to the relaxation of the protonic maltenes, the longitudinal deuteron relaxation of some deuterated maltenes in presence of the asphaltene can still be described by a single time constant, while the transverse relaxation is always characterized by two time constants.

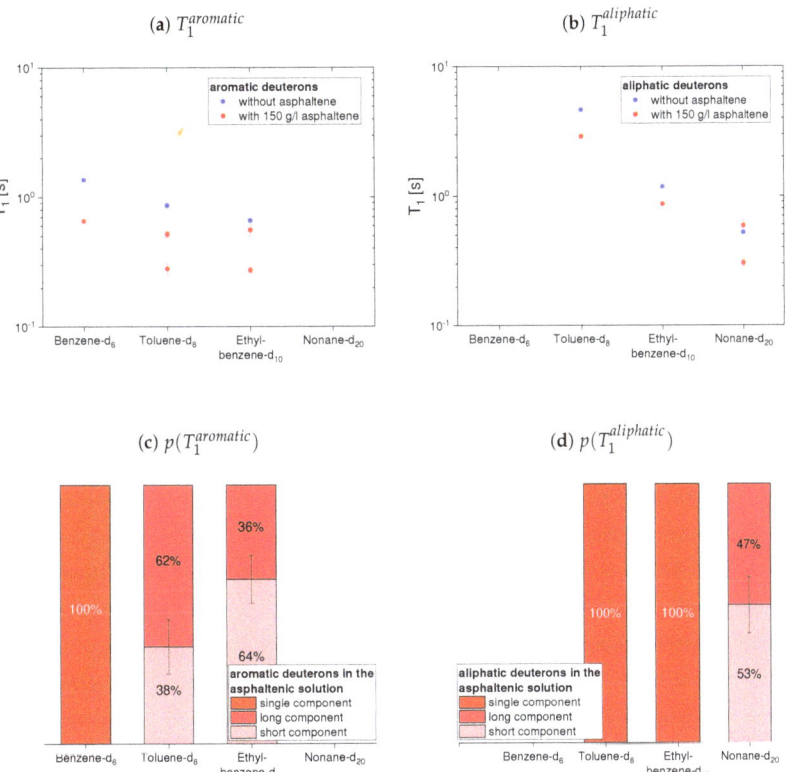

Figure 10. Top: Longitudinal relaxation time constants T_1 of the aromatic (**a**) and aliphatic (**b**) deuterons in the asphaltene-free and the asphaltenic solution. Bottom: Relative weight of the T_1 components of the aromatic (**c**) and aliphatic (**d**) deuterons in the asphaltenic solution. The x-axis distinguishes the different solutions, consisting of 5 vol% of the specified maltene and 95 vol% benzene without or with 150 g/L asphaltene.

The longitudinal and transverse relaxation time constants obtained for the aromatic deuterons in presence of the asphaltene do not differ much between benzene-d_6, toluene-d_8 and ethylbenzene-d_{10}. In contrast, a decrease of the time constants is observed for the aliphatic deuterons from toluene-d_8 via ethylbenzene-d_{10} to nonane-d_{20}. The relative weights of the two T_1 components are roughly equal, while for T_2 often the short component dominates.

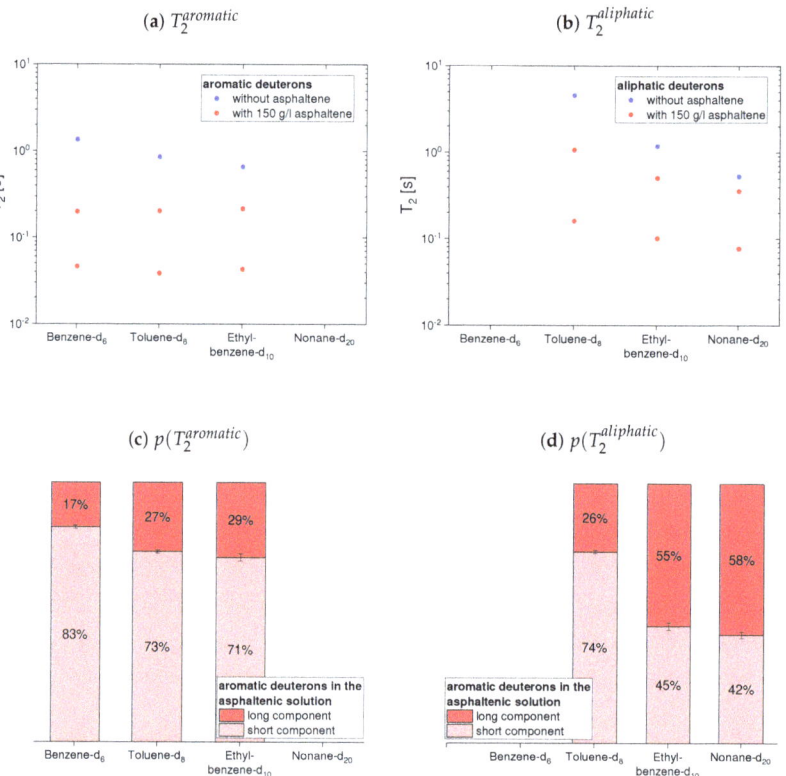

Figure 11. Top: Transverse relaxation time constants T_2 of the aromatic (**a**) and aliphatic (**b**) deuterons in the asphaltene-free and the asphaltenic solution. Bottom: Relative weight of the T_2 components of the aromatic (**c**) and aliphatic (**d**) deuterons in the asphaltenic solution. The x-axis distinguishes the different solutions, consisting of 5 vol% of the specified maltene and 95 vol% benzene without or with 150 g/L asphaltene.

In the case of the cyclic maltenes, only the longitudinal deuteron relaxation of benzene-d_6 and cyclohexane-d_{12} can be described by a single T_1. In the presence and absence of asphaltenes, the shorter time constants are mainly found for the bicyclic maltenes. Merely the long T_2 component of the aromatic cyclic maltenes does not follow this trend, as the long T_2 component of benzene-d_6 is slightly shorter than the corresponding T_2 component of naphthalene-d_8. Similar to the proton relaxation, the long T_1 component exhibits a relative weight of roughly 65%, while, except for naphthalene-d_8, a higher relative weight is found for the short T_2 component.

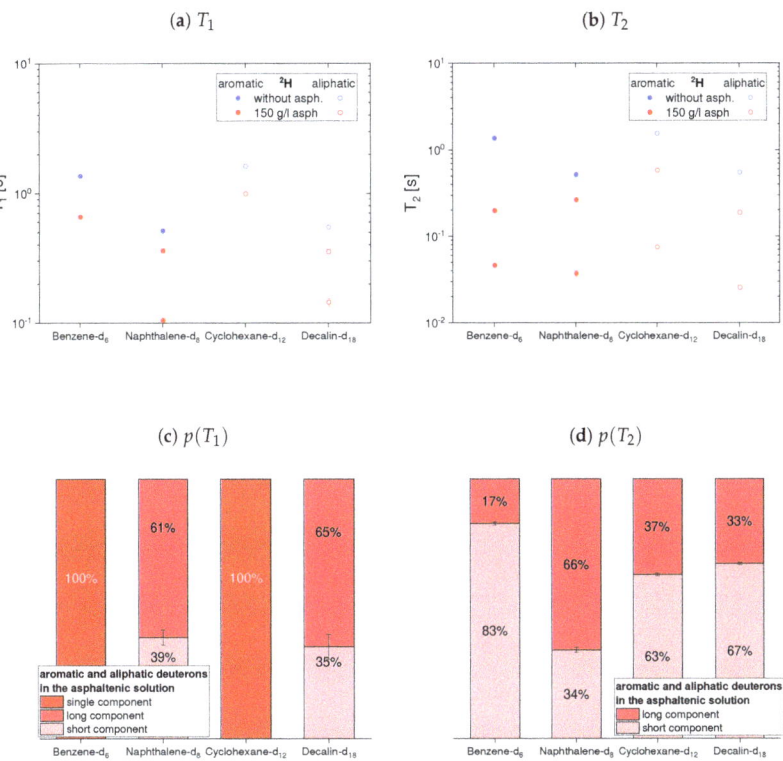

Figure 12. Top: Longitudinal (**a**) and transverse (**b**) relaxation time constants of the aromatic and aliphatic deuterons of cyclic maltenes in the asphaltene-free and the asphaltenic solution. Bottom: Relative weight of the T_1 (**c**) and T_2 (**d**) components of the maltenes' deuterons in the asphaltenic solution. The x-axis distinguishes the different solutions, consisting of 5 vol% of the specified maltene and 95 vol% benzene without or with 150 g/L asphaltene.

4.3. Mean Relaxation Rate Constants

The calculated mean relaxation rate constants (Equation (10)) for the protonic and the deuterated maltenes can be found in the Appendix C in Tables A2 and A3, respectively. In Figures 13 and 14, two different normalized mean relaxation rate constants are presented for the two maltene subsets. In the top panel of Figure 13, the maltenes mean relaxation rate constants ($\overline{R}_1$ and $\overline{R}_2$) in the asphaltenic solution are divided by the corresponding relaxation rate constants in the asphaltene-free solution to visualize the impact of the asphaltene on the maltenes rate constant describing the relaxation. The $\overline{R}_2/\overline{R}_1$ ratio of the relaxing nuclei in the different solutions is displayed in the bottom panel of Figure 13. For comparison, the diagrams contain the normalized mean relaxation rate constants of both the protonic and the deuterated maltenes, while aromatic and aliphatic nuclei are presented side by side in different plots.

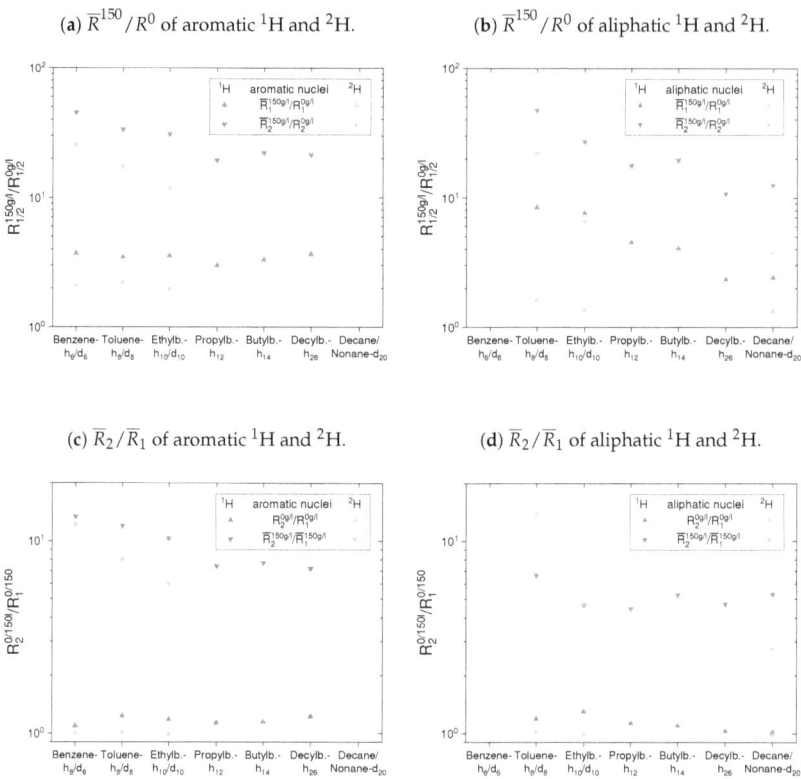

Figure 13. Top: Mean relaxation rates of the aromatic (**a**) and aliphatic (**b**) maltenes' hydrogen isotopes in the presence of the asphaltene normalized to the corresponding relaxation rates in the asphaltene-free solution. Bottom: $\overline{R}_2/\overline{R}_1$ ratio of the investigated aromatic (**c**) and aliphatic (**d**) maltenes' nuclei in the asphaltenic and asphaltene-free solution.

The maltenes $\overline{R}^{150}/R^0$ ratios in Figures 13a,b and 14a are always larger than one, i.e., the relaxation of the maltenes is shortened by the presence of asphaltene. The higher $\overline{R}^{150}/R^0$ ratios can be found for the transverse relaxation, which results in $\overline{R}_2/\overline{R}_1$ ratios, for both ^{1}H and ^{2}H, significantly larger than one in the asphaltenic solution. Furthermore, the $\overline{R}^{150}/R^0$ ratios are higher for the proton relaxation than for the deuteron relaxation. These observations are not true for cyclohexane and decalin in Figure 14 due to their already shortened transverse relaxation in absence of the asphaltene (see Appendix D).

For the first subset, the normalized proton and deuteron mean relaxation rate constants show a similar trend. The $\overline{R}^{150}/R^0$ ratio of the aromatic hydrogen isotopes is quite constant for the longitudinal relaxation, but shows a decrease from benzene-h_6/d_6 to ethylbenzene-h_{10}/d_{10} for the transverse relaxation. With a further increase of the length of the aliphatic side chain, the $\overline{R}_2^{150}/R_2^0$ ratio of the aromatic ring protons does not change any further. The aliphatic chain protons, however, show for both, the longitudinal and transverse relaxation a decrease of the $\overline{R}^{150}/R^0$ ratio with an increase of the side chain length. In principle, this is also seen for the chain deuterons, whereas the decrease is strongly damped in the case of the longitudinal mean relaxation rate constants. Note further, that the decrease of $\overline{R}_2^{150}/R_2^0$ from toluene-d_8 to ethylbenzene-d_{10} is more pronounced than in the case of the protonic counterparts.

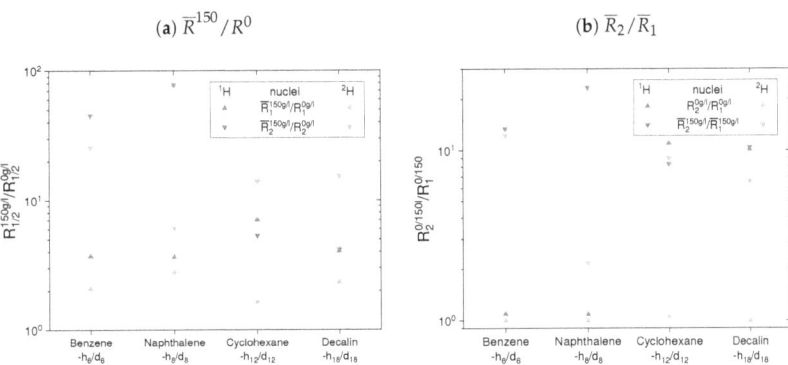

Figure 14. (a) Mean relaxation rate constants of the cyclic maltenes' hydrogen isotopes in the asphaltenic solution normalized by the corresponding relaxation rate constants in the asphaltene-free solution, and (b) $\overline{R}_2/\overline{R}_1$ ratio of the investigated cyclic maltenes' nuclei in both solutions.

The $\overline{R}_2/\overline{R}_1$ ratios show then a similar trend for the aromatic and aliphatic nuclei. In the asphaltene-free solution, the R_2/R_1 ratio differs only little from unity, slightly more for the protons than for the deuterons. In the asphaltenic solution, the maltene $\overline{R}_2/\overline{R}_1$ ratios are noticeably increased, and the highest ratio is found for benzene-h_6/d_6 and toluene-h_8/d_8, respectively. Afterward the $\overline{R}_2/\overline{R}_1$ ratio decreases for both, aromatic and aliphatic nuclei, until a plateau is reached since the $\overline{R}_2/\overline{R}_1$ ratios of butyl- and decylbenzene differ little. Furthermore, the $\overline{R}_2/\overline{R}_1$ ratio of decane's protons fits very well in this plateau. In contrast, a visible difference exists between the $\overline{R}_2/\overline{R}_1$ ratio of the deuterons of ethylbenzene-d_{10} and nonane-d_{20}. Furthermore, the decrease of the $\overline{R}_2/\overline{R}_1$ ratio between toluene-d_8 and ethylbenzene-d_{10} is more pronounced and their $\overline{R}_2/\overline{R}_1$ ratios are, in the first subset, the only ones larger than the ratios of their protonic versions.

In the second subset in Figure 14 the normalized mean relaxation rate constants of the aliphatic cyclic maltene protons differ from the trend seen so far, as the shortening of their transverse relaxation in presence of the asphaltene is, due to their already shortened relaxation in the asphaltene-free solution, similar to the shortening of the longitudinal relaxation. Their $\overline{R}_2/\overline{R}_1$ ratio in the asphaltenic solution and the asphaltene-free solution are thus almost identical. Although the comparison with the aromatic cyclic protonic maltenes is therefore limited, the $\overline{R}_2/\overline{R}_1$ ratios of cyclohexane and decalin in the asphaltenic solution are of comparable size, as the corresponding ratio of benzene. Furthermore, are the $\overline{R}_2/\overline{R}_1$ ratios of the deuterated cyclic aliphatic maltenes comparable/slightly smaller than the ratios of the protonic counterparts, which is in agreement with the overall relation of proton and deuteron $\overline{R}_2/\overline{R}_1$ observed for the first subset.

From this relation differs the $\overline{R}_2/\overline{R}_1$ ratios of naphthalene and naphthalene-d_8, with a deuteron $\overline{R}_2/\overline{R}_1$ ratio more than ten times smaller than the proton ratio. This results from a significantly lower shortening of the transverse relaxation of naphthalene-d_8 compared with naphthalene, while the shortening of the longitudinal relaxation is similar. As a result, the highest and lowest $\overline{R}_2/\overline{R}_1$ ratio of all maltenes is found for naphthalene and naphthalene-d_8, respectively.

Therefore, it is difficult to identify a trend for the influence of the asphaltene presence on the cyclic maltenes. Regarding their size, the higher proton $\overline{R}_2/\overline{R}_1$ ratio is found for the bicyclic maltenes, while in the case of the deuterated maltenes the bicyclic maltenes exhibit the lower $\overline{R}_2/\overline{R}_1$ ratio compared to the monocyclic maltenes. In terms of aromaticity, a slightly higher $\overline{R}_2/\overline{R}_1$ ratio is found for benzene-h_6/d_6 compared to the aliphatic cyclic maltenes (^{1}H and ^{2}H). However, the ratio of naphthalene-h_8/d_8 is, as mentioned before, significantly higher (^{1}H) or lower (^{2}H) than the ratios of the remaining cyclic maltene.

4.4. Diffusion

The, compared to the pulse spacing in the PGSTE sequence, long relaxation time constants of the asphaltene protons (see Table 1) cause only a small attenuation of the asphaltene signal during the diffusion measurement. Therefore, the contribution of the asphaltene protons to the obtained PGSTE signal cannot be neglected in determining the diffusion coefficients of the maltenes in the asphaltenic solution. Accordingly, the obtained proton PGSTE decays differ from a monoexponential decay in the asphaltenic solution (see Figure 15). However, the variation of the gradient strength was adjusted to obtain the diffusion coefficients of the maltenes and is therefore not sufficient to reliably fit a sum of two exponential functions to additionally extract the asphaltene's diffusion coefficient. The asphaltene's diffusion coefficient was therefore determined in the solution of 150 g/L asphaltene in benzene-d_6 and the obtained signal is displayed in Figure 16 for the aromatic (a) and the aliphatic (b) protons.

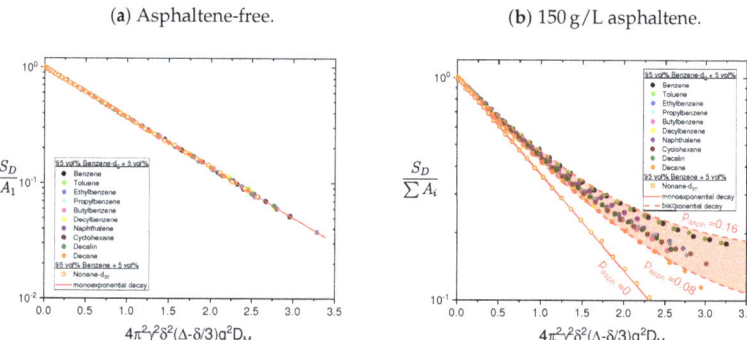

Figure 15. PGSTE signal decay of the protonic maltenes and nonane-d_{20} in the (**a**) asphaltene-free and the (**b**) asphaltenic solution. The solid line represents a monoexponential decay, while the area between the dashed lines corresponds to biexponential decays with different relative contributions $p_{asph.}$ of the asphaltene to the decay. The diffusion coefficient of the asphaltene was fixed at $D_{asph.} = 3.6 \times 10^{-11}$ m^2/s. To obtain the signal decay of nonane-d_{20} the gradient strength was varied up to 5 T/m.

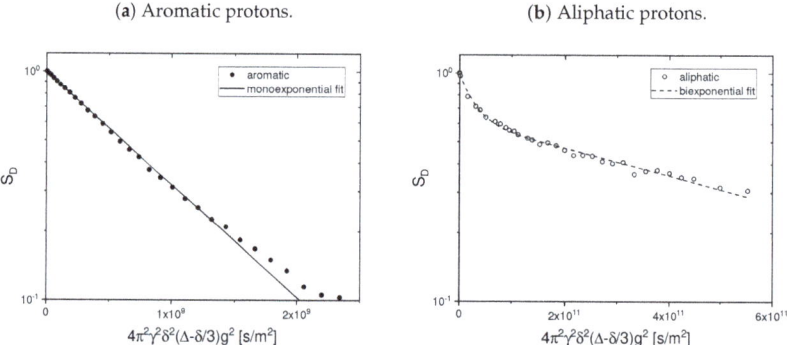

Figure 16. PGSTE signal decay of the aromatic (**a**) and aliphatic (**b**) protons in a solution of 150 g/L asphaltene in benzene-d_6. To obtain the signal decay of the aliphatic protons, the gradient strength was varied up to 10 T/m.

As already seen in the determination of the relaxation time constants (see Figure 6), the signal overlap of aromatic asphaltene protons and protons of remaining non-deuterated benzene molecules results in a superposition of the individual signal decays. By comparing the PGSTE signal decay of the aromatic and aliphatic protons in Figure 16, one notices the significantly faster decay of the aromatic protons signal. Since asphaltenes are, due to their definition, a heterogeneous class of molecules, a distribution of diffusion coefficients is not a surprising observation. However, considering the molecular weight range of asphaltenes as the heaviest part of crude oil, a diffusion coefficient of the order of 10^{-9} m^2/s, as obtained by a single-exponential fit of the aromatic protons signal decay (see Table 2), is not expected and can therefore be mainly attributed to the non-deuterated benzene molecules with high probability (For comparison, the self-diffusion coefficient of the remaining non-deuterated benzene molecules in a sample consisting of 100% benzene-d$_6$ was determined to $(1.842 \pm 0.007) \times 10^{-9}$ m^2/s). The asphaltene diffusion coefficients are therefore determined from the signal decay of the aliphatic protons, whose signal does not overlap with the signal of non-deuterated solvent molecules. The signal decay could be fitted sufficiently well by a sum of two exponential functions (see Table 2) revealing diffusion coefficients of $(3.6 \pm 0.3) \times 10^{-11}$ m^2/s and $(1.38 \pm 0.07) \times 10^{-12}$ m^2/s, respectively. These values seem reasonable in comparison with the self-diffusion coefficients of about 1×10^{-10} m^2/s found for asphaltenes in solutions at concentrations not higher than 30 g/L [50–52].

Table 2. Fit parameters used to fit the in Figure 16 shown PGSTE signal decays of aromatic and aliphatic protons in a solution of 150 g/L asphaltene in benzene-d$_6$, using Equation (9).

	Diffusion		
Parameter	Aromatic	Aliphatic	
A_1	0.993 ± 0.005	0.375 ± 0.012	38%
D_1 [m^2/s]	$(1.13 \pm 0.02) \times 10^{-9}$	$(3.6 \pm 0.3) \times 10^{-11}$	
A_2		0.617 ± 0.011	62%
D_2 [m^2/s]		$(1.38 \pm 0.07) \times 10^{-12}$	

For the gradient strengths used to determine the protonic maltenes diffusion coefficients in the asphaltenic solution, only the faster diffusion of the asphaltene has to be taken into account. The signal decays of the maltenes in the asphaltenic solution are therefore fitted with a biexponential fit according to Equation (9), with a fixed second diffusion coefficient $D_2 = 3.6 \times 10^{-11}$ m^2/s.

The signal decays of the maltenes in the absence and presence of the asphaltene are displayed in Figure 15. For comparison, the signal decays are normalized like the relaxation decays in Section 4.1, i.e., the signal is normalized to the sum of the two exponential functions amplitude and the x-axis is rescaled with the obtained maltene diffusion coefficient D_M.

In absence of the asphaltene, the decay is perfectly described by a single-exponential function, whereas in presence of the asphaltene the additional diffusion of the asphaltenes is seen in a deviation from a monoexponential decay. The relative contribution of the asphaltene to the fit $p_{\text{asph.}}$ is calculated according to Equation (11) and varies between 8% and 16%. The monoexponential decay of the deuteron signal of nonane-d$_{20}$ in the presence of the asphaltene supports the assumption that the deviation from a monoexponential decay of the protonic maltenes reflects the diffusion of the asphaltene rather than representing different diffusion coefficients of the maltenes.

The diffusion coefficients of the maltenes are listed in Table 3 together with the $D^{150g/l}/D^{0g/l}$ ratio. The ratio reflects to what extent the diffusion coefficient, and thus the translational motion of the maltenes, is diminished by the presence of the asphaltene.

The diffusion coefficients of the maltenes were mostly decreased by roughly 30% to 35% in the presence of the asphaltenes. With the minimal decrease of 20%, decane differs the most and is the maltene with the least impairment of its translational motion. This decrease is notably weaker than the similar long aliphatic chain molecule nonane-d_{20} and the phenyl ring bearing decylbenzene, which is the most restricted alkylbenzene. With a decreasing side chain length, the ratio increases to a maximum for toluene, which reflects its smaller size. The cyclic hydrocarbons are similarly affected by the presence of the asphaltene, whereby the difference between the aromatic ring molecules is slightly bigger than between the aliphatic ones. However, the overall differences between the maltenes are rather small, and no pronounced effect of the asphaltene on the translational motion of any of these molecules is observed.

Table 3. Diffusion coefficients of the maltenes in absence and presence of the asphaltene. The $D^{150g/l}/D^{0g/l}$ ratio reflects the impairment of the translational motion of the maltenes due to the presence of the asphaltene.

Maltene	$D^{0g/l}$ [10^{-9} m^2/s]	$D^{150g/l}$ [10^{-9} m^2/s]	$\dfrac{D^{150g/l}}{D^{0g/l}}$
Decalin	1.393 ± 0.005	0.949 ± 0.007	0.681 ± 0.008
Cyclohexane	1.767 ± 0.007	1.182 ± 0.006	0.669 ± 0.006
Naphthalene	1.409 ± 0.005	0.903 ± 0.007	0.641 ± 0.007
Benzene	1.856 ± 0.006	1.280 ± 0.007	0.690 ± 0.006
Toluene	1.769 ± 0.008	1.256 ± 0.005	0.710 ± 0.006
Ethylbenzene	1.652 ± 0.005	1.162 ± 0.007	0.704 ± 0.006
Propylbenzene	1.511 ± 0.004	1.049 ± 0.006	0.695 ± 0.006
Butylbenzene	1.410 ± 0.005	0.956 ± 0.008	0.678 ± 0.008
Decylbenzene	0.983 ± 0.003	0.646 ± 0.005	0.657 ± 0.006
Decane	1.465 ± 0.004	1.152 ± 0.011	0.786 ± 0.009
Nonane-d_{20}	1.604 ± 0.004	1.140 ± 0.005	0.711 ± 0.005

5. Discussion

In the presence of 150 g/L asphaltene, the longitudinal and transverse relaxation of most investigated maltenes are no longer monoexponential and can be sufficiently described by a fit of two exponential functions. For the ^{1}H measurements, the signal of the asphaltene and maltene protons overlap, and the obtained relaxation decays reflect not only the motion of the maltenes, but also of the asphaltene. From the two time constants, the shorter is often similar to the relaxation time constants found for asphaltene in benzene-d_6. However, the non-monoexponential, biexponentially fittable, relaxation decays are not only observed for the protonic maltene solutions, but also for the ^{2}H relaxation of their deuterated analogs, where the asphaltenes are not visible and the non-monoexponential decays purely reflect the maltenes dynamic.

In the current picture of the maltene–asphaltene interaction, the two relaxation time constants observed in the asphaltenic solutions could represent different relaxation environments of the maltenes, which are caused by the proximity of the maltenes to the asphaltenes. The dynamic of maltenes in the intermediate space between the asphaltene clusters is mainly affected by the viscosity of the solution, but not the direct interaction with the asphaltene structures. The longer relaxation time constant could therefore be attributed to these bulk molecules. In contrast, the molecules in close proximity to the asphaltene macroaggregates experience an additional relaxation mechanism due to paramagnetic species VO^{2+} in the asphaltenes, and a stronger restriction of their motion, due to steric hindrances or additional interactions such as π-stacking. These molecules then exhibit a shorter relaxation time than the bulk molecules farther apart from the asphaltene clusters.

The border between these two environments is not expected to be sharp, but rather a more or less uniform transition, so that a distribution of relaxation times could be expected.

For further comparison, the mean relaxation rate constants are constructed from the $T_{1,2}$ components and the corresponding relative weights. The $\overline{R}_{1,2}$ condense the relaxation of the investigated nuclei into a single parameter and ease the general comparison of the different maltene solutions. In presence of the asphaltene, the relaxations of the investigated nuclei become faster. Thereby, the transverse relaxation is more affected than the longitudinal relaxation, which was already observed before for a mixture of maltenes with a varying asphaltene concentration [19]. As mentioned above, the shortening of the maltene's relaxation can be expected due to the presence of paramagnetic species and a slowed maltene's motion, resulting from the contact and entanglement between the maltenes and the asphaltene macrostructures. The, compared to R_1, stronger increase of R_2 indicates the presence of motions with correlation times comparable or slower than the inverse of the Larmor frequency, and thus the latter effect is significant since the presence of paramagnetic species influences the longitudinal and transverse relaxation equally. As a consequence, the $\overline{R}_2/\overline{R}_1$ ratios, which reflect the strength of the impairment, are significantly larger than unity in the asphaltenic solution, while in the asphaltene-free solution $\overline{R}_2/\overline{R}_1 \approx 1$ indicates the extreme narrowing limit.

As a result of an internal motion of cyclohexane and decalin (see Appendix D), the $\overline{R}_2/\overline{R}_1$ ratios in the asphaltene-free solution are larger than unity and differ little from the ratios in the asphaltenic solution. The comparability with the other maltenes is therefore limited, since the impact of the asphaltene presence on the dynamic is not clear. Furthermore, the theory and models developed so far have assumed rigid maltenes without considering internal motions, especially not ring flips, which can result in totally different behavior. This then needs to be considered in future oil studies, in case a significant proportion of the maltenes in crude oil exhibit similar internal motions.

For the deuterated maltenes the shortening of T_2 due to the internal motion is negligible and $\overline{R}_2/\overline{R}_1$ ratios close to and significantly larger than unity are observed in the asphaltene-free and the asphaltenic solution, respectively. Comparing the bicyclic ring system with the monocyclic, the lower $\overline{R}_2/\overline{R}_1$ ratio is found for the bicyclic one. However, the decrease from the monocyclic to the bicyclic maltene is much stronger for the aromatic ring systems than for the aliphatic ones. In contrast, the ratio of the protonic aromatic bicyclic maltene (naphthalene) is larger than the ratio of benzene. As a result, the $\overline{R}_2/\overline{R}_1$ ratios of benzene and benzene-d_6 differ little, while for naphthalene and naphthalene-d_8, the ratios differ the most among all investigated maltenes. However, the results from the protonic and the deuterated maltene solutions should be compared with caution, since one compares the maltene–asphaltene dynamic (^{1}H) with the maltene dynamic (^{2}H) and in addition different relaxation mechanisms. While the ^{1}H relaxation is caused by the rotational and translational motion of the molecule (as well as the presence of free electrons), the ^{2}H relaxation results mainly from the rotational motion of the molecule (and the presence of free electrons).

Regarding the influence of the side chain length of the maltene, the observed $\overline{R}_2/\overline{R}_1$ ratio decreases with the increase of the side chain length, probably approaching a plateau for long side chains, since the ratios of butylbenzene (#$C_{aliph.}$ = 4) and decylbenzene (#$C_{aliph.}$ = 10) differ little. The ratios of decane and decylbenzene are similar, indicating that the phenyl ring is of minor importance for the restrictions the aliphatic chain experiences in presence of the asphaltene. The initial decrease of the $\overline{R}_2/\overline{R}_1$ ratio is also observed for the deuterated variants.

However, the $\overline{R}_2/\overline{R}_1$ ratios of ethyl- (aliphatic) and propylbenzene (aromatic and aliphatic) are slightly lower than approximated from the trend, which reflects their significantly different long relaxation component, compared to the rest of the maltenes. This is especially well pronounced for the long T_2 component, as it was shown in Figure 8. Due to the lack of solutions with deuterated analogs, it cannot be verified whether this is a real characteristic of the maltene dynamic.

The self-diffusion coefficients of the maltenes (including nonane-d_{20}) in the high (150 g/L) concentrated asphaltenic solution are only reduced by 30% to 35% compared to their self-diffusion coefficients in the asphaltene-free solution. Merely decane deviates slightly from this observation since its self-diffusion coefficient is reduced by about 20%. Overall, the diffusion of the different maltenes is similar affected by the asphaltene's presence, and especially no pronounced effect is found for ethyl- and propylbenzene. The impairment of the translational motion is therefore caused by the increased viscosity of the solution due to the high asphaltene concentration. If there is a prolonged interaction time between the maltenes and asphaltenes, it is still short compared to the observed time interval in the diffusion measurements.

6. Conclusions

In this work, the relaxation and diffusion of different maltenes in the presence and absence of asphaltene were investigated. The self-diffusion is equally restricted for all maltenes in presence of the asphaltene, due to the increased viscosity of the solution. In contrast to crude oil, the relaxation time constants of asphaltene in solution are considerably long and as a consequence, the ^{1}H relaxation decays in the maltene–asphaltene solution reflect the dynamic of both maltenes and asphaltenes.

In presence of the asphaltene, the maltenes' relaxation is enhanced, due to the influence of paramagnetic species, as well as the slowdown of the maltenes' motion upon contact and entanglement with the asphaltene structures. Since R_2 is more affected than R_1, the slowdown is of significant importance. The increase of R_2 and R_1 due to the presence of asphaltene is more pronounced for ^{1}H in comparison to ^{2}H, but amounts only to an increase of typical 50%. If dipolar interaction were the sole origin of relaxation, this factor should be equivalent to the square of the ratio of the gyromagnetic constant, about 42. If intramolecular contributions were the sole origin, the ratio of relaxation rate constants with and without asphaltene should be identical for ^{2}H and ^{1}H. It is therefore reasonable to assume that relaxation of ^{1}H with the unpaired electrons of the radicals contained in the asphaltenes does contribute to ^{1}H relaxation, but is not dominating.

Overall, a higher ratio of mean relaxation rate constants is often found for the "more aromatic" maltene, i.e., the shorter chained alkylbenzene or the aromatic ring systems, indicating a possible influence of aromatic maltene–asphaltene interactions, e.g., π-stacking. Since the self-diffusion coefficients of the maltenes are all similar, a possible prolonged maltene–asphaltene interaction time due to aromatic interactions is still short compared to the studied diffusion time interval. However, a definite statement about the role of aromatic interactions cannot be made, as the remarkably small mean relaxation rate constants ratio of naphthalene-d_8 shows. Hence, more detailed studies are needed.

Most maltenes show at least two distinct relaxation time constants due to internal motions of different parts, e.g., ring and chain. In presence of the asphaltene, the relaxation further becomes non-monoexponential, especially the transverse relaxation. The individual relaxations and the averaged values R_2 and R_1 differ from one maltene to another. For some maltenes, the relaxation can be influenced significantly by conformational interconversion processes. Therefore, the distribution of either T_1 or T_2, or both (in 2D experiments) of crude oil is much more complicated than conventionally assumed, and is strongly dependent on the composition of the maltenes.

This approach serves as a further step to analyze these compositions from a thorough analysis of the data, especially in two-dimensional experiments. Further modeling and density functions theory computations will help to elucidate the picture, at the same time integrating a realistic model for asphaltene aggregate structure and radical location.

Author Contributions: Conceptualization, S.S.; methodology, S.S., C.M., B.G. and K.L.; investigation, K.L.; data curation, K.L.; writing—original draft preparation, K.L.; writing—review and editing, S.S., C.M. and B.G.; visualization, K.L.; supervision, C.M., B.G. and S.S.; project administration, S.S.; funding acquisition, S.S. All authors have read and agreed to the published version of the manuscript.

Funding: This research was funded in part by Deutsche Forschungsgemeinschaft (DFG) under project number STA 511/15-1 and 15-2.

Data Availability Statement: The data presented in this study are openly available in FigShare at [10.6084/m9.figshare.15082656].

Acknowledgments: The authors thank Schlumberger Doll Research for providing the asphaltene sample and Kerstin Geyer for assistance with sample preparation.

Conflicts of Interest: The authors declare no conflict of interest.

Abbreviations

The following abbreviations are used in this manuscript:

NMR   nuclear magnetic resonance
EPR   electron paramagnetic resonance
IR   inversion recovery
CPMG Carr-Purcell-Meiboom-Gill

Appendix A. List of Chemicals and Suppliers

Table A1. List of chemicals and suppliers.

Provider	Chemical
Alfa Aesar	Decalin
Merck KGaA	Decane
Fluka Analytical	Ethylbenzene, Naphthalene
Sigma-Aldrich	Benzene, Propylbenzene, Butylbenzene, Decylbenzene, Ethylbenzene-d_{10}, Naphthalene-d_8
Carl Roth GmbH + Co. KG	Toluene, Cyclohexane
Cambridge Isotope Laboratories, Inc.	Benzene-d_6, Cyclohexane-d_{12}
Chemotrade Chemiehandelsgesellschaft Leipzig mbH & Co. KG	Cyclohexane-d_{12}, Toluene-d_8
Deutero GmbH	Decalin-d_{18}
Akademie der Wissenschaften der DDR, Zentralinstitut für Isotopen- und Strahlenforschung	Nonane-d_{20}

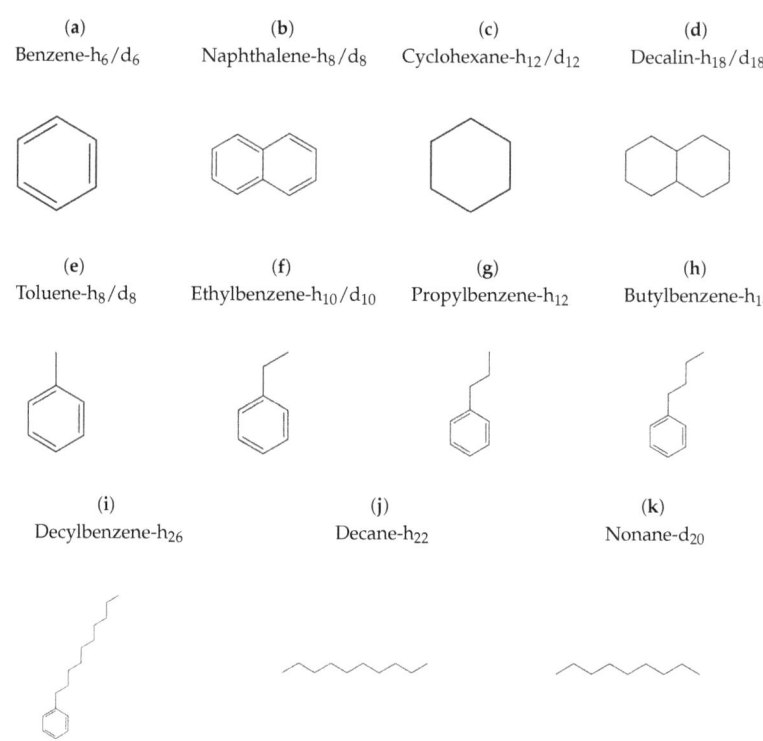

Figure A1. The chemicals used in this study include cycloalkanes (**a–d**), alkylbenzenes (**e–i**), as well as linear alkanes (**j,k**). The cycloalkanes can be distinguished in monocyclic (**a,c**) and bicyclic (**b,d**), as well as in aromatic (**a,b**) and aliphatic (**c,d**) ring systems. Protons and deuterons are not shown for clarity. The individual headings specify the used variants (^{1}H, ^{2}H). MolView was used to generate the structural formulas.

Appendix B. Spectral Resolution

(a) Longitudinal relaxation

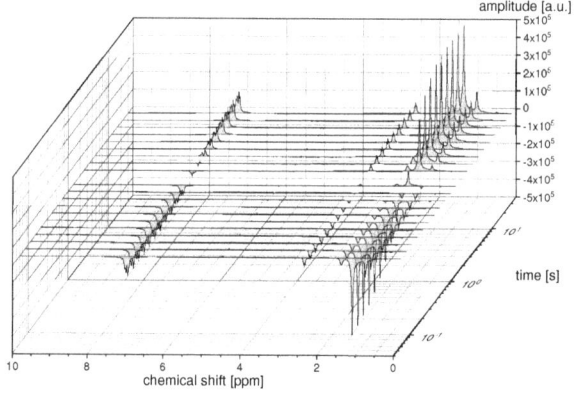

Figure A2. *Cont.*

(**b**) Transverse relaxation

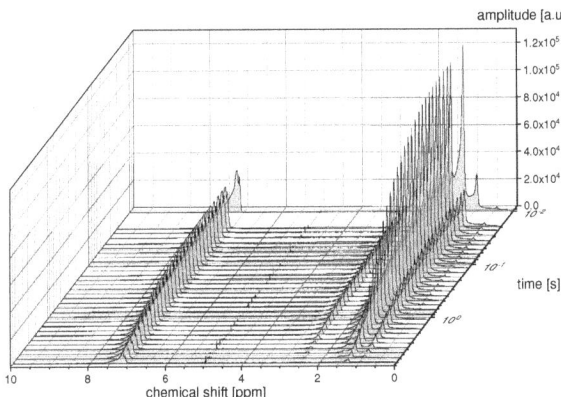

Figure A2. Longitudinal (**a**) and transverse (**b**) relaxation of decylbenzene in the asphaltene-free solution. The chemical shift axis is normalized to the dominant peak in the spectrum at 1.26 ppm, which results from the seven CH_2 groups, most apart from the phenyl ring. Due to the spectral resolution the signal from the aliphatic chain (0 ppm–3 ppm) and from the phenyl ring (6 ppm–8 ppm) are well distinguishable and the relaxation of each part can be evaluated separately.

Appendix C. Mean Relaxation Rate Constants (^{1}H and ^{2}H)

Table A2. (Mean) relaxation rate constants of the longitudinal (R_1) and transverse (R_2) relaxation of the protonic maltenes in absence and presence of the asphaltene.

Protonic Maltene	Without Asphaltene		With 150 g/L Asphaltene	
	R_1 [s^{-1}]	R_2 [s^{-1}]	$\overline{R}_1$ [s^{-1}]	$\overline{R}_2$ [s^{-1}]
Decalin	0.272(2)	2.74(8)	1.10(1)	11.4(2)
Cyclohexane	0.1437(4)	1.576(8)	1.012(9)	8.4(2)
Naphthalene	0.1686(5)	0.184(1)	0.616(9)	14.3(3)
Benzene	0.1140(3)	0.1243(8)	0.421(2)	5.64(5)
Toluene (ring)	0.1296(4)	0.156(2)	0.448(3)	5.35(5)
Toluene (CH$_3$)	0.176(1)	0.210(4)	1.492(9)	9.9(4)
Ethylbenzene (ring)	0.1210(3)	0.142(1)	0.428(4)	4.40(7)
Ethylbenzene (chain)	0.1704(7)	0.222(5)	1.30(2)	6.0(2)
Propylbenzene (ring)	0.1702(5)	0.193(1)	0.508(2)	3.76(7)
Propylbenzene (chain)	0.276(2)	0.312(4)	1.24(2)	5.5(2)
Butylbenzene (ring)	0.1813(6)	0.208(1)	0.600(4)	4.61(5)
Butylbenzene (chain)	0.315(2)	0.346(3)	1.282(9)	6.7(3)
Decylbenzene (ring)	0.174(2)	0.212(2)	0.634(4)	4.55(4)
Decylbenzene (chain)	0.581(3)	0.599(3)	1.367(8)	6.5(2)
Decane	0.3686(9)	0.375(3)	0.889(8)	4.7(4)

Table A3. (Mean) relaxation rate constants of the longitudinal (R_1) and transverse (R_2) relaxation of the deuterated maltenes in absence and presence of the asphaltene.

Deuterated Maltene	Without Asphaltene		With 150 g/L Asphaltene	
	R_1 [s^{-1}]	R_2 [s^{-1}]	$\overline{R}_1$ [s^{-1}]	$\overline{R}_2$ [s^{-1}]
Decalin-d$_{18}$	1.827(5)	1.82(2)	4.27(4)	28.0(6)
Cyclohexane-d$_{12}$	0.617(2)	0.647(4)	1.011(7)	9.1(1)
Naphthalene-d$_8$	1.953(8)	1.94(2)	5.42(7)	11.8(8)
Benzene-d$_6$	0.735(2)	0.733(2)	1.534(5)	18.8(2)
Toluene-d$_8$ (ring)	1.162(3)	1.167(4)	2.56(3)	20.4(3)
Toluene-d$_8$ (CH$_3$)	0.2162(7)	0.2192(8)	0.348(2)	4.83(4)
Ethylbenzene-d$_{10}$ (ring)	1.523(4)	0.854(4)	3.00(3)	17.9(4)
Ethylbenzene-d$_{10}$ (chain)	0.854(4)	0.847(5)	1.168(5)	5.6(2)
Nonane-d$_{20}$	1.928(7)	1.89(2)	2.56(2)	7.06(9)

Appendix D. Fast Transverse Relaxation of Cyclohexane and Decalin

A biexponential fit of the transverse relaxation decay of the decalin protons in the asphaltene-free solution results in one T_2 component similar to T_1, and one T_2 component 20 times shorter than T_1. In addition, the time constant of the transverse relaxation of the cyclohexane protons is by a factor of ten shorter than T_1. These observations contradict the expected $T_1 = T_2$ in the extreme narrowing limit. This short T_2 is a result of an internal motion present in cyclohexane and decalin, known as ring flip [53], which is the interconversion between equivalent ring shapes of cyclic conformers.

This interconversion can be described as chemical exchange between nuclei at two different sites A and B, with the respective resonance frequencies ω_A and ω_B and transverse relaxation times T_{2A} and T_{2B}. The exchange takes place at the exchange rate $k_{ex} = k_A + k_B = \tau_{ex}^{-1}$. In case of a fast exchange, a single resonance line is observed (e.g., cyclohexane at room temperature), while two lines can be observed for a slow exchange (e.g., cyclohexane at, for instance, $-77.9\,^\circ$C [54]). In the presence of a chemical exchange, the transverse relaxation time constant obtained by the CPMG pulse sequence depends on the echo time, since the dephasing rate is affected when a nucleus changes its precession frequency from ω_A to ω_B during the exchange from site A to B. If the echo time is much smaller than the exchange time, the dephasing resulting from the exchange is negligible, while in the opposite case it is maximal, and a shortened T_2 can be observed [55,56]. The transverse relaxation rate constant in the limit of $\tau_E \to \infty$ can be computed as follows:

$$\lim_{\tau_E \to \infty} R_2 = \frac{(T_{2A} + T_{2B})}{2 T_{2A} T_{2B}} + \frac{k_{ex}}{2} - \Gamma \tag{A1}$$

$$\Gamma = \Re\left[\left(\frac{(T_{2B} - T_{2A})}{2 T_{2A} T_{2B}} + \frac{k_{ex}}{2}(p_B - p_A)\right)^2 - \frac{\Delta\omega^2}{4} + k_{ex}^2 p_A p_B \right.$$
$$\left. + i\Delta\omega\left(\frac{(T_{2B} - T_{2A})}{2 T_{2A} T_{2B}} + \frac{k_{ex}}{2}(p_B - p_A)\right)\right]^{1/2}$$

with $p_A + p_B = 1$, where p_A and p_B denote the equilibrium population of spins in sites A and B. Ring flips occur in the chair conformation of cyclohexane and the *cis* conformation of decalin, but not its *trans* conformation due to steric constraints. Since

$\tau_e = 1$ ms $\gg \tau_{ex}(C_6H_{12}) \approx 14$ µs (see Table A4), the shortening of T_2 can be expected. Furthermore, the chair conformation of cyclohexane is slightly more stable than the twist-boat conformation, so that the first is the dominant conformation at room temperature [53] and only one time constant for the relaxation is observed. The biexponential decay of decalin then simply results from the coexistence of a mixture of cis and trans decalin. Since the dynamic of the ring flip does not differ much for the protonic and the deuterated variants, see, for example, the exchange rate of C_6H_{12} and C_6D_{12} in Table A4, the shortened T_2 is in principle also observable for cyclohexane-d_{12} and decalin-d_{18}. Due to the more than six times smaller gyromagnetic ratio of ^{2}H, the chemical shift difference diminishes, and the shortening of T_2 becomes negligible at the used magnetic field strength, as one can see in Table A4 by comparing the observed T_1 with the short limit of T_2.

Table A4. Estimation of the short limit of T_2 of cyclohexane-h_{12}/-d_{12} in case of $\tau_E \to \infty$, according to Equation (A1). Since the asphaltene-free solution should fulfill the extreme narrowing limit, where $T_1 = T_2$, $T_{2A} = T_{2B} \approx T_1$ is assumed for the transverse relaxation time constants of the nuclei at the different sites. The exchange rates correspond to a sample temperature of ~ 20 °C and were taken from references [57] (C_6H_{12}) and [58] (C_6D_{12}) (determined for cyclohexane-h_{12}/-d_{12} in a solution of liquid crystals). The chemical shift differences were taken from references [54] ($C_6D_{11}H$ at -100 °C) and [59] (neat plastic-crystalline C_6D_{12} at -80 °C).

Parameter	Cyclohexane	Cyclohexane-d_{12}
T_1	6.96 s	1.621 s
$p_A = p_B$	0.5	0.5
k_{ex}	72 885 s^{-1}	105 390 s^{-1}
$\Delta \omega$	0.462 ppm	0.5 ppm
	871 Hz	145 Hz
$\lim_{\tau_E \to \infty} T_2$	0.364 s	1.5 s

References

1. Mullins, O.C. Petroleomics and Structure–Function Relations of Crude Oils and Asphaltenes. In *Asphaltenes, Heavy Oils, and Petroleomics*; Mullins, O.C., Sheu, E.Y., Hammami, A., Marshall, A.G., Eds.; Springer: New York, NY, USA, 2007; pp. 1–16. [CrossRef]
2. Freed, D.E. Dependence on chain length of NMR relaxation times in mixtures of alkanes. *J. Chem. Phys.* **2007**, *126*, 174502. [CrossRef]
3. Freed, D.E. Temperature and Pressure Dependence of the Diffusion Coefficients and NMR Relaxation Times of Mixtures of Alkanes. *J. Phys. Chem. B* **2009**, *113*, 4293–4302. [CrossRef]
4. Mutina, A.R.; Hürlimann, M.D. Correlation of Transverse and Rotational Diffusion Coefficient: A Probe of Chemical Composition in Hydrocarbon Oils. *J. Phys. Chem. A* **2008**, *112*, 3291–3301. [CrossRef]
5. Benamsili, L.; Korb, J.P.; Hamon, G.; Louis-Joseph, A.; Bouyssiere, B.; Zhou, H.; Bryant, R.G. Multi-dimensional Nuclear Magnetic Resonance Characterizations of Dynamics and Saturations of Brine/Crude Oil/Mud Filtrate Mixtures Confined in Rocks: The Role of Asphaltene. *Energy Fuels* **2014**, *28*, 1629–1640. [CrossRef]
6. Gracheva, I.N.; Gafurov, M.R.; Mamin, G.V.; Biktagirov, T.B.; Rodionov, A.A.; Galukhin, A.V.; Orlinskii, S.B. ENDOR study of nitrogen hyperfine and quadrupole tensors in vanadyl porphyrins of heavy crude oil. *Magn. Reson. Solids Electron. J.* **2016**, *18*, 16102.
7. Yakubov, M.R.; Milordov, D.V.; Yakubova, S.G.; Abilova, G.R.; Sinyashin, K.O.; Tazeeva, E.G.; Borisova, U.U.; Mironov, N.A.; Morozov, V.I. Vanadium and paramagnetic vanadyl complexes content in asphaltenes of heavy oils of various productive sediments. *Pet. Sci. Technol.* **2017**, *35*, 1468–1472. [CrossRef]
8. Malhotra, V.M.; Graham, W.R. Characterization of P.R. Spring (Utah) tar sand bitumen by the EPR technique: Free radicals. *Fuel* **1983**, *62*, 1255–1264. [CrossRef]
9. Tannous, J.H.; de Klerk, A. Quantification of the Free Radical Content of Oilsands Bitumen Fractions. *Energy Fuels* **2019**, *33*, 7083–7093. [CrossRef]
10. Alili, A.S.; Siddiquee, M.N.; de Klerk, A. Origin of Free Radical Persistence in Asphaltenes: Cage Effect and Steric Protection. *Energy Fuels* **2020**, *34*, 348–359. [CrossRef]

11. Stapf, S.; Ordikhani-Seyedlar, A.; Ryan, N.; Mattea, C.; Kausik, R.; Freed, D.E.; Song, Y.Q.; Hürlimann, M.D. Probing Maltene–Asphaltene Interaction in Crude Oil by Means of NMR Relaxation. *Energy Fuels* 2014, *28*, 2395–2401. [CrossRef]
12. Stapf, S.; Ordikhani-Seyedlar, A.; Mattea, C.; Kausik, R.; Freed, D.E.; Song, Y.Q.; Hürlimann, M.D. Fluorine tracers for the identification of molecular interaction with porous asphaltene aggregates in crude oil. *Microporous Mesoporous Mater.* 2015, *205*, 56–60. [CrossRef]
13. Ordikhani-Seyedlar, A.; Neudert, O.; Stapf, S.; Mattea, C.; Kausik, R.; Freed, D.E.; Song, Y.Q.; Hürlimann, M.D. Evidence of Aromaticity-Specific Maltene NMR Relaxation Enhancement Promoted by Semi-immobilized Radicals. *Energy Fuels* 2016, *30*, 3886–3893. [CrossRef]
14. Zielinski, L.; Hürlimann, M.D. Nuclear Magnetic Resonance Dispersion of Distributions as a Probe of Aggregation in Crude Oils. *Energy Fuels* 2011, *25*, 5090–5099. [CrossRef]
15. Korb, J.P.; Louis-Joseph, A.; Benamsili, L. Probing Structure and Dynamics of Bulk and Confined Crude Oils by Multiscale NMR Spectroscopy, Diffusometry, and Relaxometry. *J. Phys. Chem. B* 2013, *117*, 7002–7014. [CrossRef]
16. Vorapalawut, N.; Nicot, B.; Louis-Joseph, A.; Korb, J.P. Probing Dynamics and Interaction of Maltenes with Asphaltene Aggregates in Crude Oils by Multiscale NMR. *Energy Fuels* 2015, *29*, 4911–4920. [CrossRef]
17. Gizatullin, B.; Gafurov, M.; Rodionov, A.; Mamin, G.; Mattea, C.; Stapf, S.; Orlinskii, S. Proton–Radical Interaction in Crude Oil—A Combined NMR and EPR Study. *Energy Fuels* 2018, *32*, 11261–11268. [CrossRef]
18. Gizatullin, B.; Gafurov, M.; Vakhin, A.; Rodionov, A.; Mamin, G.; Orlinskii, S.; Mattea, C.; Stapf, S. Native Vanadyl Complexes in Crude Oil as Polarizing Agents for In Situ Proton Dynamic Nuclear Polarization. *Energy Fuels* 2019, *33*, 10923–10932. [CrossRef]
19. Zielinski, L.; Saha, I.; Freed, D.E.; Hürlimann, M.D.; Liu, Y. Probing Asphaltene Aggregation in Native Crude Oils with Low-Field NMR. *Langmuir* 2010, *26*, 5014–5021. [CrossRef]
20. Speight, J.G.; Moschopedis, S.E. On the Molecular Nature of Petroleum Asphaltenes. In *Advances in Chemistry*; American Chemical Society: Washington, DC, USA, 1982; Volume 195 (Chemistry of Asphaltenes), pp. 1–15. [CrossRef]
21. Gawrys, K.L.; Spiecker, P.M.; Kilpatrick, P.K. The Role of Asphaltene Solubility and Chemical Composition on Asphaltene Aggregation. *Pet. Sci. Technol.* 2003, *21*, 461–489. [CrossRef]
22. Hemmingsen, P.V.; Silset, A.; Hannisdal, A.; Sjöblom, J. Emulsions of Heavy Crude Oils. I: Influence of Viscosity, Temperature, and Dilution. *J. Dispers. Sci. Technol.* 2005, *26*, 615–627. [CrossRef]
23. Tharanivasan, A.K.; Svrcek, W.Y.; Yarranton, H.W.; Taylor, S.D.; Merino-Garcia, D.; Rahimi, P.M. Measurement and Modeling of Asphaltene Precipitation from Crude Oil Blends. *Energy Fuels* 2009, *23*, 3971–3980. [CrossRef]
24. Abragam, A. *The Principles of Nuclear Magnetism*; Clarendon Press Oxford University Press: Oxford, UK, 1983.
25. Kimmich, R. *NMR-Tomography, Diffusometry, Relaxometry*, 1st ed.; Springer: Berlin/Heidelberg, Germany, 1997. [CrossRef]
26. Hwang, L.P.; Freed, J.H. Dynamic effects of pair correlation functions on spin relaxation by translational diffusion in liquids. *J. Chem. Phys.* 1975, *63*, 4017. [CrossRef]
27. Chen, J.J.; Hürlimann, M.; Paulsen, J.; Freed, D.; Mandal, S.; Song, Y.Q. Dispersion of T1 and T2 Nuclear Magnetic Resonance Relaxation in Crude Oils. *ChemPhysChem* 2014, *15*, 2676–2681. [CrossRef] [PubMed]
28. Lozovoi, A.; Hürlimann, M.; Kausik, R.; Stapf, S.; Mattea, C. High Temperature Fast Field Cycling Study of Crude Oil. *Diffus. Fundam.* 2017, *29*, 1–4.
29. Gilli, G.; Gilli, P. *The Nature of the Hydrogen Bond: Outline of a Comprehensive Hydrogen Bond Theory*, 1st ed.; IUCr Monographs on Crystallography; Oxford University Press: Oxford, UK, 2009.
30. Sinnokrot, M.O.; Valeev, E.F.; Sherrill, C.D. Estimate of the Ab Initio Limit for π-π Interactions: The Benzene Dimer. *J. Am. Chem. Soc.* 2002, *124*, 10887–10893. [CrossRef] [PubMed]
31. Sinnokrot, M.O.; Sherrill, C.D. High-Accuracy Quantum Mechanical Studies of π-π Interactions in Benzene Dimers. *J. Phys. Chem. A* 2006, *110*, 10656–10668. [CrossRef]
32. Martinez, C.R.; Iverson, B.L. Rethinking the term "pi-stacking". *Chem. Sci.* 2012, *3*, 2191–2201. [CrossRef]
33. Spiecker, P.; Gawrys, K.L.; Kilpatrick, P.K. Aggregation and solubility behavior of asphaltenes and their subfractions. *J. Colloid Interface Sci.* 2003, *267*, 178–193. [CrossRef]
34. Gray, M.R.; Tykwinski, R.R.; Stryker, J.M.; Tan, X. Supramolecular Assembly Model for Aggregation of Petroleum Asphaltenes. *Energy Fuels* 2011, *25*, 3125–3134. [CrossRef]
35. da Costa, L.M.; Stoyanov, S.R.; Gusarov, S.; Tan, X.; Gray, M.R.; Stryker, J.M.; Tykwinski, R.; de Carneiro, J.W.M.; Seidl, P.R.; Kovalenko, A. Density Functional Theory Investigation of the Contributions of π-π Stacking and Hydrogen-Bonding Interactions to the Aggregation of Model Asphaltene Compounds. *Energy Fuels* 2012, *26*, 2727–2735. [CrossRef]
36. Santos Silva, H.; Alfarra, A.; Vallverdu, G.; Bégué, D.; Bouyssiere, B.; Baraille, I. Asphaltene aggregation studied by molecular dynamics simulations: role of the molecular architecture and solvents on the supramolecular or colloidal behavior. *Pet. Sci.* 2019, *16*, 669–684. [CrossRef]
37. Zhang, Y.; Siskin, M.; Gray, M.R.; Walters, C.C.; Rodgers, R.P. Mechanisms of Asphaltene Aggregation: Puzzles and a New Hypothesis. *Energy Fuels* 2020, *34*, 9094–9107. [CrossRef]
38. Kertesz, M. Pancake Bonding: An Unusual Pi-Stacking Interaction. *Chem. A Eur. J.* 2019, *25*, 400–416. [CrossRef]
39. Castellano, O.; Gimon, R.; Soscun, H. Theoretical Study of the σ-π and π-π Interactions in Heteroaromatic Monocyclic Molecular Complexes of Benzene, Pyridine, and Thiophene Dimers: Implications on the Resin–Asphaltene Stability in Crude Oil. *Energy Fuels* 2011, *25*, 2526–2541. [CrossRef]

40. Alvarez-Ramírez, F.; Ruiz-Morales, Y. Island versus Archipelago Architecture for Asphaltenes: Polycyclic Aromatic Hydrocarbon Dimer Theoretical Studies. *Energy Fuels* **2013**, *27*, 1791–1808. [CrossRef]
41. Mousavi, M.; Fini, E.H. Non-Covalent π-Stacking Interactions between Asphaltene and Porphyrin in Bitumen. *J. Chem. Inf. Model.* **2020**, *60*, 4856–4866. [CrossRef] [PubMed]
42. Wang, P.; Dong, Z.j.; Tan, Y.q.; Liu, Z.y. Investigating the Interactions of the Saturate, Aromatic, Resin, and Asphaltene Four Fractions in Asphalt Binders by Molecular Simulations. *Energy Fuels* **2015**, *29*, 112–121. [CrossRef]
43. Ahmadi, M.; Chen, Z. Insight into the Interfacial Behavior of Surfactants and Asphaltenes: Molecular Dynamics Simulation Study. *Energy Fuels* **2020**, *34*, 13536–13551. [CrossRef]
44. Chang, C.L.; Fogler, H.S. Stabilization of Asphaltenes in Aliphatic Solvents Using Alkylbenzene-Derived Amphiphiles. 1. Effect of the Chemical Structure of Amphiphiles on Asphaltene Stabilization. *Langmuir* **1994**, *10*, 1749–1757. [CrossRef]
45. Chang, C.L.; Fogler, H.S. Stabilization of Asphaltenes in Aliphatic Solvents Using Alkylbenzene-Derived Amphiphiles. 2. Study of the Asphaltene-Amphiphile Interactions and Structures Using Fourier Transform Infrared Spectroscopy and Small-Angle X-ray Scattering Techniques. *Langmuir* **1994**, *10*, 1758–1766. [CrossRef]
46. Qiao, P.; Harbottle, D.; Li, Z.; Tang, Y.; Xu, Z. Interactions of Asphaltene Subfractions in Organic Media of Varying Aromaticity. *Energy Fuels* **2018**, *32*, 10478–10485. [CrossRef]
47. Sheu, E.Y.; Tar, M.M.D.; Storm, D.A.; DeCanio, S.J. Aggregation and kinetics of asphaltenes in organic solvents. *Fuel* **1992**, *71*, 299–302. [CrossRef]
48. Oh, K.; Deo, M.D. Near Infrared Spectroscopy to Study Asphaltene Aggregation in Solvents. In *Asphaltenes, Heavy Oils, and Petroleomics*; Mullins, O.C., Sheu, E.Y., Hammami, A., Marshall, A.G., Eds.; Springer: New York, NY, USA, 2007; pp. 469–488. [CrossRef]
49. Goual, L.; Sedghi, M.; Zeng, H.; Mostowfi, F.; McFarlane, R.; Mullins, O.C. On the formation and properties of asphaltene nanoaggregates and clusters by DC-conductivity and centrifugation. *Fuel* **2011**, *90*, 2480–2490. [CrossRef]
50. Norinaga, K.; Wargardalam, V.J.; Takasugi, S.; Iino, M.; Matsukawa, S. Measurement of Self-Diffusion Coefficient of Asphaltene in Pyridine by Pulsed Field Gradient Spin-Echo 1H NMR. *Energy Fuels* **2001**, *15*, 1317–1318. [CrossRef]
51. Östlund, J.A.; Andersson, S.I.; Nydén, M. Studies of asphaltenes by the use of pulsed-field gradient spin echo NMR. *Fuel* **2001**, *80*, 1529–1533. [CrossRef]
52. Kawashima, H.; Takanohashi, T.; Iino, M.; Matsukawa, S. Determining Asphaltene Aggregation in Solution from Diffusion Coefficients As Determined by Pulsed-Field Gradient Spin-Echo 1H NMR. *Energy Fuels* **2008**, *22*, 3989–3993. [CrossRef]
53. Eliel, E.L.; Wilen, S.H.; Mander, L.N. *Stereochemistry of Organic Compounds*, 1st ed.; Wiley-Interscience: Hoboken, NJ, USA, 1994; pp. 686–690.
54. Anet, F.A.L.; Bourn, A.J.R. Nuclear Magnetic Resonance Line-Shape and Double-Resonance Studies of Ring Inversion in Cyclohexane-d11. *J. Am. Chem. Soc.* **1967**, *89*, 760–768. [CrossRef]
55. Jen, J. Chemical exchange and NMR-T2 relaxation. *Adv. Mol. Relax. Process.* **1974**, *6*, 171–183. [CrossRef]
56. Sobol, W.T. A complete solution to the model describing Carr-Purcell and Carr-Purcell-Meiboom-Gill experiments in a two-site exchange system. *Magn. Reson. Med.* **1991**, *21*, 2–9. [CrossRef]
57. Burnell, E.E.; de Lange, C.A.; Dong, R.Y. Ring inversion in cyclohexane: A textbook example. *Liq. Cryst.* **2019**, *47*, 1965–1974. [CrossRef]
58. Poupko, R.; Luz, Z. Dynamic deuterium NMR in liquid crystalline solvents: Ring inversion of cyclohexane-d12. *J. Chem. Phys.* **1981**, *75*, 1675–1681. [CrossRef]
59. McGrath, K.J.; Weiss, R.G. Rate of chair-to-chair interconversion of cyclohexane-d12 in its neat plastic crystalline phase. *J. Phys. Chem.* **1993**, *97*, 2497–2499. [CrossRef]

Article

Impact of Xylose on Dynamics of Water Diffusion in Mesoporous Zeolites Measured by NMR

Madison L. Nelson [1,2], Joelle E. Romo [2], Stephanie G. Wettstein [2] and Joseph D. Seymour [1,2,*]

[1] Department of Physics, Montana State University, Bozeman, MT 59717, USA; madison.nelson1@montana.edu
[2] Department of Chemical and Biological Engineering, Montana State University, Bozeman, MT 59717, USA; jojoromo@gmail.com (J.E.R.); stephanie.wettstein@montana.edu (S.G.W.)
* Correspondence: jseymour@montana.edu

Abstract: Zeolites are known to be effective catalysts in biomass converting processes. Understanding the mesoporous structure and dynamics within it during such reactions is important in effectively utilizing them. Nuclear magnetic resonance (NMR) T_2 relaxation and diffusion measurements, using a high-power radio frequency probe, are shown to characterize the dynamics of water in mesoporous commercially made 5A zeolite beads before and after the introduction of xylose. Xylose is the starting point in the dehydration into furfural. The results indicate xylose slightly enhances rotational mobility while it decreases translational motion through altering the permeability, K, throughout the porous structure. The measurements show xylose inhibits pure water from relocating into larger pores within the zeolite beads where it eventually is expelled from the bead itself.

Keywords: NMR diffusometry; zeolites; heterogeneous catalysis sugar conversion; biomolecules

1. Introduction

Pulsed gradient spin echo (PGSE), or pulsed field gradient (PFG) nuclear magnetic resonance (NMR) is a preeminent method for characterization of transport and structure in porous media systems [1,2]. Despite the fact that application to nanoporous systems is challenging, due to the small structural length dimensions which generate complex rotational and translational molecular dynamics over a hierarchy of scales, significant characterization of systems such as zeolites has been attained [3–6]. The application of zeolites in catalytic conversion of biomass to fuel and chemical products is an area of growing application [7], and recent research has shown that zeolite beads have the potential to catalyze sugar to furan dehydration reactions [8]. Studies of water molecular dynamics in zeolites in the presence of biomolecules by NMR have been limited to solid state NMR spectroscopy, PGSE NMR using a single displacement observation time [9], and PGSE NMR to study water in zeolites [10,11] while solid state NMR has also been used to study solvents in zeolites [12]. Here displacement time-dependent PGSE NMR was applied to study the impact of xylose on water dynamics in zeolites for heterogeneous catalysis of sugars to furans [8,13].

Zeolites are known for their chemical and thermal stability, versatility [14], and have been widely used in biomass conversion reactions [15–18] including xylose dehydration to furfural. For example, research by Gao et al. found xylose dehydration reactions with ZSM-5 resulted in furfural yields of 51.5% in an aqueous system [17]. Other researchers have looked at the use of powdered silicoaluminophosphates (SAPOs), a class of small-pore zeolites, in various solvent systems to maximize furfural production from xylose, achieving moderate yields [15]. In order to improve catalyst recovery, Romo et al. used dual-layered zeolite beads (versus powdered zeolites) to convert xylose to furfural and achieved yields of up to 45%, indicating zeolite beads have the potential for sugar upgrading [8]. Although microporous zeolites beads are promising for sugar dehydration reactions, the zeolite pore size can create a diffusion limited system. This is particularly true for substates such as sugars, which have large kinetic diameters.

Commercially available Linde Type A (LTA) zeolite beads consist of ~3 μm crystallites made up of ~5Å molecular cages, which are compressed with binder into 3 mm beads that are 86% microporous [8]. Molecular transport in zeolite systems of this type of structure have been modeled as porous media with periodic permeable inclusions [6,19,20]. This results in an effective diffusivity $\frac{1}{D_{eff}} = \frac{1}{D_o} + \frac{1}{Kl}$ where D_o is the molecular diffusion within a pore structure and K is the permeability, reflecting transport resistance between pore structures separated by length scale l [19,20]. In the zeolite system studied here, D_o is the diffusion within the zeolite crystal and K the permeability at the zeolite crystal grain interfaces within the 3 mm bead and D_{eff} the NMR measured diffusion.

Diffusion measurements by NMR are obtained by application of magnetic field gradient pulses which attenuate the voltage signal due to magnetization dephasing caused by random diffusive motion. The measured signal normalized by the signal with no gradient is given by $E(g, \Delta) = \frac{S(g,\Delta)}{S(0,\Delta)} = \exp\left[-\gamma^2 g^2 \delta^2 D\left(\Delta - \frac{\delta}{3}\right)\right]$, where γ is the gyromagnetic ratio, g is the gradient amplitude, δ is the gradient duration, and Δ is the gradient separation, which is the time the nuclei are allowed to displace. PGSE NMR thus measures the time-dependent effective diffusivity $D(\Delta)$, which characterizes the length scale of the restricted diffusion dynamics of a fluid in a pore at short times, as [2,21,22].

$$D(\Delta) = D_o \left[1 - \frac{4}{9\sqrt{\pi}} \frac{S}{V} (D_o \Delta)^{1/2}\right] \tag{1}$$

Here D_0 is the free liquid diffusion, Δ the PGSE NMR displacement time for the spins, and $S/V = 3/R$ the surface to volume ratio of a spherical pore of characteristic length scale radius R. While the normalized form of the measured signal, $E(g, \Delta)$, factors out T_1 spin-lattice and T_2 spin–spin magnetization relaxation effects, the measured signal is weighted by the T_2 relaxation if T_2 times are present which are less than the PGSE echo time, as the signal from those spins are fully relaxed before being encoded for diffusive motion. T_2 relaxation is due to dipolar coupling of the NMR active spins, ^{1}H protons in the experiments conducted in this work, and interaction with solid surfaces in porous media. Longer T_2 relaxation times occur when the dipolar coupling is averaged out by rotational diffusion, as in liquids, and shorter when rotational mobility is restricted.

2. Results

T_2 relaxation measurement of water xylose solution in bulk has a large peak at 1167.1 ms due to water and a small peak at 126 ms from the xylose. T_2 relaxation of pure water and of the 20% wt. xylose solution in the 5A bead indicate two primary populations of relaxation behavior as shown in Figure 1. The fast relaxing, short T_2 relaxation populations are at 0.619 ms for water and 0.752 ms for the xylose solution. These sub millisecond relaxation times demonstrate the significant restriction of rotational mobility and interactions of primarily water and the zeolite surfaces (kinetic diameter 2.7 Å) [23] in the zeolite micropores, since the xylose (kinetic diameter 6.8 Å) [24] is too large to be within those pores. The presence of the xylose generates an increase in the T_2 relaxation time. The slower relaxing, longer T_2 populations in the zeolite beads at 8.300 ms for pure water and 9.100 ms for the xylose solution are due to the molecules in the inter-crystalline mesopores of the bead. In this more mobile population, the presence of the xylose induces a slight increase in the rotational mobility of the protons in the system. In the pure water in zeolite system there is a small peak at 806 ms which is associated with water in macro pores or leaking to the bead surface, while this population is suppressed in the xylose solution system. Due to only 20% of the solution being xylose, the signal is primarily from water. Spectral resolution is not possible in the zeolite beads due to the signal broadening caused by the restrictions of the solution within the zeolite, and the T_2 relaxation distributions show little signal changes after the introduction of xylose. Therefore, the NMR signal obtained will be primarily attributed to water within the system.

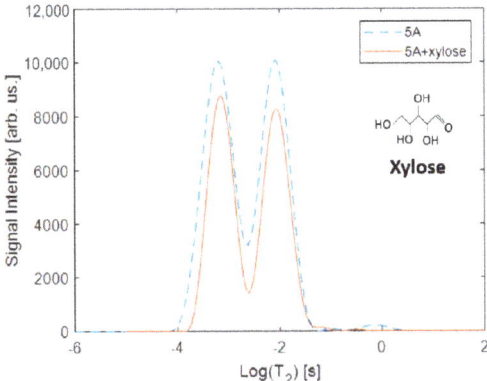

Figure 1. T_2 distributions comparing the core zeolite Grace 5A beads before and after the addition of 20% wt. xylose to the water permeating through the system. In the 5A beads with water there are two large populations at $T_2 = 6.193 \times 10^{-4}$ s and 8.300×10^{-3} s and a small population at longer $T_2 = 0.806$ s. The presence of xylose eliminates the longer relaxation component and slightly shifts the two dominant populations to slightly longer relaxation times $T_2 = 7.518 \times 10^{-4}$ s and 9.100×10^{-3} s. The 20% wt. xylose in water solution has a large relaxation population at $T_2 = 1.167$ s and a small population at 1.260×10^{-1} s representing the biopolymer and water relaxation times (not shown). The chemical structure of xylose is shown in the upper right.

The displacement time dependent pulse gradient stimulated echo (PGStE) NMR signal attenuation data as a function of increasing pulsed gradient strength is shown in Figure 2. The data are plotted in a standard Stejskal–Tanner plot format in which the slope of the curve indicates the diffusion coefficient [2]. The data exhibits biexponential behavior with a fast and slow diffusing component. It is important to note that the ^{1}H proton signal measured comes only from the longer T_2 relaxation population since the sub millisecond relaxing populations are filtered out by the 4.32 ms echo time of the stimulated echo experiment. The slow diffusing component increases in quantity as a percentage of the total signal as the displacement observation time is increased.

This can be seen by fitting a biexponential model $p_f exp\left[-D_f x\right] + p_s exp[-D_s x]$ to the data and determining the population in the fast and slow decay regions, Figure 3. In the pure water system, there is an initial increase in the amount of fast diffusion component, and commensurate decrease in the slow diffusion population, which is associated with the water moving into larger pore regions of the beads during the 20 min of each initial short displacement time Δ experiment. After the Δ = 50 ms experiment, the signal proportion in the slow component increases. This can be attributed to a transient redistribution of the pure water into larger pore spaces followed by a loss of the water signal in the large pores due to dephasing of the signal or drainage from the large pores out of the sensitive region of the rf coil over the hours long experimental time for all the displacement times Δ. While any long-time scale transient redistribution of the water prohibits determining exchange between the fast and slow diffusion populations it does not negatively impact the assessment of the length scales associated with the diffusion dynamics. Of interest is the impact of the xylose in solution on the distribution of ^{1}H proton signal in the fast and slow diffusion populations, in that it generates a more significant loss of fast diffusion signal. The xylose maintains more liquid within the slow diffusion population than the pure water and inhibits the initial redistribution of the water into the larger pores during displacement times Δ < 50 ms. experiments.

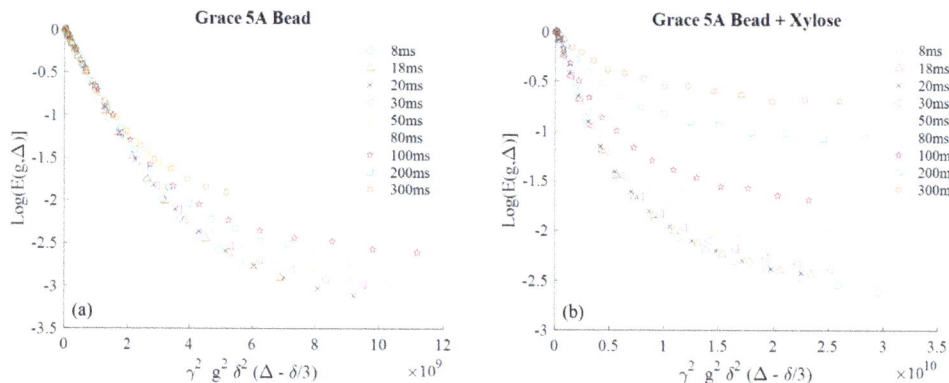

Figure 2. Stejskal–Tanner plots of Grace 5A beads (**a**) and Grace 5A beads with xylose (**b**) are shown. With increasing displacement observation time Δ from 8 to 300 ms, the attenuation of signal decreases for the pure water and xylose solution saturated beads. The gradient duration δ was 1 ms with a maximum gradient ranging from 0.500 to 1.9021 T/m in order to sample out to a similar point in the gradient domain. The data are well fit by a biexponential model with a fast and slow diffusion component $p_f exp[-D_f x] + p_s exp[-D_s x]$.

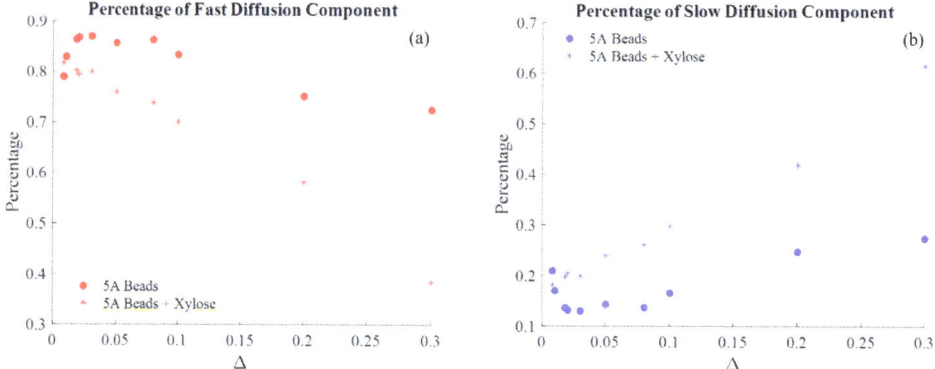

Figure 3. The population percentage of the fast (**a**) and slow (**b**) diffusion populations with varying Δ. The signal is weighted toward the slower diffusing regions before and after the addition of the xylose with increasing Δ.

The primary results of the PGStE measurement are the $D(\Delta)$ data for the fast and slow diffusion populations shown in Figure 4. The data are plotted against $\Delta^{1/2}$ so that the short displacement time data provide S/V and the long displacement time data provide the tortuosity [21,22]. The beads with pure water show an increase in diffusion coefficient for both the fast and slow component for displacement times Δ > 80 ms. This is consistent with a possible draining of the beads over the total experimental run and the loss of the fast diffusion population and precludes determination of the tortuosity, however further studies are required to determine the origin of this effect. The decrease in the fast diffusion coefficient with pure water to a value less than 1/2 that of free water indicates a diffusion length scale of the order of $l = (2D\Delta)^{1/2} = 9.49$ μm at Δ = 50 ms, representative of multiple mesopore transport. The xylose solution is much more restricted in the largest pores and has a reduction in diffusion to less than 1/5 the free water value. The slow diffusion data for the pure water in the bead displays the classic $\Delta^{1/2}$ decay predicted by theory at short times [21,22]. Determination of the length scale from Equation (1) gives R ~3.87 μm consistent with the inter-crystal mesopore length. The reduction of the slow diffusion coefficients in the xylose solution beads relative to the beads in pure water indicates a decrease in the permeability K at the zeolite crystal grain surfaces.

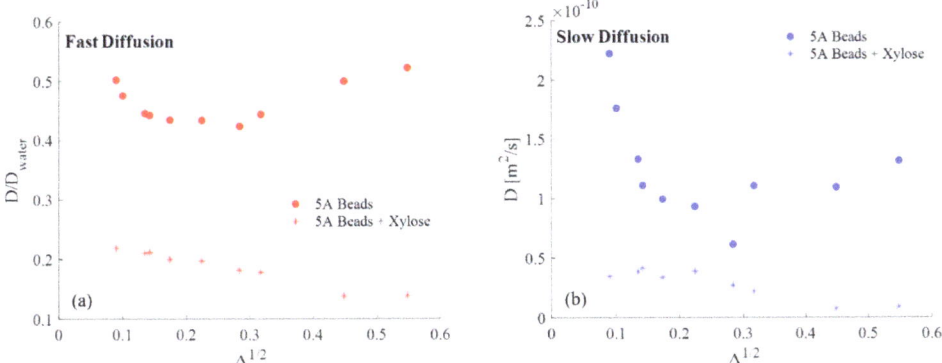

Figure 4. Diffusion coefficients calculated from the biexponential fits to the Stejskal–Tanner plots at each Δ, plotted against $\Delta^{1/2}$ for the zeolite beads with the pure water and xylose solution for the fast (**a**) and slow (**b**) diffusion component. The fast diffusion is plotted normalized by the free water diffusion coefficient 2.0×10^{-9} m^2/s. The presence of the xylose significantly decreases both the fast and slow diffusion coefficients in the beads. The increase in diffusion of pure water in the zeolite at longer Δ indicates some nonstationary water redistribution.

3. Discussion

The NMR data presented indicate the presence of a biomolecule such as xylose alters the rotational mobility of water in zeolite crystal beads slightly, while generating larger changes in translation mobility. The xylose solution alters the interplay of fluid distribution between fast and slow diffusion populations. The increase in the percentage of fluid undergoing slow diffusion in the xylose solution relative to the pure water implies that the fluid in the largest pores is more prone to drain from the beads. Any seepage from the larger pores means caution must be used in interpreting changes in the fast diffusion coefficient with displacement observation time Δ. The slow diffusion component is associated with the zeolite crystal length scale. The slow diffusion coefficients measured for the pure water and the xylose solution indicate the xylose decreases the permeability at the zeolite crystal interfaces. The xylose also makes the slow diffusion behavior less time dependent over displacement timescales from Δ = 8 to 100 ms than in pure water, consistent with a decrease in permeability restricting water translational diffusion even on shorter timescales. This demonstrates that xylose dehydration reactions in zeolite bead catalysts can be impacted by diffusion limitations, even with mesoporosity.

4. Materials and Methods

Commercially obtained Zeolite 5A beads (W. R. Grace & Co., Columbia, MD, USA) are composed of 4.2 Å molecular cages grown to ~3 μm crystallites indicated by manufacturer's data and then compressed with binder into 3 mm beads. The exact crystal size within the bead is not measured, but there is crystallinity as seen by the XRD reported in our previous manuscript [8]. The beads were first saturated with water by placing 0.3 g of beads into a 15 mL pressure tube. Approximately 4 g water and a stir bar were added to the tube and then sealed with a Teflon cap with front-seal Kalrez o-ring. The pressure tube was then placed in an oil bath set at 443 K for 10 min with a stir rate of 600 rpm. The beads were then filtered from the liquid, gently blotted with a paper towel, and then placed in the NMR tube. For the beads containing xylose, a similar method was used, but the water was replaced with a 20% wt. xylose solution.

After the heat treatment period the catalytic beads were placed in a 5 mm glass NMR tube were returned to room temperature and loaded into the NMR probe in the magnet.

Diffusion and T_2 relaxation experiments were performed in a Bruker 250 MHz superconducting magnet using a custom built high-power ^{1}H rf probe and 5 mm rf coil (Bruker, Karlsruhe). Sample temperature was controlled through the Bruker BTU system with N_2 gas flow and kept at 292 K throughout all experiments. A standard Carr–Purcell–Meiboom–Gill (CPMG) pulse sequence [1,2] was used for T_2 measurements with an echo time of τ_E = 192 µs, 10,000 echoes, a dwell time of 4 µs, and 7.4 µs 180° rf pulses at power of 100 W. Each measurement had 64 averages. The diffusion measurements were acquired using a pulsed gradient stimulated echo (PGStE) pulse sequence [1,2] with δ = 1 ms, with maximum gradients ranging from 1.9021 T/m to 0.5003 T/m dependent on the displacement time Δ value which spanned 8–300 ms. The stimulated echo pulse sequence with an echo time of τ_E = 4.32 ms was averaged 64 times.

The T_2 relaxation experimental data was processed through a Fredholm integral, also commonly described as an inverse Laplace transform method. The diffusion data were not analyzed through this method due to having only 16 echoes. This analysis technique works optimally with more echoes. The diffusion data were fit directly from the Stejskal–Tanner plots using a bi-exponential fitting process to obtain the diffusion coefficients.

5. Conclusions

NMR can provide data on the impact of biomolecules on the translational and rotational dynamics of water in zeolite beads. The significant decrease in diffusion due to decreased permeability at the zeolite crystal interfaces could impact the reaction dynamics and the catalyst performance.

Author Contributions: Conceptualization, S.G.W. and J.D.S.; methodology, M.L.N., J.E.R., S.G.W. and J.D.S.; NMR data analysis and acquisition, M.L.N. and J.D.S.; zeolite preparation, J.E.R.; writing—original draft preparation, M.L.N. and J.D.S.; writing—review and editing, M.L.N., J.E.R., S.G.W. and J.D.S.; project administration, S.G.W.; funding acquisition, S.G.W. All authors have read and agreed to the published version of the manuscript.

Funding: This research was funded by the National Science Foundation CBET, grant number 1705490 to SGW.

Institutional Review Board Statement: Not applicable.

Informed Consent Statement: Not applicable.

Data Availability Statement: The data presented in this study are available within the article.

Acknowledgments: Moise Carreon (Colorado School of Mines) for sending zeolite beads.

Conflicts of Interest: The authors declare no conflict of interest. The funders had no role in the design of the study; in the collection, analyses, or interpretation of data; in the writing of the manuscript, or in the decision to publish the results.

Sample Availability: Samples of the compounds are commercially available.

References

1. Song, Y.-Q. Novel NMR techniques for porous media research. *Magn. Reson. Imaging* **2003**, *21*, 207–211. [CrossRef]
2. Callaghan, P.T. *Translational Dynamics and Magnetic Resonance Principles of Pulsed Gradient Spin Echo NMR*; Oxford University Press: New York, NY, USA, 2011; p. 547.
3. Kärger, J.; Ruthven, D.M.; Theodorou, D.N. Diffusion Measurement by Monitoring Molecular Displacement. In *Diffusion in Nanoporous Materials*; Wiley-VCH Verlag GmbH & Co. KGaA: Weinheim, Germany, 2012; pp. 347–394.
4. Kärger, J. A study of fast tracer desorption in molecular sieve crystals. *AIChE J.* **1982**, *28*, 417–423. [CrossRef]
5. Karger, J.; Heink, W. The Propagator Representation of Molecular-Transport in Microporous Crystallites. *J. Magn. Reson.* **1983**, *51*, 1–7. [CrossRef]
6. Heinke, L.; Kärger, J. Correlating Surface Permeability with Intracrystalline Diffusivity in Nanoporous Solids. *Phys. Rev. Lett.* **2011**, *106*, 074501. [CrossRef] [PubMed]
7. Alonso, D.M.; Wettstein, S.G.; Dumesic, J.A. Bimetallic catalysts for upgrading of biomass to fuels and chemicals. *Chem. Soc. Rev.* **2012**, *41*, 8075–8098. [CrossRef] [PubMed]

8. Romo, J.E.; Wu, T.; Huang, X.; Lucero, J.; Irwin, J.L.; Bond, J.Q.; Carreon, M.A.; Wettstein, S.G. SAPO-34/5A Zeolite Bead Catalysts for Furan Production from Xylose and Glucose. *ACS Omega* **2018**, *3*, 16253–16259. [CrossRef] [PubMed]
9. Qi, L.; Alamillo, R.; Elliott, W.A.; Andersen, A.; Hoyt, D.W.; Walter, E.D.; Han, K.S.; Washton, N.M.; Rioux, R.M.; Dumesic, J.A.; et al. Operando Solid-State NMR Observation of Solvent-Mediated Adsorption-Reaction of Carbohydrates in Zeolites. *ACS Catal.* **2017**, *7*, 3489–3500. [CrossRef]
10. Valiullin, R.; Kärger, J.; Cho, K.; Choi, M.; Ryoo, R. Dynamics of water diffusion in mesoporous zeolites. *Microp. Mesoporous Mater.* **2011**, *142*, 236–244. [CrossRef]
11. Beckert, S.; Stallmach, F.; Toufar, H.; Freude, D.; Kärger, J.; Haase, J. Tracing Water and Cation Diffusion in Hydrated Zeolites of Type Li-LSX by Pulsed Field Gradient NMR. *J. Phys. Chem. C* **2013**, *117*, 24866–24872. [CrossRef]
12. Zheng, A.; Han, B.; Li, B.; Liu, S.-B.; Deng, F. Enhancement of Brønsted acidity in zeolitic catalysts due to an intermolecular solvent effect in confined micropores. *Chem. Commun.* **2012**, *48*, 6936–6938. [CrossRef] [PubMed]
13. Romo, J.E.; Bollar, N.V.; Zimmermann, C.J.; Wettstein, S.G. Conversion of Sugars and Biomass to Furans Using Heterogeneous Catalysts in Biphasic Solvent Systems. *ChemCatChem* **2018**, *10*, 4805–4816. [CrossRef]
14. Liu, F.; Huang, K.; Zheng, A.; Xiao, F.-S.; Dai, S. Hydrophobic Solid Acids and Their Catalytic Applications in Green and Sustainable Chemistry. *ACS Catal.* **2018**, *8*, 372–391. [CrossRef]
15. Bruce, S.M.; Zong, Z.; Chatzidimitriou, A.; Avci, L.E.; Bond, J.Q.; Carreon, M.A.; Wettstein, S.G. Small pore zeolite catalysts for furfural synthesis from xylose and switchgrass in a γ-valerolactone/water solvent. *J. Mol. Catal. A Chem.* **2016**, *422*, 18–22. [CrossRef]
16. Ennaert, T.; Van Aelst, J.; Dijkmans, J.; De Clercq, R.; Schutyser, W.; Dusselier, M.; Verboekend, D.; Sels, B.F. Potential and challenges of zeolite chemistry in the catalytic conversion of biomass. *Chem. Soc. Rev.* **2016**, *45*, 584–611. [CrossRef]
17. Gao, H.; Liu, H.; Pang, B.; Yu, G.; Du, J.; Zhang, Y.; Wang, H.; Mu, X. Production of furfural from waste aqueous hemicellulose solution of hardwood over ZSM-5 zeolite. *Bioresour. Technol.* **2014**, *172*, 453–456. [CrossRef]
18. Ordomsky, V.V.; van der Schaaf, J.; Schouten, J.C.; Nijhuis, T.A. The effect of solvent addition on fructose dehydration to 5-hydroxymethylfurfural in biphasic system over zeolites. *J. Catal.* **2012**, *287*, 68–75. [CrossRef]
19. Dudko, O.K.; Berezhkovskii, A.M.; Weiss, G.H. Diffusion in the presence of periodically spaced permeable membranes. *J. Chem. Phys.* **2004**, *121*, 11283–11288. [CrossRef]
20. Dudko, O.K.; Berezhkovskii, A.M.; Weiss, G.H. Time-Dependent Diffusion Coefficients in Periodic Porous Materials. *J. Phys. Chem. B* **2005**, *109*, 21296–21299. [CrossRef]
21. Mitra, P.P.; Sen, P.N.; Schwartz, L.M.; Le Doussal, P. Diffusion propagator as a probe of the structure of porous media. *Phys. Rev. Lett.* **1992**, *68*, 3555–3558. [CrossRef]
22. Sen, P.N. Time-dependent diffusion coefficient as a probe of geometry. *Concepts Magn. Reson. Part A* **2004**, *23A*, 1–21. [CrossRef]
23. Kalipcilar, H.; Bowen, T.C.; Noble, R.D.; Falconer, J.L. Synthesis and separation performance of SSZ-13 zeolite membranes on tubular supports. *Chem. Mater.* **2002**, *14*, 3458–3464. [CrossRef]
24. Antunes, M.M.; Lima, S.; Fernandes, A.; Pillinger, M.; Ribeiro, M.F.; Valente, A.A. Aqueous-phase dehydration of xylose to furfural in the presence of MCM-22 and ITQ-2 solid acid catalysts. *Appl. Catal. A Gen.* **2012**, *417–418*, 243–252. [CrossRef]

Article

Local Structures of Two-Dimensional Zeolites—Mordenite and ZSM-5—Probed by Multinuclear NMR

Marina G. Shelyapina [1,*], Rosario I. Yocupicio-Gaxiola [2], Iuliia V. Zhelezniak [1], Mikhail V. Chislov [1], Joel Antúnez-García [3], Fabian N. Murrieta-Rico [3], Donald Homero Galván [3], Vitalii Petranovskii [3] and Sergio Fuentes-Moyado [3]

- [1] Department of Nuclear Physics Research Methods, Saint-Petersburg State University, 7/9 Universitetskaya nab., St. Petersburg 199034, Russia; st022650@student.spbu.ru (I.V.Z.); mikhail.chislov@spbu.ru (M.V.C.)
- [2] Center for Scientific Research and Higher Education at Ensenada (CICESE), Ensenada, Baja California 22860, Mexico; ryocu@cnyn.unam.mx
- [3] Center for Nanoscience and Nanotechnology, National Autonomous University of Mexico (CNyN, UNAM), Ensenada, Baja California 22860, Mexico; joel.antunez@gmail.com (J.A.-G.); fabian.murrieta@uabc.edu.mx (F.N.M.-R.); donald@cnyn.unam.mx (D.H.G.); vitalii@cnyn.unam.mx (V.P.); fuentes@cnyn.unam.mx (S.F.-M.)
- * Correspondence: marina.shelyapina@spbu.ru

Academic Editor: Igor Serša
Received: 13 September 2020; Accepted: 13 October 2020; Published: 14 October 2020

Abstract: Mesostructured pillared zeolite materials in the form of lamellar phases with a crystal structure of mordenite (MOR) and ZSM-5 (MFI) were grown using CTAB as an agent that creates mesopores, in a one-pot synthesis; then into the CTAB layers separating the 2D zeolite plates were introduced by diffusion the TEOS molecules which were further hydrolyzed, and finally the material was annealed to remove the organic phase, leaving the 2D zeolite plates separated by pillars of silicon dioxide. To monitor the successive structural changes and the state of the atoms of the zeolite framework and organic compounds at all the steps of the synthesis of pillared MOR and MFI zeolites, the nuclear magnetic resonance method (NMR) with magic angle spinning (MAS) was applied. The ^{27}Al and ^{29}Si MAS NMR spectra confirm the regularity of the zeolite frameworks of the as synthesized materials. Analysis of the ^{1}H and ^{13}C MAS NMR spectra and an experiment with variable contact time evidence a strong interaction between the charged "heads" $-[N(CH_3)_3]^+$ of CTAB and the zeolite framework at the place of $[AlO_4]^-$ location. According to ^{27}Al and ^{29}Si MAS NMR the evacuation of organic cations leads to a partial but not critical collapse of the local zeolite structure.

Keywords: lamellar 2D zeolites; pillared zeolites; mordenite; ZSM-5; CTAB; NMR

1. Introduction

In recent years, much attention has been paid to the techniques of the "one-pot synthesis" for the direct production of zeolitic materials. Zeolites are undoubtedly important heterogeneous catalysts, and the number of industrial processes, in which they are used in that capacity, has been constantly increasing. The interest is mainly issued by the great opportunities these methodologies open up when developing functional zeolite based materials, such as hybrid organic-inorganic molecular sieves [1–3], hierarchical microporous-mesoporous zeolites [4–7], nanozeolites [8–10], and template-free molecular sieves [4,7,11].

Zeolites with hierarchical porous structure can also be synthesized directly, without using templates. The template-free methods are mainly based on the following strategies [7]: (i) the development of intercrystalline mesoporosity due to aggregation of nanocrystals; (ii) the emergence of intracrystalline mesoporosity, which is formed by amorphous gels that control crystallization; (iii) mesoporosity created between self-pillared two-dimensional zeolite nanosheets. The latter can be obtained by synthesizing layered zeolites in the presence of organic structure directing agents (OSDA) followed by calcination [4]. It is known that the use of various OSDA makes it possible to obtain target zeolites with specific physicochemical properties, and even novel or improved zeolitic frameworks [12]. In this sense, the physicochemical properties of zeolites are highly dependent on the synthesis procedure, including the choice of OSDA. The latter is of primary importance for the aluminum distribution, the acidic properties of the obtained material, the size and morphology of crystals, which are the key parameters for the catalysts [13–16].

One of the widely used OSDA for the synthesis of mesoporous silica and organic-inorganic layered materials is Cetyltrimethylammonium Bromide (CTAB). It is also widely applied in the synthesis of zeolites [15,17–19]. For example, under certain synthesis conditions, it is possible to grow a layered material in which inorganic layers of ZSM-5 zeolite and organic layers consisting of ordered CTAB molecules alternate. Under the conditions generally used for zeolite synthesis (100–180 °C and high pH), the CTAB molecules do not decompose, but interact strongly with the components of aluminosilicate gel. As a result of using CTAB, it was possible to direct the synthesis towards the formation of inorganic-organic microporous materials and the design of hierarchical zeolite catalysts from a plate-like zeolite precursor, which opens up new possibilities for the complex production of mesoporous zeolites. The main factor in this process is the guest-host interactions between organic surfactant and inorganic framework during the self-assembly and structure evolution development [17,19]. In such a process, a swelling-type multilamellar ECNU-7P with alternative stacking of MWW nanosheets and organic CTAB layers was successfully prepared through a dissolution–recrystallization route. This was the first time that a simple surfactant CTAB and a layered zeolite precursor could act synergistically during self-assembly. As a result, an alternative, attractive pathway opens up to current post-synthetic approaches, or to the hydrothermal syntheses of MWW nanosheets with designed surfactants. Calcined Al-ECNU-7 turned into a hierarchical zeolite catalyst, and exhibited excellent activity, selectivity and stability during the catalytic conversion of bulky molecules. The present approach would be a general methodology and would be suitable for the direct synthesis of hierarchical layered zeolites with other topologies by controlling the self-assembly of a simple surfactant and zeolite precursor. More significantly, the low cost and commercial availability of the CTAB simple surfactant makes it more promising than the complex bifunctional surfactants currently used for the preparation of industrial heterogeneous catalysts.

Despite quite numerous studies of the morphology and catalytic properties of layered zeolites obtained by self-assembling method, studies of their local structure are not so widespread, although this is a key point for understanding of the catalytic activity of materials. Nuclear magnetic resonance (NMR) is one of the most versatile experimental methods to probe the local structure [20]; besides this technique enables to obtain at the microscopic level information on dynamics of intercalated species [21,22] and is successfully applied to study organic-inorganic layered materials [22–25].

Earlier, we reported on the results of the successful synthesis of 2D ZSM-5 and mordenite [26]. The aim of this work is by applying multinuclear NMR to follow up changes in the local structure at all the stages of preparation, starting from a freshly synthesized hybrid material, in which the confinement of organic and inorganic layers is implemented, then pillaring between the layers, and finally removal of organic material during calcination.

2. Results and Discussion

In this work, layered two-dimensional (2D) zeolites with mordenite and ZSM-5 structures were prepared and studied. Further in the text and in the figures, they will be denoted by three-letter structural codes adopted by the International Zeolite Association (IZA) [27], as MOR and MFI, using additional abbreviations those are marking a certain stage of preparation. Both materials were synthesized according to the procedure described in our previous work [26]. Pillaring of the obtained materials was done in accordance with the process proposed by Na et al. [28]. As a result, our method for preparing samples included four steps: (i) obtaining of organic-inorganic hybrid lamellar zeolites by self-assembling method with addition of CTAB (and tetrapropylammonium bromide (TPABr) as OSDA for the synthesis of 2D ZSM-5): the MOR-AS and MFI-AS samples; (ii) introduction of tetraethoxysilane (TEOS) molecules into the organic layer of the interlamellar space filled with CTAB molecules: the MOR-T and MFI-T samples; (iii) hydrolysis of an organosilicon compound and the formation of pillars of amorphous SiO_2: the MOR-TH and MFI-TH samples; (iv) calcination to remove organic molecules: the MOR-P and MFI-P samples. A more detailed description of the preparation method can be found in Section 3.

2.1. X-ray Analysis

Figures 1 and 2 represent the X-ray Diffraction (XRD) patterns of both as synthetized and pillared MOR and MFI samples, respectively. As can be seen in Figures 1b and 2b, both samples exhibit typical features of the corresponding zeolite structure with an amorphous halo (range 2θ between 17–30 degrees) which is very consistent with this kind of materials [28,29]. This amorphous halo should vanish after calcination process, but it is necessary to remind the amorphous character of the pillars even after calcination could maintain or intensify this feature.

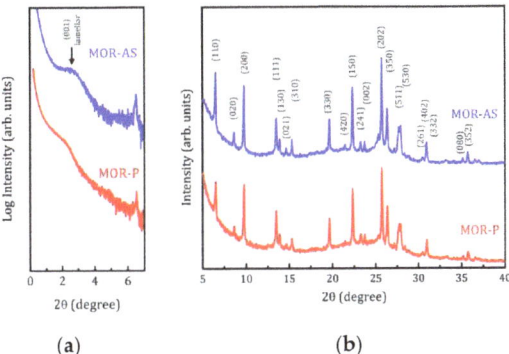

Figure 1. Small-angle (**a**) and full (**b**) XRD powder patterns for the mordenite samples as synthetized (MOR-AS) and pillared (MOR-P).

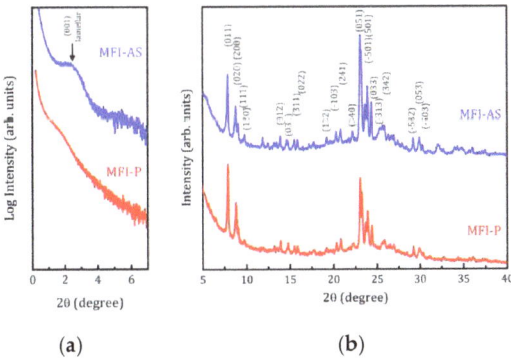

Figure 2. Small-angle (**a**) and full (**b**) XRD powder patterns for the ZSM-5 samples as synthetized (MFI-AS) and pillared (MFI-P).

Small angle X-ray scattering (SAXS) patterns shown in Figures 1a and 2a unambiguously indicate the formation of lamellar mesophases [28]. For lamellar samples, the peak at $2\theta = 2.7°$ for MOR-AS and $2\theta = 2.3°$ for MFI-AS, corresponds to the (001) reflections with interplanar distances $d = 3.2$ and 3.8 nm, respectively. For pillared samples, this peak is smoothed and shifted toward small angles ($2\theta = 2.2$ and $1.7°$ with $d = 4.0$ and 5.2 nm for MOR-P and MFI-P, respectively), which shows that the introduction of SiO_2 pillars increased the interlamellar space. The present results are consistent with our previously reported data [26]. A wider interplanar distance d distribution is a clear evidence of the random growing of pillars, that is to say, some distances can be expanded while others can be contracted. A detailed discussion of the physical structure of these zeolite plates, separated by plates consisting of organic material, is of particular interest. However, this topic is beyond the scope of the present article and will be published separately elsewhere.

2.2. SEM-EDX Studies

Figure 3 shows scanning electron microscopy (SEM) images for the initial as synthetized zeolites and the final samples after pillaring. As seen from Figure 3a,c both MOR-AS and MFI-AS have similar morphology: elongated plates up to 1 µm in length and 0.1 µm in width, combined in stacks. The pillaring does not change noticeably the morphology of the layered zeolites, see Figure 3b,d.

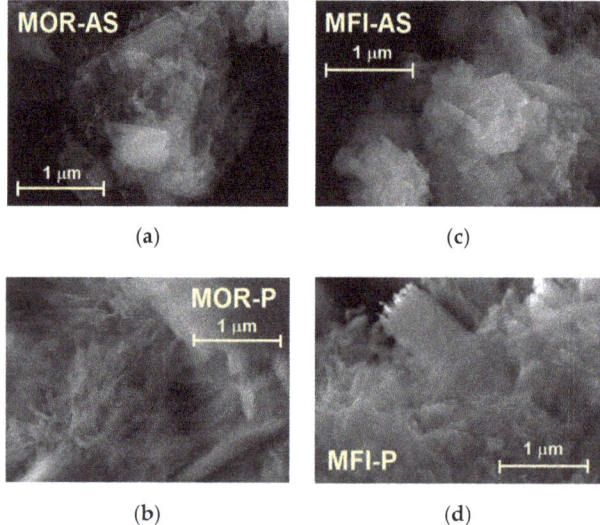

Figure 3. SEM images of the samples, as synthetized and pillared: MOR-AS (**a**), MOR-P (**b**), MFI-AS (**c**), MFI-P (**d**).

The results of the energy dispersive X-rays (EDX) elemental analysis of the as-synthesized and pillared materials are summarized in Table 1. Both MOR-AS and MFI-AS have nearly the same Si/Al ratios, 8.4 ± 0.3 and 8.8 ± 0.3, respectively. The Na/Al ratio in MFI-AS is close to unity within the experimental error, which means that all the negative charge due to partial substitution of Si for Al is compensated by Na^+. In MOR-AS an excess of positive charge (Na/Al > 1) must be balanced by Br anions. No trace of Br^- was detected in the MFI set of samples and in MOR-P. This means that both CTAB and TPABr are present only in their cationic forms, CTA^+ (hexadecyltrimethylammonium) and TPA^+ (tetrapropylammonium), respectively, balancing the dangling bonds of the zeolite layers. Such a rather nontrivial question of the coordination of charged Al tetrahedra and organic cations requires additional research, which the authors plan to carry out in the future, and the results of which will be published elsewhere.

Table 1. EDX elemental analysis of the as synthetized and pillared samples.

Sample	Na/Al	Si/Al	Br/Al
MOR-AS	1.14 ± 0.03	8.4 ± 0.3	0.24 ± 0.05
MOR-P	0.17 ± 0.06	15.2 ± 1.3	Not detected
MFI-AS	1.11 ± 0.09	8.8 ± 0.4	Not detected
MFI-P	0.34 ± 0.05	14.8 ± 1.0	Not detected

The pillaring results in an almost twofold increase in the Si/Al ratio, 15.2 ± 1.3 and 16.5 ± 1.0 for MOR-P and MFI-P, respectively, with a simultaneous decrease in sodium content, more pronounced for MOR-P. Sodium leaching is quite likely during sample processing in TEOS hydrolysis (Step 3). The role of compensating cations should eventually pass to protons, even if organic cations are involved in the hydrolyzed samples. In addition, it should be noted that all the materials are characterized by a certain inhomogeneity in the distribution of elements: there are regions with higher and lower Si/Al ratios, which is reflected in a rather large experimental error. An example of the element map distribution for the MOR-P sample is shown in Figure 4.

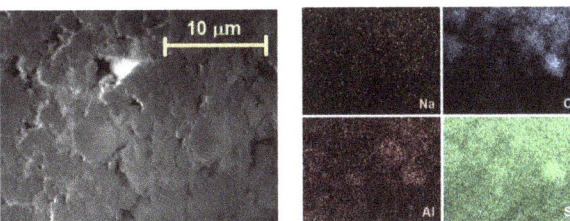

Figure 4. Element distribution maps in the MOR-P sample.

2.3. Thermal Analysis

The results of the simultaneous thermal analysis (including thermogravimetry (TG) and differential scanning calorimetry (DSC) combined with mass spectrometric analysis (MS) of the evolved gases) of the MOR-AS and MFI-AS samples are shown in Figure 5a–d, respectively.

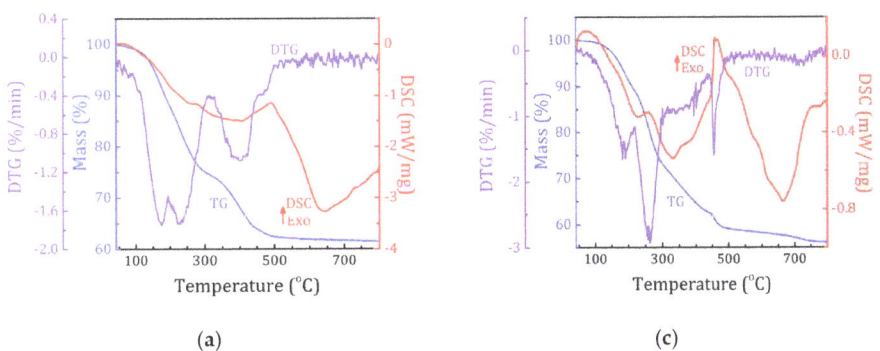

(a) (c)

Figure 5. *Cont.*

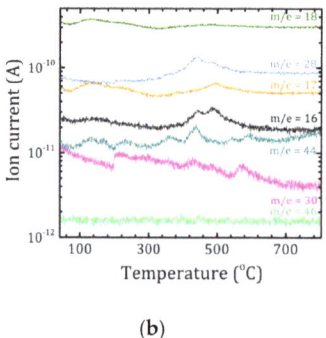 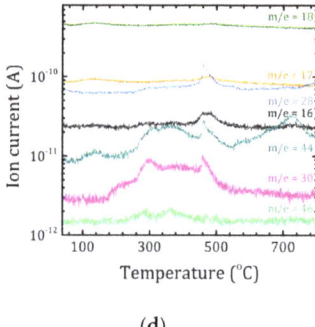

(b) (d)

Figure 5. (**a**,**c**): TG (blue), DTG (violet), DSC (red) and (**b**,**d**): ion current (various colors) curves for the MOR-AS (**a**,**b**) and MFI-AS (**c**,**d**) samples.

Both samples exhibit a rather complex mass loss. For MOR-AS, the mass loss below 300 °C is associated with water release (m/e = 16, 17, 18) from macro- and microcavities, peaks at 175 and 228 °C, respectively. The mass loss between 300 and 500 °C is related with a multistep decomposition of CTA^+ and PEG with formation of ammonium fragments, volatile low-carbon residues, e.g., ethylene (m/e = 28) and products of combustion of non-volatile high-carbon residues, CO_2 and H_2O. The peaks in the corresponding ion-current curves are accompanied by DSC peaks at 445 and 488 °C. In MFI-AS, water release occurs in one step below 220 °C, mass loss above 250 °C is associated with the decomposition of organic cations and molecules (CTA^+, TPA^+ and PEG).

2.4. NMR Study

2.4.1. ^{1}H MAS NMR and ^{13}C CP-MAS NMR

Figure 6 shows the ^{1}H MAS and ^{13}C CP/MAS (at the contact pulse duration τ_{cp} = 2 ms) NMR spectra for the studied samples of "as-synthesized" MOR-AS and MFI-AS. All the spectral lines can be attributed to the organic molecules CTAB and TPABr, the latter for the MFI-AS sample only. The reference spectra of pure CTAB and TPABr substances, simulated using the online service www.nmrdb.org [30], as well as the spatial structure of their molecules with atom labeling, are shown in the upper part of Figure 6a,b. As can be seen, the spectra for both zeolite samples exhibit typical features of the corresponding organic molecules (except the ^{1}H lines above 4.5 ppm that can be attributed to PEG and water), but all lines are broadened and shifted towards a lower magnetic field. The broadness of the spectral lines points out that the molecule mobility is frozen.

For better visualization, in Figure 6c the ^{13}C chemical shifts for all carbon atoms of CTAB in the MOR-AS and MFI-AS samples are plotted versus the numbering of carbon atoms. The data for a crystalline CTAB powder from Ref. [23], together with the simulation for a CTAB molecule are given as a comparison. The ^{13}C CTAB spectra for both MOR-AS and MFI-AS samples are very similar and correspond to immobilized rigid molecules in the all-trans conformation [23]. A chemical shift value of about 31 ppm can be attributed to C_4–C_{13} carbon atoms in the central part of the CTAB chain. A typical value of the chemical shift obtained in crystalline *n*-alkanes in the *trans*-conformation is 33 ppm [31]; the lower chemical shift of these methylene carbons by 2 ppm is usually attributed to the presence of a significant fraction of gauche conformers [23]. A higher chemical shift of carbon C_1 by 4 ppm evidences a strong interaction between the charged $-[N(CH_3)_3]^+$ head of the CTAB and the zeolite framework. This is also confirmed by a linewidth of the ^{1}H-NMR lines. The ^{1}H NMR lines broaden as one moves from the tail towards the $-[N(CH_3)_3]^+$ head of the molecule that means a decrease of the mobility of methyl groups (broad H_N peak) as compared to the CTAB tail. This is very consistent with the findings of studies on the inclusion of CTAB into a MWW type structure synthesized under basic/alkaline conditions, where it is suggested that CTAB can be included into hemicavities of

MWW through intermolecular hydrogen bonding with bridged oxygen atoms that are connected to Q^4 sites [19]. In the same way, in some NMR studies on the inclusion of Al into MCM-41 mesoporous aluminosilicates, which are synthesized using CTAB as a mesoporogen or structure directing agent, it was found that the polar head of CTAB shows a strong correlation with four-coordinated Al through electrostatic interactions between cationic ammonium-methyl head groups and tetrahedral Al (in the framework) [32]. In this sense, the interaction of the polar head of the surfactant with the surface of silica is carried out by silanol groups [33], which are very weak sites; while the interaction of CTAB with ordered aluminosilicate occurs by electrostatic interaction with the framework charge in the place where tetrahedral aluminum is present.

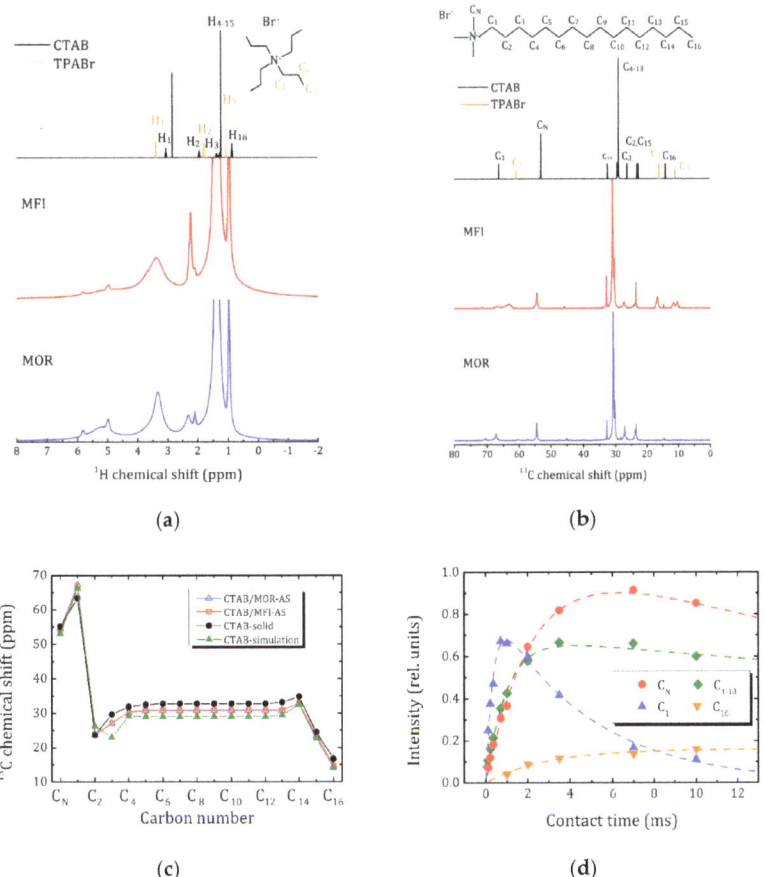

Figure 6. ^{1}H MAS NMR (**a**) and ^{13}C CP/MAS NMR at τ_{cp} = 2 ms (**b**) spectra of the studied samples MOR-AS and MFI-AS with simulated spectra of CTAB and TPABr (at the top) given for a comparison; (**c**) ^{13}C isotropic chemical shift profiles of CTAB: CTAB/MOR-AS (open triangles), CTAB/MFI-AS (open squares), crystalline powder from Ref. [23] (closed circles), simulated using www.nmrdb.org [30] (closed triangles); (**d**) integral peak intensity for C_N, C_1, C_{4-13} and C_{16} atoms of CTAB in MOR-AS versus contact pulse duration τ_{cp} (VCT experiment), dashed lines corresponds to the fitting within Equation (1).

This conclusion is in agreement with the VCT experiment. Figure 6d represents the intensity of the selected carbon peaks of CTAB in MOR-AS as a function of contact time ^{1}H-^{13}C. For C_{4-13}, the signal intensity is normalized per carbon nucleus.

The building up and loss of signal intensity during VCT can be described by the following Equation (1) [34]:

$$I = I_0 \left(1 - \frac{T_{CH}}{T_{1\rho}}\right)^{-1} \times \left[\exp\left(-\frac{\tau_{cp}}{T_{1\rho}}\right) - \exp\left(-\frac{\tau_{cp}}{T_{CH}}\right)\right] \quad (1)$$

T_{CH} determines the rising part of the intensity and represents the efficiency of the cross polarization between ^{1}H and ^{13}C nuclei and is often related to the mobility of the nuclei under study: mobile atoms have a high T_{CH} value because of the inefficiency of cross-polarization. The decay of the signal is governed by the rate of ^{1}H spin $T_{1\rho}$ relaxation. The T_{CH} and $T_{1\rho}$ parameters for the selected carbon sites of CTAB in MOR-AS, as determined from the dependencies shown in Figure 6d using Equation (1), are listed in Table 2. The C_N and the terminal methyl group (C_{16}) have a much longer T_{CH} than other carbon atoms. The terminal C_{16} group has the greatest mobility: the largest T_{CH}, and within the studied contact time range, the signal intensity does not decrease even at the longest τ_{cp} values that point out to a large $T_{1\rho}$ value that cannot be determined due to the low signal intensity. The C_1 group has the shortest T_{CH} and $T_{1\rho}$ times and hence the lowest mobility.

Table 2. Values of T_{CH} and $T_{1\rho}$ for the selected CTAB spectral lines in MOR-AS derived from ^{1}H-^{13}C CP/MAS NMR measurements.

CTAB Carbon Site	δ (ppm)	T_{CH} (ms)	$T_{1\rho}$ (ms)
C_N	54.5	2.21 ± 0.04	31 ± 13
C_1	67.3	0.37 ± 0.04	4.5 ± 0.5
C_{4-13}	30.5	0.98 ± 0.12	70 ± 30
C_{16}	14.6	2.8 ± 0.3	-

Coming back to the NMR spectral analysis, the introduction of TEOS and subsequent hydrolysis result in an essential broadening of the ^{1}H- and ^{13}C-NMR lines that is associated with a further decrease in the mobility of CTAB, see Figure 7.

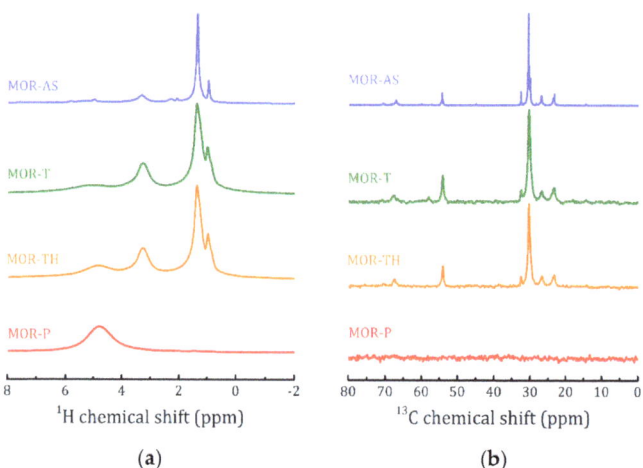

Figure 7. Cont.

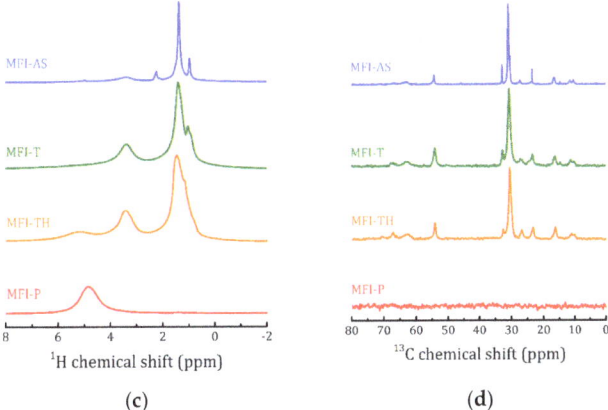

Figure 7. ^{1}H MAS NMR (**a**,**c**) and ^{13}C CP/MAS NMR at τ_{cp} = 2 ms (**b**,**d**) spectra of the studied MOR (**a**,**b**) and MFI (**c**,**d**) samples at all the preparation steps.

The line positions in MOR-T, MOR-TH, MFI-T and MFI-TH remain untouched as compared to MOR-AS and MFI-AS, respectively. Annealing results in the complete disappearance of organic matters (there are no traces of the ^{13}C-NMR signal in the MOR-P and MFI-P spectra). A broad ^{1}H line at 4.8 ppm points out to the presence of water molecules with restricted mobility.

2.4.2. ^{27}Al and ^{29}Si MAS NMR

The ^{27}Al MAS NMR spectra confirm the regularity of the zeolite frameworks of the as prepared samples, see Figure 8, the only line at about 54 ppm corresponds to Al in regular tetrahedral sites. The introduction of TEOS and the subsequent hydrolysis procedure do not perturb much the framework aluminum: the line slightly broadens and shifts at 2–3 ppm due to interaction with TEOS, see Table 3. However, the calcination procedure results in the appearance of six-coordinated extra-framework Al (the line at about 2 ppm) (20 and 27% for MOR-P and MFI-P, respectively). The presence of six-coordinated Al is often observed in protonated zeolites obtained by calcination of the ammonium form [35–38].

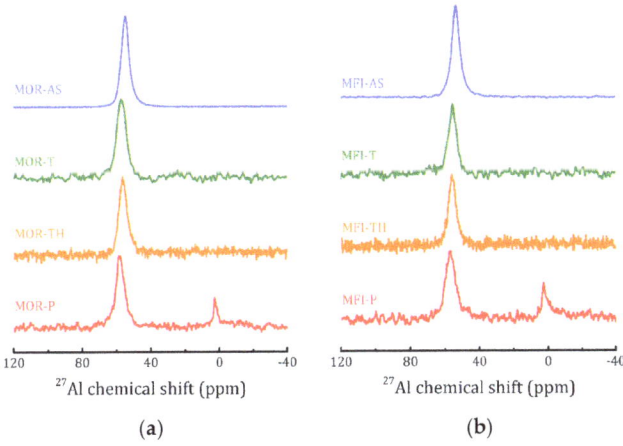

Figure 8. ^{27}Al MAS NMR spectra of the studied MOR (**a**) and MFI (**b**) samples at all the preparation steps.

Table 3. ^{27}Al NMR chemical shift (ν_0) and line width at half maximum ($\Delta\nu_{1/2}$) in the studied samples.

Sample	ν_0 (ppm)	$\Delta\nu_{1/2}$ (ppm)	Sample	ν_0 (ppm)	$\Delta\nu_{1/2}$ (ppm)
MOR-AS	54.6 ± 0.1	4.8 ± 0.1	MFI-AS	53.7 ± 0.1	4.9 ± 0.1
MOR-T	56.9 ± 0.1	5.6 ± 0.1	MFI-T	55.9 ± 0.1	4.9 ± 0.1
MOR-TH	56.3 ± 0.1	5.0 ± 0.1	MFI-TH	56.1 ± 0.1	4.7 ± 0.1
MOR-P	58.1 ± 0.1	5.7 ± 0.1	MFI-P	57.2 ± 0.1	6.2 ± 0.1
	2.6 ± 0.1	4.2 ± 0.2		1.6 ± 0.1	8.5 ± 0.1

It should be noted that the data reported in Table 3 can be used for a rough estimate of the ^{27}Al MAS NMR spectra that was done by a simple fitting by a Lorentzian line. But, as one can see from Figure 8, after hydrolysis, a shoulder at about 52 ppm appears (more pronounced for pillared samples). Such an asymmetric shape of the spectral line is issued by quadrupole interactions [39] and may point out at an increase in the quadrupole interactions due to the deformation of [AlO$_4$]$^-$ tetrahedra [40,41].

Figure 9 represents the ^{29}Si MAS NMR spectra of all the studied compounds. The spectra for the as-synthesized layered zeolites exhibit features typical for 3D zeolites.

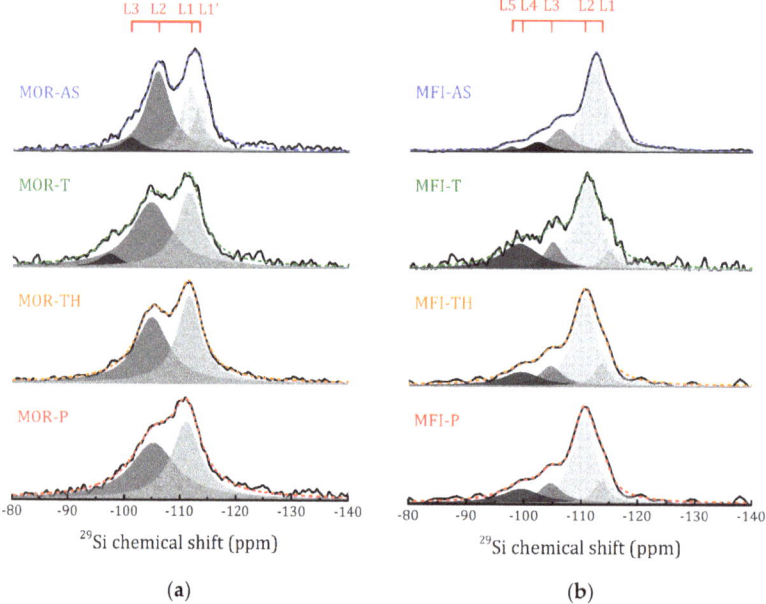

Figure 9. ^{29}Si MAS NMR spectra for the MOR (**a**) and MFI (**b**) samples at the all preparation steps. Filled patterns represent decomposition on Lorentzian functions, dashed lines represent the total fit.

For MOR-AS, the ^{29}Si spectrum was fitted by four Lorentzian lines (L1, L1', L2 and L3) that can be assigned with specific Q-type Si sites in the mordenite lattice: −113.5 and −111.6 ppm correspond to two different Q^4(0Al) sites, whereas −105.8 and −101.0 ppm can be attributed to Q^4(1Al) and Q^4(2Al), respectively [42]. Using the integrated areas of these lines, the Si/Al ratio can be estimated as 8.3 [43,44]. This is in a fair agreement with the EDX data and implicitly confirms the ^{29}Si NMR line assignment. However, rather important contributions of Q^4(1Al) and Q^4(2Al) are observed in layered mordenite, as compared with 3D mordenites with a close Si/Al ratio [45,46]. This could be due to the interaction of CTAB heads with Al tetrahedra at the interface that is truncated, resulting in a higher prevalence of Al in certain preferred sites of the zeolitic structure. In addition, one should take into account the change in the T-O-T angles for the Q^4(nAl) sites due to local structural distortions in the 2D plate. For several zeolites, including mordenite, there is an almost linear correlation between the ^{29}Si chemical shift of the $Q^4(n\text{Al})$ signal with the magnitude of T-O-T angle [47]. And finally, there may be an effect of the simultaneous action of these two factors.

To follow the changes in the ^{29}Si spectra that occur at each preparation step, in Figure 10 we plotted the parameters of individual spectral lines shown in Figure 9. As can be seen, for the mordenite sample, the introduction of TEOS (MOR-T) results in essential broadening the line width, the two lines previously attributed to $Q^4(0\text{Al})$ are merged, and a low chemical shift part of the ^{29}Si spectrum becomes more pronounced. After hydrolysis (MOR-TH) the spectrum can be perfectly fitted by two Lorentzian lines. Further, annealing (MOR-P) results in a slight line broadening and redistribution of the line intensities. The main difficulty in assigning the spectral line is caused by the partial overlapping of the $Q^4(1\text{Al})$ and $Q^4(2\text{Al})$ ^{29}Si mordenite signal with Q^4 and Q^3 of TEOS. The typical ranges of ^{29}Si chemical shift in mordenite [42,44], and TEOS [48] are shown in Figure 10a.

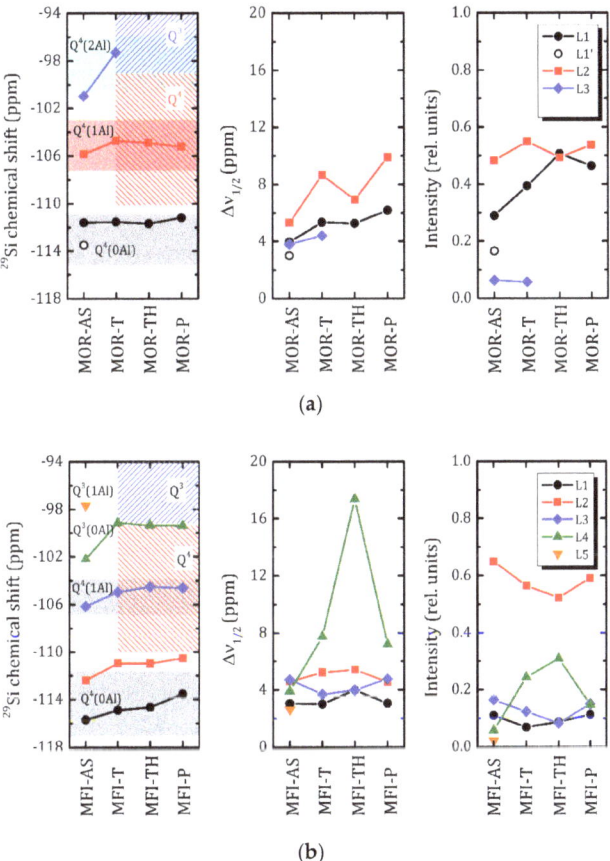

Figure 10. ^{29}Si NMR chemical shift (left), the line width at half maximum, $\Delta\nu_{1/2}$ (center) and relative integral intensities (right) of the individual Lorentzian lines for the MOR (**a**) and MFI (**b**) samples at the all preparation steps. Filled areas show the typical ranges of ^{29}Si chemical shift in zeolites (solid fill) and TEOS (hatched fill).

It should be noted that the ^{29}Si signal below 90 ppm was not detected even in CP/MAS mode (see Figure S1 in Supplementary Materials). In MOR-T, after the introduction of TEOS, this means that from the first moments a sol-gel reaction (hydrolysis and polycondensation) takes place, which triggers the formation of a three-dimensional network of Si tetrahedra in the interlayer space of the zeolite, between the CTAB molecules. This is consistent with the data for gelled TEOS reported in [48]. The next step, a targeted hydrolysis procedure, completes it (total disappearance of the Q^3 signal). A consistent increase in the intensity of the L1 line at each step of preparation implicitly points out the formation of SiO_2 oligomers. And in the pillared MOR-P signal at −111.2 ppm corresponds to the overlapping mordenite $Q^4(0Al)$ site and Q^4 of amorphous SiO_2 [49,50].

In the layered ZSM-5 zeolite, MFI-AS, the ^{29}Si spectrum can be decomposed into five Lorentzian lines (L1–L5), Figure 9b. The signals at −115.7 and −112.4 ppm can be attributed to two different $Q^4(0Al)$ sites, the remaining signals, −106.16, −102.2 and −97.7 ppm, can be assigned with $Q^4(1Al)$, $Q^3(0Al)$ and $Q^3(1Al)$, respectively [44,47,51]. The introduction of TEOS, in MFI-T, leads to significant changes in the $Q^3(0Al)$ signal: an increasing of the linewidth and integral intensity is observed, and the chemical shift is 3 ppm lower, Figure 10b. Overlapping basically triggers these changes with the Q^3 signal of TEOS. Similarly to MOR-T, an autohydrolysis and condensation also occur in MFI-T, but with the predominant formation of Q^3 structures. The subsequent hydrolysis procedure (MFI-P) leads to a further increase in the formation of Q^3, but after annealing (MFI-P), a sharp decrease in the L4 line intensity means that this is due to $Q^3(0Al)$ of the ZSM-5 zeolite. The autohydrolysis is in good agreement with the ease of hydrolysis of this type of alkoxide compounds when exposed to a small amount of water or humidity. Step 1 was carried out in the presence of water in the reaction mixture and flushing with methanol of this zeolite/organics hybrid interlayered compound under reflux did not completely remove the water. In addition, the interlaminar diffusion treatments with TEOS (Step 2) were not carried out in a controlled dry atmosphere, so the presence of internal water in the sample and ambient humidity may cause autohydrolysis of the compound even before it is treated in water at 90 °C (Step 3).

3. Materials and Methods

Both sets of materials were synthesized according to the procedure described in our previous work [26]. Pillaring of the obtained materials was done in accordance with the process proposed by Na et al. [28].

Step 1: For mordenite, the organic components (3.123 g of CTAB, 0.5205 g of polyethylene glycol (PEG) 20000) and 0.47 g of NaOH were completely dissolved in 36.3 mL of H_2O. After that 21.46 g of sodium silicate solution (25 wt% of SiO_2 and 10.6 wt% of Na_2O) were added. The obtained mixture was vigorously stirred for 20 min. Then, the solution of sodium aluminate (0.48 g of $NaAlO_2$ dissolved in 26.6 g of H_2O) was added dropwise. Finally, 26 g of a 10 wt% H_2SO_4 solution was added under vigorous stirring. The same method was used to obtain ZSM-5 zeolite, but with the addition of 2.66 g of TPABr as OSDA to other organic components.

The obtained mixtures were heated at 150 °C for 4 days in a stainless-steel autoclave with teflon coating under autogenous pressure. Then, the samples were filtered and washed with distilled water, and then washed with methanol under refluxing for 12 h at 60 °C to remove physically occluded surfactants. The resulting samples were labeled as MOR-AS and MFI-AS.

Step 2: 1.0 g of MOR-AS (or MFI-AS) sample was stirred in 5.0 g of TEOS for 12 h at 25 °C. Then samples were filtered and dried at 35 °C for 12 h. The obtained samples were labeled as MOR-T and MFI-T.

Step 3: To hydrolyze TEOS, 1.0 g of the MOR-T and MFI-T samples were stirred in 10.0 g of distilled water at 90 °C for 12 h. Washed with distilled water, filtered, and dried at 120 °C samples were labeled as MOR-TH and MFI-TH.

Step 4: Samples of MOR-TH and MFI-TH obtained after hydrolysis were calcined at 550 °C for 4 h in air to remove organic compounds. As a result, samples of pillared MOR and MFI were obtained, which were labeled MOR-P and MFI-P, respectively.

Powder XRD analysis was done on a Bruker D8 DISCOVER diffractometer using monochromatic CuK_α radiation (λ = 0.154056 nm). Diffractograms were recorded in the 2θ range of 5–40° (step width

0.0302°), where the main characteristic peaks of the MOR and MFI zeolites appear. SAXS patterns were recorded in a scan range from 0.2 to 7.0 2θ degree, step width 0.01°.

Simultaneous thermal analysis was carried out using a Netzsch STA 449 F1 Jupiter coupled with a QMS 403 Aëolos quadrupole mass spectrometer. The mass change of the samples and the composition of the evolved gases were registered. Analysis of samples was carried out in the temperature range 40–820 °C at a heating rate of 10 °C/min in an argon stream at a rate of 90 mL/min.

The morphology and elemental analysis of the samples was studied by an optical system integrated into D8 DISCOVER spectrometer (Bruker AXS, Karlsruhe, Germany) and by SEM applying Zeiss Merlin (Zeiss, Oberkochen, Germany) equipped with an EDX Oxford Instruments INCAx-act.

NMR spectra were recorded using a Bruker Avance IIIWB 400 MHz (Bruker, Karlsruhe, Germany) solid-state NMR spectrometer (operating with Topspin version 3.2) using a double-resonance 4 mm Magic Angle Spinning (MAS) probe. The operating frequencies were 400.23, 100.64, 104.28 and 79.5 MHz for ^{1}H, ^{13}C, ^{27}Al, and ^{29}Si nuclei, respectively. The rotor speed was 14 kHz. For all nuclei except ^{13}C, the direct excitation method was used. To increase the intensity of ^{13}C-NMR spectra, the cross-polarization (CP/MAS) method was applied. Variable contact time (VCT) experiments were performed with contact time τ_{cp} varied between 70 and 10,000 μs. The relaxation delay time was 5 s. Tetramethylsilane (TMS) was used as an external standard.

4. Conclusions

Mesostructured zeolite materials with the crystalline structure of MOR and MFI having Si/Al ratios equal to 8.4 and 8.8, respectively, were grown in the form of lamellar inorganic phases separated by layers of organic material. CTAB was used as an agent that creates mesopores, in a one-pot synthesis. It was shown that the mesostructured array consists of alternating lamellas of CTAB, with a thickness of ~3.5 nm, and a zeolite, with a thickness of one-unit cell along the z axis for each of the synthesized structures. Both lamellar zeolites have a similar morphology: elongated plates up to 1 μm long and 0.1 μm wide combined in stacks.

^{27}Al and ^{29}Si MAS NMR spectra confirm the regularity of the zeolite frameworks of the as-synthetized layered 2D materials: there is no extra-framework Al, ^{29}Si spectra correspond to bulk 3D MOR and MFI with broadened lines from Q^4(0Al), Q^4(1Al) and Q^4(2Al) sites.

Analysis of the ^{1}H and ^{13}C MAS NMR spectra and the VCT experiment evidence a strong interaction between the charged –[N(CH$_3$)$_3$]$^+$ heads of CTAB and the zeolite framework: the C_1 and terminal C_{16} groups of CTAB have the lowest and highest mobility, respectively. Since in the both MOR-AS and MFI-AS samples the Na/Al ratio is close to unity (a slight excess of Na$^+$ found in MOR-AS is balanced by Br$^-$ anions), CTA$^+$ and TPA$^+$ cations balance the dangling bonds of the zeolite layers.

The introduction of TEOS from the beginning leads to autohydrolysis and the formation of SiO$_2$ oligomers due to the water contained in the sample. Further targeted hydrolysis completes the formation of amorphous SiO$_2$ pillars separating the zeolite layers and holding them at fixed distances after thermal removal of the organic layers. Annealing leads to a partial drop out of Al from the zeolite frameworks (the appearance of extra-framework six-coordinated Al species). This implicitly points out that CTA$^+$ cations in the as-synthetized materials are localized near [AlO$_4$]$^-$, and the removal of organic cations leads to a partial collapse of the local structure. However, in general, the zeolite structure of the layers is preserved. After calcination the role of compensating cations should eventually pass to protons; moreover, the surface hydroxyls should balance of the dangling bonds of the zeolite layers. From this perspective the study of the inner surface of the pillared zeolites is of great interest and is actually under evaluation.

Supplementary Materials: The following is available online, Figure S1: ^{29}Si {^{1}H} CP/MAS NMR at τ_{cp} = 2 ms for the MOR-T and MFI-T samples.

Author Contributions: M.G.S. and V.P. conceived an experiment and analyzed the general data set; R.I.Y.-G. fabricated the layered two-dimensional (2D) MOR and MFI samples; I.V.Z. carried out the NMR analysis of the synthesized samples under supervision of M.G.S.; M.V.C. performed the thermal analysis; J.A.-G. and D.H.G.

analyzed XRD and SAXS data; F.N.M.-R. and S.F.-M. examined data of SEM-EDX studies. All the authors discussed the topic, helped identify specific aspects studied in this work, discussed the experimental results and contributed to the writing of the manuscript. All authors have read and agreed to the published version of the manuscript.

Funding: This research was funded by the Russian Foundation for Basic Research project No 18-53-34004 (RFBR and CITMA), and UNAM PAPIIT IN115920 Grant.

Acknowledgments: The samples were synthetized at CNyN-UNAM and studied at the Research Park of Saint Petersburg State University: Centre for X-ray Diffraction Studies, Interdisciplinary Resource Centre for Nanotechnology, Centre of Thermal Analysis and Calorimetry and Centre for Magnetic Resonance.

Conflicts of Interest: The authors declare no conflict of interest.

References

1. Wight, A.P.; Davis, M.E. Design and Preparation of Organic−Inorganic Hybrid Catalysts. *Chem. Rev.* **2002**, *102*, 3589–3614. [CrossRef] [PubMed]
2. Li, D.; Wu, Z.; Zhou, D.; Xia, Y.; Lu, X.; He, H.; Xia, Q. One-step synthesis of hybrid zeolite with exceptional hydrophobicity to accelerate the interfacial reaction at low temperature. *Microporous Mesoporous Mater.* **2019**, *280*, 195–202. [CrossRef]
3. Čejka, J.; Millini, R.; Opanasenko, M.; Serrano, D.P.; Roth, W.J. Advances and challenges in zeolite synthesis and catalysis. *Catal. Today* **2020**, *345*, 2–13. [CrossRef]
4. Roth, W.J.; Nachtigall, P.; Morris, R.E.; Čejka, J. Two-Dimensional Zeolites: Current Status and Perspectives. *Chem. Rev.* **2014**, *114*, 4807–4837. [CrossRef] [PubMed]
5. Zhang, K.; Ostraat, M.L. Innovations in hierarchical zeolite synthesis. *Catal. Today* **2016**, *264*, 3–15. [CrossRef]
6. Ge, T.; Hua, Z.; He, X.; Zhu, Y.; Ren, W.; Chen, L.; Zhang, L.; Chen, H.; Lin, C.; Yao, H.; et al. One-pot synthesis of hierarchically structured ZSM-5 zeolites using single micropore-template. *Chin. J. Catal.* **2015**, *36*, 866–873. [CrossRef]
7. Jia, X.; Khan, W.; Wu, Z.; Choi, J.; Yip, A.C. Modern synthesis strategies for hierarchical zeolites: Bottom-up versus top-down strategies. *Adv. Powder Technol.* **2019**, *30*, 467–484. [CrossRef]
8. Koohsaryan, E.; Anbia, M. Nanosized and hierarchical zeolites: A short review. *Chin. J. Catal.* **2016**, *37*, 447–467. [CrossRef]
9. Mintova, S.; Grand, J.; Valtchev, V. Nanosized zeolites: Quo Vadis? *Comptes Rendus Chim.* **2016**, *19*, 183–191. [CrossRef]
10. Margarit, V.J.; Portilla, M.T.; Navarro, M.T.; Abudawoud, R.; Al-Zahrani, I.M.; Shaikh, S.; Martínez, C.; Corma, A. One-pot co-crystallization of beta and pentasil nanozeolites for the direct conversion of a heavy reformate fraction into xylenes. *Appl. Catal. A Gen.* **2019**, *581*, 11–22. [CrossRef]
11. Moliner, M. Direct Synthesis of Functional Zeolitic Materials. *ISRN Mater. Sci.* **2012**, *2012*, 1–24. [CrossRef]
12. Moliner, M.; Rey, F.; Corma, A. Towards the Rational Design of Efficient Organic Structure-Directing Agents for Zeolite Synthesis. *Angew. Chem. Int. Ed.* **2013**, *52*, 13880–13889. [CrossRef] [PubMed]
13. Guo, G.-Q.; Long, Y.-C.; Sun, Y.-J. Synthesis of FER type zeolite with tetrahydrofuran as the template. *Chem. Commun.* **2000**, 1893–1894. [CrossRef]
14. Migliori, M.; Aloise, A.; Giordano, G. Methanol to dimethylether on H-MFI catalyst: The influence of the Si/Al ratio on kinetic parameters. *Catal. Today* **2014**, *227*, 138–143. [CrossRef]
15. Li, H.-J.; Zhou, X.-D.; Di, Y.H.; Zhang, J.-M.; Zhang, Y. Effect of Si-ATP/CTAB ratio on crystal morphology, pore structure and adsorption performance of hierarchical (H) ZSM-11 zeolite. *Microporous Mesoporous Mater.* **2018**, *271*, 146–155. [CrossRef]
16. Catizzone, E.; Migliori, M.; Mineva, T.; Van Daele, S.; Valtchev, V.; Giordano, G. New synthesis routes and catalytic applications of ferrierite crystals. Part 2: The effect of OSDA type on zeolite properties and catalysis. *Microporous Mesoporous Mater.* **2020**, *296*, 109988. [CrossRef]
17. Che, S.; Feng, J.; Che, S. An insight into the role of the surfactant CTAB in the formation of microporous molecular sieves. *Dalton Trans.* **2014**, *43*, 3612–3617. [CrossRef]
18. Wang, X.; Chen, H.; Meng, F.; Gao, F.; Sun, C.; Sun, L.; Wang, S.; Wang, L.; Wang, Y. CTAB resulted direct synthesis and properties of hierarchical ZSM-11/5 composite zeolite in the absence of template. *Microporous Mesoporous Mater.* **2017**, *243*, 271–280. [CrossRef]

19. Xu, L.; Ji, X.; Li, S.; Zhou, Z.; Du, X.; Sun, J.; Deng, F.; Che, S.; Wu, P. Self-Assembly of Cetyltrimethylammonium Bromide and Lamellar Zeolite Precursor for the Preparation of Hierarchical MWW Zeolite. *Chem. Mater.* **2016**, *28*, 4512–4521. [CrossRef]
20. Chizhik, V.I.; Chernyshev, Y.S.; Donets, A.V.; Frolov, V.V.; Komolkin, A.V.; Shelyapina, M.G. *Magnetic Resonance and Its Applications*; Springer International Publishing: Cham, Switzerland, 2014. [CrossRef]
21. Shelyapina, M.; Nefedov, D.Y.; Kostromin, A.V.; Siluykov, O.; Zvereva, I. Proton mobility in Ruddlesden-Popper phase $H_2La_2Ti_3O_{10}$ studied by ^{1}H-NMR. *Ceram. Int.* **2019**, *45*, 5788–5795. [CrossRef]
22. Shelyapina, M.; Lushpinskaya, I.P.; Kurnosenko, S.A.; Silyukov, O.I.; Zvereva, I.A. Identification of Intercalates and Grafted Organic Derivatives of $H_2La_2Ti_3O_{10}$ by Multinuclear NMR. *Russ. J. Gen. Chem.* **2020**, *90*, 760–761. [CrossRef]
23. Kharkov, B.B.; Dvinskikh, S.V. Chain dynamics of surfactants in mesoporous silica. *Phys. Chem. Chem. Phys.* **2013**, *15*, 18620–18626. [CrossRef] [PubMed]
24. Kharkov, B.B.; Corkery, R.W.; Dvinskikh, S.V. Phase Transitions and Chain Dynamics of Surfactants Intercalated into the Galleries of Naturally Occurring Clay Mineral Magadiite. *Langmuir* **2014**, *30*, 7859–7866. [CrossRef]
25. Khimyak, Y.Z.; Klinowski, J. Solid-state NMR studies of the organic template in mesostructured aluminophosphates. *Phys. Chem. Chem. Phys.* **2001**, *3*, 616–626. [CrossRef]
26. Yocupicio-Gaxiola, R.I.; Petranovskii, V.; Antúnez-García, J.; Moyado, S.F. One-pot synthesis of lamellar mordenite and ZSM-5 zeolites and subsequent pillaring by amorphous SiO_2. *Appl. Nanosci.* **2019**, *9*, 557–565. [CrossRef]
27. Baerlocher, C.; McCusker, L.B.; Olson, D.H. *Atlas of Zeolite Framework Types*, 6th ed.; Elsevier: Amsterdam, The Netherlands, 2007.
28. Na, K.; Choi, M.; Park, W.; Sakamoto, Y.; Terasaki, O.; Ryoo, R. Pillared MFI Zeolite Nanosheets of a Single-Unit-Cell Thickness. *J. Am. Chem. Soc.* **2010**, *132*, 4169–4177. [CrossRef]
29. Na, K.; Park, W.; Seo, Y.; Ryoo, R. Disordered Assembly of MFI Zeolite Nanosheets with a Large Volume of Intersheet Mesopores. *Chem. Mater.* **2011**, *23*, 1273–1279. [CrossRef]
30. Banfi, D.; Patiny, L. www.nmrdb.org: Resurrecting and Processing NMR Spectra On-line. *Chim. Int. J. Chem.* **2008**, *62*, 280–281. [CrossRef]
31. Ishikawa, S.; Kurosu, H.; Ando, I. Structural studies of n-alkanes by variable-temperature solid-state high-resolution ^{13}C-NMR spectroscopy. *J. Mol. Struct.* **1991**, *248*, 361–372. [CrossRef]
32. Janicke, M.; Landry, C.C.; Christiansen, S.C.; Kumar, D.; Stucky, G.D.; Chmelka, B.F. Aluminum Incorporation and Interfacial Structures in MCM-41 Mesoporous Molecular Sieves. *J. Am. Chem. Soc.* **1998**, *120*, 6940–6951. [CrossRef]
33. Baccile, N.; Babonneau, F. Organo-modified mesoporous silicas for organic pollutant removal in water: Solid-state NMR study of the organic/silica interactions. *Microporous Mesoporous Mater.* **2008**, *110*, 534–542. [CrossRef]
34. Voelkel, R. High-Resolution Solid-State ^{13}C-NMR Spectroscopy of Polymers [New Analytical Methods(37)]. *Angew. Chem. Int. Ed.* **1988**, *27*, 1468–1483. [CrossRef]
35. Deng, F.; Yue, Y.; Ye, C. Observation of Nonframework Al Species in Zeolite β by Solid-State NMR Spectroscopy. *J. Phys. Chem. B* **1998**, *102*, 5252–5256. [CrossRef]
36. Li, S.; Zheng, A.; Su, Y.; Fang, H.; Shen, W.; Yu, Z.; Chen, L.; Deng, F. Extra-framework aluminium species in hydrated faujasite zeolite as investigated by two-dimensional solid-state NMR spectroscopy and theoretical calculations. *Phys. Chem. Chem. Phys.* **2010**, *12*, 3895. [CrossRef]
37. Zhukov, Y.; Efimov, A.Y.; Shelyapina, M.; Petranovskii, V.; Zhizhin, E.; Burovikhina, A.; Zvereva, I. Effect of preparation method on the valence state and encirclement of copper exchange ions in mordenites. *Microporous Mesoporous Mater.* **2016**, *224*, 415–419. [CrossRef]
38. Shelyapina, M.; Krylova, E.A.; Zhukov, Y.M.; Zhukov, Y.; Rodríguez-Iznaga, I.; Petranovskii, V.; Fuentes-Moyado, S.; Iznaga, R.; Moyado, F. Comprehensive Analysis of the Copper Exchange Implemented in Ammonia and Protonated Forms of Mordenite Using Microwave and Conventional Methods. *Molecules* **2019**, *24*, 4216. [CrossRef]
39. Freude, D. Quadrupolar nuclei in solid-state nuclear magnetic resonance. In *Encyclopedia of Analytical Chemistry*; Meyers, R.A., Dybowski, C., Eds.; Wiley: New York, NY, USA, 2006; pp. 1–37.

40. Kasperovich, V.S.; Sodel', N.E.; Shelyapina, M. Nonempirical cluster calculations of the electric field gradient tensor in yttrium-aluminum garnet $Y_3Al_5O_{12}$. *Phys. Solid State* **2006**, *48*, 1684–1688. [CrossRef]
41. Shelyapina, M.; Kasperovich, V.; Wolfers, P. Electronic structure and electric-field-gradients distribution in Y3Al5O12: An ab initio study. *J. Phys. Chem. Solids* **2006**, *67*, 720–724. [CrossRef]
42. Fyfe, C.A.; Feng, Y.; Grondey, H.; Kokotailo, G.T.; Gies, H. One- and two-dimensional high-resolution solid-state NMR studies of zeolite lattice structures. *Chem. Rev.* **1991**, *91*, 1525–1543. [CrossRef]
43. Engelhardt, G.; Michel, D. *High Resolution Solid State NMR of Silicates and Zeolites*; Wiley: New York, NY, USA, 1987.
44. Brouwer, D.H.; Brouwer, C.C.; Mesa, S.; Semelhago, C.A.; Steckley, E.E.; Sun, M.P.; Mikolajewski, J.G.; Baerlocher, C. Solid-state 29Si NMR spectra of pure silica zeolites for the International Zeolite Association Database of Zeolite Structures. *Microporous Mesoporous Mater.* **2020**, *297*, 110000. [CrossRef]
45. Kato, M.; Itabashi, K.; Matsumoto, A.; Tsutsumi, K. Characteristics of MOR-Framework Zeolites Synthesized in Fluoride-Containing Media and Related Ordered Distribution of Al Atoms in the Framework. *J. Phys. Chem. B* **2003**, *107*, 1788–1797. [CrossRef]
46. Paixão, V.; Carvalho, A.P.; Rocha, J.; Fernandes, A.; Martins, F. Modification of MOR by desilication treatments: Structural, textural and acidic characterization. *Microporous Mesoporous Mater.* **2010**, *131*, 350–357. [CrossRef]
47. Klinowski, J. Recent Advances in Solid-State NMR of Zeolites. *Annu. Rev. Mater. Res.* **1988**, *18*, 189–218. [CrossRef]
48. Glaser, R.H.; Wilkes, G.L.; Bronnimann, C.E. Solid-state 29Si NMR of TEOS-based multifunctional sol-gel materials. *J. Non-Cryst. Solids* **1989**, *113*, 73–87. [CrossRef]
49. Brinker, C.; Kirkpatrick, R.; Tallant, D.; Bunker, B.; Montez, B. NMR confirmation of strained "defects" in amorphous silica. *J. Non-Cryst. Solids* **1988**, *99*, 418–428. [CrossRef]
50. Rainho, J.P.; Rocha, J.; Carlos, L.D.; Almeida, R.M. 29Si nuclear-magnetic-resonance and vibrational spectroscopy studies of SiO2–TiO2 powders prepared by the sol-gel process. *J. Mater. Res.* **2001**, *16*, 2369–2376. [CrossRef]
51. Klinowski, J.; Carpenter, T.; Gladden, L. High-resolution solid-state NMR studies of temperature-induced phase transitions in silicalite (zeolite ZSM-5). *Zeolites* **1987**, *7*, 73–78. [CrossRef]

Sample Availability: Samples of the compounds are available from the author (R.I.Y.-G.) upon request.

Publisher's Note: MDPI stays neutral with regard to jurisdictional claims in published maps and institutional affiliations.

© 2020 by the authors. Licensee MDPI, Basel, Switzerland. This article is an open access article distributed under the terms and conditions of the Creative Commons Attribution (CC BY) license (http://creativecommons.org/licenses/by/4.0/).

Article

NMR Relaxivities of Paramagnetic Lanthanide-Containing Polyoxometalates

Aiswarya Chalikunnath Venu [1], Rami Nasser Din [2], Thomas Rudszuck [3], Pierre Picchetti [1], Papri Chakraborty [1], Annie K. Powell [1,4,5,*], Steffen Krämer [2,*], Gisela Guthausen [3,6,*] and Masooma Ibrahim [1,*]

1. Karlsruhe Institute of Technology (KIT), Institute of Nanotechnology (INT), Hermann-von-Helmholtz-Platz 1, 76344 Eggenstein-Leopoldshafen, Germany; aiswarya.venu@kit.edu (A.C.V.); pierre.picchetti@kit.edu (P.P.); papri.chakraborty@kit.edu (P.C.)
2. LNCMI-EMFL, CNRS, INSA-T and UPS, Université Grenoble Alpes, Boîte Postale 166, CEDEX 9, 38042 Grenoble, France; rami.nasser-din@lncmi.cnrs.fr
3. Karlsruhe Institute of Technology (KIT), MVM-VM, Adenauerring 20b, 76131 Karlsruhe, Germany; thomas.rudszuck@kit.edu
4. Institute of Inorganic Chemistry, Karlsruhe Institute of Technology (KIT), Engesserstrasse 15, 76131 Karlsruhe, Germany
5. Institute for Quantum Materials and Technologies (IQMT), Karlsruhe Institute of Technology (KIT), Hermann-von-Helmholtz-Platz 1, 76344 Eggenstein-Leopoldshafen, Germany
6. Karlsruhe Institute of Technology (KIT), EBI-WCWT, Adenauerring 20b, 76131 Karlsruhe, Germany
* Correspondence: annie.powell@kit.edu (A.K.P.); steffen.kramer@lncmi.cnrs.fr (S.K.); gisela.guthausen@kit.edu (G.G.); masooma.ibrahim@kit.edu (M.I.)

Abstract: The current trend for ultra-high-field magnetic resonance imaging (MRI) technologies opens up new routes in clinical diagnostic imaging as well as in material imaging applications. MRI selectivity is further improved by using contrast agents (CAs), which enhance the image contrast and improve specificity by the paramagnetic relaxation enhancement (PRE) mechanism. Generally, the efficacy of a CA at a given magnetic field is measured by its longitudinal and transverse relaxivities r_1 and r_2, i.e., the longitudinal and transverse relaxation rates T_1^{-1} and T_2^{-1} normalized to CA concentration. However, even though basic NMR sensitivity and resolution become better in stronger fields, r_1 of classic CA generally decreases, which often causes a reduction of the image contrast. In this regard, there is a growing interest in the development of new contrast agents that would be suitable to work at higher magnetic fields. One of the strategies to increase imaging contrast at high magnetic field is to inspect other paramagnetic ions than the commonly used Gd(III)-based CAs. For lanthanides, the magnetic moment can be higher than that of the isotropic Gd(III) ion. In addition, the symmetry of electronic ground state influences the PRE properties of a compound apart from diverse correlation times. In this work, PRE of water ^{1}H has been investigated over a wide range of magnetic fields for aqueous solutions of the lanthanide containing polyoxometalates $[Dy^{III}(H_2O)_4GeW_{11}O_{39}]^{5-}$ (**Dy-W$_{11}$**), $[Er^{III}(H_2O)_3GeW_{11}O_{39}]^{5-}$ (**Er-W$_{11}$**) and $[\{Er^{III}(H_2O)(CH_3COO)(P_2W_{17}O_{61})\}_2]^{16-}$ (**Er$_2$-W$_{34}$**) over a wide range of frequencies from 20 MHz to 1.4 GHz. Their relaxivities r_1 and r_2 increase with increasing applied fields. These results indicate that the three chosen POM systems are potential candidates for contrast agents, especially at high magnetic fields.

Keywords: polyoxometalates; nuclear magnetic resonance imaging; paramagnetic relaxation enhancement; lanthanides; relaxivity; dysprosium; erbium

1. Introduction

Nuclear magnetic resonance (NMR) studies involving paramagnetic systems have been the subject of research in fields such as biochemistry, medicine, and material science [1,2]. The presence of an unpaired electron originating from paramagnetic molecules significantly affects the NMR behavior of the entire system [3,4]. The paramagnetic effects, which arise due to the hyperfine interactions between a nuclear spin I and the unpaired electronic spin S of the paramagnetic center, can be broadly categorized into two types. First, the NMR chemical shift range can be largely expanded over a wide spectral range, for example ppm range (δ-para) up to 100 for ^{1}H. Even more than 1000 ppm for heavier ligand atoms such as ^{13}C or ^{15}N can be observed due to the large magnetic moment of unpaired electrons [5,6]. Second, the presence of an unpaired electron causes a faster nuclear spin relaxation that leads to the enhancement of longitudinal, $R_1 = 1/T_1$, and transverse, $R_2 = 1/T_2$, nuclear spin relaxation rates. This effect is commonly called paramagnetic relaxation enhancement (PRE). In this context, Curie spin relaxation is an important contributor to the water relaxivity in complexes of certain lanthanide ions (e.g., Tb, Dy, Ho, Er) due to their high magnetic moments and short electronic relaxation times, which limit the efficiency of Solomon-Bloembergen-type relaxation. The alignment of paramagnetic ions increases with field and inverse temperature according to Curie's law. Thus, in high fields, where the Curie spin becomes large, nuclear transverse relaxation may be dominated by interaction with the Curie spin [7].

Magnetic anisotropy is one of the most important properties of the paramagnetic ions for technological applications and plays a particularly important role in the development of single-molecule magnets (SMMs) [8]. The anisotropic distribution of electrons in the 4f orbitals of lanthanide ions can be prolate (axially elongated) (Pm^{III}, Sm^{III}, Er^{III}, Tm^{III}, and Yb^{III}) or oblate (equatorially elongated) (Ce^{III}, Pr^{III}, Nd^{III}, Tb^{III}, Dy^{III}, and Ho^{III}) (see Table S1) [9]. As a result of the anisotropy in the ground state of paramagnetic ions, a Larmor frequency difference occurs between the Ln-coordinated and the bulk water molecules, which is proportional to the external magnetic field [10]. Moreover, the Ln anisotropies strongly influence both paramagnetic NMR effects of solutions containing Ln complexes, the shifts [11], and the relaxation properties of the neighboring nuclei [12].

Paramagnetic relaxation enhancement (PRE) is usually explored in the field of magnetic resonance imaging (MRI) and NMR spectroscopy for the development of contrast agents and the structural determination of biomolecules, respectively [13].

Traditional contrast agents are mainly based on a paramagnetic gadolinium metal ion due to its seven unpaired electrons, high magnetic moment, and long electron spin relaxation time (10^{-9} s). However, recent information has raised new concerns about their toxicity [14]. Alternatively, other paramagnetic lanthanides are promising candidates to be used as contrast agents and among them dysprosium, due to its asymmetric/anisotropic electronic ground state (^{6}H$_{15/2}$) and a very high magnetic moment (μ_{eff} = 10.6 μB) may improve relaxivity [3,15].

Another important factor concerning MRI technologies is the use of ultra-high magnetic field instruments to improve sensitivity and spatial resolution. This is due to the strong dependency of signal to noise ratio in relation to the applied magnetic field (B_0) [16,17]. In the past few decades, a different range of magnetic fields has been utilized depending on the various field of applications. The currently used magnetic field for clinical MRI systems is at a maximum of 11.7 T [18]. The preclinical system and the small animal imaging instruments are operated with up to 21.1 T. The maximum magnetic field of 28.2 T is used in NMR for the structural determination of small organic molecules or large biomolecules. MRI technologies with ultra-high magnetic fields require the use of new materials as contrast agents [16]. Moreover, for characterization of CAs over the entire field range of interest, results from NMR spectrometers using different magnet types need to be combined. These include fast field cycling techniques, permanent magnets, superconducting magnets and resistive high field magnets [19].

Recently, we have investigated NMR relaxivity of heterometallic high-spin molecular clusters [Fe$_{10}$Ln$_{10}$(Me-tea)$_{10}$(Me-teaH)$_{10}$(NO$_3$)$_{10}$] (lanthanides (Ln) Ln = Gd^{III}, Tb^{III}, Dy^{III}, Er^{III} and Tm^{III}) {Fe$_{10}$Ln$_{10}$} [20,21] and [Ln$_{30}$Co$_8$Ge$_{12}$W$_{108}$O$_{408}$(OH)$_{42}$(OH$_2$)$_{30}$]$^{56-}$ (Ln = Gd^{III}, Dy^{III}, Eu^{III}, and Y^{III}) {Ln$_{30}$Co$_8$} [22]. The relaxivities of {Fe$_{10}$Ln$_{10}$} and {Ln$_{30}$Co$_8$} clusters in water were measured over a wide range of ^{1}H Larmor frequencies from 10 MHz up to 1.4 GHz. The alteration of the lanthanide ions (at structurally similar sites within the same scaffold) in {Fe$_{10}$Ln$_{10}$} and {Ln$_{30}$Co$_8$} made it possible to differentiate the relaxation impacts of electronic states and molecular dynamics. The transverse relaxivity was found to increase with the field, whereas field dependence of the longitudinal relaxivities of these molecules depends on the nature of the lanthanide.

Apart from the electronic properties of the paramagnetic centers, the relaxivities depend on several other structural and dynamic features such as the number of bound water molecules, their rate of exchange with the bulk, size, clustering and aggregation, tumbling, diffusion and rotational correlation times [2,23].

MRI contrast agents increase both r_1 and r_2 to different extents depending on their nature as well as on the applied magnetic field [24,25]. Accordingly, the development of new contrast agents is also a demanding area of research, which aims to increase their efficiency and sensitivity with increasing field. Field-dependent measurements of relaxivity are thus important to characterize new potential contrast agents. In this quest, one of the approaches we took was to synthesize and analyze the NMR relaxometry of ultrahigh-spin polyoxometalates (POM)-based heterometallic clusters in aqueous solutions [22]. The usage of Ln-POM nanocomposites as host/guest assemblies for contrast agents has also been reported by other groups [26–28]. POMs, as anionic metal oxide clusters, bear many properties that make them attractive for applications in various fields such as catalysis, magnetism, imaging, optics, medicine and also have interesting electrochemical properties [29–34]. Lacunary POMs or POM ligands are usually synthesized from parent plenary precursors by removal of one or more MO$_6$ octahedra. Various types and the number of spin-coupled paramagnetic centers (d- or f-block), usually bridged via μ_2-oxo/hydroxo can be incorporated into the structurally well-defined vacant sites of the lacunary POMs. This leads to the formation of monomeric dimeric, trimeric and tetrameric assemblies [35,36]. POM ligands can be viewed as an inorganic analogue of the porphyrins (Figure 1). By analogy, a paramagnetic metal containing polyoxometalates (PM-POMs) could be designed and investigated as contrast agents. As part of our continuous research in this field, we report another approach to investigate PRE of POM molecules that contain homometallic paramagnetic lanthanides. Thus, NMR relaxometry of aqueous solutions containing POMs K$_5$[Dy(H$_2$O)$_4$GeW$_{11}$O$_{39}$]·16H$_2$O (**Dy-W$_{11}$**), K$_5$[Er(H$_2$O)$_3$GeW$_{11}$O$_{39}$]·20H$_2$O (**Er-W$_{11}$**) [37], and (NH$_2$Me$_2$)$_{13}$Na$_3$[{Er(H$_2$O)(CH$_3$COO)(P$_2$W$_{17}$O$_{61}$)}$_2$]·40H$_2$O (**Er$_2$-W$_{34}$**) [38] has been carried out at ^{1}H Larmor frequencies from 20 MHz to 1.4 GHz.

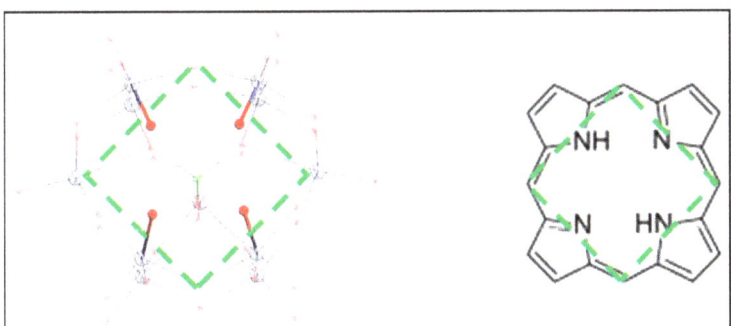

Figure 1. Analogy of monolacunary POM ligand [GeW$_{11}$O$_{39}$]$^{8-}$ with porphyrin.

Structure of the POMs

Monolanthanide-substituted Keggin-type POMs: The Keggin POMs [Dy^{III}(H$_2$O)$_4$GeW$_{11}$O$_{39}$]$^{5-}$ (**Dy-W$_{11}$**) and [Er^{III}(H$_2$O)$_3$GeW$_{11}$O$_{39}$]$^{5-}$ (**Er-W$_{11}$**) contain one monolacunary Keggin [α-GeW$_{11}$O$_{39}$]$^{8-}$ subunit and one lanthanide metal ion. This occupies the position that has been formed by loss of a W–O$_t$ group from the plenary [α-GeW$_{12}$O$_{40}$]$^{4-}$ anion, which consists of a central {GeO$_4$} tetrahedron surrounded by four vertex-sharing {W$_3$O$_{13}$} triads (Figure 2). Mass spectrometric studies were carried out to check the stability of the compounds in solution which indicated that the Ln metal ion remains attached to the POM skeleton. However, the closed three-dimensional (3D) framework architecture, that was built by connecting POM moieties via K ion linkers, collapses in solution. The two isostructural POMs are different only in their number of exchangeable aqua ligands that are coordinated to Ln ions. There are three terminal water ligands in **Er-W$_{11}$** due to the smaller ionic radius of Er^{III} ion compared to Dy^{III} ion which has four terminal water ligands [37].

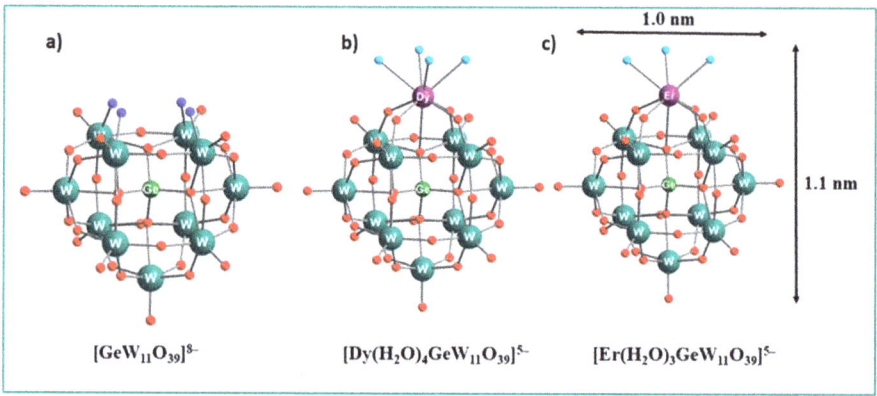

Figure 2. Ball-and-stick representation: (**a**) Keggin POM ligand [GeW$_{11}$O$_{39}$]$^{8-}$. (**b**) [Dy(H$_2$O)$_4$GeW$_{11}$O$_{39}$]$^{5-}$. (**c**) [Er(H$_2$O)$_3$GeW$_{11}$O$_{39}$]$^{5-}$. Color scheme: O, red; aqua ligand, turquoise; four oxo ligands that coordinate to paramagnetic metal ions, dark blue.

Er-substituted Wells-Dawson-type POM: The anionic unit of [{Er^{III}(H$_2$O)(CH$_3$COO)(P$_2$W$_{17}$O$_{61}$)}$_2$]$^{16-}$ (**Er$_2$-W$_{34}$**) consists of dinuclear erbium(III) core, [{Er(H$_2$O)(CH$_3$COO)}$_2$]$^{4+}$, which is sandwiched between two monolacunary Wells-Dawson [P$_2$W$_{17}$O$_{61}$]$^{10-}$ units. Each Er^{III} in these units takes up the void (monolacunary) site of [P$_2$W$_{17}$O$_{61}$]$^{10-}$ and is coordinated to the four available oxygen atoms. The two {ErP$_2$W$_{17}$O$_{61}$} units are bridged by acetate groups in the η^1:η^2:μ_2 fashion. Each Er(III) ion is eight coordinate with square antiprismatic geometry. Thus, every Er^{III} ion in these units is coordinated by four O atoms of the POM ligand, two oxygen atoms of an acetate ligand, one O atom of the other acetate ligand and one terminal aqua ligand (Figure 3). The mass spectrometry data suggest that the polyanion **Er$_2$-W$_{34}$** is fragmented into monomeric species [38]. In some cases, MS performed under vacuum and μM conditions does not necessarily detect the species in solution, but rather captures the fragments of complexes that exist in the gas phase. Furthermore, the detection probability of the parent molecule during the transition from solution to gas phase decreases due to the weaker hydrophobic and stronger electrostatic interactions among gas phase assemblies compared to the solution phase [39]. Magnetic properties of the isolated Ln(III) are summarized in Table 1.

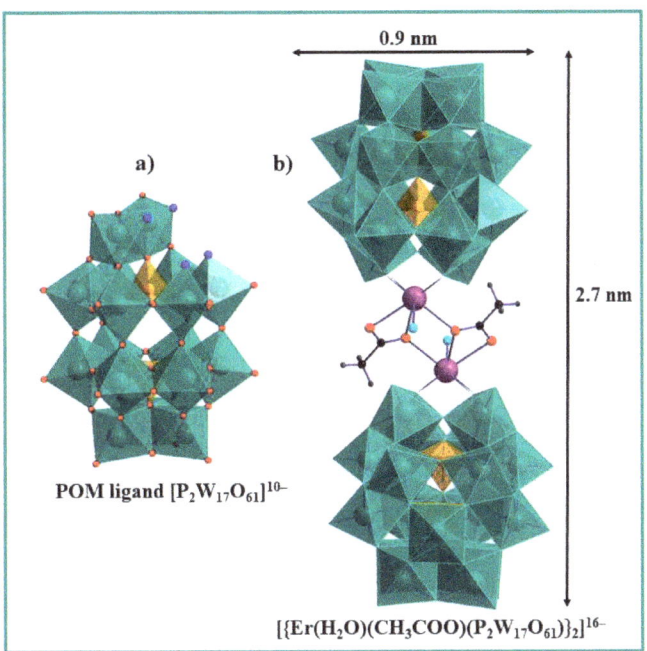

Figure 3. Combined polyhedral/ball-and-stick representation: (**a**) Wells-Dawson POM ligand $[P_2W_{17}O_{61}]^{10-}$. (**b**) $[\{Er(H_2O)(CH_3COO)(P_2W_{17}O_{61})\}_2]^{16-}$. Color scheme: O, red; aqua ligand, turquoise; four oxo ligands that coordinate to Er metal ions, dark blue; WO_6 octahedra, teal; PO_4 tetrahedra, yellow; Er, violet; C, black; H, grey.

Table 1. Magnetic properties of non-interacting lanthanide ions.

Ln(III)	Dy(III)	Er(III)
Free-ion ground state term $^{2S+1}L_J$	$^6H_{15/2}$	$^4I_{15/2}$
S	5/2	3/2
L	5	6
J	15/2	15/2
g-factor	4/3	6/5
$\mu_{eff} = gJ\mu_B\sqrt{J(J+1)}$	10.6	9.6
χT expected value for non-interacting ions per complex (cm^3Kmol^{-1})	14.7	11.5

2. Results and Discussions

2.1. Water ^1H Relaxation Measurements

To investigate the paramagnetic relaxation enhancement of the POMs $[Dy^{III}(H_2O)_4 GeW_{11}O_{39}]^{5-}$ (**Dy-W$_{11}$**), $[Er^{III}(H_2O)_3GeW_{11}O_{39}]^{5-}$ (**Er-W$_{11}$**) and $[\{Er^{III}(H_2O)(CH_3COO)(P_2W_{17}O_{61})\}_2]^{16-}$ (**Er$_2$-W$_{34}$**), longitudinal and transverse relaxivity measurements were performed at NMR Larmor frequencies from 20 MHz to 1.4 GHz. The detailed experimental procedures are explained in the previous publications [20–22,40]. For the current studies, pure crystals of POMs were dissolved in 9:1 D_2O/H_2O, and sets of five dilutions were prepared: 0-, 0.2-, 0.4-, 0.6-, 0.8- and 1-mM of complex. The investigated clusters are soluble and form aqueous solutions which remain clear after the NMR measurements have been performed (Figure S2). The dependence of T_1 and T_2 on POM dilutions were measured with different NMR magnetic fields with ^{1}H-NMR Larmor frequencies ranging from 20 MHz to 1.4 GHz. Due to the limited availability of high power (24 MW) resistive magnets ($\geq$800 MHz), only 1-, 0.6-, 0.2- and 0-mM dilutions were measured with the resistive magnet ($\geq$800 MHz). Inversion recovery, progressive saturation recovery and Carr-Purcell-Meiboom-Gill (CPMG) multi-echo pulse sequences were used to measure the T_1 and T_2, respectively. All compounds exhibit a monoexponential relaxation curve that remains unchanged, when the solution is stored for a long time (several weeks). This indicates that the solution of the clusters is stable and no phase separation occurs. Both longitudinal and transverse relaxation rates of these POM clusters were determined as a function of concentration and, at each magnetic field, a linear relation was observed, as expected for a homogenous solution.

2.2. Factors Influencing PRE

PRE properties of the studied POM metal clusters can be derived from the ^{1}H relaxation of water molecules of POM clusters in solution. PRE correlates with the structure and electronic properties of the POM metal clusters, which act as water ^{1}H relaxing agents. The water molecule in the vicinity of the paramagnetic metal center follows a different relaxation behavior which reduces both longitudinal and transverse relaxation time. PRE also depends on the dynamics. Inner and outer sphere contributions are commonly distinguished. For the former the hydration number q of the molecules is an important factor, which also determines the efficiency of a molecule as a potential contrast agent. The theoretical description is governed by many parameters, such as the effective magnetic moments of the paramagnetic metal ions, their electron spin relaxation times, $T_{i,e}$ (i = 1; 2), the tumbling rate of the complex, τ_R^{-1}, the residence time of water molecules near the paramagnetic center, τ_M, as well as diffusion processes that are relevant for outer sphere contributions. For transverse relaxivities and high magnetic fields, Curie mechanisms, i.e., interaction of the nuclear spin with the thermal average of the electron spin, need to be taken into account [41].

2.3. Stability Studies of Er_2-W_{34} in Solution

Electrospray ionization mass spectrometry (ESI-MS): Since our previously reported mass spectrometric (MS) studies on a dilute solution of **Er$_2$-W$_{34}$** have shown the fragmentation of the POM in gas phase, we decided to carry out ESI MS measurement of a diluted **Er$_2$-W$_{34}$** sample (0.2 mM taken from the NMR tube) in order to investigate their solution phase behavior (Figure 4). This was performed by further dilution in 1:1 ACN/H$_2$O to μM concentration and produced spectra similar to our reported ones [39].

As discussed earlier, MS is a harsh technique of analysis. We therefore decided to perform a comparatively soft method of analysis.

Dynamic light scattering (DLS) and Z-potential (Z-pot) measurements: DLS and Z-pot were used to study the colloidal dispersions of **Er$_2$-W$_{34}$** (5 mM in distilled water). First, the hydrodynamic diameter (D_h) of **Er$_2$-W$_{34}$** was determined both at pH 5.6, the pH at which the NMR relaxivities of POMs were studied, and at pH 7 to investigate their stability at physiologically relevant pH values (see Supplementary Information for details). As shown in Figure 5, at pH 5.6, the POMs have an average D_h of 1.2 $\pm$ 0.4 nm with a polydispersity index (PDI) of 0.147, indicating good colloidal stability in water. A comparable colloidal stability was observed at pH 7.0 (Figure 6) with a very similar hydrodynamic size (D_h = 1.3 $\pm$ 0.4 nm; PDI = 0.193). However, a second population of particles with a larger mean D_h of 176.6 $\pm$ 71.1 nm was also observed. These larger particle aggregates can most likely be explained as sodium hydroxide solution was used to adjust the pH of the particle dispersion, and the presence of positively charged Na$^+$ ions could promote aggregation between particles through electrostatic interactions [42]. The net negative charge of **Er$_2$-W$_{34}$** when dispersed in water was confirmed by Z-pot measurements. At pH 5.6, the POMs have a negative Z-pot of -9.0 ± 0.8 mV, while at neutral pH the Z-pot increased towards more negative values of -20.4 ± 0.1 mV. The more negative Z-potential value at neutral pH for **Er$_2$-W$_{34}$** can be explained by the presence of more deprotonated oxygen groups at more basic pH, which contribute to a larger number of negative charges.

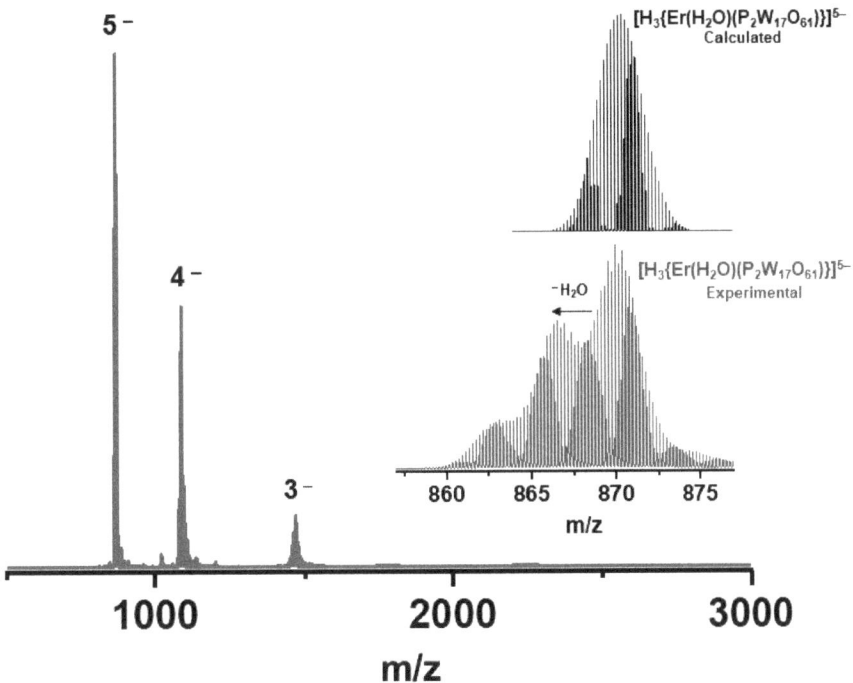

Figure 4. Negative ion electrospray ionization mass spectrometry (ESI MS) of **Er$_2$-W$_{34}$** sample having 0.2 mM concentration taken from the NMR tube and diluted further in 1:1 ACN/H$_2$O. The 5− region is expanded in the inset and the peaks are compared with their calculated isotope pattern.

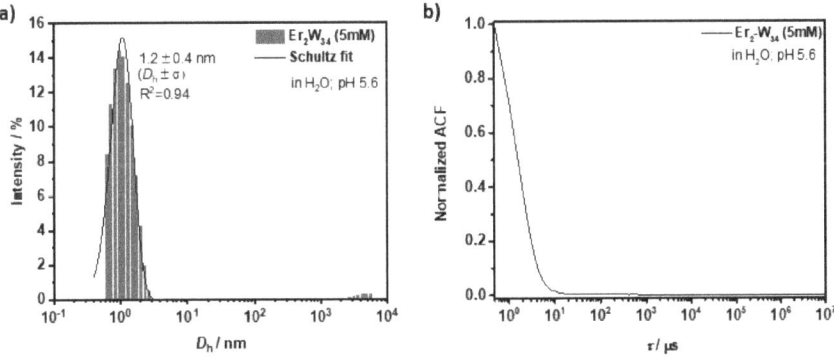

Figure 5. DLS analysis on **Er$_2$-W$_{34}$** (5 mM) at pH 5.6 in water. (**a**) Hydrodynamic diameter of **Er$_2$-W$_{34}$**. (**b**) Corresponding normalized autocorrelation function.

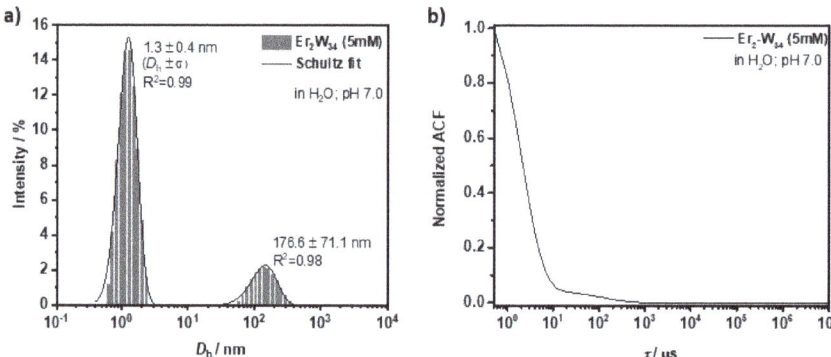

Figure 6. DLS analysis on **Er$_2$-W$_{34}$** (5 mM) at pH 7.0 in water. (**a**) Hydrodynamic diameter of **Er$_2$-W$_{34}$**. (**b**) Corresponding normalized autocorrelation function.

As the anionic units of **Er$_2$-W$_{34}$** are linked by two acetate groups, the possibility of a change in the aggregation state of the POM upon addition of acid was further investigated by DLS analysis. The acetate linker (p$K_{a,\text{acetic acid}}$ = 4.7) is expected to be completely protonated in strongly acidic media (pH 2) and loses its coordination ability, leading to a significant change in the aggregation state of the POM. To test this hypothesis, the hydrodynamic diameter was measured before (pH 5.6) and 40 min after acid addition (HCl$_{aq}$, pH 2). As shown in Figure 7, it can be observed that at pH 2 the scattering intensity of the initial population of small particles (1.2 nm) decreases greatly, while a second population of particles with a D_h of about 187 nm is formed. The formation of this second and larger population of particles can most likely be attributed to the degradation of the small particles leading to the formation of large aggregates. This data suggests that, unless the pH of the particle dispersion is acidified (pH < 4.7), the dimeric structure of **Er$_2$-W$_{34}$** is preserved in water.

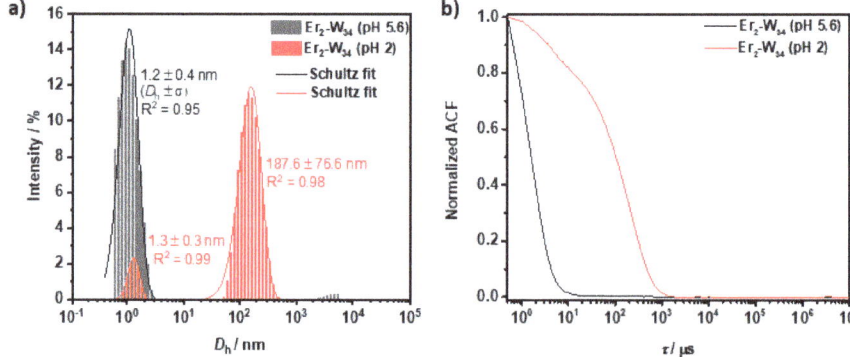

Figure 7. DLS analysis on **Er$_2$-W$_{34}$** (5 mM) at different pH values of the particle dispersion. (**a**) Hydrodynamic diameter of **Er$_2$-W$_{34}$** at pH 5.6 (gray) and at 40 min after addition of HCl$_{aq}$ (37%, 3 μL to 500 μL of sample) at pH 2. (**b**) Corresponding normalized autocorrelation function.

Fourier transform infrared (FTIR) spectroscopy: FTIR spectroscopy was carried out on dried **Er$_2$-W$_{34}$** sample, that was obtained by evaporating the aqueous solvents. FTIR spectra of the recovered **Er$_2$-W$_{34}$** show the expected absorption bands ascribed to the pristine POM. The two spectra perfectly match each other and no wavenumber shifts are observed indicating the stability of the **Er$_2$-W$_{34}$** in solution state (Figure S1).

Xylenol Orange Test: In addition to the instrumental analysis to check the integrity of **Er$_2$-W$_{34}$** in solution, a quick visual Xylenol Orange test was performed to detect the presence of free Er(III) ions. Xylenol Orange test allows an assessment of the amounts of free metal ions and free ligand in a solution of a given lanthanide containing complex. In the presence of free lanthanide ions, the xylenol orange solution undergoes a color change from orange to violet that can be visually detected [43]. No color change was observed on addition of Xylenol Orange (30 µL) to the **Er$_2$-W$_{34}$** solution (pH 5.6) which was used for NMR relaxivity studies. This indicate that the Er(III) ions do not leach out into solution. Whereas, violet color was observed upon addition of Xylenol Orange to the **Er$_2$-W$_{34}$** solution containing ca. 2 mg Er(NO$_3$)$_3$·6H$_2$O as a source of free Er(III) ions (Figure S3).

Based on the above conducted experiments (DLS, post FTIR and Xylenol Orange tests) the structural integrity of **Er$_2$-W$_{34}$** in solution can be confirmed.

2.4. Longitudinal Relaxivity r_1

The longitudinal relaxation rates $R_1(c)$ of the studied POM metal complexes show a linear dependence with concentration c, which leads to the calculation of the slope (i.e., the longitudinal relaxivity r_1). Please note that the concentration used for the relaxivity is the one of the entire POM metal complex rather than the Ln-ion. For all compounds r_1 monotonically increases with Larmor frequency (i.e., with magnetic field (Figure 8)). Among these three POM clusters (**Er$_2$-W$_{34}$**, **Dy-W$_{11}$** and **Er-W$_{11}$**) **Er$_2$-W$_{34}$** shows the highest longitudinal relaxivity.

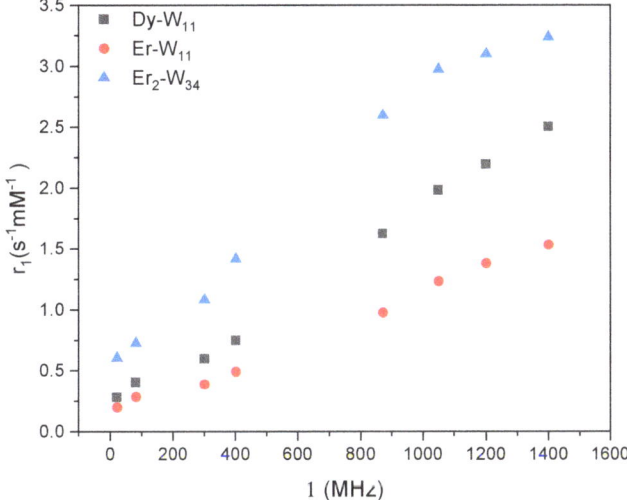

Figure 8. PRE as measured by longitudinal relaxivities r_1 of ^{1}H of water with POM metal clusters (**Dy-W$_{11}$** and **Er-W$_{11}$**, and **Er$_2$-W$_{34}$**) as a function of the Larmor frequency ν. In first approximation, the PRE in the presence of the paramagnetic metal clusters increases linearly with the applied magnetic field except for **Er$_2$-W$_{34}$**, where a levelling-off is observed at high Larmor frequencies.

Taking the data from Table 1, the magnetic moment of Dy^{III} (10.6 µ$_B$) is higher than the one of Er^{III} (9.6 µ$_B$). Thus, the higher longitudinal relaxivity of **Dy-W$_{11}$** compared with **Er-W$_{11}$** supports the influence of electron spin of the paramagnetic centers on PRE, since, in these POMs, single paramagnetic metal centers (Dy^{III} and Er^{III}) are incorporated within similar diamagnetic Keggin moieties. As already mentioned, another important factor that determines the inner sphere contribution to longitudinal relaxivity is the number of exchangeable water molecules that are directly associated with the paramagnetic centers of the complexes. Assuming fast exchange, the longitudinal relaxivity is directly proportional to the hydration number q and the residence time of the water molecules. In the case of **Dy-W$_{11}$** and **Er-W$_{11}$** POMs, Dy^{III} and Er^{III} ions have four and three exchangeable aqua ligands, respectively. The difference in the number of water ^{1}H in {Dy(H$_2$O)$_4$}$^{3+}$ and {Er(H$_2$O)$_3$}$^{3+}$ also contributes to the higher longitudinal relaxivity of **Dy-W$_{11}$** compared with that of **Er-W$_{11}$**. Other relevant properties that can account for the difference in relaxivities of **Dy-W$_{11}$** and **Er-W$_{11}$** are the ionic radii of Dy^{III} and Er^{III}, their magnetic anisotropies and the electron spin relaxation times [12].

The relaxivities of **Er$_2$-W$_{34}$** are more than doubled compared to **Er-W$_{11}$** and this ratio is enhanced towards lower frequencies. Moreover, r_1 of **Er$_2$-W$_{34}$** levels off a very high field. In order to explain these differences in relaxivity dispersion, several factors need to be considered. First, the magnetic moment of dimerized Wells-Dawson POM **Er$_2$-W$_{34}$** is twice the magnetic moment of **Er-W$_{11}$**. This can cause increase of the relaxivitivy. However, for a detailed quantitative modelling further information is needed on the correlations and couplings between the two Er^{III} ions in the dimer. Second, in the chemical structure of the **Er$_2$-W$_{34}$**, only one of the eight coordination sites of each paramagnetic Er^{III} is coordinated by a single water molecule in the solid-state, which limits the possibility of chemical exchange, and hence reduces the PRE. Third, the Wells-Dawson POM **Er$_2$-W$_{34}$** (2.7×0.9 nm in size) has a bigger molecular weight compared to **Er-W$_{11}$** (1.1×1.0 nm in size). This causes lower tumbling rates and different diffusion behavior that can explain the relative enhancement of r_1 of **Er$_2$-W$_{34}$** with respect to **Er-W$_{11}$** at low fields and the levelling-off at highest fields. However, with the available set of data, our model is rather qualitative at the current state, but makes the observed PRE behavior plausible. More systematic studies of other Ln$_2$-W$_{34}$ complexes are needed to identify and quantify the microscopic mecanisms that are responsible for the observed PRE behavior in these highly complex systems.

2.5. Transverse Relaxivity r_2

For the application of a material as an MRI contrasting agent, the significance of transverse relaxivity r_2, which causes negative image contrasts, is as important as the longitudinal relaxivity r_1. In the present studies, transverse relaxivities of the POMs also show approximately linear field dependence due to the PRE effect (Figure 9). The higher relaxivity r_2 of **Dy-W$_{11}$** compared to **Er-W$_{11}$** can be correlated with the effective magnetic moment of the paramagnetic center since r_2 is proportional to the square of the magnetic moment of the lanthanide ions. Dy^{III} has a higher magnetic moment of 10.6 μ_B than Er^{III} which has μ_{eff} = 9.6 μ_B. The comparison of **Er$_2$-W$_{34}$** and **Er-W$_{11}$** in (Figure 9) indicates that the PRE induced transverse relaxivity r_2 of the POM metal complexes also depends on the number of paramagnetic centers in the molecule. Moreover, the r_2 of **Er$_2$-W$_{34}$** starts to deviate from a linear field dependence above 300 MHz. This behavior is consistent with an increasing Curie-spin contribution to the relaxivity. This term has a quadratic field dependence and becomes enhanced for large molecules and at high magnetic field [41,44].

Measurements of r_2 in a very high magnetic field ($\geq$800 MHz) were not carried out on the investigated samples due to the large T_2 values of the complexes compared to the field fluctuations of the resistive magnet.

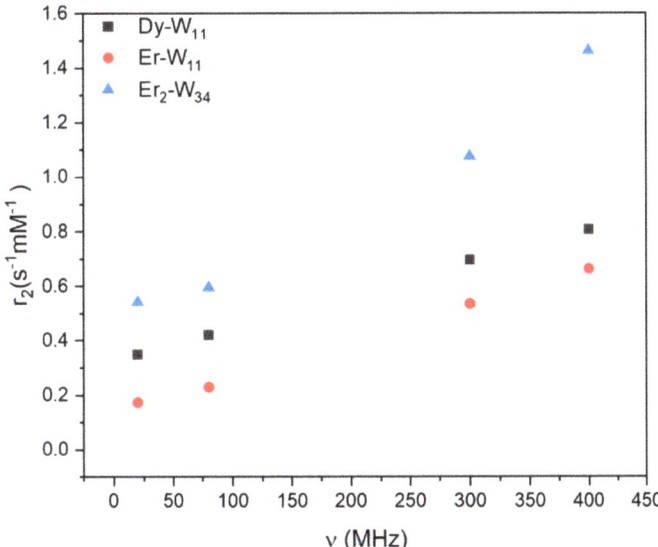

Figure 9. PRE measured by transverse relaxivities r_2 of ^{1}H of water with POM metal clusters as a function of the Larmor frequency ν. r_2 of **Dy-W$_{11}$** and **Er-W$_{11}$** show an approximately linear dependence with the applied magnetic field for Larmor frequencies up to 400 MHz. For **Er$_2$-W$_{34}$** a deviation occurs above 300 MHz, probably due to a Curie-contribution due to the large size of the molecule.

3. Experimental Section

3.1. ^{1}H—Frequencies 870–1400 MHz

High field NMR relaxivity studies were accomplished at the Laboratoire National des Champs Magnétiques Intense (LNCMI) in Grenoble equipped with 24 MW resistive magnet providing variable fields up to 35 T in a 34 mm room temperature bore. To overcome the low field homogeneity (50 ppm/mm at 1 mm off-center position) of these magnets, small sample volumes (1 mm^3) were used that were precisely centered (better than 0.2 mm) at the center of the magnet. For this purpose, a specially designed sample filling system was developed to ensure the placement of precise sample volumes at the center of capillary tubes (10 mm length and 1 mm inner diameter). Moreover, single scan inversion recovery measurements were used to reduce the impact of fast field fluctuations (up to 50 ppm) on the measurement. For the experiment, a specially designed single-resonance ^{1}H-NMR probe was used enabling in situ tuning of NMR frequencies between 850 MHz and 1.4 GHz. Data acquisition and data analysis were performed by using a home-built variable-frequency NMR spectrometer covering Larmor frequency up to 2 GHz and specially designed data acquisition software. As resistive magnets provide access to variable fields (field ramp rate of 5 T/min), the relaxation rates for each concentration were measured at all the frequencies 870 MHz (20.4 T), 1050 MHz (24.7 T), 1200 MHz (28.2 T), and 1400 MHz (33 T) using the same magnet.

3.2. ^{1}H—Frequencies 20–400 MHz

The relaxation measurements at lower magnetic fields were measured on commercially available NMR instruments. The experimental details are described in [20,21] to which we refer here.

4. Conclusions

Water ^{1}H relaxivities are a determining factor for assessing the effectiveness of MRI contrast agents. Often, paramagnetic systems with higher relaxivities result in images with better contrast. PRE of water has been investigated over a large range of magnetic fields for aqueous solutions for the paramagnetic POM clusters **Dy-W$_{11}$**, **Er-W$_{11}$** and **Er$_2$-W$_{34}$**. The r_1 and r_2 values are the fitted slopes of R_1 and R_2 as a function of POM concentration, respectively. As a basis for interpretation, we compare the PRE of the three compounds. POMs with paramagnetic Dy^{III} and Er^{III} ions have very short electronic relaxation times compared to their Gd^{III} analogs and are thus efficient relaxing agents at intermediate and high magnetic fields. The short electronic relaxation time is due to the highly anisotropic electronic ground state of these ions. In addition to the total number of unpaired electrons, magnetic moment and the anisotropic nature of paramagnetic ions prove to be key for the development of efficient CAs for high field MR imaging. There is a need to track these effects through systematic experimental studies on the relaxation properties of paramagnetic compounds.

Supplementary Materials: The following are available online, Figure S1: Comparison of FTIR spectra of Er2-W34, Figure S2: Pictures of NMR tubes containing Er2-W34 solution after NMR relaxivity studies, Figure S3: Pictures of NMR tubes: Upon addition of xylenol orange solution to the Er2-W34 solution after NMR relaxivity studies. Table S1. Magnetic properties of isolated lanthanide ions

Author Contributions: All the authors contributed to this work. G.G., A.K.P., S.K. and M.I. conceived and designed the experiments. A.C.V., T.R. and R.N.D. performed the relaxivity measurements and analysis. P.P. performed DLS and Z-pot studies. P.C. performed the mass spectrometry. The manuscript was written with contributions from all authors. All authors have checked and approved the manuscript.

Funding: M.I. and A.K.P. acknowledge funding by the Helmholtz Gemeinschaft through the program Science and Technology of Nanosystems (STN) and program oriented funding phase four (PoF IV). A.C.V. acknowledges KIT for PhD grant. Diverse financial support of DFG is highly acknowledged for Pro^{2}NMR @ KIT. COST Actions MultiComp (CA15107) and MOLSPIN (CA15128). R.N.D and S.K. acknowledge support by the French National Research Agency in the framework of the "Investissements d'avenir" program (ANR-15-IDEX-02). We acknowledge support of the LNCMI-CNRS, member of the European Magnetic Field Laboratory (EMFL). P.P. acknowledges the DFG for support. P.C. thanks Alexander von Humboldt foundation for her postdoctoral research fellowship. The APC was funded by KIT publication fund.

Data Availability Statement: Not applicable.

Conflicts of Interest: The authors declare no conflict of interest.

Sample Availability: Samples of the compounds K$_5$[Dy(H$_2$O)$_4$GeW$_{11}$O$_{39}$]·16H$_2$O (**Dy-W$_{11}$**), K$_5$[Er(H$_2$O)$_3$GeW$_{11}$O$_{39}$]·20H$_2$O (**Er-W$_{11}$**), and (NH$_2$Me$_2$)$_{13}$Na$_3$[{Er(H$_2$O)(CH$_3$COO)(P$_2$W$_{17}$O$_{61}$)}$_2$] ·40H$_2$O (**Er$_2$-W$_{34}$**) [{Er^{III}(H$_2$O)(CH$_3$COO)(P$_2$W$_{17}$O$_{61}$)}$_2$]$^{16-}$ (**Er$_2$-W$_{34}$**) are available from the authors.

References

1. Pell, A.J.; Pintacuda, G.; Grey, C.P. Paramagnetic NMR in solution and the solid state. *Prog. Nucl. Magn. Reson. Spectrosc.* **2019**, *111*, 1–271. [CrossRef]
2. Bertini, I.; Luchinat, C.; Parigi, G. *Solution NMR of Paramagnetic Molecules: Applications to Metallobiomolecules and Models*, 1st ed.; Elsevier Science: Amsterdam, The Netherlands, 2001.
3. Pintacuda, G.; John, M.; Su, X.C.; Otting, G. NMR structure determination of protein—Ligand complexes by lanthanide labeling. *Acc. Chem. Res.* **2007**, *40*, 206–212. [CrossRef]
4. Helm, L. Relaxivity in paramagnetic systems: Theory and mechanisms. *Prog. Nucl. Magn. Reson. Spectrosc.* **2006**, *49*, 45–64. [CrossRef]
5. Novotný, J.; Sojka, M.; Komorovsky, S.; Nečas, M.; Marek, R. Interpreting the paramagnetic NMR spectra of potential Ru(III) metallodrugs: Synergy between experiment and relativistic DFT calculations. *J. Am. Chem. Soc.* **2016**, *138*, 8432–8445. [CrossRef]
6. Invernici, M.; Trindade, I.B.; Cantini, F.; Louro, R.O.; Piccioli, M. Measuring transverse relaxation in highly paramagnetic systems. *J. Biomol. NMR* **2020**, *74*, 431–442. [CrossRef]

7. Caravan, P.; Greenfield, M.T.; Bulte, J.W.M. Molecular factors that determine Curie spin relaxation in dysprosium complexes. *Magn. Reson. Med.* **2001**, *922*, 917–922. [CrossRef]
8. Gao, C.; Genoni, A.; Gao, S.; Jiang, S.; Soncini, A.; Overgaard, J. Observation of the asphericity of 4f-electron density and its relation to the magnetic anisotropy axis in single-molecule magnets. *Nat. Chem.* **2020**, *12*, 213–219. [CrossRef]
9. Rinehart, J.D.; Long, J.R. Exploiting single-ion anisotropy in the design of f-element single-molecule magnets. *Chem. Sci.* **2011**, *2*, 2078–2085. [CrossRef]
10. Norek, M.; Peters, J.A. MRI contrast agents based on dysprosium or holmium. *Prog. Nucl. Magn. Reson. Spectrosc.* **2011**, *59*, 64–82. [CrossRef]
11. Suturina, E.A.; Mason, K.; Geraldes, C.F.G.C.; Kuprov, I.; Parker, D. Beyond Bleaney's theory: Experimental and theoretical analysis of periodic trends in lanthanide-induced chemical shift. *Angew. Chem. Int. Ed.* **2017**, *56*, 12215–12218. [CrossRef]
12. Suturina, E.A.; Mason, K.; Geraldes, C.F.G.C.; Chilton, N.F.; Parker, D.; Kuprov, I. Lanthanide-induced relaxation anisotropy. *Phys. Chem. Chem. Phys.* **2018**, *20*, 17676–17686. [CrossRef]
13. Lauffer, R.B. Paramagnetic metal complexes as water proton relaxation agents for NMR imaging: Theory and design. *Chem. Rev.* **1987**, *87*, 901–927. [CrossRef]
14. Rogosnitzky, M.; Branch, S. Gadolinium-based contrast agent toxicity: A review of known and proposed mechanisms. *BioMetals* **2016**, *29*, 365–376. [CrossRef] [PubMed]
15. Hamer, A.M.; Livingstone, S.E. The magnetic moments and electronic spectra of lanthanide chelates of 2-Thenoyltrifluoroacetone. *Transit. Met. Chem.* **1983**, *304*, 298–304. [CrossRef]
16. Moser, E.; Laistler, E.; Schmitt, F.; Kontaxis, G. Ultra-high field NMR and MRI-the role of magnet technology to increase sensitivity and specificity. *Front. Phys.* **2017**, *5*, 1–15.
17. Leone, L.; Ferrauto, G.; Cossi, M.; Botta, M.; Tei, L. Optimizing the relaxivity of MRI probes at high magnetic field strengths with binuclear Gd^{III} complexes. *Front. Chem.* **2018**, *6*, 1–12. [CrossRef]
18. Nowogrodzki, A. The world's strongest MRI machines are pushing human imaging to new limits. *Nature* **2018**, *563*, 24–26. [CrossRef] [PubMed]
19. Kowalewski, J.; Fries, P.H.; Kruk, D.; Odelius, M.; Egorov, A.V.; Krämer, S.; Stork, H.; Horvatić, M.; Berthier, C. Field-dependent paramagnetic relaxation enhancement in solutions of Ni(II): What happens above the NMR proton frequency of 1 GHz? *J. Mag. Reson.* **2020**, *314*, 106737. [CrossRef] [PubMed]
20. Machado, J.R.; Baniodeh, A.; Powell, A.K.; Luy, B.; Krämer, S.; Guthausen, G. Nuclear magnetic resonance relaxivities: Investigations of ultrahigh-spin lanthanide clusters from 10 MHz to 1.4 GHz. *ChemPhysChem* **2014**, *15*, 3608–3613. [CrossRef] [PubMed]
21. Guthausen, G.; Machado, J.R.; Luy, B.; Baniodeh, A.; Powell, A.K.; Krämer, S.; Ranzinger, F.; Herrling, M.P.; Lackner, S.; Horn, H. Characterisation and application of ultra-high spin clusters as magnetic resonance relaxation agents. *Dalt. Trans.* **2015**, *44*, 5032–5040. [CrossRef]
22. Ibrahim, M.; Krämer, S.; Schork, N.; Guthausen, G. Polyoxometalate-based high-spin cluster systems: A NMR relaxivity study up to 1.4 GHz/33 T. *Dalt. Trans.* **2019**, *48*, 15597–15604. [CrossRef]
23. Caravan, P. Strategies for increasing the sensitivity of gadolinium based MRI contrast agents. *Chem. Soc. Rev.* **2006**, *35*, 512–523. [CrossRef]
24. Caravan, P.; Ellison, J.J.; McMurry, T.J.; Lauffer, R.B. Gadolinium(III) chelates as MRI contrast agents: Structure, dynamics, and applications. *Chem. Rev.* **1999**, *99*, 2293–2352. [CrossRef]
25. Pellico, J.; Ellis, C.M.; Davis, J.J. Nanoparticle-based paramagnetic contrast agents for magnetic resonance imaging. *Contrast Media Mol. Imaging* **2019**, *2019*, 1845637. [CrossRef]
26. Elistratova, J.; Akhmadeev, B.; Korenev, V.; Sokolov, M.; Nizameev, I.; Ismaev, I.; Kadirov, M.; Sapunova, A.; Voloshina, A.; Amirov, R.; et al. Aqueous solution of triblock copolymers used as the media affecting the magnetic relaxation properties of gadolinium ions trapped by metal-oxide nanostructures. *J. Mol. Liq.* **2019**, *296*, 111821. [CrossRef]
27. Pizzanelli, S.; Zairov, R.; Sokolov, M.; Mascherpa, M.C.; Akhmadeev, B.; Mustafina, A.; Calucci, L. Trapping of Gd(III) ions by keplerate polyanionic nanocapsules in water: A ^1H fast field cycling NMR relaxometry study. *J. Phys. Chem. C* **2019**, *123*, 18095–18102. [CrossRef]
28. Carvalho, R.F.S.; Pereira, G.A.L.; Rocha, J.; Castro, M.M.C.A.; Granadeiro, C.M.; Nogueira, H.I.S.; Peters, J.A.; Geraldes, C.F.G.C. Lanthanopolyoxometalate-silica core/shell nanoparticles as potential MRI contrast agents. *Eur. J. Inorg. Chem.* **2021**, 3458–3465. [CrossRef]
29. Wu, Y.; Bi, L. Research progress on catalytic water splitting based on polyoxometalate/semiconductor composites. *Catalysts* **2021**, *11*, 524. [CrossRef]
30. Li, N.; Liu, J.; Dong, B.; Lan, Y. Polyoxometalate-based compounds for photo- and electrocatalytic applications. *Angew. Chem.* **2020**, *132*, 20963–20977. [CrossRef]
31. Ibrahim, M.; Mereacre, V.; Leblanc, N.; Wernsdorfer, W.; Anson, C.E.; Powell, A.K. Self-assembly of a giant tetrahedral 3 d-4 f single-molecule magnet within a polyoxometalate system. *Angew. Chem. Int. Ed.* **2015**, *54*, 15574–15578. [CrossRef]
32. Ibrahim, M.; Peng, Y.; Moreno-Pineda, E.; Anson, C.E.; Schnack, J.; Powell, A.K. Gd 3 triangles in a polyoxometalate matrix: Tuning molecular magnetocaloric effects in {$Gd_{30}M_8$} polyoxometalate/cluster hybrids through variation of M^{2+}. *Small Struct.* **2021**, *2*, 2100052. [CrossRef]

33. Anyushin, A.V.; Kondinski, A.; Parac-Vogt, T.N. Hybrid polyoxometalates as post-functionalization platforms: From fundamentals to emerging applications. *Chem. Soc. Rev.* **2020**, *49*, 382–432. [CrossRef] [PubMed]
34. Wang, K.Y.; Bassil, B.S.; Lin, Z.; Römer, I.; Vanhaecht, S.; Parac-Vogt, T.N.; Sáenz De Pipaón, C.; Galán-Mascarós, J.R.; Fan, L.; Cao, J.; et al. Ln_{12}-containing 60-tungstogermanates: Synthesis, structure, luminescence, and magnetic studies. *Chem. Eur. J.* **2015**, *21*, 18168–18176. [CrossRef]
35. Zhao, J.; Li, Y.; Chen, L.; Yang, G. Research progress on polyoxometalate-based transition-metal–rare-earth heterometallic derived materials: Synthetic strategies, structural overview and functional applications. *Chem. Commun.* **2016**, *52*, 4418–4445. [CrossRef] [PubMed]
36. Wang, W.; Izarova, N.V.; Van Leusen, J.; Kögerler, P. Polyoxometalates with separate lacuna sites. *Chem. Commun.* **2020**, *56*, 14857–14860. [CrossRef] [PubMed]
37. Ibrahim, M.; Mbomekallé, I.M.; de Oliveira, P.; Baksi, A.; Carter, A.B.; Peng, Y.; Bergfeldt, T.; Malik, S.; Anson, C.E. Syntheses, crystal structure, electrocatalytic, and magnetic properties of the monolanthanide-containing germanotungstates $[Ln(H_2O)_nGeW_{11}O_{39}]^{5-}$ (Ln = Dy, Er, n = 4,3). *ACS Omega* **2019**, *4*, 21873–21882. [CrossRef] [PubMed]
38. Ibrahim, M.; Baksi, A.; Peng, Y.; Al-zeidaneen, F.K.; Mbomekall, I.M.; De Oliveira, P.; Anson, C.E. Synthesis, characterization, electrochemistry, photoluminescence and magnetic properties of a dinuclear erbium(III)-containing monolacunary dawson-type tungstophosphate: $[\{Er(H_2O)(CH_3COO)(P_2W_{17}O_{61})\}_2]^{16-}$. *Molecules* **2020**, *25*, 4229. [CrossRef]
39. Bich, C.; Baer, S.; Jecklin, M.C.; Zenobi, R. Probing the hydrophobic effect of noncovalent complexes by mass spectrometry. *J. Am. Soc. Mass Spectrom.* **2010**, *21*, 286–289. [CrossRef]
40. Ibrahim, M.; Rudszuck, T.; Kerdi, B.; Krämer, S.; Guthausen, G.; Powell, A.K. Comparative NMR relaxivity study of polyoxometalate-Based clusters $[Mn_4(H_2O)_2(P_2W_{15}O_{56})_2]^{16-}$ and $[\{Dy(H_2O)_6\}_2Mn_4(H_2O)_2(P_2W_{15}O_{56})_2]^{10-}$ from 20 MHz to 1.2 GHz. *Appl. Magn. Reson.* **2020**, *51*, 1295–1305. [CrossRef]
41. Peters, J.A.; Huskens, J.; Raber, D.J. Lanthanide induced shifts and relaxation rate enhancements. *Prog. Nucl. Magn. Reson. Spectrosc.* **1996**, *28*, 283–350. [CrossRef]
42. Pigga, J.M.; Teprovich, J.A.; Flowers, R.A.; Antonio, M.R.; Liu, T. Selective monovalent cation association and exchange around keplerate polyoxometalate macroanions in dilute aqueous solutions. *Langmuir* **2010**, *26*, 9449–9456. [CrossRef] [PubMed]
43. Barge, A.; Cravotto, G.; Gianolio, E.; Fedeli, F. How to determine free Gd and free ligand in solution of Gd chelates. A technical note. *Contrast Media Mol. Imaging* **2006**, *1*, 184–188. [CrossRef] [PubMed]
44. Gueron, M. Nuclear relaxation in macromolecules by paramagnetic ions: A novel mechanism. *J. Magn. Reson.* **1975**, *19*, 58–66. [CrossRef]

Article

Real-Time Monitoring Polymerization Reactions Using Dipolar Echoes in ^{1}H Time Domain NMR at a Low Magnetic Field

Rodrigo Henrique dos Santos Garcia [1], Jefferson Gonçalves Filgueiras [2,3], Luiz Alberto Colnago [4] and Eduardo Ribeiro de Azevedo [5,*]

1. Instituto de Química de São Carlos, Universidade de São Paulo, CP 369, São Carlos 13660-970, SP, Brazil; rodrigogarciaquimico@yahoo.com.br
2. Instituto de Química, Universidade Federal Fluminense, Outeiro de São João Batista, Niterói 24020-007, RJ, Brazil; jgfilgueiras@yahoo.com
3. Instituto de Física, Universidade Federal do Rio de Janeiro, CP 68528, Rio de Janeiro 21941-972, RJ, Brazil
4. Embrapa Instrumentação, Rua XV de Novembro, 1452, São Carlos 13560-970, SP, Brazil; luiz.colnago@embrapa.br
5. Instituto de Física de São Carlos, Universidade de São Paulo, CP 369, São Carlos 13660-970, SP, Brazil
* Correspondence: azevedo@ifsc.usp.br

Abstract: ^{1}H time domain nuclear magnetic resonance (^{1}H TD-NMR) at a low magnetic field becomes a powerful technique for the structure and dynamics characterization of soft organic materials. This relies mostly on the method sensitivity to the ^{1}H-^{1}H magnetic dipolar couplings, which depend on the molecular orientation with respect to the applied magnetic field. On the other hand, the good sensitivity of the ^{1}H detection makes it possible to monitor real time processes that modify the dipolar coupling as a result of changes in the molecular mobility. In this regard, the so-called dipolar echoes technique can increase the sensitivity and accuracy of the real-time monitoring. In this article we evaluate the performance of commonly used ^{1}H TD-NMR dipolar echo methods for probing polymerization reactions. As a proof of principle, we monitor the cure of a commercial epoxy resin, using techniques such as mixed-Magic Sandwich Echo (MSE), Rhim Kessemeier—Radiofrequency Optimized Solid Echo (RK-ROSE) and Dipolar Filtered Magic Sandwich Echo (DF-MSE). Applying a reaction kinetic model that supposes simultaneous autocatalytic and noncatalytic reaction pathways, we show the analysis to obtain the rate and activation energy for the epoxy curing reaction using the NMR data. The results obtained using the different NMR methods are in good agreement among them and also results reported in the literature for similar samples. This demonstrates that any of these dipolar echo pulse sequences can be efficiently used for monitoring and characterizing this type of reaction. Nonetheless, the DF-MSE method showed intrinsic advantages, such as easier data handling and processing, and seems to be the method of choice for monitoring this type of reaction. In general, the procedure is suitable for characterizing reactions involving the formation of solid products from liquid reagents, with some adaptations concerning the reaction model.

Keywords: time-domain NMR; dipolar echoes; polymerization reaction; epoxy resin; autocatalytic reaction

1. Introduction

Industries increasingly benefit from the use of process analytical technology (PAT) for production control and quality assurance. The usual goal is a comprehensive understanding and thus better control of manufacturing processes. Thus, predictability and reliability of quality should be incorporated into the process; consequently, in the ideal scenario it should be monitored online or in-situ [1,2].

In online or in-situ monitoring it is often necessary to track structural and dynamical changes in real-time. PAT in polymers can be achieved by several physical techniques, among them high- and low-resolution NMR, infrared spectroscopy (FTIR) and Differential Scanning Calorimetry (DSC) techniques [3,4].

Among the NMR methods, ^{1}H TD-NMR at a low magnetic field is particularly useful for studying molecular processes where chemical specificity is not necessary. Indeed, because of its high sensitivity to molecular mobility, TD-NMR can detect dynamical chances associated to thermal transitions [5–8]. In addition, the technique allows to determine other important information such as the ratio between rigid and mobile segments, sizes of molecular clusters in emulsions [9], determination of the degree of crystallinity and crystallite sizes in semicrystalline systems [10], gelation rates in polymer gels [11] and crosslinking and entanglement in elastomers [12–14], among others [15–17].

^{1}H TD-NMR at a low magnetic field is extensively used to determine solid fractions in food. The Solid Fat Content (SFC) [9] method and its fusion curve are used as one of the most important quality parameters for fats. The measured values are used for controlling hydrogenation, interesterification (enzymatic by the enzyme lyase or chemically catalyzed by alkaline metal) and blending, as well as the level of solid fat necessary for the development of new products or replacement of raw material.

^{1}H TD-NMR has also been used to monitor polymer crystallization in real-time via T2 measurements. For instance, it was first demonstrated by C. Hertlein et al. [18], who studied the crystallization kinetics of polymer S-Poly(propylene) (sPP), Polyethylene-co-(PEcO) and Poly(ε-caprolactone) (PεCL). In these cases, the reduction of the T2 values, which occurs due the decrease of the polymer chain mobility upon crystallization, is monitored to obtain the crystallization kinetics parameters. The values obtained by ^{1}H TD-NMR usually show good agreement with dilatometry and X-ray scattering.

The advancement of ^{1}H TD-NMR methods made possible many other applications. Litvinov and co-workers [19] used the FID and Solid Echo pulse sequences to determine phase composition (mobile fraction, semi-rigid or rigid) and molecular mobility of Nylon 6 fibers. The results satisfactory agreed with results obtained with X-ray diffraction and ^{13}C NMR- spectroscopy. Maus [20] used mixed-MSE-CPMG (a combination of the mixed-MSE and Carr–Purcell–Meiboom–Gill pulse sequences) to investigate the solid/liquid ratio and crystallinity of (syndiotactic poly(propylene) (sPP). Maus also investigated the online monitoring of polymer crystallization kinetics and obtained results in good agreement with dilatometric measurements reported for the same sample.

Another example of the real-time monitoring of polymer crystallization was provided by Faria and co-workers [21], who used a dipolar filtered MSE (DF-MSE) pulse sequence [8,22] to monitor one-dimensional crystallization in poly(3-(2′-ethylhexyl)thiophene) (P3EHT) films. In this case, the crystallization does not start from the melt, but from a solid amorphous state, so only little changes in the shape of the ^{1}H TDNMR signal are observed. Using a dipolar filter pulse sequence in conjunction with MSE (DF-MSE), these subtle changes could be monitored in order to observe the crystallite growth which was explained using a modified Avrami model for one-dimensional crystallization.

Oztop [23] determined the crystalline/amorphous fraction obtained with the NMR signal acquired with the MSE pulse sequences of various sugars (glucose, sucrose, etc.) to develop a basic work for a reliable quality-control method. It was demonstrated that the crystallinity of powdered sugars could be predicted using TD-NMR and confirmed the linear dependence of crystallinity with the second moment of the NMR line shape.

There are also many examples showing the real time monitoring of processes using time-domain ^{1}H double quantum NMR in a low magnetic field [11,18,24]. As ^{1}H double quantum coherences are only created in the presence of ^{1}H-^{1}H dipolar coupling, the double quantum evolution acts as a filter, keeping only the signal from segments presenting mobility restrictions (rigidified). Hence, the growth of rigidified structures can be monitored in real time. For instance, Saalwächter and co-workers studied gelation in flexible polymer systems [11] and the real time observation of polymer network formation [24]. Another example is the work by Valentin et. al., who successfully used the double quantum filter pulse sequence to selectively detect rigid components and provides a direct probing of the product formation during the cure of an epoxy resin [25].

In this article we used ^{1}H TD-NMR dipolar echo pulse sequences, shown in Figure 1, for real time monitoring of reactions associated to the cure of a commercial epoxy resin. The cure of a thermorigid resin is a complex process that can be defined as the change in the chemical and physical properties of a given resin/hardener formulation. Because of the formation of a product with rigid molecular segments, understanding the mechanisms and kinetics of cure reactions is essential for a better knowledge of structure–property relationships. The article is organized as follows. First, we discuss the main features related to the dipolar echo pulse sequences mixed-MSE and RK-ROSE. We follow with a discussion about the changes in the signal profile during the polymerization reaction and present the basic data processing needed to extract the parameter used to characterize the rection kinetic. We also discuss a specific reaction model used for characterizing the epoxy cure and establish the meaning of the NMR data in terms of the kinetic parameters of this model. Then, we use the model to fit NMR data acquired with mixed-MSE [26] and RK-ROSE [27] and demonstrate that these experiments are able to bring reliable estimations of the kinetic rate and activation energy of the reactions. Last, we propose the use of the DF-MSE pulse sequence as a simple method for characterizing the reaction, showing its main features and advantages.

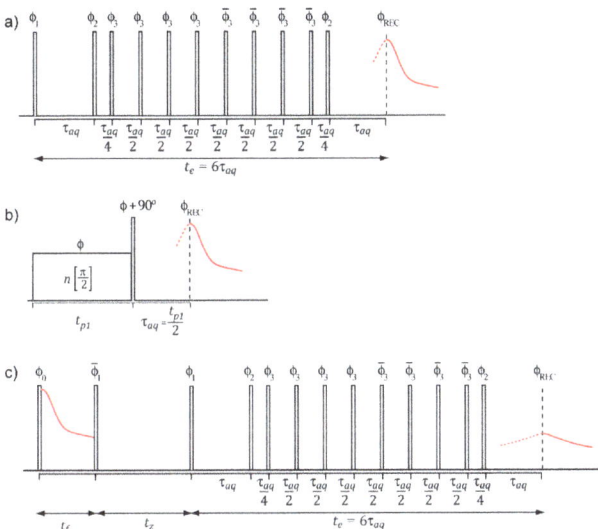

Figure 1. Schemes of the NMR pulse sequences used in this article. (a) Mixed-Magic Sandwich Echo (mixed-MSE). (b) Rhim–Kessemeier (RK). (c) Dipolar Filtered Magic Sandwich Echo (DF-MSE). The phase cycling used in the pulse sequence are provided in references [24,7]. REC: refer to the receiver phase.

2. Results

2.1. Signal Profiles during a Polymerization Reaction

The idea behind using ^{1}H TD-NMR to probe chemical reactions with the formation of solid products is to follow the evolution of signals from rigid and mobile segments during the reaction. However, ^{1}H NMR signals arising from rigid molecular segments have short decay times ($\lesssim$50 µs) due to the strong ^{1}H-^{1}H dipolar coupling. Thus, to detect full NMR signals from these segments, it is necessary to start the signal acquisition right after the excitation pulse. This is barely achieved in most of the commercial probeheads, which have typical dead times in the range of 5–30 µs. Thus, a simple $\pi/2$—acquisition scheme (denoted here as FID acquisition) frequently implies in the loss of a considerable part of the signal from solid components. One strategy to avoid this dead time issue is to use probeheads with the lowest possible quality factor, but this usually implies in lower sensitivity and requires higher rf power for excitation. An alternative is to use a dipolar echo pulse sequence, capable of refocusing the ^{1}H-^{1}H dipolar coupling, to produce an echo right out of the dead time region where the solid component is recovered [28]. However, the recovery efficiency of the solid components signals is not 100%, depending on echo times, interpulse delays, pulses errors, etc. Moreover, the shape of the signal can also be affected by the use of a specific pulse sequence or by the experimental set-up [20]. Therefore, different methods can be chosen to obtain better echo recovery efficient or shape accuracy, depending on desired application.

Here we rely on two basic pulse sequences to refocus the signal from rigid components. The first one is an approach based on the experiments developed by Rhim and Kassemeier in the early 1970s. It consists of applying a continuous wave pulse of duration t_p followed by a delay τ and a hard $\pi/2$ pulse 90° phase shifted with respect to the CW pulse [28]. Recently, the Rhim and Kassemeier pulse sequence was used without interpulse delays and had a power setup chosen to maximize the signal recovery of solid components at expense of some signal distortion due to the magnitude mode acquisition. This dipolar echo pulse sequence refocuses the dipolar coupling at $\tau_{acq} = t_p/2$ after the second pulse and was referred to as Rhim and Kessemeier Radiofrequency Optimized Solid-Echo (RK-ROSE) [27]. The second pulse sequence is the traditional mixed Magic Sandwich Echo (mixed-MSE—Figure 1a) method [26,29]. In this method, a sequence of properly phased pulses refocuses the ^{1}H-^{1}H dipolar couplings at the same time as eliminating the interference of linear spin interactions such as magnetic field inhomogeneities, chemical shielding, heteronuclear dipolar interactions and local susceptibility variations. Because it can be acquired in the phase mode, the shape analysis of the mixed-MSE echo provide a more reliable way to estimate dipolar coupling second moment, which is particularly important for applications relies on the analysis of the signal shape [30,31]. However, because of the large number of pulses and interpulse delays, the minimum echo time is limited in mixed-MSE, which may compromise the recovery efficiency of signals arising from rigid segments.

As already mentioned, both RK-ROSE and mixed-MSE pulse sequences allow for differentiation between signals from rigid and mobile segments. Thus, both methods can be used for probing the emergence of rigid segments and/or the disappearing of mobile ones. This is the case of the curing reaction of the epoxy resin, where the formation of the solid products occurs at the expense of the liquid reagents. During such processes, an epoxy resin is converted into crosslinked thermosetting networks, and the thermosetting polymer properties depend on the extent of the chemical reactions that occur during cure and the resin morphology [32]. Briefly, epoxy resin cure goes from a liquid state to a gel point, then turns to rubber, and finally reaches the vitrification point, where it is converted to glass [32,33]. The effect of these processes on the RK-ROSE and mixed-MSE dipolar echoes is shown in Figure 2, where normalized half-echoes signals acquired at several reaction times are presented. The RK-ROSE and mixed-MSE correspond to two different reactions, but with the same relative amount of resin and hardener and at the same temperature. The general behavior of the signals is similar for both type of experiments. The first signal is acquired only after the homogenization of the reagents (~2 min) and temperature equilibration (~10 min). At shorter reaction times (10 min), the signal is comprised of a slow decaying signal associated to ^{1}H nuclei in the mobile segments of the reagents. As the reaction takes place (reaction time of 200 min, Figure 2b), the slow decaying fraction of the signal decreases while the fast-decaying fraction increase. When the reaction finishes (reaction time of 450 min in Figure 2a), a fast-decaying signal is observed, as is typical for rigid segments.

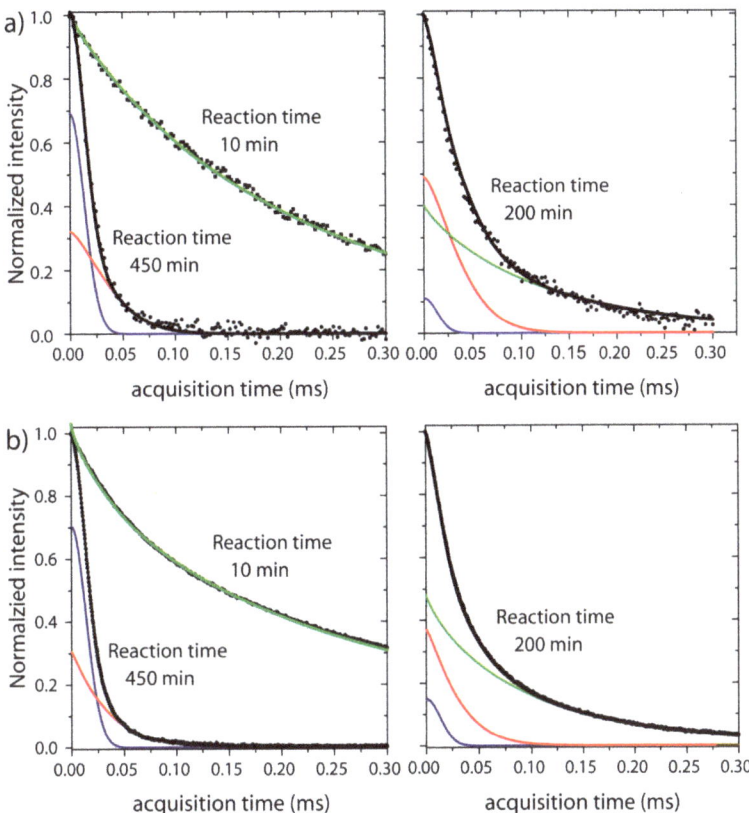

Figure 2. Example of deconvolution of dipolar echoes signals obtained with mixed-MSE (**a**) and RK-ROSE (**b**) pulse sequences using Equation (1). The signals shown in (**a**,**b**) were acquired at the indicated reaction times during the cure of the epoxy resin at 293 K. Note that the differences in the signal to noise between the mixed-MSE and RK-ROSE are due to the differences in the number of scans and echo times as shown in the methods section. Both pulse sequences have similar efficiencies at this set-up condition.

To provide a more quantitative analysis, we refer to the work by Mauss et. al. [20] who decomposed the mixed-MSE signal from a semicrystalline polymer above its glass transition temperature T_g in three components, associated to the crystalline (rigid), amorphous (highly mobile) and interfacial (with intermediate mobility) regions. The signals from rigid molecular segments can be well described by the so-called Abragam function, i.e., a multiplication between a gaussian and a sinc function (Van Vleck theory [34]). The signals from intermediate and highly mobile segments are described by modified exponential functions (stretched or compressed), but with different time constant and shape parameters [20]. In summary, the NMR signal can be represented by the following function:

$$\frac{s(t)}{s(0)} = f_r e^{-(\frac{t}{T_{2r}})^2} \frac{\sin(bt)}{bt} + f_i e^{-(\frac{t}{T_{2i}})^{\nu_i}} + f_m e^{-(\frac{t}{T_{2m}})^{\nu_m}} \qquad (1)$$

where f_r, f_i and f_m are the fractions associated to rigid, intermediate and mobile components, respectively. T_{2r}, T_{2i}, T_{2m}, ν_i and ν_m are the respective shape parameters. It is worth to stress that in a low field NMR the signal has significant contribution from static field inhomogeneities. Thus, despite the signal fitting being performed within the 0 – 300 µs acquisition time window, the shape parameters of mobile components may still have a minor contribution from static field inhomogeneities.

Here we will use the same model to describe the ^{1}H NMR signal, but with the signal from the rigid segments assigned to the solid polymer formed during the cure and the mobile component accounting for the liquid reagents. The segments with intermediate mobility are attributed to polymer chains with restricted mobility in the solid product, for instance, chains in the solid to liquid interface, chain segments with local mobility such side groups or remaining small segments such as oligomers.

Figure 2 (right) illustrates the fit of normalized signals acquired at 293 K with RK-ROSE and mixed-MSE 10 min and 450 min after the polymerization started. For RK-ROSE the absolute value of the signals is shown (magnitude mode acquisition), while for mixed-MSE the real part of phase adjusted signals are exhibited (phase mode acquisition). After 10 min of reaction, the signal is well adjusted using only the last term of Equation (1) (green curve, $T_{2m} = (0.24 \pm 0.01)$ ms, $\nu_m = 0.72 \pm 0.05$, $f_m = 1.00 \pm 0.02$ for RK-ROSE and $T_{2m} = (0.21 \pm 0.01)$ ms, $\nu_m = 0.92 \pm 0.02$, $f_m = 0.99 \pm 0.03$ for mixed-MSE), showing that there is no significant rigid phase at this reaction time and temperature, i.e., the signal is only associated to the reagent. On the other hand, after 450 min of reaction only rigid components are observed in the signal. As shown in Figure 2, this fast component can be adjusted using the first and the second terms of Equation (1), i.e., a gaussian-type decay (blue curve, $T_{2r} = (0.0135 \pm 0.0002)$ ms, $f_r = 0.70 \pm 0.02$ for RK-ROSE and $T_{2r} = (0.0135 \pm 0.0002)$ ms, $f_r = 0.69 \pm 0.02$ for mixed-MSE) and a compressed exponential (red curve, $T_{2i} = (0.035 \pm 0.002)$ ms, $\nu_i = 1.20 \pm 0.05$, $f_i = 0.30 \pm 0.02$ for RK-ROSE and $T_{2i} = (0.045 \pm 0.001)$ ms, $\nu_i = 1.50 \pm 0.05$, $f_i = 0.31 \pm 0.02$ for mixed-MSE). The signal profile does not change for reaction times longer than 450 min (not shown), suggesting that the cure is already completed. As discussed, the existence of an intermediate mobility phase in the final product can be related to restricted local mobility in the solid phase, for instance due to side chain motions or other local segmental reorientations, or remaining oligomers. Thus, the product fraction can be associated to the sum of the rigid and intermediate mobility components, which will be referred simply as solid fraction, i.e., $f_S = f_r + f_i$. Moreover, we should point out that in both experiments, mixed-MSE and RK-ROSE, there is an underestimation of the rigid fraction because the efficient of the pulse sequences in recovering signals from rigid segments is not 100%. This effect can be corrected as suggested elsewhere [31]. Here this is completed by analyzing the dipolar echoes (acquired with mixed-MSE or RK-ROSE) the simple FID signals acquired after the end of the reaction. This FID signal was shifted by the dead time of our spectrometer (12 µs), and then we perform a joint fit of the dipolar echoes and the corresponding FID signals. In this joint fit, the shape parameters in Equation (1) are shared to impose the FID and the dipolar echo with the same shape. Because the amplitudes are free parameters, the fitting provides independent values of f_r, f_i and f_m for the dipolar echo and the FID signals. Thus, it is possible to calculate the ratio between the amplitude parameters associated to dipolar echoes and the FID to estimate a correction factor α, which gives how much of the solid signals is lost due to the acquisition with mixed-MSE or RK-ROSE. For the experiments conducted as a function of temperature, this procedure was performed at each temperature. This correction factor was used to correct the values of $f_S = f_r + f_i$ in all data presented here.

Figure 2 (left) illustrates the fit of normalized signals acquired 200 min after mixing the reagents. In this case, the best fit is achieved using all three components of Equation (1) (blue curve, $T_{2r} = (0.0135 \pm 0.0002)$ ms, $f_r = 0.15 \pm 0.01$, red curve, $T_{2i} = (0.035 \pm 0.002)$ ms, $\nu_i = 1.2 \pm 0.1$, $f_i = 0.37 \pm 0.03$, green curve, $T_{2m} = (0.10 \pm 0.05)$ ms, $\nu_m = 0.84 \pm 0.05$, $f_m = 0.48 \pm 0.05$ for RK-ROSE and blue, $T_{2r} = (0.0135 \pm 0.0002)$ ms, $f_r = 0.11 \pm 0.02$, dark yellow, $T_{2i} = (0.044 \pm 0.004)$ ms, $\nu_i = 1.5 \pm 0.1$, $f_i = 0.49 \pm 0.03$, magenta, $T_{2m} = (0.12 \pm 0.04)$ ms, $\nu_m = 0.90 \pm 0.05$, $f_m = 0.40 \pm 0.06$ for mixed-MSE), as shown individually in Figure 2b.

Changes in the molecular dynamics of the mobile component throughout the reaction can be observed monitoring the T_{2m} values. T_{2m} values decrease for longer reaction times, suggesting an average slowdown of molecular motions in the mobile phase. This is somewhat expected because the formation of rigid components increases the local molecular constraints, reducing the molecular mobility of the remaining mobile segments. Indeed, T_{2m} has been extensively used to probe reaction kinetics using ^{1}H TD-NMR [1]. However, kinetic models for describing polymerization reactions mostly rely on monitoring changes in the reagent and product concentrations. Thus, a more straightforward analysis can be completed using f_r, f_i and f_m.

2.2. Autocatalyzed and Non-Catalyzed Reaction: Relationship between NMR Parameters and Reaction Kinetic Parameters

To obtain the kinetic parameters of the polymerization reaction we first need to establish the meaning of the NMR data in terms of the kinetic parameters. As mentioned, in the curing reaction of epoxy resins, the formation of the solid products occurs at the expense of the liquid reagents. Thus, f_m and $f_s = f_r + f_i$ can be associated to the relative reagent, $A \propto f_m$, and product, $P \propto f_r + f_i$, concentrations. Therefore, assuming a model for the polymerization reaction, it would be possible to obtain the kinetic parameters by fitting the reaction time dependence of f_m or $f_r + f_i$.

Epoxy cure is usually described in terms of noncatalyzed or autocatalytic single step reaction models [14]. The analysis of such reactions using DSC makes it possible to distinguish the contribution of noncatalyzed and autocatalyzed paths by performing dynamic or isothermal experiments, respectively [33,35]. Here, the experiment is essentially isothermal, so we would expect an autocatalytic reaction to prevail. Nonetheless, despite the external temperature of the sample being kept constant, a relatively large sample volume (~0.5 cm^3) is used, making it difficult to completely avoid internal temperature gradients due to the exothermal character of the reaction, so noncatalytic paths cannot be ruled out. Thus, we build upon a model described in reference [36] that assumes autocatalytic and noncatalytic paths occurring simultaneously. In this model, the reagent and product concentrations are given by:

$$A(t) = (A_0 + q) \frac{A_0}{A_0 + q e^{[k_c(A_0+q)t]}}; \quad P(t) = 1 - A(t) \qquad (2)$$

Here t is the reaction time, A_0 is the initial reagent concentration and k_c is the average reaction rate at a given temperature. The parameter q assumes different meaning for autocatalyzed and noncatalyzed paths, being equal to the initial product concentration, $q = P_0$ for autocatalyzed and $q = \frac{k_0}{k_c}$ for noncatalyzed reactions. k_0 is the initial reaction rate.

As already discussed, one may establish a direct correlation between the reagent concentration and the mobile component of the signal, $A \propto f_m$, as well as between the product concentration and the solid component, $P \propto f_s = f_r + f_i$. Thus, the curves of f_m or f_s as a function of the reaction time could be fitted to obtain A_0, k_c and q. However, for higher temperatures one might observe a slow decaying component in the signal that is not associated to the reagent, but to mobile segments in the product. Fortunately, this component can be identified as a mobile fraction kept constant after the reaction is finished. Moreover, it can be taken into account by adding a constant term, f_∞, in the fitting function. Hence, the reaction time dependence of f_m can be fitted by:

$$f_m(t) = (f_0 + q) \frac{f_0}{f_0 + q e^{[k_c(f_0+q)t]}} + f_\infty; \quad f_s(t) = 1 - f_m(t) \qquad (3)$$

with k_c and q assumed as free fitting parameters and f_0 obtained as the initial and final f_∞ mobile fractions.

2.3. Reaction Kinetic Parameters Extracted from Dipolar Echoes

In order to probe the dependence of the mobile fraction on the reaction time, we acquired a series of dipolar echoes during the epoxy cure. The dipolar echoes were acquired using mixed-MSE and RK-ROSE in different batches, but with the same proportion of resins and hardeners as well as the same temperature. The signal acquisition started 15 min after mixing the reagents to assure temperature equilibration in the sample. The resulting curves of f_m versus the reaction time, $f_m(t)$, were fitted using Equation (3) to obtain k_c and q. The fitting procedure used to obtain these parameters was: first, we manually deconvolute the mixed-MSE and RK-ROSE signals at the shortest and the longest reaction times to obtain the shape and amplitude parameters f_r, f_i, f_m, T_{2r}, T_{2i}, T_{2m}, v_i and v_m. Note that this fitting procedure already provides the values for f_0 and f_∞. Then, these values were used as input values for an automated fit procedure using the origin lab software to obtain the parameter values for each reaction time.

Figure 3a shows the plot of the mobile fractions $f_m(t)$ as a function of the reaction time obtained from mixed-MSE and RK-ROSE data at 313 K. Even though the curves obtained by mixed-MSE and RK-ROSE correspond to different batches of samples, the k_c values obtained are similar, since the reactions were performed at the same temperatures. In both cases the values obtained for q are close to zero. As mentioned, for autocatalytic reaction the q parameter is equal to the initial product concentration ($q = P_0$). Because the initial product concentration should be quite small, the lower q obtained suggests the reaction is predominantly of autocatalytic, as expected for an isothermal process. The values of f_∞ are different from zero, indicating that part of the segments remain mobile after the end of the reaction. As mentioned, this corresponds to molecular segments in the product, such as pendant groups, chain ends and oligomers, which contributes as intermediate mobility components at the lower temperature (293 K, see Figure 3), but have their mobility increased at 313 K. This remaining mobile component is more evident in the mixed-MSE data due to its ability of refocusing field inhomogeneity effects. We shall consider that even though these segments can contribute to a significant fraction of the signal, they do not hinder the analysis of the reaction kinetics, as they remain constant after the reaction is completed.

Figure 3b shows the plot of the mobile fractions, $f_m(t)$, as a function of the reaction time, obtained from mixed-MSE and RK-ROSE data at 293 K, 313 K, 333 K and 353 K. The curve fit using Equation (3) provides the k_c values at each temperature, as shown in the insets. The temperature dependence of k_c is shown in the Arrhenius plot of Figure 3c. Activation energies of $E_a = (47 \pm 4)$ kJ/mol and $E_a = (52 \pm 4)$ kJ/mol were obtained from the RK-ROSE and mixed-MSE data, respectively. Such values for the activation energies are in good agreement with literature values for similar samples [25,35].

2.4. Reaction Kinetics Parameters from Dipolar Filtered Magic Sandwich Echo (DF-MSE)

In the last section we showed how dipolar echo pulse sequences can be used to extract kinetic parameters. Nevertheless, performing the signal deconvolution to obtain the mobile, intermediate and solid fractions for each reaction time and temperature can be a quite tedious task and add significant fitting errors to the data. A possible strategy to avoid this is to acquire the data using a pulse sequence that selects only signals from mobile or rigid components or allow to separate these signals in a proper manner. This approach permits an intensity-only analysis, requiring minimal processing effort. For instance, any dipolar filter pulse sequence that suppresses the signal from the rigid phase [37–39] can be used to obtain an echo arising only from the mobile components. The amplitude of such an echo, normalized by the corresponding full echo signal, can be plotted as a function of the reaction time to obtain curves similar to those shown in Figure 3b, but without further need of data processing. It is also possible to use a pulse sequence to suppress the mobile phase reaching only the signal from the rigid components, which could also be monitored as a function of the reaction time to probe the polymerization reaction. The most common filter to keep the signal from the rigid phase is the double quantum filter, which only selects signals from dipolar coupled spins, i.e., stiffened segments [40]. Such an approach was already used by Valentin and co-workers to probe epoxy polymerization reactions [25].

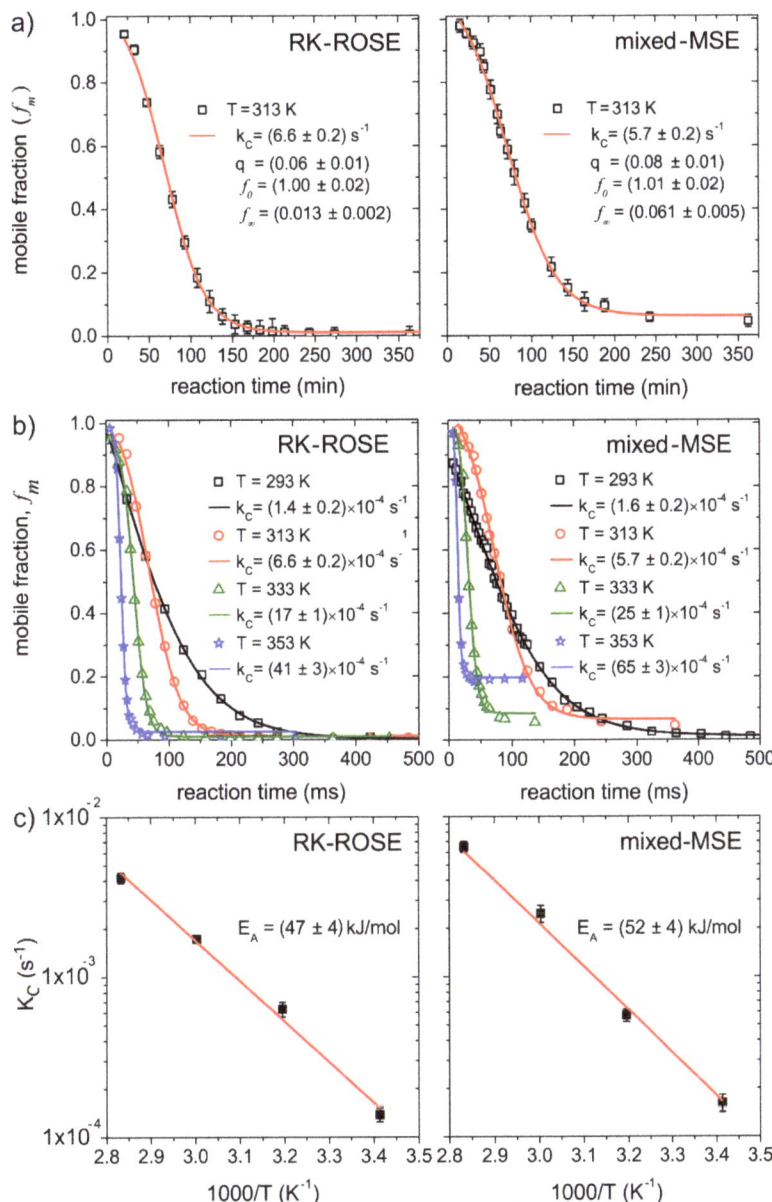

Figure 3. (a) Mobile fractions, as function of the reaction time during the epoxy cure, obtained from the deconvolution of signals acquired by the mixed-MSE (right) and RK ROSE (left) pulse sequences at 313 K. (b) Same as in (a) for 293 K, 313 K, 333 K and 353 K. (c) Arrhenius plot of the k_c values estimated from the fits shown in (b). The error bars were omitted in (b), but they are in the same order as those shown in (a). The meaning of k_c, q, f_0, f_∞ are describe in the main text.

Here we discuss an approach based on a simple pulse sequence named Dipolar Filtered Magic Sandwich Echo (DF-MSE); see Figure 1 and reference [8] for details. In its simplest form, DF-MSE is comprised by a Goldman–Shen dipolar filter of duration t_f [38,39]

followed by the mixed-MSE pulse sequence. At the shorter possible filter time t_{f0} (here 3.5 μs) the Goldman–Shen period has no effect, so a standard mixed-MSE echo is obtained. As already discussed, this signal may contain contributions from both the solid and mobile components with the solid component signal somewhat reduced due to the finite efficiency of the pulse sequence, i.e., $S_{DF-MSE}(t_{f0} \sim 0) = S_m + \alpha S_s$, with $\alpha \leq 1$ accounting for the reduction in the echo intensity associated to the solid component. Note that α does not change with reaction time. As the filter time is increased the signals from rigid segments are progressively attenuated by the Goldman–Shen dipolar filter, while the signal from the mobile ones is detected without attenuation as long as $T_{2m} \gg t_f$. Thus, the detected signal is comprised by the full signal from the mobile components and an attenuated signal from the rigid components. The attenuation can be taken into account considering a factor depending on the Goldman–Shen filter time, i.e., $S_{DF-MSE}(t_f) = S_m + \beta_{GS}(t_f)\alpha S_s$. If the filter time is long enough $\beta_{GS}(t_f)$ becomes equal to zero, meaning the signals from rigid components are suppressed. This limit is easily identified by the absence of fast decaying signals in the DF-MSE echo. Thus, the ratio between the DF-MSE echo intensities at long and short Goldman-Shen filter times becomes:

$$f_{DF-MSE}(t_f) = \frac{S_{DF-MSE}(t_f)}{S_{DF-MSE}(0)} = \frac{S_m}{S_m + \alpha S_s} + \frac{\beta_{GS}(t_f)\alpha S_s}{S_m + \alpha S_s} = \gamma f_m + f_{GS}(t_f) \quad (4)$$

Therefore, for a filter time adjusted such as $\beta_{GS}(t_f) = 0$, the f_{DF-MSE} fraction has the same behavior of f_m concerning the dependence with the reaction time. Thus, it can also be fitted using Equation (3) to obtain the reaction kinetic parameters.

Figure 4a show f_{DF-MSE} fractions acquired with filter times of t_f = 50 μs, 100 μs, 200 μs, 400 μs as a function of the reaction times for the epoxy cure carried out at 313 K. The curves show similar decay shapes, but different plateau values. Another feature observed in Figure 4 is the reduction of the initial f_{DF-MSE} fractions as the filter time increases. This is a result of the relative short T_{2m} values, so at longer filter times part of the mobile component signals is also filtered out by the Goldman–Shen pulse sequence. This attenuation of the mobile signals is taken into account the γ factor in Equation (4). Due to the partial filtering of the solid components at shorter filter times, i.e., t_f = 50 μs and 100 μs, the plateau is associated to both the $f_{GS}(t_f)$ contribution and the remaining mobile component f_∞ as observed in the mixed-MSE experiments. At longer filter times, so $\beta_{GS}(t_f) = 0$, the plateau value is only related to f_∞, but it can assume a different value because of the partial attenuation of the mobile component. These features can be observed in Figure 4a, where a progressive decrease of the plateau value and initial fractions are observed as the filter time increases. Fitting the curves using Equation (3), we obtained almost the same values for the kinetic rate k_c, showing that the terms f_∞ and γ suffice to take into account the features discussed above. Despite not being necessary, in order to assure a situation with $\beta_{GS}(t_f) = 0$ and to minimize the attenuation of mobile component signal, we used t_f = 200 μs in the DF-MSE experiments to probe the epoxy cure.

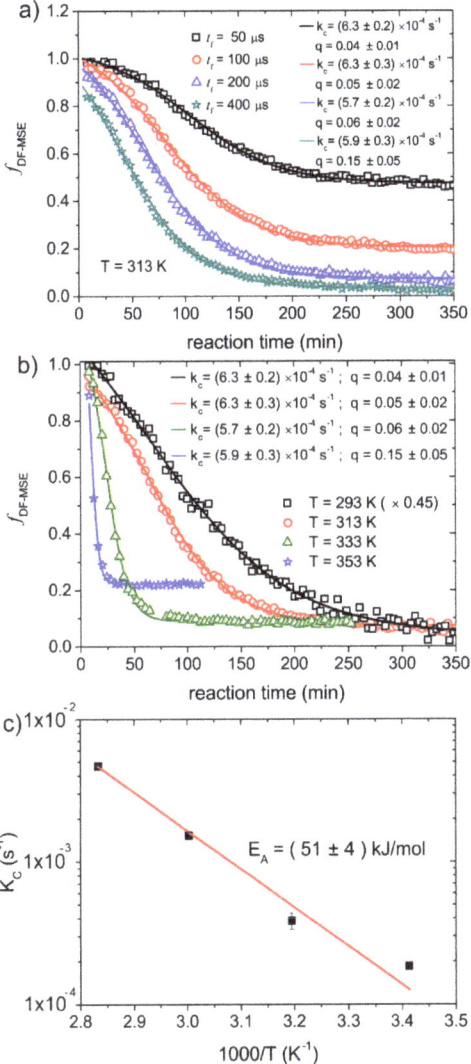

Figure 4. (a) Mobile fractions as function of the reaction time during the epoxy cure, calculated from DF-MSE experiments with different filter times. Best fit curves using Equation (3), as well as the obtained values for k_C and q, are also shown. The values obtained for f_0 are all equal to 1.0 ± 0.1. (b) Mobile fractions as a function of the reaction time during the epoxy cure at different temperatures calculated from DF-MSE experiments. For better visualization, the curve for 293 K was scaled up by a factor of 0.45. Best fit curves using Equation (3), as well as the obtained values for k_C and q, are also shown. The values obtained for f_0 at all temperatures using the first mixed-MSE echo are equal to 1.01 ± 0.01. The drop in f_{DF-MSE} due to the partial suppression of the mobile phase was taken into account by the γ factor in Equation (3). (c) Arrhenius plot of the k_c values obtained from the fits shown in (b).

Figure 4b shows the f_{DF-MSE} fractions as a function of reaction time for epoxy polymerization reactions carried out at 293 K, 313 K, 333 K and 353 K. The increase of the plateau values with temperature is associated with the gain of molecular mobility in the molecular segments of the product, as already discussed in the case of the mixed-MSE experiments. The fits using Equation (3) are also shown with the fitting parameters presented as inset. The k_c as a function of temperature are shown in the Arrhenius plot of Figure 4c. An activation energy of $E_a = (51 \pm 5)$ kJ/mol is obtained. This value is in excellent agreement with those obtained from the mixed-MSE and RK-ROSE data and also with the values found in the literature [25].

3. Discussions

In summary, our analysis showed that mixed-MSE, RK-ROSE and DF-MSE can be used to describe reactions with formation of solid products. The use of DF-MSE is particularly advantageous, since no extensive processing is required. Indeed, the normalization process can be completed in a fully automated way to display the result while the reaction is in progress, making possible a real-time monitoring of this type of reactions. DF-MSE can be used to monitor reactions when there is a formation of a solid product from liquid reagents. We should also mention that other type of dipolar filters, such as as the Magic and Polarization Echo (MAPE) pulse sequence [37], for mobile fraction determination, or double quantum filter, for rigid fraction determination, can also be used to acquire the data, so the same analysis presented here can be conducted. However, the simplicity of the Goldman–Shen filter used in the current DF-MSE version makes the data acquisition also quite straightforward.

4. Materials and Methods

Epoxy resin Araldite® with cure time of 90 min was acquired in the local trade of São Carlos-SP, Brazil. The two components (resin/hardener) were first mixed in the proportion of 10:8 (m/m) for its homogenization (about 120 s), as indicated by the manufacturer and was subsequently placed to the sample in the 10 mm NMR tube.

The spectrometer used was the Bruker Minispec ND mq-20 operating at ^{1}H frequency of 20 MHz (0.47 T) with a 10 mm probehead (dead time of 11.6 µs). A $\pi/2$ pulse length of 2.4 µs and acquisition time of 5 ms were typically used. The typical decay constant time due to static field inhomogeneity in the spectrometer was about 1.5 ms, so all data processing involving signal deconvolution was restricted to the first 300 µs of the signals. The recycle delays were ~5 s, based on the longitudinal relaxation time (T_1) determined with the inversion recovery pulse sequence [17]. Variable temperature experiments were carried out using the BVT 3000 temperature controller (Bruker). A previous calibration of the sample temperature was performed by placing a thermocouple immersed in silicon oil in the sample position and relating the thermocouple and the BVT indications. RK-ROSE experiments were performed with a long pulse duration of 24 µs (τ_{aq} in Figure 1 equal to the equipment dead time of 12 µs) and 32 scans. Mixed-MSE experiements were performed with an echo time of 100 µs and 16 scans.

5. Conclusions

We presented an extensive analysis about the usage of dipolar echo pulse sequences at a low magnetic field for monitoring polymerization reactions with the formation of solid-products from liquid reagents. As proof of principle, we probed the curing reactions of epoxy resins using three different ^{1}H TD-NMR dipolar based methods, i.e., mixed-MSE, RK-ROSE and DF-MSE.

We showed how mixed-MSE, RK-ROSE data can be processed to bring information about the changes in the product and reagents concentration during the reaction. Using a model assuming autocatalytic and noncatalytic reaction pathways, we establish the relationship between reaction and NMR parameters, so its kinetics could be followed and characterized from NMR data. Using this procedure, we determined an activation energy of about 50 kJ/mol for the cure of a commercial epoxy resin.

In the last part of the article, we applied the DF-MSE method as a straightforward approach for characterizing these types of reactions. By monitoring a simple normalized intensity as a function of the reaction time, we obtained a curve representing the consumption of the reagents during the process. Using a specific reaction model, this curve was fitted to obtain the reaction rates and to estimate the activation energy by monitoring reactions at different temperatures. In the specific case of the epoxy resin cure, we used the aforementioned model and estimated an activation energy of (51 ± 4) kJ/mol, in agreement with the values obtained using mixed-MSE and RK-ROSE as well as with values reported in the literature.

Author Contributions: Conceptualization, L.A.C., R.H.d.S.G., J.G.F. and E.R.d.A.; Methodology, R.H.d.S.G. and E.R.d.A.; formal analysis, R.H.d.S.G., J.G.F. and E.R.d.A.; investigation, R.H.d.S.G.; resources, L.A.C. and E.R.d.A.; writing—original draft preparation, R.H.d.S.G.; writing—review and editing, L.A.C., J.G.F. and E.R.d.A.; supervision, L.A.C. and E.R.d.A.; funding acquisition, L.A.C. and E.R.d.A. All authors have read and agreed to the published version of the manuscript.

Funding: This research was funded by the Brazilian Agencies FAPESP Grants 2019/13656-8, 2021/12694-3, 2017/24465-3, CNPq Grants 302866/2017-5, 303753/2018-8 and CAPES Finance Code 001.

Institutional Review Board Statement: Not applicable.

Informed Consent Statement: Not applicable.

Conflicts of Interest: The authors declare no conflict of interest.

References

1. Dalitz, F.; Cudaj, M.; Maiwald, M.; Guthausen, G. Process and Reaction Monitoring by Low-Field NMR Spectroscopy. *Prog. Nucl. Magn. Reson. Spectrosc.* **2012**, *60*, 52–70. [CrossRef]
2. Colnago, L.A.; Andrade, F.D.; Souza, A.A.; Azeredo, R.B.V.; Lima, A.A.; Cerioni, L.M.; Osán, T.M.; Pusiol, D.J. Why Is Inline NMR Rarely Used as Industrial Sensor Challenges and Opportunities. *Chem. Eng. Technol.* **2014**, *37*, 191–203. [CrossRef]
3. Hardis, R. Cure Kinetics Characterization and Monitoring of an Epoxy Resin for Thick Composite Structures. Ph.D. Thesis, Iowa State University, Ames, IA, USA, 2012; 55p. [CrossRef]
4. Ghaemy, M.; Barghamadi, M.; Behmadi, H. Cure Kinetics of Epoxy Resin and Aromatic Diamines. *J. Appl. Polym. Sci.* **2004**, *94*, 1049–1056. [CrossRef]
5. Papon, A.; Montes, H.; Hanafi, M.; Lequeux, F.; Guy, L.; Saalwächter, K. Glass-Transition Temperature Gradient in Nanocomposites: Evidence from Nuclear Magnetic Resonance and Differential Scanning Calorimetry. *Phys. Rev. Lett.* **2012**, *108*, 065702. [CrossRef]
6. Sturniolo, S.; Saalwächter, K. Breakdown in the Efficiency Factor of the Mixed Magic Sandwich Echo: A Novel NMR Probe for Slow Motions. *Chem. Phys. Lett.* **2011**, *516*, 106–110. [CrossRef]
7. Diuk Andrade, F.; Newson, W.R.; Bernardinelli, O.D.; Rasheed, F.; Cobo, M.F.; Plivelic, T.S.; DeAzevedo, E.R.; Kuktaite, R. An Insight into Molecular Motions and Phase Composition of Gliadin/Glutenin Glycerol Blends Studied by 13C Solid-State and 1H Time-Domain NMR. *J. Polym. Sci. Part B Polym. Phys.* **2018**, *56*, 739–750. [CrossRef]
8. Filgueiras, J.G.; da Silva, U.B.; Paro, G.; D'Eurydice, M.N.; Cobo, M.F.; DeAzevedo, E.R. Dipolar Filtered Magic-Sandwich-Echoes as a Tool for Probing Molecular Motions Using Time Domain NMR. *J. Magn. Reson.* **2017**, *285*, 47–54. [CrossRef]
9. Cobo, M.F.; Deublein, E.J.; Haber, A.; Kwamen, R.; Nimbalkar, M.; Decker, F. TD-NMR in Quality Control: Standard Applications. In *Modern Magnetic Resonance*; Webb, G., Ed.; Springer: Cham, Switzerland, 2018; pp. 1–18. [CrossRef]
10. Räntzsch, V.; Haas, M.; Özen, M.B.; Ratzsch, K.F.; Riazi, K.; Kauffmann-Weiss, S.; Palacios, J.K.; Müller, A.J.; Vittorias, I.; Guthausen, G.; et al. Polymer Crystallinity and Crystallization Kinetics via Benchtop 1H NMR Relaxometry: Revisited Method, Data Analysis, and Experiments on Common Polymers. *Polymer* **2018**, *145*, 162–173. [CrossRef]
11. Seiffert, S.; Oppermann, W.; Saalwächter, K. Hydrogel Formation by Photocrosslinking of Dimethylmaleimide Functionalized Polyacrylamide. *Polymer* **2007**, *48*, 5599–5611. [CrossRef]
12. Malmierca, M.A.; González-Jiménez, A.; Mora-Barrantes, I.; Posadas, P.; Rodríguez, A.; Ibarra, L.; Nogales, A.; Saalwächter, K.; Valentín, J.L. Characterization of Network Structure and Chain Dynamics of Elastomeric Ionomers by Means of 1H Low-Field NMR. *Macromolecules* **2014**, *47*, 5655–5667. [CrossRef]
13. Vieyres, A.; Pérez-Aparicio, R.; Albouy, P.A.; Sanseau, O.; Saalwächter, K.; Long, D.R.; Sotta, P. Sulfur-Cured Natural Rubber Elastomer Networks: Correlating Cross-Link Density, Chain Orientation, and Mechanical Response by Combined Techniques. *Macromolecules* **2013**, *46*, 889–899. [CrossRef]
14. Saalwächter, K.; Chassé, W.; Sommer, J.U. Structure and Swelling of Polymer Networks: Insights from NMR. *Soft Matter* **2013**, *9*, 6587–6593. [CrossRef]
15. Conway, T.F.; Earle, F.R. Nuclear Magnetic Resonance for Determining Oil Content of Seeds. *J. Am. Oil Chem. Soc.* **1963**, *40*, 265–268. [CrossRef]

16. Todt, H.; Guthausen, G.; Burk, W.; Schmalbein, D.; Kamlowski, A. Time-Domain NMR in Quality Control: Standard Applications in Food. *Mod. Magn. Reson.* **2008**, *1*, 1739–1743. [CrossRef]
17. Vold, R.L.; Waugh, J.S.; Klein, M.P.; Phelps, D.E. Measurement of Spin Relaxation in Complex Systems. *J. Chem. Phys.* **1968**, *48*, 3833–3834. [CrossRef]
18. Hertlein, C.; Saalwächter, K.; Strobl, G. Low-Field NMR Studies of Polymer Crystallization Kinetics: Changes in the Melt Dynamics. *Polymer* **2006**, *47*, 7216–7221. [CrossRef]
19. Litvinov, V.M.; Penning, J.P. Phase Composition and Molecular Mobility in Nylon 6 Fibers as Studied by Proton NMR Transverse Magnetization Relaxation. *Macromol. Chem. Phys.* **2004**, *205*, 1721–1734. [CrossRef]
20. Maus, A.; Hertlein, C.; Saalwächter, K. A Robust Proton NMR Method to Investigate Hard/Soft Ratios, Crystallinity, and Component Mobility in Polymers. *Macromol. Chem. Phys.* **2006**, *207*, 1150–1158. [CrossRef]
21. Faria, G.C.; Duong, D.T.; da Cunha, G.P.; Selter, P.; Strassø, L.A.; Davidson, E.C.; Segalman, R.A.; Hansen, M.R.; DeAzevedo, E.R.; Salleo, A. On the Growth, Structure and Dynamics of P3EHT Crystals. *J. Mater. Chem. C* **2020**, *8*, 8155–8170. [CrossRef]
22. Fernandes, H.; Filgueiras, J.G.; DeAzevedo, E.R.; Lima-Neto, B.S. Real Time Monitoring by Time-Domain NMR of Ring Opening Metathesis Copolymerization of Norbornene-Based Red Palm Olein Monomer with Norbornene. *Eur. Polym. J.* **2020**, *140*, 110048. [CrossRef]
23. Grunin, L.; Oztop, M.H.; Guner, S.; Baltaci, S.F. Exploring the Crystallinity of Different Powder Sugars through Solid Echo and Magic Sandwich Echo Sequences. *Magn. Reson. Chem.* **2019**, *57*, 607–615. [CrossRef]
24. Kovermann, M.; Saalwächter, K.; Chassé, W. Real-Time Observation of Polymer Network Formation by Liquid- and Solid-State NMR Revealing Multistage Reaction Kinetics. *J. Phys. Chem. B* **2012**, *116*, 7566–7574. [CrossRef]
25. Martin-Gallego, M.; González-Jiménez, A.; Verdejo, R.; Lopez-Manchado, M.A.; Valentin, J.L. Epoxy Resin Curing Reaction Studied by Proton Multiple-Quantum NMR. *J. Polym. Sci. Part B Polym. Phys.* **2015**, *53*, 1324–1332. [CrossRef]
26. Fechete, R.; Demco, D.E.; Blümich, B. Chain Orientation and Slow Dynamics in Elastomers by Mixed Magic-Hahn Echo Decays. *J. Chem. Phys.* **2003**, *118*, 2411–2420. [CrossRef]
27. Garcia, R.H.S.; Filgueiras, J.G.; DeAzevedo, E.R.; Colnago, L.A. Power-Optimized, Time-Reversal Pulse Sequence for a Robust Recovery of Signals from Rigid Segments Using Time Domain NMR. *Solid State Nucl. Magn. Reson.* **2019**, *104*, 101619. [CrossRef]
28. Pines, A.; Rhim, W.K.; Waugh, J.S. Homogeneous and Inhomogeneous Nuclear Spin Echoes in Solids. *J. Magn. Reson.* **1972**, *6*, 457–465. [CrossRef]
29. Matsui, S. Suppressing the Zero-Frequency Artifact in Magic-Sandwich-Echo Proton Images of Solids. *J. Magn. Reson.* **1992**, *98*, 618–621. [CrossRef]
30. Schäler, K.; Achilles, A.; Bärenwald, R.; Hackel, C.; Saalwächter, K. Dynamics in Crystallites of Poly(ε-Caprolactone) as Investigated by Solid-State NMR. *Macromolecules* **2013**, *46*, 7818–7825. [CrossRef]
31. Kurz, R.; Achilles, A.; Chen, W.; Schäfer, M.; Seidlitz, A.; Golitsyn, Y.; Kressler, J.; Paul, W.; Hempel, G.; Miyoshi, T.; et al. Intracrystalline Jump Motion in Poly(Ethylene Oxide) Lamellae of Variable Thickness: A Comparison of NMR Methods. *Macromolecules* **2017**, *50*, 3890–3902. [CrossRef]
32. Manson, J.A.; Sperling, L.H. *Polymer Blends and Composites*; Plenum Press: New York, NY, USA, 1976.
33. Vyazovkin, S.; Sbirrazzuoli, N. Kinetic Methods to Study Isothermal and Nonisothermal Epoxyanhydride Cure. *Macromol. Chem. Phys.* **1999**, *200*, 2294–2303. [CrossRef]
34. Abragam, A. *The Principles of Nuclear Magnetism*; Oxford University Press: New York, NY, USA, 1961; Volume 1.
35. Sourour, S.; Kamal, M.R. Differential Scanning Calorimetry of Epoxy Cure: Isothermal Cure Kinetics. *Thermochim. Acta* **1976**, *14*, 41–59. [CrossRef]
36. Odian, G. *Principles of Polymerization*, 4th ed.; John Wiley & Sons: Hoboken, NJ, USA, 2004.
37. Demco, D.E.; Johansson, A.; Tegenfeldt, J. Proton Spin Diffusion for Spatial Heterogeneity and Morphology Investigations of Polymers. *Solid State Nucl. Magn. Reson.* **1995**, *4*, 13–38. [CrossRef]
38. Zhang, S.; Mehring, M. A Modified Goldman-Shen NMR Pulse Sequence. *Chem. Phys. Lett.* **1989**, *160*, 644–646. [CrossRef]
39. Goldman, M.; Shen, L. Spin-Spin Relaxation in LaF$_3$. *Phys. Rev.* **1966**, *144*, 321–331. [CrossRef]
40. Saalwächter, K. Proton Multiple-Quantum NMR for the Study of Chain Dynamics and Structural Constraints in Polymeric Soft Materials. *Prog. Nucl. Magn. Reson. Spectrosc.* **2007**, *51*, 1–35. [CrossRef]

Article

Compact NMR Spectroscopy for Low-Cost Identification and Quantification of PVC Plasticizers

Anton Duchowny and Alina Adams *

Institut für Technische und Makromolekulare Chemie, RWTH Aachen University, Templergraben 55, 52056 Aachen, Germany; Duchowny@itmc.rwth-aachen.de
* Correspondence: Alina.Adams@itmc.rwth-aachen.de; Tel.: +49-241-8026-428

Abstract: Polyvinyl chloride (PVC), one of the most important polymer materials nowadays, has a large variety of formulations through the addition of various plasticizers to meet the property requirements of the different fields of applications. Routine analytical methods able to identify plasticizers and quantify their amount inside a PVC product with a high analysis throughput would promote an improved understanding of their impact on the macroscopic properties and the possible health and environmental risks associated with plasticizer leaching. In this context, a new approach to identify and quantify plasticizers employed in PVC commodities using low-field NMR spectroscopy and an appropriate non-deuterated solvent is introduced. The proposed method allows a low-cost, fast, and simple identification of the different plasticizers, even in the presence of a strong solvent signal. Plasticizer concentrations below 2 mg mL^{-1} in solution corresponding to 3 wt% in a PVC product can be quantified in just 1 min. The reliability of the proposed method is tested by comparison with results obtained under the same experimental conditions but using deuterated solvents. Additionally, the type and content of plasticizer in plasticized PVC samples were determined following an extraction procedure. Furthermore, possible ways to further decrease the quantification limit are discussed.

Keywords: plasticizer; PVC; identification; quantification; non-deuterated solvent; low-field NMR spectroscopy

Citation: Duchowny, A.; Adams, A. Compact NMR Spectroscopy for Low-Cost Identification and Quantification of PVC Plasticizers. *Molecules* **2021**, *26*, 1221. https://doi.org/10.3390/molecules26051221

Academic Editor: Igor Serša

Received: 9 February 2021
Accepted: 22 February 2021
Published: 25 February 2021

Publisher's Note: MDPI stays neutral with regard to jurisdictional claims in published maps and institutional affiliations.

Copyright: © 2021 by the authors. Licensee MDPI, Basel, Switzerland. This article is an open access article distributed under the terms and conditions of the Creative Commons Attribution (CC BY) license (https://creativecommons.org/licenses/by/4.0/).

1. Introduction

The amount of plastics produced worldwide has been increasing steadily in recent decades, with poly(vinyl chloride) (PVC) being the third most produced polymer after polyethylene and polypropylene [1]. PVC products are widely used in many fields of application including consumer products, construction, and packaging materials as well as medical devices. Concomitant with the growth of polymer production also comes a significant increase in the amount of used additives such as antioxidants [2], organic peroxides [3], and plasticizers. In particular, plasticizers account for about one third of the additives [4]. Forecasts predict a rise in the global demand for plasticizers to about 9.75 million tons in 2024 [5]. Plasticizers play an essential role in almost all formulations of polymer products, being especially important for PVC. The plasticizer content in PVC ranges from small amounts up to about 80 wt% for various industrial products [6,7]. Plasticizers are usually larger molecules with molar masses between 200 and 500 g/mol which have bulky or long side groups and serve the purpose to improve the flexibility of the PVC products by lowering the glass transition T_g of the pure polymer which is about 82 °C.

Due to economic and technical reasons, plasticizers are, in most cases, simply mixed with the polymer material [4]. In contrast to inner plasticizers, these external plasticizers are not chemically bound to the polymer chain and generally tend to migrate out of the product over time or in the presence of solvents. The detected leaching is strongly dependent on the experimental conditions and on the type and amount of plasticizer [8–14].

As a consequence, developing suitable mathematical models for predicting plasticizer leaching under various experimental conditions is a very challenging task, as also shown by the failure of accelerated tests to predict its long-term migration behavior [15,16]. The modeling is even more complicated for the plasticizer loss in the presence of solvents. This is due to a combined interplay between the migration of the plasticizer in the surrounding liquid and the ingress of the liquid itself into the PVC product at diffusivity rates which strongly depend on the concentration of the plasticizer inside the PVC and on the type of solvent [17].

However, a precise prediction of the behavior of a particular plasticizer under working conditions of a PVC product is of key importance for an improved understanding of the two major issues accompanying the plasticizer loss: 1) the deterioration of the original performance of the PVC products [18–21] and 2) possible environmental and health risks due to the toxicity of phthalate plasticizers [22–25]. Due to this, risk assessments and regulations have been introduced concerning the usage of plasticizers in products designed to be in contact with human skin or groceries [26–28]. This also led to the development of novel strategies to reduce the plasticizer loss and to design alternative and phthalate-free plasticizers [4,29–31].

Only very few studies exist today about the migration of these novel plasticizers and their health risks [14,32,33]. A reliable assessment of the above-mentioned risks requires identifying the type of plasticizer inside a PVC product and quantifying its release under particular experimental conditions. This in turn will help in designing plasticizers with improved properties. In addition, simple, cost-efficient, and reliable ways are needed for controlling, on a regular basis, how far the legal regulations are indeed respected.

Various analytical techniques are nowadays applied in the identification and/or quantification of plasticizers in PVC. They include Fourier-transform infrared (FT-IR) spectroscopy, mass spectrometry (MS), liquid and gas chromatography (LC and GC), thermogravimetric analysis, terahertz spectroscopy, and solid- and liquid-state nuclear magnetic resonance (NMR) spectroscopy [34–41]. Aside from FT-IR, LC, and GC-MS, a recent publication compared the analytical performance of liquid-state NMR spectroscopy conducted at a high magnetic field of 500 MHz and identified NMR as being a primer method able to precisely discriminate all investigated plasticizers [37]. This result is further supported by the identification of seven plasticizers in medical devices followed by quantification of their concentrations by adhering to a defined measuring procedure of the high-field NMR method in deuterated solvents [42]. Despite being, nowadays, an indispensable method for structural determination in chemistry, the applicability of high-field liquid-state NMR for the identification and eventual quantification of plasticizers in PVC was, up to now, restricted to a few dedicated studies largely from academia [38,42]. This is because the high-field NMR devices are expensive, need to be operated by skilled personnel in special facilities, and require the usage of deuterated solvents, which are more costly than the corresponding non-deuterated solvents. Hence, high-field NMR is usually not the method of choice for low-cost routine analysis in industry, in a medical environment, and even in academia.

However, the development of compact NMR instruments with open and closed geometries has opened new perspectives in many fields of activities [43–51]. Such NMR devices working at a low magnetic field, in the range of 40 to 60 MHz, are commercially available at low prices and can be operated by non-experts. Having a small size and being light, they can be placed in a synthesis laboratory on a bench or under a fume hood, near a production line, or in a corner in a hospital. Furthermore, a large variety of experimental NMR methods, including one-dimensional ^{1}H and ^{13}C spectroscopy, 2D spectroscopy, and relaxation, are readily implemented to work at low-field NMR. As a consequence, compact low-field NMR has become an excellent alternative to high-field NMR for a large variety of investigations. In particular, low-field NMR spectroscopy is well suited for the detailed structural characterization of small molecules and reaction monitoring [44,46]. The application of low-field NMR spectroscopy for the study of larger and more complex molecules has started more recently. This is because the analysis of the corresponding spectra is far more challenging than for smaller molecules due to a stronger overlap of the resonances than in the high field. Comparisons with the corresponding liquid spectra from high-field NMR and/or a combination with chemometrics or other analytical methods are useful and often the needed method for a reliable signal identification and assignment [45,52–55].

In this work, the applicability of low-field NMR spectroscopy for the identification and quantification of PVC plasticizers is, for the first time, investigated and demonstrated with the help of five different plasticizers (Figure 1). In addition, a novel experimental protocol is proposed for gaining the needed information in a fast way and without the need for deuterated solvents. Moreover, possible methodological and hardware improvements to further lower the quantification limit are discussed. These traits make the introduced approach particularly interesting as an alternative to high-field NMR and for routine quality control in various environments. Furthermore, it could be used for the identification and quantification of additives for other polymer materials.

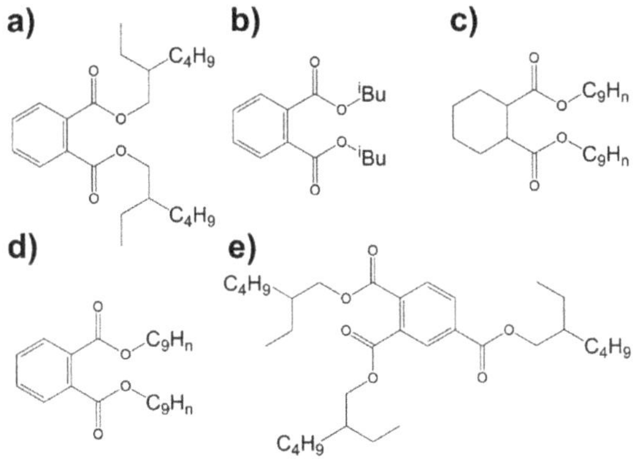

Figure 1. Structure of the investigated plasticizers in the current study: (**a**) diethylhexyl phthalate (DEHP), (**b**) diisobutyl phthalate (DIBP), (**c**) diisononyl cyclohexane-1,2-dicarboxylate (DINCH), (**d**) diisononyl phthalate (DINP), (**e**) tris(2-ethylhexyl) trimellitate (TOTM).

2. Results and Discussions

2.1. ^{1}H NMR Spectroscopy

^{1}H NMR spectroscopy is an appropriate analytical method to analyze PVC plasticizers as it can differentiate between the signals given by various functional groups and their intensities are directly related to the amount of plasticizer inside the NMR tube. According to results acquired at high-magnetic fields in deuterated solvents [42,56], the ^{1}H NMR spectra of various plasticizers contain peaks in the aromatic region around 7 ppm, between 3 and 4 ppm for the α-CH$_2$ groups next to the ester bond, and at around 1 ppm for aliphatic CH$_2$, while the CH$_3$ chain ends appear at around 0.8 ppm. An exception to this is DINCH which has a cyclohexene dicarboxylic acid core instead of phthalic acid and consequently shows no peak in the aromatic region of the spectrum. These features show that various plasticizers can be well discriminated by using characteristic proton resonances.

One can thus anticipate that a discrimination of plasticizers would also be possible using low-field NMR spectroscopy even if the resonances would be less separated. Furthermore, given that the characteristic proton resonances are located in a spectral range above 2.5 ppm, one could argue that non-deuterated solvents, which give signals outside the range of interest, can be used to acquire the proton spectra instead of deuterated solvents. One example of such solvent is n-hexane which is reported to be a good solvent for plasticizers [57] and gives proton signals under 1.3 ppm, at positions largely independent of the presence of other solvents [58]. A methodology using non-deuterated solvents is especially attractive in the view of the strongly reduced costs as compared to the case when deuterated solvents are employed for routine analyses. It is also more appropriate for investigating plasticizer leaching under real conditions.

Typical proton low-field NMR spectra of various plasticizers dissolved in deuterated chloroform as well as in non-deuterated hexane are depicted in Figure 2a,b and in full scale in Figure S1. One can clearly observe from Figure 2a that the spectral region below 3 ppm is very crowded, and the signals are largely similar among the various plasticizers, except for DIBP. Nevertheless, this is not an impediment as the signals above 3 ppm are well separated and can be used for the purpose of identification and quantification. This observation is in agreement with the results previously reported [42,56]. Exactly these spectral features are also advantageous when a plasticizer is dissolved in a non-deuterated solvent. This is demonstrated in Figure 2b by the typical low-field ^{1}H NMR spectra of the investigated plasticizers in non-deuterated n-hexane for a plasticizer concentration of 10 vol.% (94.4–103.9 mg mL^{-1}) and in Figure S2 for varying concentrations down to 0.1 vol.% (0.97 mg mL^{-1}). As expected, the solvent shows strong signals at lower ppm values. Obviously, the resonances below 2 ppm in Figure 2b are covered by the strong n-hexane peak and, thus, cannot be used for quantification. However, the spectra acquired in deuterated chloroform and non-deuterated n-hexane are highly similar, being above 3 ppm, except the residual chloroform peak and differences in chemical shift due to solvent effects. Thus, the aromatic and α-CH$_2$ regions can be used to identify and quantify PVC plasticizers. Even at a plasticizer concentration as low as 1 vol.% and below, characteristic resonances of the various plasticizers can be identified in this spectral range in the presence of the non-deuterated solvent which poses no impediments by its strong signal in the low-ppm region (Figures S2 and S3). Moreover, the ^{1}H low-field NMR spectra depicted in Figure 2 conveniently allow the identification of the various plasticizers with the help of the specific spectral features, despite the used magnetic field being a factor 10 lower than in [42].

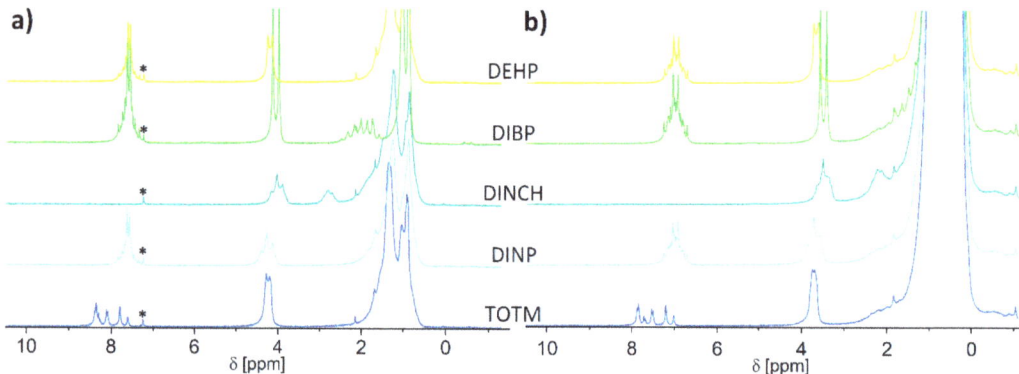

Figure 2. 40 MHz ^{1}H NMR spectra of the investigated plasticizers at concentrations of 10 vol.% in (**a**) deuterated chloroform and (**b**) non-deuterated n-hexane. Spectra have been referenced to the residual deuterated chloroform (signal marked with asterisk at 7.26 ppm) and n-hexane (0.8 ppm) peaks, respectively. The difference in the chemical shift of ~0.5 ppm in the signals in the two subfigures can be attributed to solvent effects. Most of the spectra are zoomed in on for a better view of the signals above 2 ppm.

The aromatic spectral region around 7 ppm indicates that one can easily differentiate DINCH from all the other plasticizer types by the lack of signals in this region. Furthermore, TOTM shows a completely different spectral pattern at this position compared to all the other investigated plasticizers. This specific signal can be used for its identification. The observed spectral pattern is because TOTM is a derivate of trimellitic acid instead of phthalic acid, which has three rather than two carboxylic acid side groups. This leads to the strong splitting of the aromatic peak. Additionally, as TOTM possesses one more substituent at the aromatic ring, less hydrogen atoms contribute to the aromatic signal. DINP, DIBP, and DEHP show similar peak shapes in this spectral region, which makes their identification impossible when using only this signal. However, due to the structural differences in their aliphatic side chains, the peaks between 3 and 4 ppm have different shapes and intensities compared to the signals around 7 ppm. The specific characteristics of the three plasticizers in this region can also be used for their identification. In particular, DIBP shows the most distinctive and intense signal. Its peak at 3.5 ppm has a doublet with a J-coupling constant of 6.5 Hz due the single proton bound to the tertiary carbon atom at the β-position of the isobutylic group. DEHP also shows a doublet, which is shifted towards the lower field by 0.19 ppm and shows a significantly lower J-coupling constant of 4.5 Hz when compared to the DIBP doublet.

DINCH and DINP are mixtures of plasticizers with several iso-nonyl chains, causing the signals in this spectral region to be relatively broad. Despite having a similar shape, the signal of DINCH is shifted to the higher field by 0.23 ppm, allowing further differentiating them.

From the above findings, it becomes clear that the analysis of the peak positions, their shapes, and their intensities in the aromatic and ester regions of the ^{1}H spectrum delivers enough information to identify each of these five plasticizers. This type of identification, however, could be more complicated, especially at a low magnetic field, when investigating mixtures of various plasticizers due to the overlap of the resonances of interest. In this case, the use of ^{13}C spectra may be needed (see Section 2.2).

Following the identification of the plasticizer type, quantifying its concentration is the next step. Proton NMR spectroscopy is generally well suited for this purpose as the integral of a signal is directly proportional to the number of protons contributing to it and, through that, is directly proportional to the concentration of the investigated sample. The procedure applied in [42] uses an internal reference compound in a coaxial tube inserted into the 5 mm NMR tube containing the plasticizer solution to be measured. The concentration of the used reference compound was a priori calibrated with a plasticizer solution of known concentration. This procedure is common in liquid-state NMR and can be also implemented at low fields but leads to a reduced signal intensity due to the decreased volume of the sample of interest and possibly to peak distortions as well.

To overcome these issues, we generated an external calibration curve by correlating the integral value of the signal of interest to the plasticizer's concentration. This procedure was applied for all investigated plasticizers in the whole range of studied concentrations. Figure 3a shows, exemplarily for DINP, the dependence between the known plasticizer concentration and the corresponding integral of the two signals, which could, in principle, be used for chemical identification. The integral of the aromatic peak (around 7 ppm) shows a linear behavior with the plasticizer concentration. The ester peak at around 3.5 ppm, however, shows a slight offset from the ideal linear trend which becomes more pronounced on a logarithmic scale with decreasing concentrations. This can be explained by an additional signal given by the solvent's ^{13}C satellite peaks which, at a lower plasticizer concentration, start to overlap with this spectral region. ^{1}H-^{13}C decoupling techniques come standard with modern compact NMR spectrometer but were not available for the 40 MHz instrument used for this study. If a quantification with the aromatic region is not applicable—like in the case of DINCH—the utilization of a different suitable solvent such as benzene or chloroform would move the strong solvent peaks away from the region of 3.5 ppm, hence abolishing overlapping of solvent and ester peaks. Alternatively, spectral deconvolution techniques could be applied at this point to numerically decrease the error introduced by overlap. Although powerful, spectral deconvolution requires the operator to have the know-how and experience in that field. To keep the proposed quantification methodology simple, we determined the peak prominence in addition to the integral to serve as a compensation for overlapping peaks.

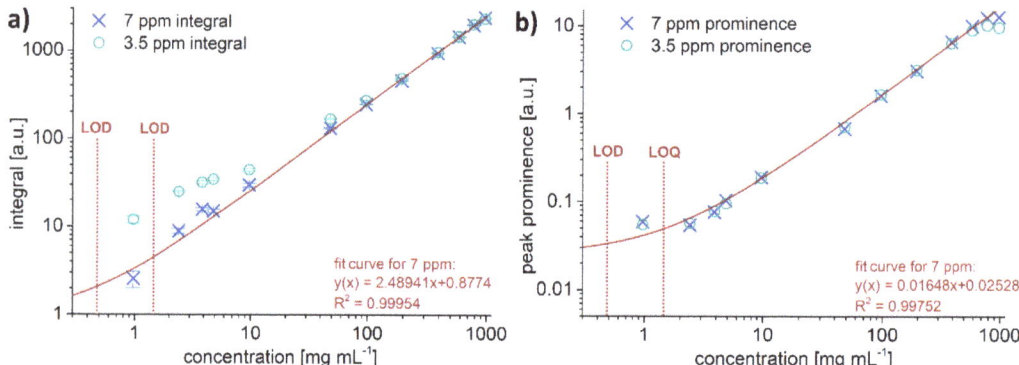

Figure 3. (a) Integral and (b) peak prominence of the ester (3.5 ppm) and aromatic (7 ppm) peak regions of DINP solutions using non-deuterated n-hexane as a solvent. Concentrations range from 0.97 to 970 mg mL^{-1}. The continuous lines are the fit results for the signal at 7 ppm showing the excellent linear correlation between the NMR integral (a) or the peak prominence (b) and the plasticizer concentration as demonstrated by the R^2 value. The apparent curvature of the linear fit is due to the log–log scaling of the axes.

Figure 3b shows that the peak prominences follow a linear trend with the exception of the data point of the lowest concentration. This is below the determined limit of quantification for DINP measured with four scans and thus is strongly influenced by noise. Additionally, the data points for the peak prominence of the 80 and 100 vol.% solutions also do not follow the linear trend. This can be explained by the high, sirup-like viscosity of the pure plasticizer, which causes the Free Induction Decay (FID) to decay more rapidly, resulting in broader, instead of taller, peaks. This behavior is also reflected in the samples' spin–spin relaxation times as exemplarily shown for TOTM and DIBP in Figure S4. Thus, for quantification purposes, the aromatic peak region around 7 ppm is preferable. If not applicable, like in the case of DINCH, the peak prominence around 3.5 ppm can be utilized as an alternative as it also shows a strongly linear correlation with the plasticizer concentration.

The determined limit of detection (LOD) and limit of quantification (LOQ) values of every plasticizer are highly influenced by the peak structure in the NMR spectrum. Hence, they vary for every individual plasticizer and the used ppm range for this purpose. Consequently, both limits are higher for TOTM in Table 1, caused by the molecular features discussed above. These features effectively lower the signal-to-noise ratio for this plasticizer in the aromatic spectral range, which was used for the determination of the LOD and LOQ. However, when analyzing the ester region of the spectrum, the signal intensity of TOTM would be higher compared to the other plasticizers as there are three, rather than two, ester groups in this molecule.

Table 1. Detection and quantification limits of examined plasticizers dissolved in non-deuterated n-hexane and measured at 40 MHz within 1 min. Both limits are given in concentrations for the analyzed solution and in amounts a PVC material would need to contain to achieve these concentrations after a solvent extraction. The aromatic peaks' integral was used to determine detection and quantification limits for all plasticizers except DINCH, where the ester peaks' prominence at 3.5 ppm was selected.

	DINP	DIBP	DEHP	TOTM	DINCH
LOD [mg mL^{-1}]	0.48	0.42	0.57	1.52	0.63
LOQ [mg mL^{-1}]	1.45	1.25	1.70	4.58	1.90
LOD [wt% in PVC]	0.96	0.83	1.13	3.05	1.27
LOQ [wt% in PVC]	2.89	2.49	3.39	9.15	3.80

2.2. ^{13}C NMR Spectroscopy

Nowadays, liquid-state ^{13}C NMR spectroscopy at a high magnetic field is often the method of choice in the structural characterization of complex and larger molecules owing to its lack of homonuclear coupling and broader signal dispersion compared to ^{1}H NMR. Liquid-state ^{13}C spectra of various plasticizers recorded at a high magnetic field are reported in various sources but, to our knowledge, never in a systematic way towards comparison and quantification and never at a low magnetic field. Therefore, we investigated, for the first time, the applicability of low-field ^{13}C NMR liquid-state spectroscopy for the identification of plasticizers and quantification of their concentration.

Due to the low natural abundance of ^{13}C, the acquisition of the spectra for obtaining a reasonable signal-to-noise ratio lasts much longer than the corresponding ^{1}H spectra. For investigating how far ^{13}C spectroscopy at a low field strength is applicable to everyday practice, the acquisition time of recording the spectra was set to around 32 min. This was achieved by accumulating 128 scans with a repetition delay of 15 s. Figure 4a exemplarily depicts the ^{13}C low-field NMR spectra of all studied plasticizers in n-hexane at a concentration of 60 vol.%. In addition, Figure S5 shows the acquired ^{13}C spectra of DIBP, ranging from 10 to 100 vol.%.

While n-hexane shows three distinct signals at around 14, 23, and 32 ppm (marked with asterisks in Figures 4a and S5), all the other signals belong to the plasticizers. They show a large dispersion over almost 180 ppm and all are well observable, even under the used experimental conditions. Furthermore, the signals of all plasticizers are outside the range where dissolved PVC would have its own signals (from about 44 to 49 ppm and from about 55 to about 58 ppm) [38]. This means that one could perform the measurements directly on the dissolved plasticized PVC sample without the need for removing the polymer or performing extraction studies.

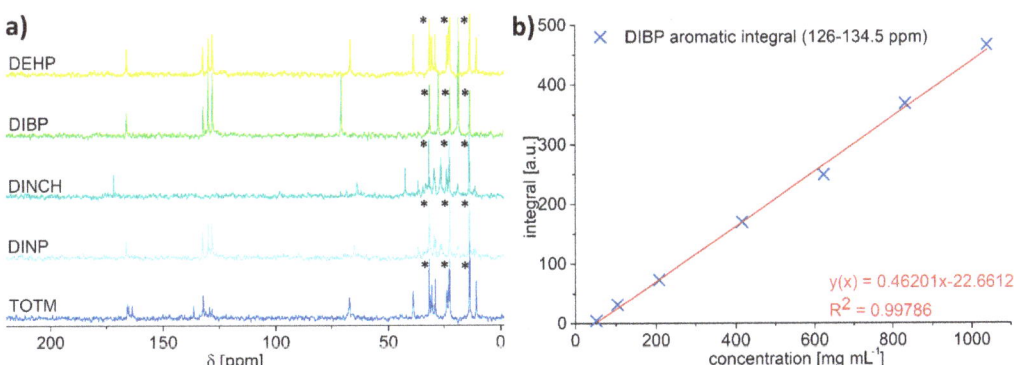

Figure 4. (a) ^{13}C low-field NMR spectra recorded with a plasticizer concentration of 60 vol.% in non-deuterated n-hexane (signals marked with asterisks). The asterisks correspond to the signals from n-hexane. (b) Integral versus concentration plot showing excellent linear behavior as illustrated by the aromatic region of DIBP.

Analyzing the spectral range of the carboxylic carbon atom between 160 and 175 ppm already allows identifying DINCH and TOTM. This is because the ring structure of these plasticizers differs from the phthalate-based ones. As a result, DINCH shows a peak at 171.54 ppm and TOTM shows three peaks at 163.49, 165.02, and 165.61 ppm, whereas the other plasticizers exhibit one peak at roughly 166 ppm. As these differences in the carboxylic peak region arise from differences in the aromatic core of these plasticizers, the same conclusion can be drawn when analyzing the aromatic part of the spectrum between 127 and 137 ppm. DINCH shows no signals in this region, whereas TOTM exhibits a more complex peak structure compared to the signals of the other three molecules.

DEHP, DIBP, and DINP all possess different aliphatic side chains. Therefore, these components can be identified by analyzing the signal of the carbon atom at the α-position of the side group between 60 and 71 ppm. Here, DIBP shows the highest chemical shift with 70.66 ppm. DEHP, on the other hand, has its peak at 66.87 ppm, followed by DINP which has the smallest chemical shift in this spectral region and shows two peaks at 64.72 and 65.05 ppm. In theory, DINCH and DINP should have an identical peak pattern here. However, as both chemicals are mixtures with different iso-nonyl side chains, this is not true. In this case, DINP only shows one signal for the α-position of the side chain and no signal around 38–43 ppm where an aliphatic tertiary carbon atom would appear. This fact suggests that the examined DINP contains more n-nonyl side chains, whereas for DINCH, the aliphatic groups are a mixture of several iso-nonyl groups.

As a conclusion, the investigated plasticizers can easily be distinguished from others even at a low magnetic field strength with a simple comparison of the ^{13}C spectra. However, this will not be feasible during the half an hour measuring time when very low amounts are present. Furthermore, their quantification using ^{13}C NMR is challenging. The relaxation times T_1 of ^{13}C nuclei usually have higher values compared to ^{1}H. This circumstance and the low ^{13}C natural abundance of 1.1% translate to longer measuring times. Moreover, ^{13}C liquid-state spectra are, in most cases, recorded using shorter recycle delays than those dictated by the ^{13}C longitudinal relaxation times T_1. Thus, the recorded spectrum is not quantitative and the signal integral will be also affected by the nuclear Overhauser effect (NOE) enhancement. This means that the peaks' integral does not necessarily correspond to the amount of contributing ^{13}C nuclei from the molecule. The signal intensity of the observed carbon nucleus will be boosted depending on the amount of hydrogen nuclei close to it as these can transfer their nuclear polarization to the carbon. Therefore, the spectra shown in Figure 4 are not quantitative in the way ^{1}H spectra are, meaning that, e.g., if one peak in the spectrum is double in integral compared to a second, it is not necessarily the case that this signal is produced by twice as many nuclei.

However, following the same procedure as applied for proton spectra and using the same series of samples, the correlation between the integrals of particular ^{13}C signals and the known plasticizer concentrations can be investigated. A linear behavior between the plasticizer's concentration and the peak integral is obtained for various signals as exemplarily depicted for DIBP (Figure 4b). These curves can then be employed for any further quantification of the plasticizer concentration. The use of only 128 scans enables identification of plasticizers at concentrations as low as 50 mg mL^{-1}, but they are, however, not enough for a reliable quantification of concentrations under 100 mg mL^{-1} due to the high noise level. Both values could be further improved at the cost of longer measurement times.

2.3. Test of the Proposed Method

To test the reliability of the proposed method for the identification and quantification of plasticizers using ^{1}H low-field spectroscopy in the presence of non-deuterated solvents and with the help of external calibration, plasticizer extraction experiments with five plasticized PVC samples with unknown histories using both deuterated and non-deuterated solvents were performed. Each extraction, as described and validated in the literature [57], was repeated three times using deuterated chloroform and non-deuterated n-hexane. Figure 5 shows the obtained ^{1}H spectra.

Identification of the plasticizer in the samples is convenient, even at a low magnetic field strength, as the spectra of the pure components were available from the calibration step. Given the specific spectral features of each plasticizer, the results from Figure 5 indicate that all samples contain only one type of plasticizer. In particular, the samples 1 to 3 can be identified as samples containing DINCH, while the samples 4 and 5 can be identified as samples containing DINP. Integrating the spectrum in the same ppm region as employed for the calibration curve yields the plasticizer concentration in the sample solution in the tube. The obtained results are shown in Table 2.

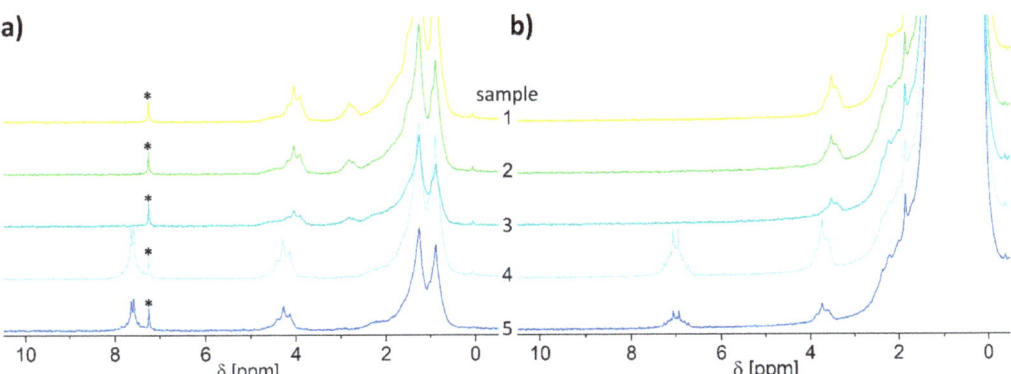

Figure 5. ^{1}H spectra of the unknown PVC samples 1–5 after the extraction in (**a**) CDCl$_3$ (residual non-deuterated solvent marked with asterisks) and (**b**) non-deuterated n-hexane.

Table 2. Results of the extraction experiment. The determined concentration of the extraction solution has been converted to wt.% of plasticizer the PVC initially had, assuming all of the plasticizer had been extracted. Each extraction step was performed 3 times for each investigated PVC sheet by using samples from different positions in the sheet.

Sample	Identified Plasticizer Type	Determined Plasticizer Content [wt.%] from CDCl3 Extraction		Determined Plasticizer Content [wt.%] from n-hexane Extraction	
1	DINCH	38.49	±1.93	42.69	±1.64
2	DINCH	31.68	±1.83	34.14	±1.19
3	DINCH	15.92	±3.63	17.97	±2.60
4	DINP	40.94	±0.10	43.26	±3.98
5	DINP	23.85	±0.95	22.64	±1.38

Since DINCH does not contain aromatic structures, its signal around 3.5 ppm was used for analysis. This ppm range shows some overlap with the hexane peak as the concentration of the plasticizer in the extraction liquid is relatively low. In order to increase the accuracy of the DINCH quantification, we determined the peak prominence and full-width-half-maximum (FWHM) to compensate for the hexane overlap.

The results in Table 2 would indicate, at a first glance, a large standard deviation of the analytical measurement. Re-measuring the same samples yielded a highly similar spectrum and plasticizer content. The relative error of the integral between several measurements of the same sample tube with plasticizer concentrations between 5 and 30 mg mL^{-1} is about 0.5–0.7%, which is much lower than the errors in Table 2. The reason for the detected differences is due to a heterogeneous distribution of the plasticizer in the PVC sheets. The fact that samples taken in the middle of the sheet contained less plasticizer compared to the edges indicates that the plasticizer has already migrated to the outside of the polymer structure. A comparison of ^{1}H-NMR spectra measured at the low field and at 400 MHz on the same sample tubes validated the fact that samples taken from the same PVC sheet indeed contained different amounts of plasticizer. This further indicates that the precision of the measurement is high, but the local plasticizer content in different areas of the PVC sheet varies.

Ascertaining the plasticizer content gravimetrically to cross-check the NMR results yielded a plasticizer content between 7.2 and 11 wt.% lower compared to the extraction method. This result suggests that either the plasticizer was not fully extracted, or the solvent could not completely be removed from the PVC sample after several days of vacuum drying. A second extraction step with just enough solvent to cover the sample, however, showed no plasticizer signal at all, even with a larger number of scans. Kastner et al. [14] reported complex interactions between the solvent and plasticized polymer, indicating that gravimetric analysis, though simple in preparation and execution, is not suitable for plasticizer quantification without further methodical modifications. Results of the gravimetric analysis are shown in Table S1 and S2.

2.4. How to Further Improve the Low-Field NMR Identification and Quantification

The most accessible method of identifying and quantifying plasticizers in PVC products, which offers reasonable precision with a lower consumption of time, costs, and workforce, was aimed for. Possibilities to improve this method are plentiful but they will add an additional step in the sample preparation, costs, or the time needed for an analysis. The easiest improvement would be the implementation of ^{13}C-decoupling in the ^{1}H spectra. This option comes standard in today's benchtop NMR spectrometers with a carbon channel but was not available with the 40 MHz instrument used in this study as it is one of the first-generation devices. ^{13}C-decoupling would have a noticeable effect especially on the n-hexane peak and would vastly enhance the spectral resolution at low plasticizer concentrations. Figures S2 and S3 exemplarily visualize this effect on various plasticizers. The ^{13}C satellite peaks of hexane appear between 2 and 2.5 ppm as well as below 0 ppm. At low plasticizer concentrations, the satellite peaks have a similar intensity to the analyte and increase the width of the already pronounced n-hexane peak.

Another option to improve the analysis outcome in terms of identification followed by quantification would be to increase the signal-to-noise ratio (SNR). For the extraction method, we used plenty of solvent compared to the mass of PVC. Hence, the plasticizer concentration of the extract was low (1–3 wt.%). The easiest way to improve the SNR would be to decrease the amount of solvent used for extraction. However, in this case, one runs the risk of an incomplete extraction. Alternatively, one could let the solvent evaporate after the extraction time and re-solve the plasticizer with just enough solvent to fill the sensitive region of the NMR spectrometer. In this way, the concentration and therefore the signal strength of the examined sample can drastically be increased which facilitates the plasticizer identification. Having a relative vapor pressure of a maximum of 60 Pa (DINP) compared to 162 hPa for n-hexane (209 hPa for chloroform) at 20 °C makes it unlikely to lose a significant amount of plasticizer during solvent evaporation. A second alternative to increase the SNR for given experimental conditions in terms of the magnetic field strength and temperature is by increasing the number of scans. This in turn leads to an increase in the experimental time. Nevertheless, in this way, the LOD and LOQ can be further decreased.

Furthermore, benefits can be gained by eliminating the peak overlap. This can be achieved by using deuterated solvents such as d-chloroform, which we used in the solvent extraction method as an alternative to n-hexane. Most of these, however, are more costly and thus the choice of solvents is limited. Technical n-hexane is around 10–15 times less expensive than d-chloroform, which can be considered as one of the low-priced deuterated solvents. As the results of this work have shown, there is no drawback in terms of identification and quantification of plasticizer solutions if n-hexane is being utilized. However, deuterated solvents can be beneficial if, e.g., mixtures of multiple plasticizers are present, making their identification more challenging. As shown in Figure 6a, all three spectral regions of interest (aliphatic, ester, and aromatic) are analyzable and separated by the baseline when d-chloroform is being used as a solvent.

Alternatively, both the signal-to-noise ratio and peak overlap can be improved by increasing the magnetic field strength of the NMR device. In order to keep all the benefits benchtop NMR offers for routine analysis, we compared spectra acquired with the Spinsolve 40 Carbon, which was used in this study, with those acquired using a Spinsolve 60 ULTRA working at a magnetic field of 60 MHz, rather than 40 MHz. The effect of increasing the magnetic field strength is shown in Figure 6b where a sample containing a DIBP concentration of 16.67 mg mL^{-1} is measured on both devices with four scans and a

15 s repetition delay. The result indicates that the 60 MHz spectrometer delivers baseline separated peaks with the n-hexane solution, which is a great benefit compared to the 40 MHz spectrum. When extracting the plasticizer with deuterated chloroform, however, upgrading to the 60 MHz device hardly shows to be beneficial.

According to the regulation (EC) No 1907/2006 of the European Parliament and of the European Council [59], a relevant plasticizer present in a polymer by more than 0.1% per weight must undergo a chemical safety assessment. Such a plasticizer content in a PVC product leads to a concentration of 0.05 mg of plasticizer solved in 1 mL of n-hexane after the solvent extraction has been performed, as described in this work. As depicted in Table 1, this concentration is lower than the LOQ achieved with the 1 min measuring time by ^{1}H NMR spectroscopy. In order to determine concentrations at this legal threshold with the low-field NMR method, the number of scans should theoretically be increased to around 400 in order to increase the SNR. Measuring a 0.05 mg mL^{-1} solution of DIBP with 256, 512, 1024, and 2048 scans with a 7 s repetition time led to the result that the LOQ could only be reached with 1024 scans in combination with an advanced baseline correction. Nevertheless, Figure S6 shows that distinct qualitative spectral features are still noticeable without implementing more sophisticated methods. Thus, slightly adjusting the straightforward low-field NMR method proposed in this work with further improvements discussed in this section can meet the criteria given by the European Union, even for less experienced users.

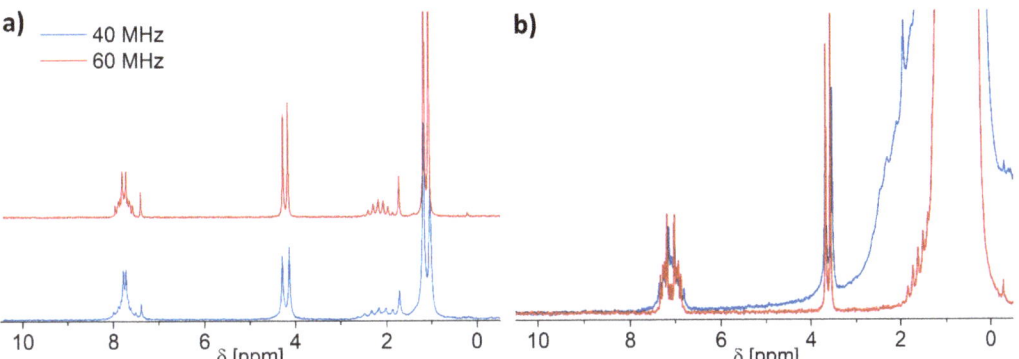

Figure 6. 16.67 mg mL^{-1} DIBP solution in (**a**) CDCl$_3$ and (**b**) n-hexane measured at 40 and 60 MHz. The left graphic shows a stacked plot as the differences between the spectra are hardly visible when displayed superimposed.

2.5. Low-Field NMR versus Conventional High-Field NMR

Given that low-field NMR hardware is, by far, more affordable compared to high-field NMR hardware and the costs for maintenance, extra personnel, and facilities are negligible, the results shown in the previous sections indicate that the proposed low-field NMR analysis is a low-cost alternative for the study of PVC plasticizers. The analysis costs at a low field can be further decreased with the use of non-deuterated solvents. However, limits of identification and quantification for plasticizers are higher when comparing our approach with the reported results in [42], which employed high-field NMR.

With the data provided in [42], a minimum plasticizer content in PVC was calculated, which can still be detected by high-field NMR after solvent extraction within 20 min of the measuring time, and we compared this value to the reported low-field NMR results. The minimum plasticizer content in PVC detected at the low field within 1 min, a measuring time which is 20 times lower than in the high field, is only higher by a factor 3 compared to the high-field value (0.336 and 0.96 wt%). This factor 3 relates primarily to the combination of two effects: the size of the sensitive volume and the amount of solvent used for the extraction step.

Advantageous for the low-field method is the larger sensitive volume (0.4 mL) compared to the high-field NMR magnet which results in higher absolute plasticizer contents at identical concentrations. More precisely, the high-field NMR magnet has a sensitive volume of 0.2 mL, which is further decreased to 0.167 mL due to the employed coaxial insert [42]. Furthermore, this work successfully lowered the detectable plasticizer content by reducing the amount of solvent in the extraction step to the lowest ratio reported in the literature. This leads to the phenomenon that the LOD of a plasticizer solution at the high field is lower by a factor of roughly 1000 (42 µg mL^{-1} at the high field versus 48 mg mL^{-1} at the low field for DINP), but the minimum detectable plasticizer content in PVC is only lower by a factor of 3.

3. Materials and Methods

3.1. Samples

All solvents and plasticizers investigated in this study were purchased from Sigma-Aldrich and used without further purification. Dilution series with plasticizer concentrations ranging from 0.1 to 100 vol.% (0.97–1039 mg mL^{-1}) were prepared with deuterated chloroform and non-deuterated hexane using Eppendorf pipettes. A total volume of the plasticizer/solvent mixture of 0.5 mL was then filled into a standard 5 mm NMR tube. ^{1}H-NMR spectroscopy measurements revealed that no detectable amount of additive from the pipette's plastic tip was extracted during the time of filling the utilized solvents into the NMR tubes. The NMR tube was then tightly sealed to prevent solvent evaporation during measurements and storage. In addition, the tube's filling level was marked after filling to serve as a control feature.

Plasticized PVC samples with unknown histories were used to test the proposed procedure and for identifying the plasticizer type and quantifying its amount with the help of solvent extraction procedures. For this, each available PVC sample was cut in small pieces and carefully weighted. Extraction was conducted at room temperature by using 130–300 mg PVC and 2.6–6 mL (roughly 20 times the amount of the PVC sample) of deuterated chloroform or non-deuterated hexane for 24 h according to [57]. Proton NMR spectra were recorded for the extracted solutions to obtain the plasticizer concentration c_{sample}. The total mass of extracted plasticizer, as defined by equation (2), can be calculated by transposing equation (1). This value is then required to determine the mass percentage of plasticizer in the PVC sheet. Following the extraction step, the sample sheets were dried under vacuum for 24 h and weighted again to additionally analyze the amount of plasticizer loss gravimetrically.

$$c_{sample} = \frac{m_{plasticizer}}{m_{solvent} + m_{plasticizer}} \quad (1)$$

$$m_{plasticizer} = \frac{m_{solvent} * c_{sample}}{1 - c_{sample}} \quad (2)$$

3.2. NMR Experiments

The NMR experiments were performed on a Magritek Spinsolve 40 Carbon (Figure 7) working at a frequency of 43 MHz for protons and 11 MHz for ^{13}C and at a constant magnet temperature of 28 °C. For selected samples, ^{1}H spectra were measured also using a Magritek Spinsolve 60 ULTRA working at a frequency of 60 MHz for protons and at a constant temperature of 26.5 °C. Before each measurement, the magnetic field was shimmed to a linewidth of less than 0.5 Hz at half peak height using a 90/10 D$_2$O/H$_2$O sample for 40 MHz and a linewidth of less than 0.3 Hz at half peak height was achieved at 60 MHz using a 95/5 D$_2$O/H$_2$O sample as specified by the manufacturer. ^{1}H and ^{13}C spectra as well as ^{1}H spin–lattice relaxation times T_1 were measured for dilution series for every plasticizer. All ^{1}H spectra were acquired using 4 scans and the device's standard repetition time of 15 s, which is longer than the corresponding $5\times T_1$, with T_1 being between 1 and 2 s at most for all plasticizers. Thus, the total measuring time for a spectrum was 1 min.

This number of scans was chosen to keep the experimental time short. This is especially of interest if a daily analysis routine with large sample quantities is planned.

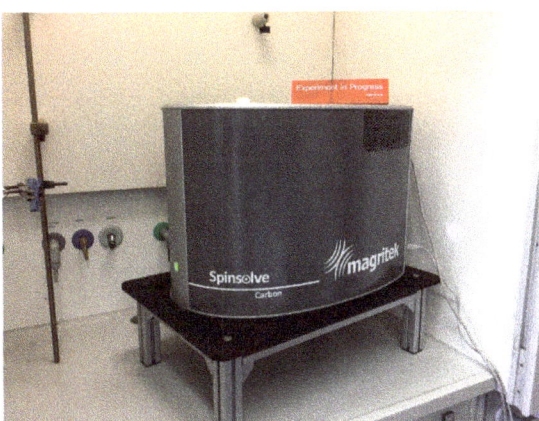

Figure 7. Photo of the utilized Spinsolve placed inside a fume hood in a chemistry laboratory.

All spectra were referenced to the distinct solvent peak present in the mixtures. T_1 were acquired with the inversion recovery method, accumulating 4 scans at inversion times ranging from 1 ms to 5 s. Knowing the samples' relaxation times permits the manual reduction in the repetition time without suffering a signal intensity loss. However, for the purpose of simplicity and the easy application of the method for non-NMR users, we decided to employ the standard repetition time of 15 s implemented in the Spinsolve software. All ^{13}C spectra were measured using proton decoupling during acquisition by accumulating 128 scans also with a repetition time of 15 s.

Due to reproducibility concerns, all NMR measurements conducted for the calibration were performed 5 times for all plasticizers. Plasticizer signals in another spectrum region different to that of the solvent were integrated for quantifying the plasticizer content in a particular sample.

For determining the limit of detection (LOD) and the limit of quantification (LOQ) [60] of the low-field NMR method, phase and baseline corrections as well as a line broadening of 0.3 Hz were applied to the spectra. The peak integral of the aromatic peak region was plotted against the solution's concentration and then a linear fit curve was used to determine the LOD and LOQ. Knowing the instrument's sensitive volume allows calculating the corresponding weight percentages of a plasticizer in a PVC material in order to detect or quantify it by the solvent extraction method.

4. Conclusions

This work evaluates, for the first time, the applicability of ^{1}H and ^{13}C low-field NMR spectroscopy for the identification of various PVC plasticizers in solution and the quantification of their amount. While the standard way of conducting such studies is by dissolving the plasticizers in expensive deuterated solvents, our work demonstrates that the same information can be obtained with the help of suited non-deuterated solvents, given that they have their signals outside the range of interest. Furthermore, the same non-deuterated solvents can also be used for the extraction step of plasticizers from PVC products.

The identification of plasticizers was conducted with the help of specific ^{1}H and ^{13}C resonances, which are well separated even at the low field. These signals were then used for quantification purposes in conjunction with a priori established correlation curves. The correlation curves were established between the integral of the signals of interests in both the ^{1}H and ^{13}C and the known concentration of plasticizer in solution. ^{1}H NMR spectroscopy showed to be the most promising tool in terms of the minimum plasticizer content to be identified and quantified in a short period of time due to the much higher SNR than ^{13}C NMR. More precisely, low-field ^{1}H spectroscopy enables the identification and the quantification of plasticizer concentrations as low as 2 mg mL^{-1} in solution, corresponding to ~3 wt% in a PVC product, within one minute of the measurement time. The suitability of the proposed method using non-deuterated solvents is demonstrated by comparisons with the spectra recorded on the same plasticizer dissolved in a suited deuterated solvent and by identifying the type of plasticizer extracted from the PVC of samples with unknown histories and quantifying its amount.

Analyzing the minimum plasticizer content in PVC products requested by the European Union is achievable, if the straightforward approach discussed in this work is refined with the methods discussed in Section 2.4. Thus, the combination of low-field NMR with non-deuterated solvents offers a very cost-effective, yet powerful method complementary to high-field NMR, LC, FTIR, or GC-MS, making it well suited for routine quality analysis of large numbers of plasticized PVC samples.

Supplementary Materials: The following are available online, Figure S1: 40 MHz ^{1}H NMR spectra of the investigated plasticizers at concentrations of 10 vol.% in (a) deuterated chloroform and (b) non-deuterated n-hexane without zoom. Spectra have been referenced to the residual deuterated chloroform (signal marked with asterisk at 7.26 ppm) and n-hexane (0.8 ppm) peaks, respectively. Figure S2: 40 MHz ^{1}H NMR spectra of DINP in non-deuterated hexane at varying concentrations. All spectra have been referenced to the signal of the n-hexane (0.8 ppm) peak. The specific resonances at around 7 ppm can be observed even at concentrations as low as 0.1 vol.% with only 4 scans. Figure S3: 40 MHz ^{1}H NMR spectra of all investigated plasticizers in non-deuterated n-hexane at a concentration of 1 vol. %. All spectra have been referenced to the signal of the n-hexane (0.8 ppm) peak. Figure S4: ^{1}H-NMR (a) T_1 and (b) T_2 relaxation times of TOTM and DIBP at all investigated concentrations measured at 40 MHz. For each concentration, the reported relaxation times are the average of three measurements. Figure S5: ^{13}C spectra of DIBP at various concentrations recorded at 40 MHz. The asterisks correspond to the signals from n-hexane. Figure S6: ^{1}H NMR spectra of a 0.05 mg mL^{-1} DIBP solution in non-deuterated n-hexane in a stacked plot. SNRs at 3.5 ppm are 5.6, 8.5, and 8.7 for 512, 1024, and 2048 scans, respectively. SNRs at 7 ppm are 6.1, 7.0, and 7.5 for 512, 1024, and 2048 scans, respectively. Table S1: Results of gravimetric analyses of unknown PVC samples used for solvent extraction discussed in Section 2.3. Table S2: Mean results from each sample displayed in Table S1 and individual results from CDCl$_3$ and n-hexane extraction.

Author Contributions: Conceptualization, A.A.; investigation, A.D., A.A.; writing—original draft preparation, A.D.; writing—review and editing, A.A.; supervision, A.A. All authors have read and agreed to the published version of the manuscript.

Funding: This research received no external funding.

Data Availability Statement: The data presented in this study are available on request from the corresponding author.

Conflicts of Interest: The authors declare no conflict of interest.

References

1. Andrady, A.L.; Neal, M.A. Applications and societal benefits of plastics. *Philos. Trans. R. Soc. B Biol. Sci.* **2009**, *364*, 1977–1984. [CrossRef] [PubMed]
2. Plastic antioxidants market projected to reach US$2.11 billion by 2022. *Addit. Polym.* **2018**. [CrossRef]
3. Organic peroxide market forecast to be worth US$1.20 billion by 2025. *Addit. Polym.* **2018**. [CrossRef]
4. Rahman, M.; Brazel, C. The plasticizer market: An assessment of traditional plasticizers and research trends to meet new challenges. *Prog. Polym. Sci.* **2004**, *29*, 1223–1248. [CrossRef]
5. Global demand for plasticizers continues to rise. *Addit. Polym.* **2017**, 10–11. [CrossRef]
6. Wilkes, C.E.; Summers, J.W.; Daniels, C.A.; Berard, M.T. PVC Handbook, Hanser, Munich, Cincinnati. In *PVC Formulary*; Wypych, G., Ed.; ChemTec Publishing: Scarborough, ON, Canada, 2005.
7. Wypych, G. *PVC Formulary*, 3rd ed.; ChemTec Publishing: Scarborough, ON, Canada, 2020.

8. Bernard, L.; Cueff, R.; Bourdeaux, D.; Breysse, C.; Sautou, V. Analysis of plasticizers in poly (vinyl chloride) medical devices for infusion and artificial nutrition: Comparison and optimization of the extraction procedures, a pre-migration test step. *Anal. Bioanal. Chem.* **2015**, *407*, 1651–1659. [CrossRef]
9. Messadi, D.; Vergnaud, J.M. Plasticizer transfer from plasticized PVC into ethanol–water mixtures. *J. Appl. Polym. Sci.* **1982**, *27*, 3945–3955. [CrossRef]
10. Messadi, D.; Taverdet, J.L.; Vergnaud, J.M. Plasticizer migration from plasticized poly (vinyl chloride) into liquids. Effect of several parameters on the transfer. *Ind. Eng. Chem. Prod. Res. Dev.* **1983**, *22*, 142–146. [CrossRef]
11. Demir, A.P.T.; Ulutan, S. Migration of phthalate and non-phthalate plasticizers out of plasticized PVC films into air. *J. Appl. Polym. Sci.* **2012**. [CrossRef]
12. Kovačić, T.; Mrklić, Ž. The kinetic parameters for the evaporation of plasticizers from plasticized poly (vinyl chloride). *Thermochim. Acta* **2002**, *381*, 49–60. [CrossRef]
13. Reddy, N.N.; Mohan, Y.M.; Varaprasad, K.; Ravindra, S.; Vimala, K.; Raju, K.M. Surface treatment of plasticized poly (vinyl chloride) to prevent plasticizer migration. *J. Appl. Polym. Sci.* **2010**, *115*, 1589–1597. [CrossRef]
14. Kastner, J.; Cooper, D.G.; Marić, M.; Dodd, P.; Yargeau, V. Aqueous leaching of di-2-ethylhexyl phthalate and "green" plasticizers from poly (vinyl chloride). *Sci. Total. Environ.* **2012**, *432*, 357–364. [CrossRef] [PubMed]
15. Linde, E.; Gedde, U. Plasticizer migration from PVC cable insulation—The challenges of extrapolation methods. *Polym. Degrad. Stab.* **2014**, *101*, 24–31. [CrossRef]
16. Ekelund, M.; Azhdar, B.; Hedenqvist, M.; Gedde, U. Long-term performance of poly (vinyl chloride) cables, Part 2: Migration of plasticizer. *Polym. Degrad. Stab.* **2008**, *93*, 1704–1710. [CrossRef]
17. Taverdet, J.L.; Vergnaud, J.M. Modelization of matter transfers between plasticized PVC and liquids in case of a maximum for liquid-time curves. *J. Appl. Polym. Sci.* **1986**, *31*, 111–122. [CrossRef]
18. Audouin, L.; Dalle, B.; Metzger, G.; Verdu, J. Thermal Aging of Plasticized PVC. 11. Effect of Plasticizer Loss on Electrical and Mechanical Properties. *J. Appl. Polym. Sci.* **1992**, *45*, 2097–2103. [CrossRef]
19. Ekelund, M.; Edin, H.; Gedde, U. Long-term performance of poly(vinyl chloride) cables. Part 1: Mechanical and electrical performances. *Polym. Degrad. Stab.* **2007**, *92*, 617–629. [CrossRef]
20. Jakubowicz, I.; Yarahmadi, N.; Gevert, T. Effects of accelerated and natural ageing on plasticized PVC. *Polym. Degrad. Stab.* **1999**, *66*, 415–421. [CrossRef]
21. Yu, Q.; Selvadurai, A. Mechanical behaviour of a plasticized PVC subjected to ethanol exposure. *Polym. Degrad. Stab.* **2005**, *89*, 109–124. [CrossRef]
22. Latini, G.; Ferri, M.; Chiellini, F. Materials Degradation in PVC Medical Devices, DEHP Leaching and Neonatal Outcomes. *Curr. Med. Chem.* **2010**, *17*, 2979–2989. [CrossRef]
23. Mersiowsky, I. Long-term fate of PVC products and their additives in landfills. *Prog. Polym. Sci.* **2002**, *27*, 2227–2277. [CrossRef]
24. Abb, M.; Heinrich, T.; Sorkau, E.; Lorenz, W. Phthalates in house dust. *Environ. Int.* **2009**, *35*, 965–970. [CrossRef] [PubMed]
25. Yost, E.E.; Euling, S.Y.; Weaver, J.A.; Beverly, B.E.; Keshava, N.; Mudipalli, A.; Arzuaga, X.; Blessinger, T.; Dishaw, L.; Hotchkiss, A.; et al. Hazards of diisobutyl phthalate (DIBP) exposure: A systematic review of animal toxicology studies. *Environ. Int.* **2019**, *125*, 579–594. [CrossRef] [PubMed]
26. European Commission. *Commission Implementing Decision (EU) 2017/1210 of 4 July 2017 on the Identification Of Bis(2-Ethylhexyl) Phthalate (DEHP), Dibutyl Phthalate (DBP), Benzyl Butyl Phthalate (BBP) and Diisobutyl Phthalate (DIBP) as Substances of very High Concern According to Article 57(F) of Regulation (EC) No 1907/2006 of the European Parliament and of the Council (Notified Under Document C(2017) 4462) (Text with EEA Relevance), OPOCE, Brussels EU/2017/4462*; Publications Office of the European Union: Luxembourg, 2017.
27. Sazan, P.; Karin, A.; Orna, C.; Bert-Ove, L.; Ana, P.P.; Stefania, V. *European Union Summary Risk Assessment Report-Bis (2-ethylhexyl) Phthalate (DEHP)*; European Chemicals Bureau: Ispra, Italy, 2008.
28. Allanou, R.; Munn, S.J.; Hansen, B.G. *European Union Risk Assessment Report Dibutyl Phthalate*; European Chemicals Bureau: Ispra, Italy, 2009.
29. Navarro, R.; Perrino, M.P.; Tardajos, M.G.; Reinecke, H. Phthalate Plasticizers Covalently Bound to PVC: Plasticization with Suppressed Migration. *Macromolecules* **2010**, *43*, 2377–2381. [CrossRef]
30. Greco, A.; Brunetti, D.; Renna, G.; Mele, G.; Maffezzoli, A. Plasticizer for poly(vinyl chloride) from cardanol as a renewable resource material. *Polym. Degrad. Stab.* **2010**, *95*, 2169–2174. [CrossRef]
31. Chiellini, F.; Ferri, M.; Morelli, A.; Dipaola, L.; Latini, G. Perspectives on alternatives to phthalate plasticized poly(vinyl chloride) in medical devices applications. *Prog. Polym. Sci.* **2013**, *38*, 1067–1108. [CrossRef]
32. Bui, T.T.; Giovanoulis, G.; Cousins, A.P.; Magnér, J.; Cousins, I.T.; de Wit, C.A. Human exposure, hazard and risk of alternative plasticizers to phthalate esters. *Sci. Total. Environ.* **2016**, *541*, 451–467. [CrossRef]
33. Weiss, J.M.; Gustafsson, Å.; Gerde, P.; Bergman, Å.; Lindh, C.H.; Krais, A.M. Daily intake of phthalates, MEHP, and DINCH by ingestion and inhalation. *Chemosphere* **2018**, *208*, 40–49. [CrossRef]
34. Wypych, G. (Ed.) 15 Specialized Analytical Methods in Plasticizer Testing. In *Handbook of Plasticizers*, 3rd ed.; ChemTec Publishing: Scarborough, ON, Canada, 2017; pp. 661–669.
35. Marcilla, A.; Garcia, S.; Garcia-Quesada, J. Migrability of PVC plasticizers. *Polym. Test.* **2008**, *27*, 221–233. [CrossRef]

36. Gimeno, P.; Thomas, S.; Bousquet, C.; Maggio, A.-F.; Civade, C.; Brenier, C.; Bonnet, P.-A. Identification and quantification of 14 phthalates and 5 non-phthalate plasticizers in PVC medical devices by GC–MS. *J. Chromatogr. B* **2014**, *949–950*, 99–108. [CrossRef]
37. Bernard, L.; Bourdeaux, D.; Pereira, B.; Azaroual, N.; Barthelemy, C.; Breysse, C.; Chennell, P.; Cueff, R.; Dine, T.; Eljezi, T.; et al. Analysis of plasticizers in PVC medical devices: Performance comparison of eight analytical methods. *Talanta* **2017**, *162*, 604–611. [CrossRef] [PubMed]
38. Barendswaard, W.; Litvinov, V.M.; Souren, F.; Scherrenberg, R.L.; Gondard, C.; Colemonts, C. Crystallinity and Microstructure of Plasticized Poly (vinyl chloride). A 13C and 1H Solid State NMR Study. *Macromolecules* **1999**, *32*, 167–180. [CrossRef]
39. Giachet, M.T.; Schilling, M.R.; McCormick, K.; Mazurek, J.; Richardson, E.; Khanjian, H.; Learner, T. Assessment of the composition and condition of animation cels made from cellulose acetate. *Polym. Degrad. Stab.* **2014**, *107*, 223–230. [CrossRef]
40. Adams, A.; Kwamen, R.; Woldt, B.; Grass, M. Nondestructive Quantification of Local Plasticizer Concentration in PVC by (1)H NMR Relaxometry. *Macromol. Rapid Commun.* **2015**, *36*, 2171–2175. [CrossRef]
41. Sommer, S.; Koch, M.; Adams, A. Terahertz Time-Domain Spectroscopy of Plasticized Poly (vinyl chloride). *Anal. Chem.* **2018**, *90*, 2409–2413. [CrossRef] [PubMed]
42. Genay, S.; Feutry, F.; Masse, M.; Barthelemy, C.; Sautou, V.; Odou, P.; Decaudin, B.; Azaroual, N. Armed Study Group. Identification and quantification by (1)H nuclear magnetic resonance spectroscopy of seven plasticizers in PVC medical devices. *Anal. Bioanal. Chem.* **2017**, *409*, 1271–1280. [CrossRef] [PubMed]
43. Mitchell, J.; Gladden, L.F.; Chandrasekera, T.C.; Fordham, E. Low-Field Permanent Magnets for Industrial Process and Quality Control. *Prog. Nucl. Magn. Reson. Spectrosc.* **2014**, *76*, 1–60. [CrossRef] [PubMed]
44. Blümich, B.; Singh, K. Desktop NMR and Its Applications from Materials Science to Organic Chemistry. *Angew. Chem. Int. Ed.* **2018**, *57*, 6996–7010. [CrossRef] [PubMed]
45. Parker, T.; Limer, E.; Watson, A.; Defernez, M.; Williamson, D.; Kemsley, E.K. 60MHz 1H NMR spectroscopy for the analysis of edible oils. *TrAC Trends Anal. Chem.* **2014**, *57*, 147–158. [CrossRef]
46. Dalitz, F.; Cudaj, M.; Maiwald, M.; Guthausen, G. Process and reaction monitoring by low-field NMR spectroscopy. *Prog. Nucl. Magn. Reson. Spectrosc.* **2012**, *60*, 52–70. [CrossRef]
47. Adams, A. Analysis of solid technical polymers by compact NMR. *TrAC Trends Anal. Chem.* **2016**, *83*, 107–119. [CrossRef]
48. Kwamen, R.; Blumich, B.; Adams, A. Estimation of Self-Diffusion Coefficients of Small Penetrants in Semicrystalline Polymers Using Single-Sided NMR. *Macromol. Rapid Commun.* **2012**, *33*, 943–947. [CrossRef]
49. Duffy, J.; Urbas, A.; Niemitz, M.; Lippa, K.; Marginean, I. Differentiation of fentanyl analogues by low-field NMR spectroscopy. *Anal. Chim. Acta* **2019**, *1049*, 161–169. [CrossRef] [PubMed]
50. Adams, A.; Piechatzek, A.; Schmitt, G.; Siegmund, G. Single-sided Nuclear Magnetic Resonance for condition monitoring of cross-linked polyethylene exposed to aggressive media. *Anal. Chim. Acta* **2015**, *887*, 163–171. [CrossRef]
51. Adams, A. Non-destructive analysis of polymers and polymer-based materials by compact NMR. *Magn. Reson. Imaging* **2019**, *56*, 119–125. [CrossRef]
52. Höpfner, J.; Ratzsch, K.-F.; Botha, C.; Wilhelm, M. Medium Resolution 1 H-NMR at 62 MHz as a New Chemically Sensitive Online Detector for Size-Exclusion Chromatography (SEC-NMR). *Macromol. Rapid Commun.* **2018**, *39*, e1700766. [CrossRef] [PubMed]
53. Singh, K.; Blümich, B. Compact low-field NMR spectroscopy and chemometrics: A tool box for quality control of raw rubber. *Polymer* **2018**, *141*, 154–165. [CrossRef]
54. Grootveld, M.; Percival, B.; Gibson, M.; Osman, Y.; Edgar, M.; Molinari, M.; Mather, M.L.; Casanova, F.; Wilson, P.B. Progress in low-field benchtop NMR spectroscopy in chemical and biochemical analysis. *Anal. Chim. Acta* **2019**, *1067*, 11–30. [CrossRef]
55. Chakrapani, S.B.; Minkler, M.J.; Beckingham, B.S. Low-field 1H-NMR spectroscopy for compositional analysis of multicomponent polymer systems. *Anal. Chim. Acta* **2019**, *144*, 1679–1686. [CrossRef]
56. Garg, B.; Bisht, T.; Ling, Y.-C. Sulfonated graphene as highly efficient and reusable acid carbocatalyst for the synthesis of ester plasticizers. *RSC Adv.* **2014**, *4*, 57297–57307. [CrossRef]
57. Bernard, L.; Décaudin, B.; Lecoeur, M.; Richard, D.; Bourdeaux, D.; Cueff, R.; Sautou, V. Analytical methods for the determination of DEHP plasticizer alternatives present in medical devices: A review. *Talanta* **2014**, *129*, 39–54. [CrossRef] [PubMed]
58. Gottlieb, H.E.; Kotlyar, V.; Nudelman, A. NMR chemical shifts of common laboratory solvents as trace impurities. *J. Org. Chem.* **1997**, *62*, 7512–7515. [CrossRef] [PubMed]
59. European Commission. *REGULATION (EC) No 1907/2006 OF THE EUROPEAN PARLIAMENT AND OF THE COUNCIL of 18 December 2006 Concerning the Registration, Evaluation, Authorisation and Restriction of Chemicals (REACH), Establishing a European Chemicals Agency, Amending Directive 1999/45/EC and Repealing Council Regulation (EEC) No 793/93 and Commission Regulation (EC) No 1488/94 as well as Council Directive 76/769/EEC and Commission Directives 91/155/EEC, 93/67/EEC, 93/105/EC and 2000/21/EC*, OPOCE; European Commission: Brussels, Belgium, 2006.
60. Bharti, S.K.; Roy, R. Quantitative 1H NMR spectroscopy. *Trends Anal. Chem.* **2012**, *35*, 5–26. [CrossRef]

Article

Enhanced Resolution Analysis for Water Molecules in MCM-41 and SBA-15 in Low-Field T_2 Relaxometric Spectra

Grzegorz Stoch * and Artur T. Krzyżak

Department of Fossil Fuels, AGH University of Science and Technology, A. Mickiewicza Av., 30-059 Cracow, Poland; akrzyzak@agh.edu.pl
* Correspondence: greg.stoch@outlook.com

Abstract: Mesoporous silica materials are the subjects for relaxometric NMR studies in which we obtain information on the properties of molecules in confined geometries. The signal analysis in such investigations is generally carried out with the help of the Inverse Laplace Transform (ILT), which is accompanied by a regularization procedure. The appropriate selection of the regularization method may positively affect the resolution of the spectrum and the essence of the final conclusions. In this work, we examined the MCM-41 and SBA-15 model systems in various saturation states, using L-Curve regularization for relaxation spectra based on our own version of the fast fast ILT implementation. In a single relaxometric spectrum, the water contributions from the internal volume in the pores and between the silica particles were identified, which allowed us to trace the dynamics of the corresponding drying trends during the removal of water from the sample as a function of total water saturation.

Keywords: low field NMR; Inverse Laplace Transform; L-Curve regularization; confined liquid; relaxometry; drying process

Citation: Stoch, G.; Krzyżak, A.T. Enhanced Resolution Analysis for Water Molecules in MCM-41 and SBA-15 in Low-Field T_2 Relaxometric Spectra. *Molecules* **2021**, *26*, 2133. https://doi.org/10.3390/molecules 26082133

Academic Editor: Igor Serša

Received: 1 March 2021
Accepted: 6 April 2021
Published: 8 April 2021

Publisher's Note: MDPI stays neutral with regard to jurisdictional claims in published maps and institutional affiliations.

Copyright: © 2021 by the authors. Licensee MDPI, Basel, Switzerland. This article is an open access article distributed under the terms and conditions of the Creative Commons Attribution (CC BY) license (https:// creativecommons.org/licenses/by/ 4.0/).

1. Introduction

Water molecules trapped in silica mesoporous materials behave differently than free water. This property can be used to study model systems with a developed surface, and the results might be extended to real systems. As shown by Grünberg et al. [1] with the help of ^{1}H MAS NMR spectra, we are able to identify different water contributions in such materials due to different chemical environments for the surface water and the water from the pore's interior space. What makes this identification feasible is standard Fourier transform (FT) methodology that splits the overall signal into groups of spins rotating with different Larmor frequencies, embodied in the frequency-domain spectrum.

A similar goal was achieved in our previous work [2] by means of a different tool and methodology where a series of a low-field ^{1}H NMR measurements and time-domain analysis led to conclusions consistent with frequency domain analysis made earlier by Grünberg et al. [1]. Contributions were obtained using a combination of ILT together with the sample's weight monitoring through a series of measurements at different water saturation. However, due to the line broadening inherent for ILT regularization, we have not been able to break down a single spectrum into the desired components, contrary to what is common in frequency domain methodology. Although NMR relaxometry is an established tool for the characterization of porous materials and its results are confirmed by independent measurements (e.g., using the gas adsorption and micro porosimetry methods [3,4]), its resolution is still a matter of progress.

In this article, we check whether the improved resolution of our transform will affect the ability to separate contributions and the ability to track their evolution in individual spectra for MCM-41 and SBA-15 nanoparticle systems. The resolution improvement was achieved through the implementation of the ILT algorithm from scratch and a revised regularization methodology. Before using it, the usefulness of the algorithm was assessed by comparing it with the one used so far in previous works. In this article, we use the term "ILT" as a convenient label to refer to the exponential nature of a signal, but the mathematically correct formulation for this is: the Fredholm problem of the first kind with an exponential kernel [5].

1.1. Importance of a Low Field, Bulk and Surface Signal from within a Pore

The FID (Free Induction Decay) signal is described by the relaxation time given [6,7] by

$$\frac{1}{T_2^*} = \frac{1}{T_2^B} + \frac{1}{T_2^S} + \frac{1}{T_2^P} \tag{1}$$

where T_2^B characterizes *free water*, T_2^S is the surface relaxation term and T_2^P is the influence of inhomogeneity of magnetic field on the signal that is of microscopic internal origin and might be additionally caused by external (macroscopic) field gradient, so that $\frac{1}{T_2^P} = \frac{1}{T_2^{act}} + \frac{1}{T_2^D}$ (where: T_2^{act} is the actual refocusable term, and T_2^D is the component describing diffusion caused by external field gradient G).

Application of the CPMG sequence [8] removes refocusable component T_2^{act}, which converts T_2^* to T_2 by reducing T_2^P in Equation (1) to the diffusion term T_2^D, which depends on the echo time t_E [6] through the expression

$$\frac{1}{T_2^D} = D\frac{(\gamma G t_E)^2}{12} \tag{2}$$

This term can be minimized using short echo time t_E, which, together with the small B_0 field magnitude [9], makes the last term in Equation (1) usually neglected, and we recorded the decay of the spin echo envelope, given by the effective expression in Equation (3).

$$\frac{1}{T_2} = \frac{1}{T_2^B} + \frac{1}{T_2^S} \tag{3}$$

Brownstein and Tarr showed that the surface term in Equation (3) is given by the expression where geometrical details are reduced to a simple relationship between the pore's surface S and its volume V:

$$\frac{1}{T_2^S} = \rho \frac{S}{V} \tag{4}$$

which for the cylindrical shape of the pore of radius r takes the form:

$$\frac{1}{T_2^S} = \rho \frac{2}{r} \tag{5}$$

They considered relaxing spins within a single pore interior, experiencing a diffusion effect [10], where spin-to-surface diffusion time is significantly shorter than the spin relaxation time T_2^S. Within this time, all molecules interact with the surface [1,11] and their fast exchange with the interior volume makes the entire magnetization for the pore uniform (so-called fast diffusion regime). This work concerns the results obtained in the fast exchange mode, and the expectation of separating the contributions T_2^B and T_2^S for materials with such small pores does not seem justified. Rather, we would expect to see coarser differences like those between the inside of the pores and their closest and other neighborhoods.

For cylindrical pores, Equation (5) is the basic way to differentiate the pore size based on the T_2 spectrum and thus identify the individual water contributions. Small-sized pores correspond to shorter relaxation times, and larger pores to proportionally longer T_2, therefore, in favorable but rare circumstances, the spectrum is a set of separate lines on the logarithmic T_2 scale.

1.2. Our Mesoporous System and Experimental

The measured system was in the form of a powder of nanoparticles filled with demineralized water, which was dried in subsequent steps. The initial saturation was close to 100% and a small water film was left on the surface itself as a marker. The spaces we considered fundamental for water placement are shown in Figure 1. First, this is the water in pores we expect to identify in our spectra toward shorter relaxation times as unresolved according to internal intra pore water and surface water. Clusters of nanoparticles form a kind of dense gel in which a sort of external pore is formed between the clusters of nanoparticles, giving rise to a somewhat longer T_2 together with water particles between the pores (III—inter-pore water). Finally, all of the free water including that left intentionally on the sample's surface was characterized with the longest T_2 time ~2 s.

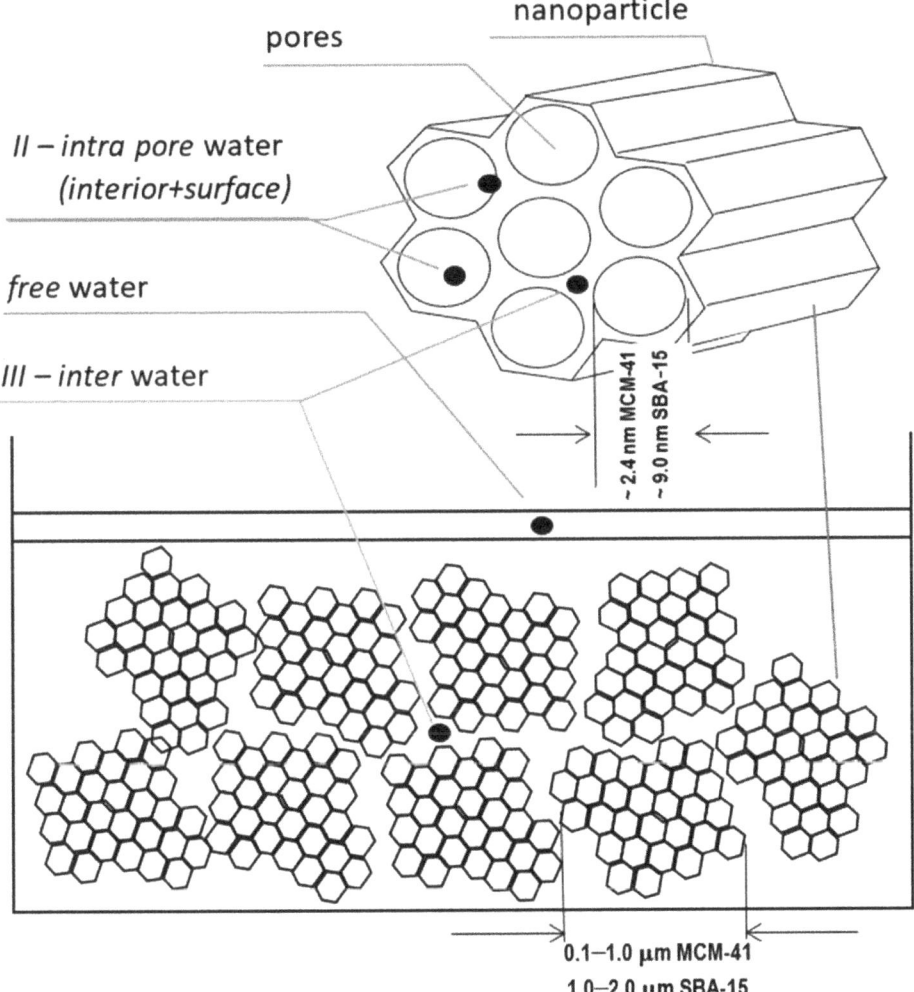

Figure 1. The considered locations for water placement: free water, water inside the pores (as the bulk and surface water), in-between the pores and nanoparticles.

The material for sample preparation was provided by Sigma Aldrich, had a structure of hexagonally arranged cylindrical pores, diameters ranging from 2.1 to 2.7 nm, average particle size of 0.1–1.0 μm for MCM-41 and 7.0 to 10.0 nm diameter, and 1–2 μm average particle size for SBA-15.

In the first step, the full saturation of the sample was obtained by adding 6.97 g of water to 0.99 g of the weighed dry powder. After completing the measurements for full saturation, the sample was dried at 80 °C in subsequent steps, each lasting 30–60 min. At each step, the sample was weighed (Radwag analytical balance, ±1 mg) and relaxation measurements repeated. Mass measurements enabled the determination of sample saturation states in the subsequent steps of the experiment based on mass balance.

Prior to hydration, the porous properties of silica were characterized by adsorption of N_2, from where the pores' surface area $S_{BET} = 1001$ m^2/g and their total volume $V_{tot}^{0.99} = 0.981$ cm^3/g per mass unit were obtained. At the end of the experiment, the characterization of the dried sample was repeated. A few percent decrease in the area of pores and volume was noted, which can be attributed to the slight hydrolysis effect [12]. However, unchanged diagram of the respective isotherm suggests [2] that the inner structure of the sample was retained in the filling and then in the successive drying processes. The nitrogen sorption isotherms at -196 °C were obtained by gas volumetry using a Micrometrics ASAP2020 analyzer in the relative pressure range of 10^{-4} to 0.99. The ^{1}H NMR relaxation signal measurements were performed on a low-field 0.05 T Magritek Rock Core Analyzer with a 29 mm probe at 30 °C using the CPMG $t_E = 60$ μs sequence with pulse length of 10 μs and 50,000 echoes.

Spectra were obtained from the collected signals using the ILT transform. The practical difference between the ILT and the more widely used Fourier transform (FFT) is the numerical instability inherent in the nature of the former as opposed to the latter [13]. Regularization is a necessary modification of the initial system of ILT equations so that this system does not become numerically ill-conditioned due to the presence of perturbations of various origins, first of all, experimental noise [14]. Regularization requires finding the so-called lambda parameter, which specifies the degree of regularization using separate criteria and is independent of the original ILT problem. These criteria (regularization algorithms) are subject to continuous improvement and the value of the lambda parameter affects the resolution of the ILT spectrum.

Our data were analyzed using the TNT-NN algorithm [15], which, being much faster than the classic Lawson–Hanson version [16], has a positive effect on the possibilities and reliability of computationally expensive regularization. We used the L-Curve method [17], which is less conservative than that of the RCA Toolbox Package used so far [18] and at the same time safer due to the risk of under-regularization compared to Generalized Cross Validation (GCV) [19,20]. Both L-Curve and GCV methods are well-established in a wide range of applications, in particular, L-Curve has been used successfully in tomography [21], the reconstruction of paintings [22], geoscience [23], and many other disciplines. For the test, we generated a pattern with which ILT fundamentally does not work well (sharp edges) to see the possibilities of the algorithm on especially demanding tasks. A comparison with the method we used so far [18] shows enhanced resolution for the new approach in Figure 2, both for data without noise and for SNR = 600. The former presents the exact reconstruction of the seven Kronecker delta patterns by the new algorithm, while the latter shows typical distortions for it, but still much smaller than in our standard approach.

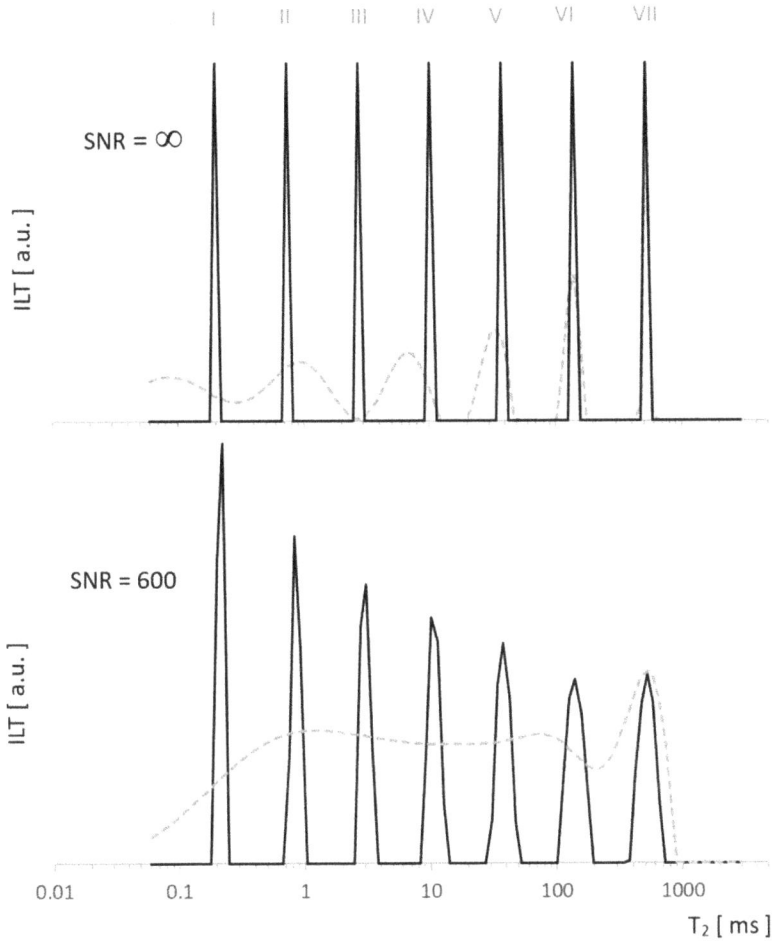

Figure 2. Spectra of the test signal with T_2 distribution as seven Kronecker deltas. Obtained using ILT with L-Curve regularization (solid line) vs. other regularization method (dashed line) for noiseless data (upper figure) and for SNR = 600 (lower figure). Significant differences in the resolution resulted from the use of different regularization algorithms.

2. Results and Discussion

Spectra T_2 obtained with L Curve regularization are shown in Figure 3 for MCM 41 and in Figure 4 for SBA-15. Intensities for different locations found from spectra along a series of water concentrations are presented in Tables 1 and 2, and then visualized in Figures 5 and 6 for MCM-41 and SBA-15, respectively. By intensity, we understand here as the sum of values in the range containing the maximum of the line (a value with some calculation error in the case of broad, overlapping lines). Respective line positions are summarized in Tables A1 and A2 in Appendix A for both samples.

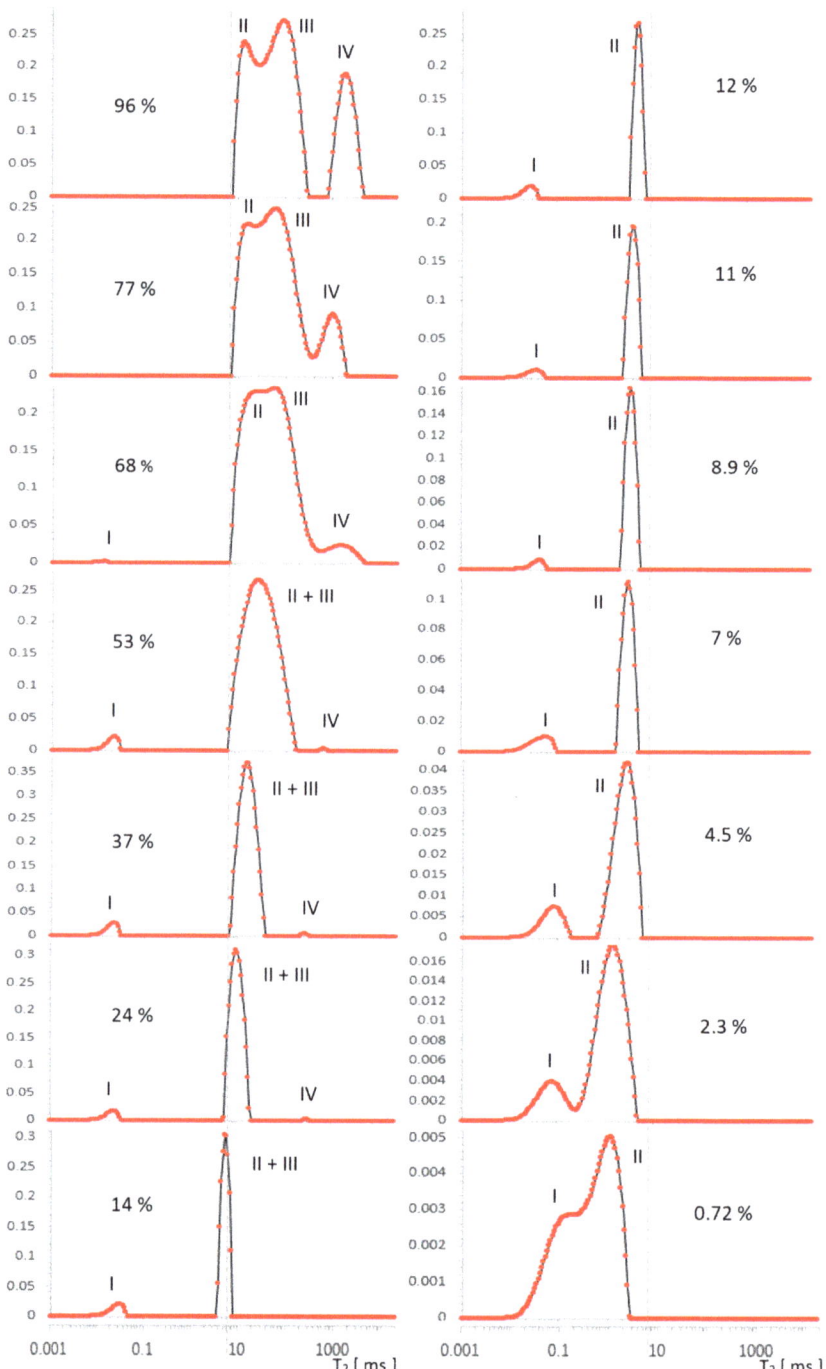

Figure 3. Evolution of T$_2$ spectrum for MCM-41, obtained in [a.u.] using ILT transform, along a series of water saturations. I—OH groups, II—intra pore water, III—inter pore water, IV—free water. The vertical axes represent the values of the ILT transform in [a.u.] units.

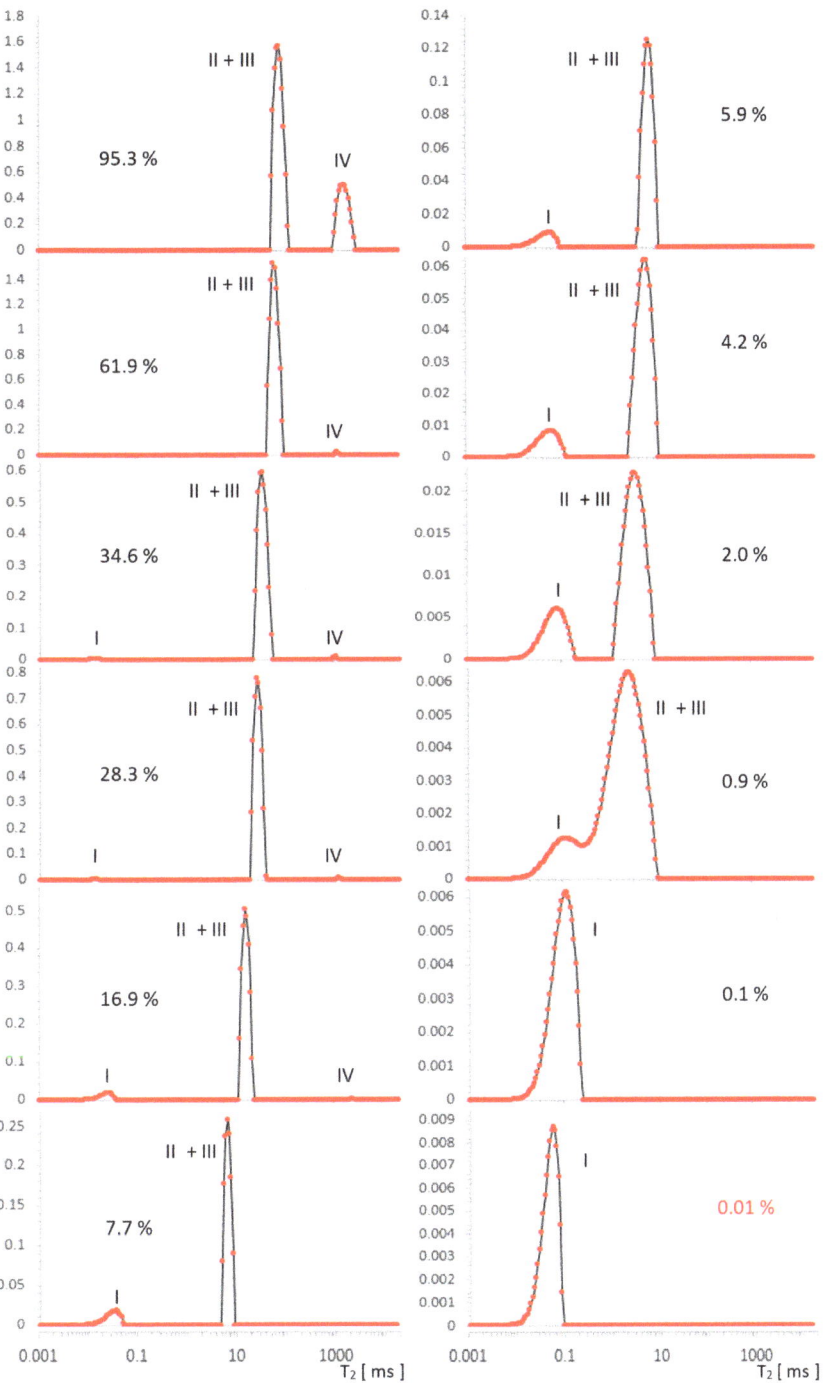

Figure 4. Evolution of the T$_2$ spectrum for SBA-15, obtained in [a.u.] using ILT transform, along a series of water saturations. I—OH groups, II—intra pore water, III—inter pore water, IV—free water. The vertical axes represent the values of the ILT transform in [a.u.] units.

Table 1. Contributions to the signal from different water locations in MCM-41 obtained from the respective integrals of the ILT spectra, expressed in [a.u.] for: I—OH groups, II—intra pore water, III—inter pore water, IV—free water.

%	I [a.u.]	II [a.u.]	III [a.u.]	IV [a.u.]
96.0	0	3.0638	5.7096	2.3942
77.0	0	2.8497	5.4366	1.1180
68.0	0.0085	3.0152	4.9602	0.4101
53.0	0.1658		6.8461	0.0108
37.0	0.2163		4.6720	0.0161
24.0	0.1417		3.0997	0.0062
14.0	0.1891	1.9050	-	-
12.0	0.1749	1.6437	-	-
11.0	0.1134	1.4223	-	-
8.9	0.0666	1.2053	-	-
7.0	0.1423	0.9463	-	-
4.5	0.1101	0.6165	-	-
2.3	0.0746	0.3472	-	-
0.7	0.0562	0.1077	-	-

Table 2. Contributions to the signal from different water locations in SBA-15 obtained from the respective integrals of the ILT spectra, expressed in [a.u.] for: I—OH groups, II + III—intra- and inter-pore water, IV—free water.

%	I [a.u.]	II + III [a.u.]	IV [a.u.]
95.3	-	10.5503	4.1646
61.9	-	9.3590	0.0429
34.6	0.0057	4.0409	0.0163
28.3	0.0028	4.4826	0.0124
16.9	0.1509	2.7507	0.0027
7.7	0.1768	1.2683	-
5.9	0.1116	0.9836	-
4.2	0.1268	0.7019	-
2.0	0.1003	0.3405	-
0.9	0.0270	0.1540	-
0.1		0.1012	-
~0.0		0.0967	-

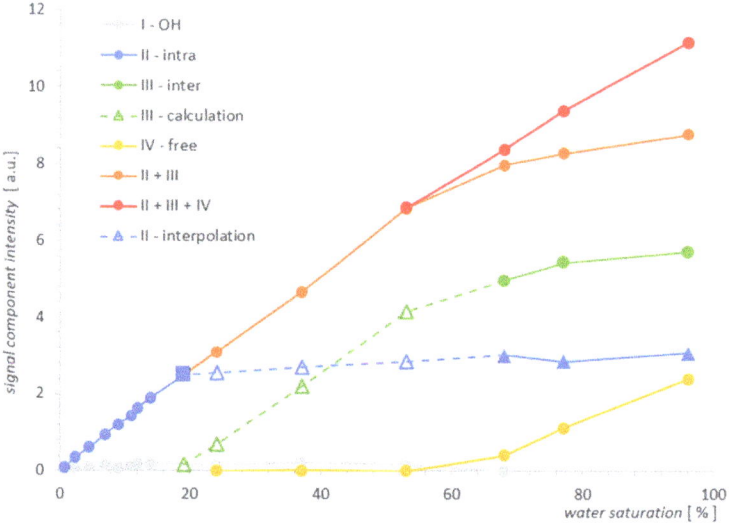

Figure 5. Individual contributions of water in MCM-41 as a function of total water saturation. The square point denotes the water saturation at which the inter water III vanishes, as seen from the comparison of the two spectra for MCM-41 and SBA-15 where both line widths are comparable.

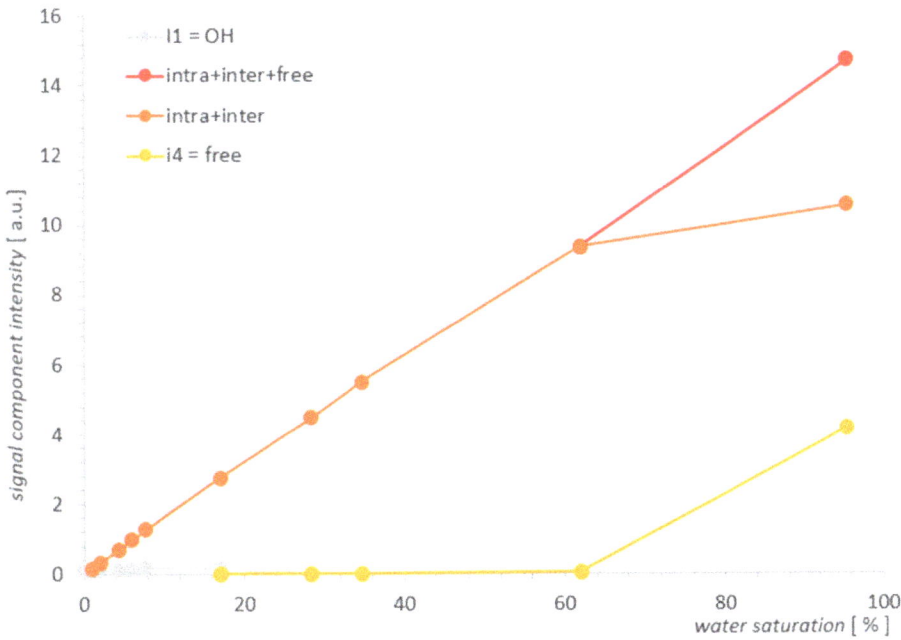

Figure 6. Individual contributions of water in SBA-15 as a function of water saturation.

2.1. Results for MCM-41

In Figure 3, four types of lines can be seen, the intensities of which change as a function of the decreasing water concentration traced in the range 96–0.72%. For the initial maximum concentration of 96%, we distinguished the following contributions in the spectrum: IV—free water around 2000 ms, III—inter water at ~100 ms, II—intra pore water at ~20 ms, and at around 0.06 ms, we had OH groups, which were better visible for lower concentrations.

With drying, the contribution from free water IV noticeably and quickly decreased, and from water inter III, it was relatively slower. It can be expected that the line from water intra II in the pores will begin to disappear when the outer layers are completely removed (i.e., at the latest). Free water IV disappeared around 24% and the line from water inter III completely overlapped that of the water intra II line in pores at a concentration of 53%. This conglomerate of lines II + III, however, loses its intensity successively further, as the water concentration decreases, and it manifests itself essentially by narrowing its width in the T_2 dimension.

Parallel to the effect of intensity decrease, there is a visible drift of the center of gravity of the complex line II + III toward the shorter T_2. This drift does not correspond to the actual translation and in this sense is apparent as it can be seen that the left border of the line remained stationary in a wide range of concentrations, which suggests that the component with a shorter T_2 in the observed sum remains constant in this range or changes little. Therefore, the only cause of the apparent maximum drift seems to be the change in component intensity with a longer T_2. Altogether, this corresponds to common-sense intuition that water is removed from the sample in a specific order, starting with the geometrically outermost (and perhaps less bound) layers, which could be, for example, a water inter III layer or free water IV as a marker on the surface. On the basis of further analysis, we will argue that this intuition, while essentially correct, is not entirely accurate in this case.

Due to the overall decrease in water concentration, individual signals in the spectral series also decrease, correspondingly reducing the SNR. The consequences of this fact can be observed independently in the form of increased values of the regularization parameter with a decrease in SNR, found in our analysis for each of the spectra separately using the L-Curve methodology (Figure A1, Appendix A). Such behavior is expected and consistent with ILT properties, and its observation allowed us to control the consistency of the analysis. The decrease in SNR ultimately results in substantial line broadening, an effect that is fundamental to the ILT and is opposite to that caused by water removal. It is noticeable in our spectra, especially at lower concentrations from the value of about 7%, but its direct influence was only a broadening of the spectrum, without affecting the intensity of its individual components, as presented in Table 1 or Table 2.

In the immediate vicinity of the value of 0.06 ms, there was a signal from strongly bound silanol OH groups, which is a separate problem that has already been studied using 2D relaxometry elsewhere [2,24]. It is also the area of possible ILT artifacts due to the time t_E used in the measurements and, due to this value itself, limits the effectiveness of the analysis in this area. As can be seen from Tables 1 and 2, the intensity of OH I with a change in water concentration slightly oscillated around small values for no apparent reason, which we attributed to the instability of ILT in this area superimposed on a constant and small value of the real signal. This behavior is systematic over the entire range of water concentrations, with the exception of the first two spectra for the strongest signals at the highest concentration, which dominate the other contributions' intensities for longer T_2 times. The signal mentioned does not have any influence on the results of this work, nor is it directly related to its topic, therefore we only note here and hereafter its presence, origin, and behavior in the spectra. The spectrum even extends down to 0.02 ms, which is the result of signal extrapolation using ILT in the sense of fitting procedure.

The change in the width of a complex line II + III in the ILT spectrum is one of the manifestations of the change in its intensity and this fact, combined with the basic knowledge for both samples, was used to plot the evolution of the components as a function of hydration in Figure 5. It follows from the characteristics provided that the pore diameters, although different, are of the same order in both samples. However, they essentially differed in the size distribution of the particles themselves, which are clusters of nanoparticles. The size distribution of these clusters for MCM-41 varied from very small 0.1 μm to the order of magnitude larger (1 μm), while for SBA-15, it remained on the same order of magnitude of 1–2 μm (see Figure 1). This created more variations in the water inter III distribution for MCM-41 and led to significantly different spectral line widths compared to SBA-15 in the state of full saturation. Such differences can actually be observed in the T_2 spectra in Figures 3 and 4. On the other hand, we know that both MCM-41 and SBA-15 have comparable pore sizes and similar pore dispersion, so for water intra II there were similar widths of spectral lines at similar T_2's. The above premise makes it easy to find the saturation value at which the complex line II + III is devoid of its water inter III component. This is roughly the first saturation toward its decreasing values at which the lines' widths for both samples are comparable at similar T_2. The spectra for which this condition holds are shown in Figure 7: at a concentration of 14% for MCM-41 and 5.9% for SBA-15, these are spectra of almost identical shape and line width. Thus, starting with the saturations mentioned toward their decreasing values, we dealt only with water intra II in the pores in both samples, respectively.

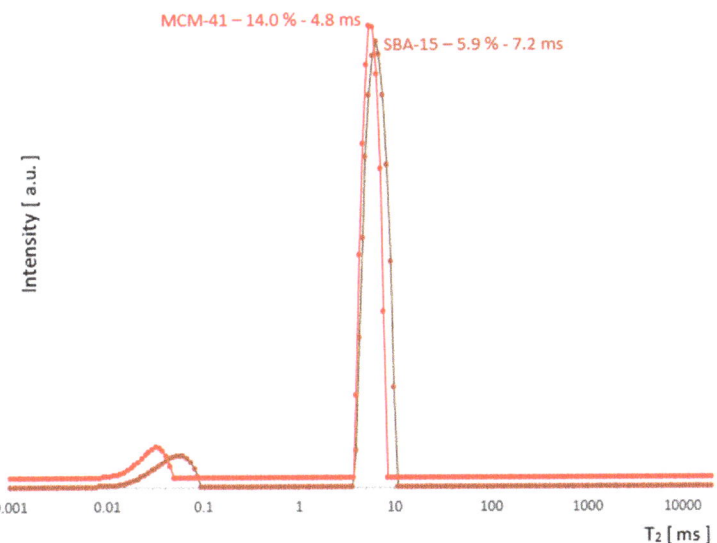

Figure 7. The saturation range for 'water only in the pores' (after removal water inter III) begins with the saturation at which the lines for MCM-41 and SBA-15 are similar in width.

In light of the above, we know that the water inter III vanishing point for MCM-41 is somewhere between a saturation of 14% and 24%, and we can find the respective signal value for water intra II by interpolation, marked by the square in Figure 5. It is worth noting that the signal at this point does not differ much from its value when the sample is fully saturated, and also when the lines are still split at 77%, we can conclude that the water intra II leaves the MCM-41 sample very slowly with a change in saturation, and the process accelerates dramatically only below 19%. Summarizing the above, the slowly-changing linear interpolation in the area of overlapping spectral lines II and III seems to be a sufficiently accurate approximation for water intra II, as illustrated by the blue dashed line in Figure 5. In the next step, from the simple balance, one can also obtain a plot of the contribution of water inter III in the remaining hydration range (marked with a green dashed line in Figure 5), using the summarized II + III contribution intensity obtained previously from the spectra.

2.2. Results for SBA-15

In Figure 4, we can see three types of lines, the intensity of which changes as a function of decreasing water concentration in the range 95.3–0.1% (presented ~0.01% is on the boundary of our accuracy). In the spectrum for the initial maximum concentration of 95.3%, overlapping contributions intra II and inter III can be suspected and are similar to MCM-41 free water IV of about 2000 ms. For MCM-41, the contribution of OH I groups with an artifact superimposed near the beginning of the T_2 timescale occurs at lower concentrations in SBA-15, which is also attributed to the hydration signal dominating these contributions.

The free water IV contribution decreased significantly around 16.9%, and over the entire saturation range, it seemed to decrease slightly faster with a saturation decrease than in MCM-41.

Following the analogy of MCM-41, here it is also logical to expect intra II and inter III contributions to be present, although starting with maximum hydration, the data seem to contradict this: as an equivalent, we see a well-defined single line across the entire range of saturation. However, this lack of structure turns out to be apparent: for 95.3%, the line position at 77 ms in SBA-15 is, we believe, not accidentally close to the average value of $T_2^{II+III} = 55$ ms of the split line positions in MCM 41 for the respective water II and water III contributions, therefore, this line is also complex in SBA-15 and includes these contributions. The position of this line in SBA-15 is between position values for lines II and III in MCM 41 and is relatively narrow, which we attributed to the small spread of the possible water inter III positions in this sample (also consistent with the manufacturer's information on the small particle size spread).

The individual contributions for SBA-15 are summarized in Table 2 and plotted in Figure 6; these are mainly contributions of water inter II and water intra III. Unlike the MCM-41 case, there is no convincing evidence in the data itself that could help separate overlapping spectral lines.

2.3. Discussion

Comparing the results visualized in Figure 5 and in Figure 6 for both samples, the obvious conclusion is that the analytical possibilities extended at the starting point had some effect for MCM-41, but for SBA-15, it proved to be insufficient. The information from the SBA-15 measurements and analysis played an important complementary role in the search for the water inter III elimination point in MCM-41, which was the foundation for determining the remaining contributions as the function of saturation and in the form presented in Figure 5 for MCM-41. However, drawing useful conclusions from the standalone T_2 spectra for SBA-15 proved to be unfeasible and the resolution improvement too small. As the comparison of SBA-15 and MCM-41 cases suggests, the successful application of the described method to other cases seems to depend on the relative development of the closed surfaces formed by the outer surfaces of adjacent pore aggregates relative to the pore surface.

Dependencies of individual contributions of water on water saturation for MCM-41 in Figure 5 illustrate, as it seems, convincingly, the actual process of water removal from the sample, obtained using T_2 NMR relaxometry. As the results show, this process takes place simultaneously and not sequentially as one might think, and as we assumed in our previous analysis and measurements [2]. However, the statement made there as an assumption, we get in this approach as a result. The water inter III and free water IV leave the sample simultaneously in the entire range of their presence, having different slopes with respect to the saturation axis in Figure 5. It seems that such a parallel transfer also applies to water from inside the pores intra II, although it happens much slower with a change in saturation and as shown in Figure 5, it accelerates rapidly for the water intra II contribution only at 19%, when the remaining contributions are completely removed. The shape of the curves of the whole process suggests approximately an exponential dependence and a box-like transfer model, in which the movement of water to successive compartments occurs continuously in the entire saturation domain. However, the verification of such a model would require more cases of resolved lines than were available in the prepared set of saturations for MCM-41.

3. Conclusions

The main motivation here was to explore the feasibility of analyzing water contributions in a single spectrum and to see if this would add new information from the point of view of a similar experiment performed previously [2]. The premise was the enormous progress that has been made in recent years in the solution of NNLS systems for Laplace analysis and the possibility of the independent implementation of appropriate algorithms. As the example for MCM-41 shows, T_2 analysis of the contributions in a single spectrum may in principle be feasible and lead to useful results. Rather rough approximations have been used in MCM-41 when analyzing partially overlapping lines, so there is room for further improvement here. In this way, we obtained a picture of the process in Figure 5, which seems to describe the distributions of individual contributions of water in MCM-41 as a function of changes in the total water saturation. The SBA-15 analysis showed the classic and fundamental problems with resolution related to the inherent need for ILT spectra regularization, and this is the case where high resolution methods should be used.

The aforementioned MAS methodology is and will probably remain the gold standard in this type of research due to the excellent differentiation of signals as a function of the chemical environment and both the phenomenal sensitivity and stable and unambiguous results of the analytical procedures (Fourier analysis).

Author Contributions: Conceptualization, G.S. and A.T.K.; methodology, G.S.; software, G.S.; validation, A.T.K. and G.S.; formal analysis, G.S.; investigation, A.T.K.; resources, A.T.K.; data curation, A.T.K.; writing—original draft preparation, G.S.; writing—review and editing, G.S.; visualization, G.S.; supervision, A.T.K.; project administration, A.T.K.; funding acquisition, A.T.K. All authors have read and agreed to the published version of the manuscript.

Funding: This research was funded by the Polish National Center for Research, grant number STRATEGMED2/265761/10/NCBR/2015.

Institutional Review Board Statement: Not applicable.

Informed Consent Statement: Not applicable.

Data Availability Statement: The data presented in this study are available on request from the corresponding author.

Acknowledgments: This work was supported by the Polish National Center for Research (grant no. STRATEGMED2/265761/10/NCBR/2015). The authors would like to kindly thank I. Habina for discussion and help with the experiments.

Conflicts of Interest: The authors declare no conflict of interest.

Appendix A

Table A1. Line position for MCM-41 in series of water saturations. I—OH groups, II—intra pore water, III—inter pore water, IV—free water.

%	T_2^I [ms]	T_2^{II} [ms]	T_2^{III} [ms]	T_2^{IV} [ms]
96.0	-	16.6275	95.6983	1931.4715
77.0	-	15.9511	90.8495	940.6998
68.0	0.0131	15.6531	96.1959	1595.4480
53.0	0.0212		37.7980	643.4950
37.0	0.0215		16.9998	262.8572
24.0	0.0230		3.0997	305.1463
14.0	0.0289	5.7398	-	-
12.0	0.0259	5.0470	-	-
11.0	0.0345	4.0882	-	-
8.9	0.0427	3.6844	-	-
7.0	0.0515	3.2950	-	-
4.5	0.0809	2.6328	-	-
2.3	0.0840	1.5408	-	-
0.7	0.1091	0.9608	-	-

Table A2. Line position for SBA-15 in series of water saturations. I—OH groups, II—intra pore water, III—inter pore water, IV—free water.

%	T_2^I [ms]	T_2^{II+III} [ms]	T_2^{IV} [ms]
95.3	-	77.3073	1672.1316
61.9	-	63.1377	1215.8725
34.6	0.0133	33.2756	1008.1585
28.3	0.0121	26.6767	1240.6753
16.9	0.0231	15.8747	2322.2875
7.7	0.0320	7.2114	-
5.9	0.0482	6.6127	-
4.2	0.0554	5.5516	-
2.0	0.0755	3.6012	-
0.9	0.1154	2.5859	-
0.1		0.0984	-
~0.0		0.0500	-

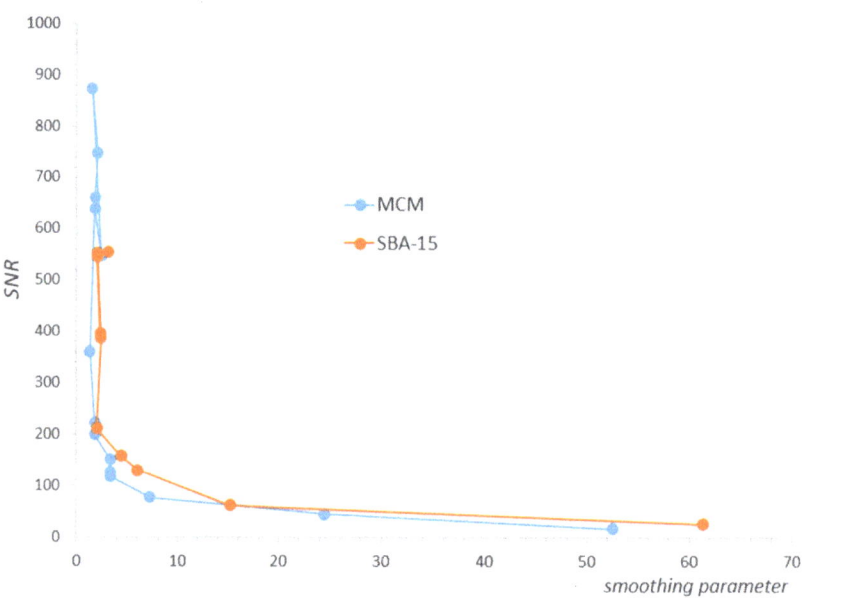

Figure A1. The regularization parameter for ILT transform increases as SNR (Signal-To-Noise-Ratio) decreases.

References

1. Grünberg, B.; Emmler, T.; Gedat, E.; Shenderovich, I.; Findenegg, G.H.; Limbach, H.H.; Buntkowsky, G. Hydrogen bonding of water confined in mesoporous silica MCM-41 and SBA-15 studied by 1H solid-state NMR. *Chem. Eur. J.* **2004**, *10*, 5689–5696. [CrossRef]
2. Krzyżak, A.T.; Habina, I. Low field 1H NMR characterization of mesoporous silica MCM-41 and SBA-15 filled with different amount of water. *Microporous Mesoporous Mater.* **2016**, *231*, 230–239. [CrossRef]
3. Stingaciu, L.R.; Weihermuller, L.; Haber-Pohlmeier, S.; Stapf, S.; Vereecken, H.; Pohlmeier, A. Determination of pore size distribution and hydraulic properties using nuclear magnetic resonance relaxometry: A comparative study of laboratory methods. *Water Resour. Res.* **2010**, *46*, W11510. [CrossRef]
4. Schmidt, R.; Stocker, M.; Hansen, E.; Akporiaye, D.; Ellestad, O.H. MCM-41: A model system for adsorption studies on mesoporous materials. *Microporous Mater.* **1995**, *3*, 443–448. [CrossRef]
5. Fordham, E.J.; Venkataramanan, L.; Mitchell, J.; Valori, A. What are, and what are not, Inverse Laplace Transforms. *Diffus. Fundam.* **2017**, *29*, 1–8.
6. Kleinberg, R.L.; Horsfield, M.A. Transverse relaxation processes in porous sedimentary rock. *J. Magn. Reson.* **1990**, *88*, 9–19. [CrossRef]
7. Mohnke, O.; Yaramanci, U. Pore size distributions and hydraulic conductivities of rocks derived from Magnetic Resonance Sounding relaxation data using multi-exponential decay time inversion. *J. Appl. Geophys.* **2008**, *66*, 73–81. [CrossRef]
8. Meiboom, S.; Gill, D. Modified spin-echo method for measuring nuclear relaxation times. *Rev. Sci. Instrum.* **1958**, *29*, 688–691. [CrossRef]
9. Straley, C.; Rosini, D.; Vinegar, H.; Tutunjijan, P.; Morris, C. Core analisis by low-field NMR. *Log Anal.* **1997**, *38*, 84–93.
10. Brownstein, K.R.; Tarr, C.E. Importance of classical diffusion in NMR studies of water in biological cells. *Phys. Rev. A* **1979**, *19*, 2446–2453. [CrossRef]
11. Buntkowsky, G.; Breitzke, H.; Adamczyk, A.; Roelofs, F.; Emmler, T.; Gedat, E.; Grunberg, B.; Xu, Y.; Limbach, H.-H.; Shenderovich, I.; et al. Structural and dynamical properties of guest molecules confined in mesoporous silica materials revealed by NMR. *Phys. Chem. Chem. Phys.* **2007**, *9*, 4843–4853. [CrossRef]
12. Pérez, L.L.; van Eck, E.R.H.; Melián-Cabrera, I. On the hydrothermal stability of MCM-41. Evidence of capillary tension-induced effects. *Microporous Mesoporous Mater.* **2016**, *220*, 88–98. [CrossRef]
13. Miller, K. Least squares methods for ill-posed problems with a prescribed bound. *SIAM J. Math. Anal.* **1970**, *1*, 52–74. [CrossRef]
14. Tikhonov, A.N.; Glasko, V.B. Use of the regularization method in non-linear problems. *USSR Comput. Math. Math. Phys.* **1965**, *5*, 93–107. [CrossRef]
15. Myre, J.M.; Frahm, E.; Lilja, D.J.; Saar, M.O. TNT-NN: A Fast Active Set Method for Solving Large Non-Negative Least Square Problems. *Procedia Comput. Sci.* **2017**, *108C*, 755–764. [CrossRef]
16. Lawson, C.L.; Hanson, R.J. Solving Least Squares Problems. In *Englewood Cliffs*; Prentice-Hall: Hoboken, NJ, USA, 1974.

17. Hansen, P.C.; O'Leary, D.P. The use of the L-curve in the regularization of discrete ill-posed problems. *SIAM J. Sci. Comput.* **1993**, *14*, 1487–1503. [CrossRef]
18. Magritek. NMR Rock Core Analyzer: RCA Toolbox version 4.25. In *Magritek Software User Manual*; Magritek: Wellington, New Zealand, 2016.
19. Golub, G.H.; Heath, M.T.; Wahba, G. Generalized cross-validation as a method for choosing a good ridge parameter. *Technometrics* **1979**, *21*, 215–223. [CrossRef]
20. Wahba, G. Practical approximate solutions to linear operator equations when the data are noisy. *SIAM J. Numer. Anal.* **1977**, *14*, 651–667. [CrossRef]
21. Kaufman, L.; Neumaier, A. PET regularization by envelope guided gradients. *IEEE Trans. Med Imaging* **1996**, *15*, 385–389. [CrossRef]
22. Luan, F.; Lee, C.; Choi, J.H.; Jung, H.-K. A Comparison of Regularization Techniques for Magnetoencephalography Source Reconstruction. *IEEE Trans. Magn.* **2010**, *46*, 3209–3212. [CrossRef]
23. Chen, L.Y.; Chen, J.T.; Hong, H.-K.; Chen, C.H. Application of Cesàro mean and the L-curve for the deconvolution problem. *Soil Dyn. Earthq. Eng.* **1995**, *14*, 361–373. [CrossRef]
24. Krzyżak, A.; Mazur, W.; Matyszkiewicz, J.; Kochman, A. Identification of Proton Populations in Cherts as Natural Analogues of Pure Silica Materials by Means of Low Field NMR. *J. Phys. Chem. C* **2020**, *124*, 5225–5240. [CrossRef] [PubMed]

Article

MAS-NMR of [Pyr$_{13}$][Tf$_2$N] and [Pyr$_{16}$][Tf$_2$N] Ionic Liquids Confined to Carbon Black: Insights and Pitfalls

Steffen Merz [1], Jie Wang [2], Petrik Galvosas [2,*] and Josef Granwehr [1,3]

[1] Fundamental Electrochemistry (IEK-9), Institute of Energy and Climate Research, Forschungszentrum Juelich, 52425 Juelich, Germany; s.merz@fz-juelich.de (S.M.); j.granwehr@fz-juelich.de (J.G.)
[2] MacDiamid Institute for Advanced Materials and Nanotechnology, School of Chemical and Physical Sciences, Victoria University Wellington, Wellington 6140, New Zealand; jie.wang@vuw.ac.nz
[3] Institute of Technical and Macromolecular Chemistry, RWTH Aachen University, 52056 Aachen, Germany
* Correspondence: Petrik.Galvosas@vuw.ac.nz

Abstract: Electrolytes based on ionic liquids (IL) are promising candidates to replace traditional liquid electrolytes in electrochemical systems, particularly in combination with carbon-based porous electrodes. Insight into the dynamics of such systems is imperative for tailoring electrochemical performance. In this work, 1-Methyl-1-propylpyrrolidinium bis(trifluoromethylsulfonyl)imide and 1-Hexyl-1-methylpyrrolidinium bis(trifluoromethylsulfonyl)imide were studied in a carbon black (CB) host using spectrally resolved Carr-Purcell-Meiboom-Gill (CPMG) and 13-interval Pulsed Field Gradient Stimulated Echo (PFGSTE) Magic Angle Spinning Nuclear Magnetic Resonance (MAS-NMR). Data were processed using a sensitivity weighted Laplace inversion algorithm without non-negativity constraint. Previously found relations between the alkyl length and the aggregation behavior of pyrrolidinium-based cations were confirmed and characterized in more detail. For the IL in CB, a different aggregation behavior was found compared to the neat IL, adding the surface of a porous electrode as an additional parameter for the optimization of IL-based electrolytes. Finally, the suitability of MAS was assessed and critically discussed for investigations of this class of samples.

Keywords: [Pyr$_{13}$][Tf$_2$N]; [Pyr$_{16}$][Tf$_2$N]; MAS; CPMG; 13-interval PGSTE; VXC72 carbon black; diffusion-NMR; Ionic liquids

Citation: Merz, S.; Wang, J.; Galvosas, P.; Granwehr, J. MAS-NMR of [Pyr$_{13}$][Tf$_2$N] and [Pyr$_{16}$][Tf$_2$N] Ionic Liquids Confined to Carbon Black: Insights and Pitfalls. *Molecules* **2021**, *26*, 6690. https://doi.org/10.3390/molecules26216690

Academic Editor: Igor Serša

Received: 25 September 2021
Accepted: 26 October 2021
Published: 5 November 2021

Publisher's Note: MDPI stays neutral with regard to jurisdictional claims in published maps and institutional affiliations.

Copyright: © 2021 by the authors. Licensee MDPI, Basel, Switzerland. This article is an open access article distributed under the terms and conditions of the Creative Commons Attribution (CC BY) license (https://creativecommons.org/licenses/by/4.0/).

1. Introduction

The investigation of ionic mobility plays a crucial role for the understanding and quantification of electrochemical systems such as batteries, fuel cells, or electrolyzers [1]. Important parameters include transference numbers, diffusion, and aggregation in liquid- and ionic liquid-based electrolytes or ion mobility in ion-conducting solid electrolytes [2]. The investigation of species inside porous matrices, which are of importance in metal-air batteries, or in solid ceramic or hybrid ceramic-polymer electrolytes for solid-state batteries is particularly challenging [3]. In both cases, ions move in different environments or across interfaces; hence, multiple environments must be distinguished to obtain a comprehensive description of dynamic processes.

Nuclear magnetic resonance (NMR) is exquisitely sensitive to the structure, environment, and dynamics of nuclei with non-zero nuclear spin [4–9]. Generally speaking, spectroscopic information provides structural and electronic information about the immediate molecular surroundings of a nucleus, while relaxation measurements describe local dynamics over a range of time scales, averaged over milliseconds to seconds. Slow processes can further be characterized using exchange experiments, which allow the identification of exchanging species and their exchange time constants [10,11]. Mobility on length scales of μm can be quantified using complementary pulsed field gradient (PFG) NMR experiments [12,13]. Solid samples show anisotropic spin interactions, which can be averaged out by magic angle spinning (MAS) [14,15]. Equally, line broadening caused by differences of magnetic susceptibility between materials in multi-component samples, such as liquids in porous media, can be reduced by MAS. The combination of MAS and PFG would be very attractive to retain the high spectral resolution for the identification of different species and their immediate environment and to correlate this information with long-range mobility in a similar way, as it was established for liquids and homogeneous liquid mixtures with the diffusion ordered spectroscopy (DOSY) experiment [16,17]. However, the combination of MAS and PFG NMR is not yet routinely employed.

Electrochemical systems are heterogeneous by nature, containing multiple components with potentially different states of matter. The study of ionic mobility in electrochemical systems would clearly benefit from a combination of MAS and PFG; however, the involved sample preparation and experimental protocols combined with limited access to, and limitations of, suitable hardware have prevented its wider application so far. However, careful experiment design may obviate these challenges. Limitations may include, in some cases, the minimum number of spins required; but experimental challenges may not be limited to this aspect [18]. In fact, electrochemical systems often contain ions in high concentrations, which imply sufficient sensitivity. On the other hand, solid-state samples and liquids in small pores or at interfaces often show short T_2 relaxation time constants, which may place constraints on the available time for applying pulsed field gradients. This can only be compensated by increased gradient amplitudes within given hardware capabilities, beyond which the method will fail. Furthermore, heterogeneous samples may be altered by fast spinning, either due to centrifugation or mechanical stress [19]. Therefore, care has to be taken because various parameters of the measuring system, e.g., spinning speed, gradient strength, or rotor filling, can become major sources of errors or influence the results due to altered sample structure. To avoid running long experiments that are eventually futile, the experimental constraints must be derived in advance from reference experiments, mapping out the available parameter space.

Carbon black (CB) is the generic name for a variety of finely divided carbon pigments originating from either pyrolysis or incomplete thermal decomposition of hydrocarbons under inert or oxygen-depleted atmosphere [20,21]. During the process, particles agglomerate to turbostratic layers. These units condense to stacks, which contain edge atoms that are reactive and adhere into chain-like structures [22]. Despite possessing roughly the same parallel and equidistant stacks of layer planes as graphite, these kinds of carbon lack a three-dimensional ordering on larger scales. In general, CB is a good electron conductor and is, therefore, commonly employed for various electrochemical applications including fuel cells [23,24], supercapacitors [25], sensors [26,27], or batteries [28,29], e.g., as additive in lithium-ion batteries (LIB) [30,31] or for gas diffusion electrodes in metal-air batteries [32].

Ionic liquids (IL) are Coulomb fluids entirely comprised of ions, with a melting point, by definition, below 373 K. They exhibit a higher conductivity than organic or aqueous electrolytes, are able to withstand voltages up to 5 V, offer large thermal and chemical stability, and possess a negligible vapor pressure [33]. ILs can be tailored by combining different anions and cations to meet individual requirements; hence, they are intensively investigated as possible replacements for aqueous electrolytes and are employed for a wide range of electrochemical applications [34–36]. In ILs, polarization, π-π, dipole-dipole, solvophobic, and van der Waals interactions (short-range) act together with Coulombic electrostatic forces (long-range). These inter- and intramolecular interdependencies can cause nanostructuring and complex IL dynamics in bulk, at surfaces, and under confinement [37–39]. Furthermore, since strong electrostatic interactions influence the dynamics, ILs behave distinctly differently than molecular liquids.

If confined to carbon matrices, the ordering of ILs depends mostly on the chemical structure and morphology of the confining matrix, its pore size distribution, and on the size of the IL anions and cations [40–43]. It has been demonstrated that a certain lateral ordering of an IL develops, which, e.g., enhances lubricity by forming crystal-like structures on ionophilic surfaces, depending on the polarity of the matrix [44–46]. This has also been shown for carbon surfaces under an applied potential [40,47,48].

ILs based on pyrrolidinium cations (Pyr) have the potential to outperform organic electrolytes and possess the advantage that the length of the alkyl chain can be used to adapt the properties of the neat IL [49,50]. Depending on the length, n, of the alkyl chain, this class of ILs can be split into short-chain ($1 \leq n \leq 4$) and long-chain ($n \geq 5$) subgroups. The latter is characterized by weaker ionic interactions. For side chains with $n \geq 4$, the ILs form nanostructures and segregate into nonpolar domains driven by van der Waals interactions [51]. The anions and the aromatic part of the cations of these domains can form a 3D network of ionic channels and facilitate distinct nano-domains, which lead to dynamic heterogeneities [52–54]. In NMR, transverse relaxation time constants (T_2) are strongly influenced by entropic changes, e.g., molecular structuring and correspondingly altered dynamics, whereas diffusion measurements are particularly sensitive regarding the clustering of ILs. Furthermore, external electric and magnetic fields can change the configuration of aggregates [55–58].

In an earlier study, we investigated the dynamics of 1-Methyl-1-propylpyrrolidiniumbis (trifluoromethylsulfonyl)imide ([Pyr$_{13}$][Tf$_2$N]) confined to CB using PGSTE-NMR in combination with T_1 relaxation time measurements [59]. We found evidence for a quasi-stationary IL film coating the carbon surface with a lubricating effect on the layer directly above. This layer showed an increased long-range mobility, causing an increased overall motion along the carbon surface [59]. However, identification of different environments using variations of their spectral signature was impeded by low resolution due to susceptibility variations inside the porous carbon environment or by incompletely averaged anisotropic interactions [60]. Such a line-broadening effect could be mitigated by using MAS [61].

In this paper, we are investigating the suitability and limitations of PFG NMR in combination with MAS for the study of ionic mobility in materials for electrochemical applications. While both solid superionic lithium conductors and electrolytes in porous electrode materials are of interest, the focus here is on the latter, as it exhibits somewhat reduced demands on the hardware in terms of pulse length or gradient strength, allowing for a more sensitive characterization of limitations. Nonetheless, the results can also serve as a feasibility assessment of experiments on solid electrolytes. The effects of confinement to CB on the dynamics of [Pyr$_{1n}$][Tf$_2$N] with different alkyl side chain lengths of $n = 3$ and $n = 6$ were investigated. We used a 13-interval stimulated echo PFG method and T_2 measurements in combination with MAS in order to establish a protocol to describe the behavior of IL confined to a CB matrix. We further report on certain pitfalls associated with MAS NMR, e.g., the electrical conductivity of CB under pressure [62–64]. However, an exhaustive interpretation of interactions between ILs and CB is outside the scope of this paper since a much larger set of systematic experiments in combination with numerical simulations would be required.

2. Results and Discussion

2.1. IL Loading Dependence of ^{1}H NMR Spectra

Figure 1 shows spectrally resolved maps of apparent ^{1}H transverse relaxation time constants, $T_{2,\text{app}}$, for [Pyr$_{13}$][Tf$_2$N] (Figure 1a) and [Pyr$_{16}$][Tf$_2$N] (Figure 1d) as neat ILs and loaded into the pore space of CB (Figure 1b,c,e,f), which was filled at two different fractions. NMR measurements of the neat ILs were recorded statically (no MAS spinning), while the IL-loaded CB samples were measured under MAS at 5 kHz.

The ^{1}H NMR spectra of the neat ILs showed chemical shift resolution of the different resonances, yet the resolution was not sufficient to also resolve J-couplings. While for [Pyr$_{16}$][Tf$_2$N] the lack of resolution was consistent with the measured $T_{2,\text{app}}$, for [Pyr$_{13}$][Tf$_2$N], a distribution of chemical shifts seemed to additionally cause a somewhat asymmetric broadening. [Pyr$_{16}$][Tf$_2$N] did not show such an asymmetry, but two sets of spectra with similar intensities were observed, both of which were clearly caused by the same IL but shifted by about 0.07 ppm with respect to each other.

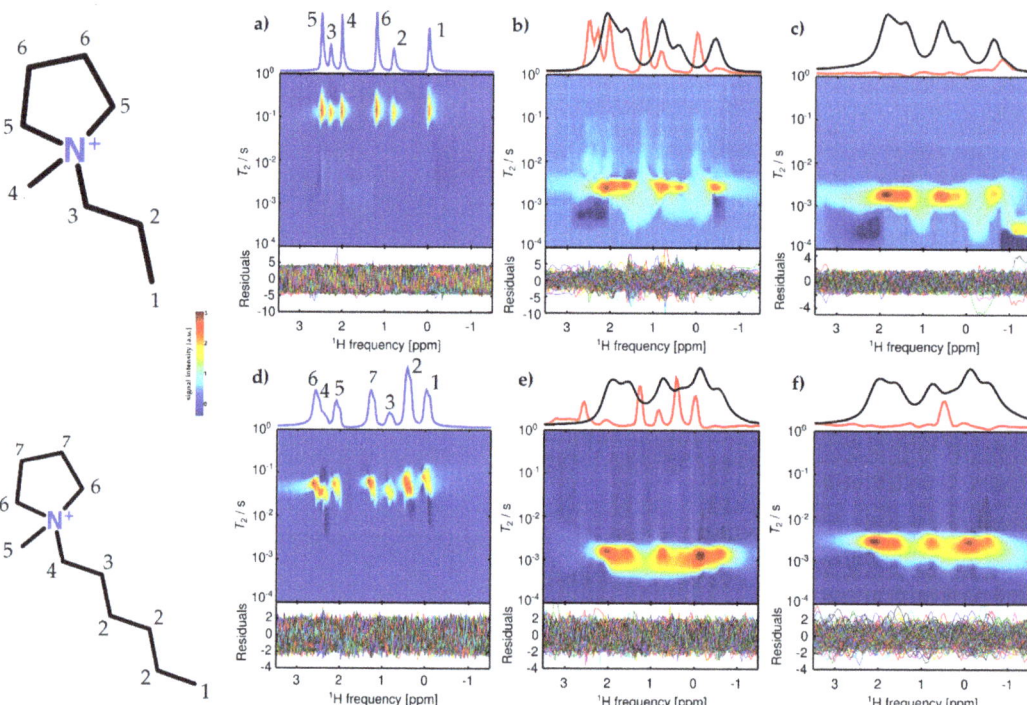

Figure 1. Pseudocolor plot of [Pyr$_{13}$] (**a**–**c**) and [Pyr$_{16}$] (**d**–**f**) $T_{2,\text{app}}$ relaxation time distribution vs. ^{1}H NMR frequency: (**a**) neat [Pyr$_{13}$][Tf$_2$N], (**b**) sample with CB and S_{IL} = 0.2 of [Pyr$_{13}$][Tf$_2$N], (**c**) sample with CB and S_{IL} = 1 of [Pyr$_{13}$][Tf$_2$N], (**d**) neat [Pyr$_{16}$][Tf$_2$N], (**e**) sample with CB and S_{IL} = 0.2 of [Pyr$_{16}$][Tf$_2$N], (**f**) sample with CB and S_{IL} = 1 [Pyr$_{16}$][Tf$_2$N]. The colormap is scaled with the square root of the amplitude to amplify features with low intensity. At the top of the panels of the neat IL, the sum projection onto the frequency axis is shown (blue). At the top of the IL in CB panels, the integral of the main feature (black; integrated from 1.3 ms to 5.0 ms for (**b**), from 0.36 ms to 2.5 ms for (**c**), from 0.7 ms to 3.2 ms for (**e**), and from 0.9 ms to 4.5 ms for (**f**)) and the integral of the slowly relaxing region (red; integrated from 5.6 ms to 180 ms for (**b**), from 4.0 ms to 100 ms for (**c**), from 3.6 ms to 56 ms for (**e**), and from 5.6 ms to 320 ms for (**f**)) are shown. For the [Pyr$_{13}$][Tf$_2$N] samples, the red spectrum is scaled by a factor of 15 compared to the black spectrum. For [Pyr$_{16}$][Tf$_2$N], the red spectrum is scaled by a factor of 100 compared to the black spectrum. Since the narrow signal represented by the red spectrum extends into the main relaxation mode, the signal intensity of the red species represents a lower limit in terms of contributing spins. The weighted residuals are shown at the bottom of each panel.

Confining both ILs to CB caused a broadening of all individual spectral features, driven by susceptibility differences between different phases, by local ring currents at the carbon surface, and by local magnetic fields induced by the electrically conductive matrix. This behavior is in line with the dynamics of molecular liquids in porous media [65,66]. Under confinement for small IL-loading fractions (Figure 1b,e) the main feature of the spectra of [Pyr$_{13}$][Tf$_2$N] and [Pyr$_{16}$][Tf$_2$N] shifted ≈ 0.5 ppm upfield. These shifts were primarily caused by ring current effects of the aromatic groups on the surface of the carbon matrix [67]. For the CB samples with S_{IL} = 1 (Figure 1c,f), the spectra of the ILs shifted upfield by approximately 0.4 ppm and 0.6 ppm for [Pyr$_{13}$][Tf$_2$N] and [Pyr$_{16}$][Tf$_2$N], respectively.

The samples with S_{IL} = 0.2 loading of the CB pore space also showed additional spectral features compared to the samples with high IL loading for both ILs. At low pore space loading, a less rapidly relaxing second species with a spectrum very similar to the neat IL, showing a considerably broadened $T_{2,\text{app}}$, could be observed. The spectra, integrated along the relaxation time dimension, are shown in the top panels of Figure 1 b,c,e,f. Since the position of the main relaxation component varied for the investigated samples, the integration limits were individually adjusted (see caption of Figure 1). Despite its low intensity, this feature could be clearly identified due to the chemical shift difference and the increased resolution afforded by MAS. The absence of such a separate narrow component for a sample with its pore space fully loaded is consistent with a previous report on MAS NMR of a similar system with a carbonaceous matrix fully saturated with [Pyr$_{13}$][Tf$_2$N] [61].

2.2. ^{1}H-$T_{2,app}$ Time Distributions

The $T_{2,app}$ distribution of the neat [Pyr$_{13}$][Tf$_2$N] was dominated by one mode with a mean relaxation time constant of 155 ms. All peaks showed a $T_{2,app}$ in the range between 130 ms to 160 ms, where all distributions showed a weak trend towards shorter $T_{2,app}$ for the broadened upfield side of each line. In addition, weak features were visible at a very short relaxation time below about 10 ms that pointed towards exchange processes. The fairly broad width in combination with the exchange features indicated an altered mobility of the spins compared to free Brownian motion, as observed in many molecular liquids. Self-aggregation of IL cations represents one possible explanation. Although it was found that the fraction of aggregated cations was small for [Pyr$_{13}$], the combination with exchange on the order of or even below milli-seconds may cause an underestimation of the fraction of aggregated species [68].

The neat [Pyr$_{16}$][Tf$_2$N] showed a faster $T_{2,app}$ mean relaxation of 72 ms, indicating a reduced mobility of the moieties driven by charge or viscosity effects, with [Pyr$_{16}$][Tf$_2$N] showing a 60% higher viscosity compared to [Pyr$_{13}$][Tf$_2$N] [69]. The values of $T_{2,app}$ vary for the different resonances. The resonance at around 0.5 ppm showed a decrease of $T_{2,app}$ towards 44 ms, whereas for the signal at about 1.4 ppm, $T_{2,app}$ increased to 100 ms. This indicated that the local mobility of each functional group, rather than collective motion of the IL cation, was the dominant dynamic process responsible for transverse relaxation, although a more complex correlated motion also could not be excluded [70].

All resonances of the [Pyr$_{16}$] cation showed a frequency-dependent $T_{2,app}$ that decreased in an upfield direction. This manifested itself in two superimposed spectra shifted by 0.07 ppm with different $T_{2,app}$; hence, mixing between the two configurations occurred on the order of the transverse relaxation rate. Slower mixing compared to [Pyr$_{13}$][Tf$_2$N] was also supported by distinctively broader $T_{2,app}$ distributions for each resonance of [Pyr$_{16}$][Tf$_2$N] compared to [Pyr$_{13}$][Tf$_2$N]. The reason for the two species could be a distribution of aggregated and separated [Pyr$_{16}$] cations. The more slowly relaxing downfield shifted spectrum, corresponding to faster dynamics of the respective species, was probably caused by free [Pyr$_{16}$] cations, while the faster relaxing upfield shifted resonances were due to aggregated [Pyr$_{16}$]. The exchange rates were slower than for [Pyr$_{13}$], confirming the hypothesis of stronger, longer-lasting aggregation for [Pyr$_{16}$] [68].

The apolarity of the ILs increased with alkyl chain length. Therefore, [Pyr$_{16}$][Tf$_2$N] had a stronger tendency to form micelles and hemi-micelles in bulk solutions compared to [Pyr$_{13}$][Tf$_2$N], leading to longer rotational correlation time constants and correspondingly reduced $T_{2,app}$. The static samples showed a behavior that can be related to a different aggregation behavior for [Pyr$_{13}$] and [Pyr$_{16}$], which was reported previously [68]. For [Pyr$_{13}$], only weak aggregation with faster exchange was found, pointing at exchange processes that were not fully averaged on the time scale of the echo time, $\tau_E = 200$ µs, rather than imperfect shimming as the origin of the line shape asymmetry.

It was reported that for longer alkyl chain lengths in [Pyr$_{1n}$][Tf$_2$N] ILs, the interatomic N–N distance decreases, the ion pair molecular volume increases, and the Mulliken charge on atoms slightly decreases. A particularly drastic change in most parameters was found at $n = 4$ [49,68]. Since the N–N distance of the cation and anion decreases for $n > 4$, the ionic couple is packed more densely and the dipolar moment drops, caused by a decrease in charge density on N that leads to decreasing $T_{2,app}$ for [Pyr$_{16}$][Tf$_2$N] because the denser packing restricts the movement of molecules. The amphiphilic properties of the pyrrolidinium cation can cause self-assembling even in bulk solutions, leading to alkyl chain length-dependent, self-organized structures [68]. Exchange between these meso-domains, micelles, and hemi-micelles can cause smeared $T_{2,app}$ distributions.

If confined to the CB matrix, the overall spectral resolution decreases and the $T_{2,app}$ main mode of [Pyr$_{13}$][Tf$_2$N] decreases to 3 ms for the S_{IL} = 0.2 sample and to 2 ms for the S_{IL} = 1 sample. For [Pyr$_{16}$][Tf$_2$N], the $T_{2,app}$ main mode decreased to 1.3 ms for S_{IL} = 0.2 and to 2.5 ms for S_{IL} = 1. The ratio of $T_{2,app}$ between neat IL and under confinement was on the same order of magnitude for both IL. The $T_{2,app}$ distributions for S_{IL} = 0.2 [Pyr$_{13}$][Tf$_2$N] were in a range of 1 ms to 5 ms, whereas [Pyr$_{16}$][Tf$_2$N] showed a distribution of 0.4 ms to 3 ms. This needs to be considered in the discussion of the diffusion measurements, since signal contributions with $T_{2,app}$ < 2 ms will be considerably affected by T_2 weighting. For S_{IL} = 1, the range of $T_{2,app}$ for [Pyr$_{13}$][Tf$_2$N] was between 0.6 ms and 4 ms. $T_{2,app}$ for S_{IL} = 1 [Pyr$_{16}$][Tf$_2$N] confined to CB ranges from 0.4 ms to 3 ms with no dependence on chemical shift. The drastic drop in $T_{2,app}$ compared to the neat ILs was caused by an ordering of the IL inside the carbon matrix and at the carbon surface equally for both ILs. This behavior might further be driven either by a separation of anion and cation below a critical pore diameter or slit distance, or by changes in conformation of the [Tf$_2$N] anion between cisoid (cis) and transoid (trans) and, thus, a change in Coulombic forces between [Pyr$_{1n}$]$^+$ and [Tf$_2$N]$^-$ [71].

The $T_{2,app}$ of the two ILs approached a similar value at high IL loading. A possible explanation could be an increased exchange between all different pore space environments at increased pore space loading. If the ILs are preferably located on the CB surface due to its high ionophilicity, a higher degree of pore space filling leads, on average, to a reduced pathway length for exchange between different sample environments. An analogous hypothesis was phrased previously based on NMR experiments on static samples [59].

The weighted residuals did not show systematic features, yet this may have been reinforced by the approximation procedure used here, where the weighting matrix was determined iteratively. While underregularization was avoided with this procedure, over-regularization may occur in situations where systematic residuals are present due to the unsuitability of the exponential kernel for certain exchange features or oscillations in the echo amplitude evolution [59]. A more accurate method that provides more faithful results would be an independent noise determination based on multiple repetitions of the experiment, followed by a noise analysis for each data point [59]. Since this was an initial study establishing the general applicability of the experimental protocol for this class of multi-component materials, noise analysis was deferred to future research.

Consistent with the shorter $T_{2,app}$, the decrease in spectral resolution was more pronounced for [Pyr$_{16}$][Tf$_2$N] than for [Pyr$_{13}$][Tf$_2$N], which might have been caused by the difference in viscosity and a possible change in cis-trans conformation of [Tf$_2$N]$^-$ that, in turn, affected [Pyr$_{1n}$]$^+$. Furthermore, restricted geometries can have significant effects on second-order or kinetic phase transitions, e.g., glass transition temperature, which can lead to a semi-solidification of the IL and, thus, coincide with a decrease in $T_{2,app}$. Since the line broadening was mainly a T_2 effect, MAS NMR can be considered successful for the investigated samples.

The experiments of IL loaded into CB showed a pronounced loading fraction dependence. For both ILs at low pore space loading, two spectral signatures with different $T_{2,app}$ could be distinguished. While both had the general features of the respective neat IL, one showed a considerable relaxation broadening and an upfield shift of about 0.5 ppm, while the other corresponded approximately to the spectrum of the neat IL, although with a different relaxation behavior that can only be explained by exchange. The fast relaxing, broad feature was strongly dominant in terms of intensity, with an integral that was about two orders of magnitude higher. Its linewidth could be fully explained by T_2 relaxation. No residual anisotropic interactions or distributions of chemical shifts were necessary. Such a feature was explained in the past by near-surface IL with restricted mobility [45,46,48]. The absence of a considerable chemical shift distribution indicated that different environments that may contribute to this signal were sampled by the ions on a time scale that was fast compared to the measured $T_{2,app}$. The neat-like IL signal in CB was not observed before in static NMR experiments. It was, however, similar to a feature that was observed at a particular IL loading level, with sufficient IL to fully cover the CB surface and then left, in addition, some surplus IL that showed a very high mobility on top of this surface layer [59]. Despite a similar loading fraction of the pore space, the looser CB packing in the current work appeared to have prevented the long-range mobility observed in [59]; yet data were consistent with the existence of an additional mobile IL fraction in the partially loaded CB samples as well.

The observation that the linewidth of the mobile species was narrower than what would be suggested by $T_{2,app}$. At least in some parts of the 2D map in Figure 1 there is a strong indication for chemical exchange with the less mobile species. Relaxation and detection of the echo were chronologically separated; hence, narrower signals than suggested by $T_{2,app}$ can originate from species that were relaxing in one environment and then exchanged prior to detection into an environment with slower relaxation. The observation of elongated distributions of $T_{2,app}$ down to the mode of the main species confirmed such a hypothesis [59]. Such an exchange behavior also indicated that this species was not artificially created by centrifugation due to MAS, but that mobile and less mobile IL species were in close proximity and not separated due to centrifugal forces. When considering the spectrum of the narrow species of the [Pyr$_{16}$] cation, it did not split into two separate species as the neat IL, indicating that within the pore space no aggregates of IL cations were formed.

An unexpected feature was the increase of the main relaxation mode for [Pyr$_{16}$][Tf$_2$N] from $S_{IL} = 0.2$ to $S_{IL} = 1$, while $T_{2,app}$ further decreased for [Pyr$_{13}$][Tf$_2$N]. When considering the simple model of two environments for the IL, one on the CB surface and one more bulk-like, then an increase of the loading past the level where the CB surface was fully covered (approximately at $S_{IL} = 0.2$) would indicate an increase of an averaged $T_{2,app}$, as was observed for [Pyr$_{16}$][Tf$_2$N]. In contrast, [Pyr$_{13}$][Tf$_2$N] showed a qualitatively different behavior, indicating that this simple model was not valid. An alternative explanation may be suggested by considering a 3D network with ionic channels forming in the neat IL [51]. Such a 3D network is structurally competing with the 2D configuration on the surface of the IL, as discussed above. There will be a transition region in between. Since [Pyr$_{16}$][Tf$_2$N] shows a more stable, slower exchanging aggregation behavior in the bulk, its 3D configuration may be more stable. At the same time, [Pyr$_{13}$][Tf$_2$N] has a lower viscosity; hence, diffusional motion could cause an enlargement of the transition region. Within the pore network of the CB, the formation of a bulk-like environment with a concomitant increase of a mean $T_{2,app}$ may, thus, be prevented for $S_{IL} = 1$ [Pyr$_{13}$][Tf$_2$N].

This may have far reaching consequences when optimizing IL-based electrolytes, since one of the main aims of IL engineering for electrochemical purposes is to enhance the mobility of the electrolyte and prevent the formation of low-mobility aggregates [59]. The result obtained here suggests that the dynamic properties of such an electrolyte in contact with an electrode cannot be implied based on the properties of the neat electrolyte alone. In addition to modifying the IL structure and the selection of a suitable electrolyte salt, IL electrolytes also offer the possibility for engineering their physical properties by altering the electrode surface. Similar hypotheses have been made based on theoretical considerations [72,73].

2.3. Diffusion Measurements

The spectrally resolved diffusion data confirmed and refined the $T_{2,app}$ data for the different samples. For all samples, exchange features were visible (Figure 2). These were apparent from the negative contributions that could not be removed without creating additional, systematic residuals [74]. On the other hand, only the neat [Pyr$_{16}$][Tf$_2$N] sample showed spectrally separable diffusion variations. The upfield shifted spectrum in Figure 2d showed a slightly slower diffusion coefficient of $D_{eff} = 6.9 \times 10^{-13}$ m^2 s^{-1} as compared to 8.7×10^{-13} m^2 s^{-1} for the downfield component. Considering the slow exchange between the two environments, as evidenced by the $T_{2,app}$ data, such a differentiation was expected with a mixing time in the PFG experiment of $\Delta = 0.1$ s. All the other spectra showed considerably faster exchange (Figure 2a–c,e–f), leading to exchange mixing during the diffusional mixing time Δ and coalescence of the effective diffusion coefficient in a single mode. In addition, the slower diffusion of the upfield-shifted component was consistent with its assignment to aggregated species, as suggested by $T_{2,app}$ data.

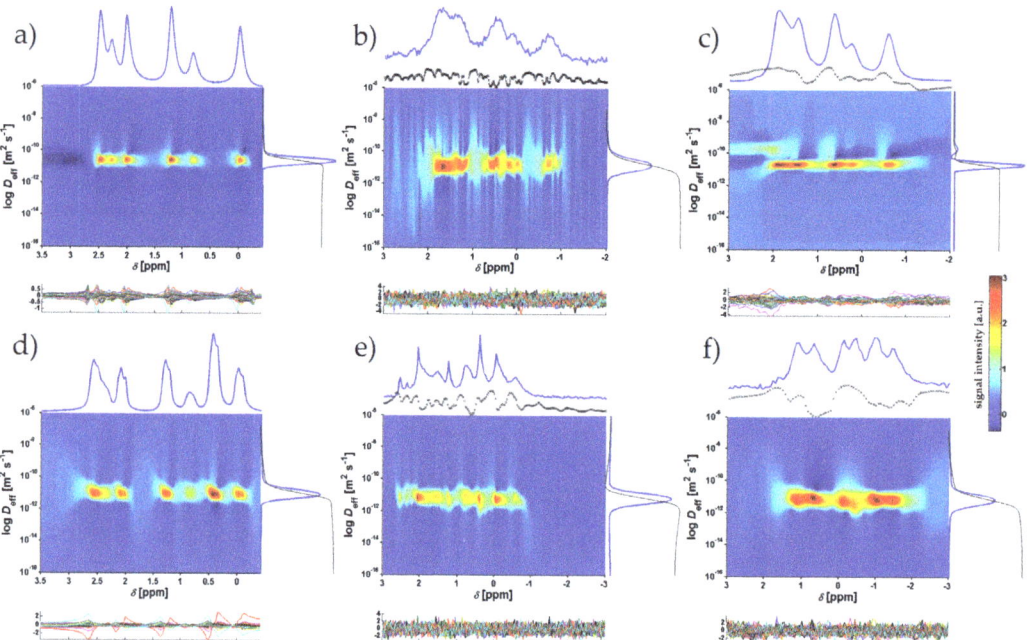

Figure 2. Pseudocolor plots of [Pyr$_{13}$] (**a**–**c**) and [Pyr$_{16}$] (**d**–**f**) diffusion coefficient distributions ($\Delta = 0.1$ s, $\delta = 0.0015$ s) vs. ^{1}H NMR frequency: (**a**) neat [Pyr$_{13}$][Tf$_2$N], (**b**) sample with CB and $S_{IL} = 0.2$ [Pyr$_{13}$][Tf$_2$N], (**c**) sample with CB and $S_{IL} = 1$ [Pyr$_{13}$][Tf$_2$N], (**d**) neat [Pyr$_{16}$][Tf$_2$N], (**e**) sample with CB and $S_{IL} = 0.2$ [Pyr$_{16}$][Tf$_2$N], (**f**) sample with CB and $S_{IL} = 1$ [Pyr$_{16}$][Tf$_2$N]. The colormap is scaled with the square root of the amplitude to amplify features with low intensity except for panel (**b**), where due to the low signal-to-noise ratio only noise was enhanced by this procedure. The sum projection onto the frequency axis (blue) and the integral of the slowly relaxing regions (dotted black) are shown on top of each panel and the sum projection onto the diffusion axis (blue) and the cumulative sum (black) to its right. The residuals are shown at the bottom of each panel.

The observed exchange features differed considerably between the samples. While the $S_{IL} = 0.2$ [Pyr$_{13}$][Tf$_2$N] sample (Figure 2b) showed a low signal-to-noise ratio and exchange features were difficult to assess, for $S_{IL} = 1$ [Pyr$_{13}$][Tf$_2$N] (Figure 2c) considerable exchange was visible. It showed a downfield-shifted broad feature with a poor spectral resolution despite MAS. Such a species was observed previously with static samples [59] and could be caused by IL in the immediate vicinity of the CB surface near crystallite edges, in slit-shaped cavities, or on amorphous regions [21]. Due to the short T_2 of this feature, only [Pyr$_{13}$] cations that were in a more slowly relaxing environment at the upper end of $T_{2,app}$ during encoding realistically contributed to this signal.

The sample $S_{IL} = 0.2$ [Pyr$_{16}$][Tf$_2$N] (Figure 2e) in CB produced spectrally resolved features, yet showed complete mixing of the diffusion coefficient during Δ. Finally, the $S_{IL} = 1$ [Pyr$_{16}$][Tf$_2$N] (Figure 2f) in CB sample showed broad exchange features that could not be easily assigned without spectral simulations that considered all the different environments, yet there was only one main mode in the diffusion dimension, also consistent with mixing of the different features during Δ, as already shown by the spectrally resolved CPMG data.

2.4. Technical Aspects

Primarily for the CPMG data but less prominent also for the diffusion experiments, the residuals of the Laplace inversion were considerably higher than the random noise level of the measurements. This observation contrasts the results of our previous study using a standard diffusion probe for static samples, where residuals above the random noise level were only observed for experiments with significant exchange contributions (see supporting info of [59]). In these cases, the systematic residuals could be traced back to features that could not be faithfully reproduced with the kernel chosen for the inversion. In the current case, however, the residuals were significantly higher. In addition, they showed typical features of multiplicative noise [59], as shown for the neat [Pyr$_{13}$][Tf$_2$N] sample in Figure 3. When plotting the standard deviation of the residuals vs. frequency, the maximum appears at positions with maximum slope of the resonances, which is characteristic for multiplicative phase noise. In contrast, if features cannot be fitted with a particular kernel or if the inversion was overregularized, then the maximum residuals occur at the peak maxima. Multiplicative noise is not a fundamental problem of the method. The $T_{2,app}$ data used for Figure 1a,d was recorded without sample spinning and without pulsed field gradients. Therefore, it was possible to reduce the impact of multiplicative noise for the neat [Pyr$_{13}$] sample to some degree by using adhesive tape to more rigidly fix the MAS rotor in the probe. In principle, it should be possible to obtain the same data quality as in [59] and this work may indicate the scope for future improvements in experimental design and hardware, especially for MAS PFG NMR. Partial remedy is possible by slightly overcoupling the resonator to reduce its quality factor or by working with reduced field gradient strength. While both measures are quite suitable for liquid or IL samples in porous media, it does limit the applicability for the investigation of solid electrolyte materials. In particular for solid Li ion conductors, chemical shift differences are small and high spinning frequencies are desirable for optimum resolution, while at the same time, strong PFGs are necessary to limit their duration and maximize the range of diffusion coefficients that can be distinguished towards low mobility. Therefore, for a wide applicability beyond proof-of-concept work, further development of the experimental method is desirable.

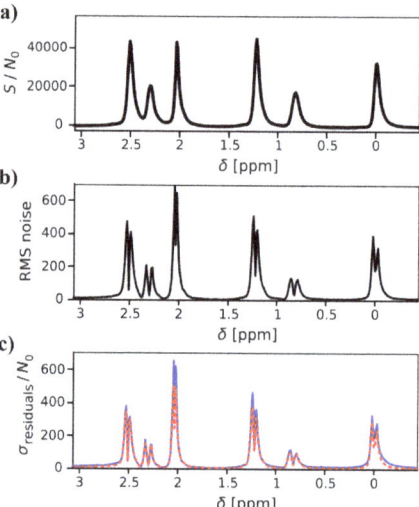

Figure 3. Illustration of multiplicative noise in ^{1}H NMR $T_{2,app}$ data of neat [Pyr$_{13}$][Tf$_2$N]. (**a**) ^{1}H NMR spectrum of first echo. The signal intensity is normalized to the noise standard deviation N_0 in the absence of any signal. Notice the very high signal-to-noise ratio, which is a prerequisite for the observation of multiplicative noise. (**b**) Estimation of root-mean-square (RMS) noise level of the above spectrum. The shape is characteristic for multiplicative phase noise, with a maximum at positions with maximum first derivative of the signal [75]. (**c**) Standard deviation of the residuals of the ILT fit (blue) and estimation of the standard deviation based on the assumption of multiplicative phase noise, as depicted in (**b**) (red dashed).

In addition to random multiplicative noise, which was a major concern mainly for the neat IL samples, CPMG data of IL loaded CB samples also showed pronounced, spectrally dependent oscillations of the echo amplitudes with longer periods than the well-known oscillations between even and odd echoes [76]. Fast relaxing components were affected more strongly by such oscillations. The effect can be observed best in the $T_{2,app}$ data of the [Pyr$_{13}$][Tf$_2$N]-loaded CB samples at the upfield and the downfield ends of the main species, where intensive negative features are observed. The spectral signature of these features suggests an origin that is not mainly dependent on hardware limitations, but has a more fundamental reason. Protons of [Pyr$_{13}$] cations in direct proximity of CB surfaces are more strongly affected. The interaction of the oscillating magnetic field caused by the NMR pulses with the conducting CB surface leads to a local field enhancement or shielding as well as a phase shift. However, the heterogeneity of the CB structure prevents the more long-range formation of eddy currents; therefore, only IL in the immediate vicinity of the surface is affected. The resulting local phase shifts cannot be corrected by phase corrections, as not all of the sample is affected, necessarily leading to negative contributions in spectrally resolved relaxation measurements. The oscillations may provide information about electrical properties of the porous host, suggesting that more in-depth investigations of these effects may be promising.

For an optimum extraction of available information in data with non-uniform noise, data analysis schemes using weighted data are called for [77,78], as applied here for the CPMG data presented in Figure 1. The oscillating features of the echo decay were also treated as noise, since the exponential kernel used for the Laplace inversion is not suitable to fit oscillations. To extract the full information contained in the data, it may be worthwhile to explore more sophisticated data analysis protocols in the future.

2.5. Influence of Centrifugal Forces

A simple estimate of the centrifugal forces acting on the IL at a MAS spinning rate of 5 kHz and a pore radius where pores remain saturated can be made based on the relationship [19]

$$P = \pi^2 \nu^2 r_r^2 \rho \tag{1}$$

where P is the pressure (Pa), ν is the spinning rate (Hz), r_r is the radius of the rotor or insert (m), and ρ is the sample density (kg m^{-3}). The pressure exerted on the IL for the given experimental parameters is 434 kPa and 467 kPa for [Pyr$_{16}$][Tf$_2$N] and [Pyr$_{13}$][Tf$_2$N], respectively. A pressure of P = 33 kPa can be assumed for the CB ($\rho \approx 100$ kg m^{-3}). For pores with a circular cross-section the corresponding pore radius can be calculated using [79]

$$r_c = \frac{2\gamma \cos\sigma}{P} \tag{2}$$

where σ is the contact angle and γ the surface tension of 0.0302 J m^{-2} and 0.0353 J m^{-2} for [Pyr$_{16}$][Tf$_2$N] and [Pyr$_{13}$][Tf$_2$N], respectively [80–82]. Assuming a good wetting of the carbon surface by the ILs with $\sigma < 60°$ yields $r_c > 69$ nm for [Pyr$_{16}$][Tf$_2$N] and $r_c > 75$ nm for [Pyr$_{13}$][Tf$_2$N]. Since pore size distributions from r = 1 nm to 60 nm have been reported for Vulcan® carbon black XC 72 and XC 72R [83–88], a centrifugal effect on the ILs is not expected because $r_c > r$.

2.6. Influence of Sample Preparation and NMR Parameters on MAS Measurements

Since CBs can be employed either as insulating or conductive additives depending on the density, both the filling fraction of CB pore space and the exerted pressure on the CB during rotor loading determine the electrical properties of the system. Electrical percolation in CBs is a known phenomenon, where a certain electrical percolation threshold determines the formation of π-electron conducting networks [89]. By increasing the graphitic character of the surface, the electrical conductivity of CBs increases, whereas for CBs possessing a low surface area, the electrical conductivity correlates with the surface chemistry [90]. For Vulcan® carbon black XC 72, the electrical conductivity increases linearly up to 2.7 S cm^{-1} at a pressure of 2000 kPa [90]. While spinning, the centrifugal forces will cause an additional compaction of the sample and, if the sample is a solid/liquid mixture, an electrically conductive slurry originates.

The sample rotation in the external magnetic field B_0 can cause eddy currents, which can slow down the rotor speed [91]. Therefore, filling and compacting CB into MAS rotors is the first delicate step that can impact the tuning and matching of the system to a point where the sample becomes immeasurable. For the system under study, an increase in the CB electrical conductivity to $\sigma \approx 0.43$ S cm^{-1} can be assumed for a spinning rate of 5 kHz. The electrical conductivity of the ILs is negligible at 0.0015 S cm^{-1} and 0.005 S cm^{-1} for [Pyr$_{16}$][Tf$_2$N] and [Pyr$_{13}$][Tf$_2$N], respectively. Thus, the compaction pressure exerted on the CB during sample preparation determines the conductive properties of the systems, whereas the increase in conductivity for the applied CB can be neglected for a spinning rate of 5 kHz.

CB samples are largely heterogeneous, based on their intrinsic chemical structure and their preparation. Therefore, chemically identical nuclei can experience different local environments and, in addition to anisotropic magnetic susceptibility that can be averaged out by MAS, variations of isotropic chemical shifts or local susceptibilities throughout the sample may occur, which can cause signal broadening.

Adding liquid to a porous, solid matrix will cause the formation of air bubbles and trapped air in dead pores (unconnected pores). These air pockets can decrease the spectral resolution and additionally alter the magnetic susceptibility of the samples. Therefore, an accurate filling and saturation procedure is imperative for proper MAS diffusion measurements.

Another source that can possibly alter diffusion measurements is the Lorentz force exerted on charges due to rapidly switched electrical currents in the gradient coils, possibly causing phase shifts, additional signal attenuation, artificial diffraction patterns in the diffusion data, loss of tuning and matching, or even a rotor crash [91–93]. The force and the resulting vibrations are a function of the applied current, the gradient wave form, its duration and its rise and fall time, the magnetic field, and the density of coil turns. The vibrations cause further phase shifts in the resulting NMR spectra, which can lead to artificial features in the ILT data. For the system under study, limiting the gradient strength to half maximum eliminated vibrational effects on the sample.

3. Materials and Methods

3.1. Sample Preparation and Characterization

The 1-methyl-1-propylpyrrolidiniumbis(trifluoromethylsulfonyl)imide ([Pyr$_{13}$][Tf$_2$N]) (water content < 100 ppm, density = 1.42 g cm^{-3} at 303.15K, melting point = 255.15K) and 1-hexyl-1-methylpyrrolidiniumbis(trifluoromethylsulfonyl)imide ([Pyr$_{16}$][Tf$_2$N]) (water content <100 ppm, density = 1.32 g cm^{-3} at 293.15K, melting point = 278.15K) were both purchased from Iolitec (Germany) and used as received (values taken from [69,94]). Both ILs were degassed under vacuum for 24 h prior to measurements.

The neat [Pyr$_{16}$][Tf$_2$N] and [Pyr$_{13}$][Tf$_2$N] were loaded into 30 µL, 4 mm HR-MAS cylindrical disposable inserts (Bruker, Germany) with an inner radius of 1 mm using a micro syringe (for details on the insert, see [93]). Each insert was sealed with a cap and screw and placed inside a cylindrical 4 mm MAS rotor (Bruker, Leipzig, Germany).

Vulcan® carbon black XC 72 (furnace black) powder (Cabot Corporation, Boston, MA, USA) with a BET (N$_2$) surface area of A_{BET} = 218 m^2 g^{-1}, primary particle size of 30 nm, and a density of 264 kg m^{-3} was grinded using an agate mortar. Since grinding alters the bead shape of Vulcan® XC72, the resulting CB after grinding is comparable to the powdered Vulcan® XC72R with an average particle size of 50 nm, a density of 96 kg m^{-3}, and A_{BET} = 222 m^2 g^{-1}. A more detailed description of the physical properties of Vulcan® XC72 and XC72R is given in [83–88,95,96]. A total pore volume of V_{total} = 0.76 cm^3 g^{-1} was determined by N$_2$ adsorption using an Autosorb iQ 2 (Quantachrome, Boynton Beach, FL, USA) equipped with a cryocooler (CTI-cryogenics, Waltham, MA, USA). A mean pore radius of r = 7 nm was estimated using the relationship $2V_{total}/A_{BET}$.

Approximately 6 mg of grinded CB was loaded and gently compacted into the 30 µL disposable HR-MAS insert, achieving a bulk density of ≈0.2 g cm^{-3}, assuming a particle density of 1.8 g cm^{-3}.

ILs were added to the CB using a microliter syringe (Hamilton, Reno, NV, USA). For each of the two ILs, two different IL volumes were applied to fill a certain fraction of the CB pore space. Since CB was filled into the MAS rotor in the form of a packed bed rather than pressed to a pellet, as in our previous work [59], the exact volume fraction was difficult to determine. As a rough estimate, about 20% of the pore space was filled in one set of samples (IL saturation level S_{IL} = 0.2) and the full pore space in the other set (S_{IL} = 1), while care had been taken that no free IL was present, as confirmed by the absence of corresponding features in the NMR data. The IL was applied dropwise, achieving the different saturation levels of S_{IL} = 0.2 (1 drop) and S_{IL} = 1 (5 drops) both for [Pyr$_{16}$][Tf$_2$N] and [Pyr$_{13}$][Tf$_2$N]. Afterwards the insert was placed under vacuum for 1 h using a vacuum desiccator until all ionic liquid was absorbed by the CB. The mass added was determined by means of NMR saturation profiles, yielding an amount of 1.84 mg (0.37mg/drop) [Pyr$_{16}$][Tf$_2$N] and 1.04 mg (0.21mg/drop) [Pyr$_{13}$][Tf$_2$N] for the fully saturated samples.

3.2. ^{1}H-NMR Measurements

All ^{1}H-NMR MAS experiments were performed using a 9.4T Bruker Avance spectrometer with a 1.5 T m^{-1} micro 2.5 imaging gradient system, equipped with a Bruker 4-mm ^{1}H/^{13}C HR-MAS probe. A probe temperature of 303.15 K was maintained for all measurements. The apparent transverse relaxation time constants ($T_{2,app}$) were monitored using the Carr-Purcell-Meiboom-Gill (CPMG) method [97]. For diffusion measurements, a 13-interval, stimulated echo PGSTEBP Bruker pulse sequence was employed, as described in [98,99] following the approach of Stejskal and Tanner [13] using a gradient pulse duration of δ = 0.0021 s, observation time of Δ = 0.1 s, 32 gradient steps, T_1 relaxation delay RD = 2.5 s, and a radio frequency pulse duration of 0.006 ms for 90° pulses. Measurements of the neat ILs were performed under static conditions, whereas for ILs confined to CB a MAS spinning speed of 5 kHz was applied.

No external or internal reference compound was employed for all measurements. Analogously to [100], the ^{1}H peak position of the terminal alkyl group protons served as 0 ppm reference for [Pyr$_{16}$][Tf$_2$N] and [Pyr$_{13}$][Tf$_2$N].

3.3. Data Analysis

For the 13-interval PGSTE method, the attenuation of the echo amplitude is given by [99]

$$\ln\frac{S(g_a)}{S(0)} = -\gamma^2 D[\delta^2\left(4\Delta + 6\tau - \frac{2}{3}\delta\right)g_a^2 + 2\tau\delta(\delta_1 - \delta_2)g_a g_0 + \frac{4}{3}\tau^3 g_0^2] \quad (3)$$

where S is the signal intensity, γ is the gyromagnetic ratio (2.6751 × 10^8 rad s^{-1} T^{-1} for ^{1}H), g_0 is the background (internal) gradient, g_a is the applied gradient, δ is the duration of the applied gradient, δ_1 is the time interval between the first rf-pulse and the start of g_a, and δ_2 is the time interval following the end of g_a until the second rf-pulse, where for all measurements $\delta_1 = \delta_2$. Since $\delta_1 = \delta_2$, the second term in Equation (3) disappears while the third term can be eliminated by normalizing the signal with pulsed gradients applied with the signal $S(g_a = 0) = S(0)$ without pulsed gradients. Δ is the interval between the second and the third gradient pulses, τ is the time between the rf-pulses, and D is the self-diffusion coefficient [101–103].

Both the 13-interval PGSTE and the CPMG data were spectrally resolved by Fourier transforming the NMR raw data along the transient dimension. The distributions of $T_{2,app}$ and D_{eff} were obtained by performing a regularized inversion with an exponential kernel without a non-negativity constraint [75,104]. Parameterization was done as described in [78] without manually fine-tuning the parameters.

CPMG experiments showed initial oscillations of the echo amplitude due to pulse imperfections and off-resonance effects, which are partially caused by the electrically conductive porous CB matrix that induces a distribution of the amplitude, B_1, of the oscillating radio frequency field used to excite the spins. This effect was reduced by using only even echoes for data processing [76]. In addition, slower oscillations of the echo amplitudes were apparent as well, which caused problems with parametrization of the regularized inversion. To avoid overregularization of data points that were not affected by these oscillations, the oscillations were considered as noise and data were weighted accordingly [77]. Since, as detailed in the discussion, data were also affected by multiplicative noise, the data weighting was determined iteratively [75]. First, using the residuals of an initial inversion with unweighted data, smoothed over three echoes, an estimate for the weight matrix was obtained. This procedure was repeated once using the residuals of an inversion with weighted data. Further iterations only led to marginal additional changes. For the PGSTE data, due to the lower signal-to-noise ratio that reduced the impact of multiplicative noise and the constant timing of the experiment that prevented oscillations from pulse imperfections, the sensitivity was mostly limited by additive white noise, and inversion was done using unweighted data.

In the figures, unless stated otherwise, the $T_{2,app}$ and the D_{eff} coefficient distributions are represented as square root scaled color maps with conservation of the sign to enhance weaker spectral features.

4. Conclusions

We investigated the suitability of MAS NMR with and without PFG for the investigation of ionic liquids in electrically conductive porous CB as a model system for an IL-based electrolyte in contact with a porous electrode. Using CPMG and diffusion experiments, it was possible to achieve spectral resolutions consistent with apparent T_2 relaxation time constants. The increased resolution afforded by MAS allowed the identification of weak, narrow spectral features of the investigated IL at partial loading of the CB host. These were caused by IL cations with a high local mobility; yet fast exchange indicated that a spatially distributed species was responsible for these features rather than a reservoir created by centrifugal force exerted by MAS sample spinning.

When comparing the local mobility of neat IL with IL in CB, significant differences were observed. In particular, the relaxation and the associated spectral properties varied to an extent that suggested that bulk IL studies were ill-suited for the optimization of electrolyte properties in a porous electrode environment. A porous host system may alter the aggregation behavior of ILs and thereby provide an additional degree of freedom to optimize transport in IL electrolyte-based electrochemical systems. One example is the existence of highly mobile IL on top of a more strongly bound surface layer, as suggested in earlier work [59] and supported here by evidence of exchange.

While overall the samples showed a very good signal-to-noise ratio, the signal was affected by multiplicative noise. Part of the problem may have been caused by the ionic and conductive nature of the sample, with ions interacting with electromagnetic fields and an electrically conductive porous host, leading to partial shielding of the sample. This led to a stronger than usual inhomogeneity of the exciting radio frequency field amplitude B_1 across the sample, causing oscillations of the echo amplitude that are usually common for imperfectly adjusted pulse lengths. For the data analysis, these oscillations were treated as noise, which allowed an estimate of the sensitivity for individual data points and, consequently, Laplace inversion with weighted data. Thereby, signal contributions with very fast relaxation were suppressed since no reliable differentiation between fast relaxing and oscillating features was possible within the scope of this study. While artifacts and spurious signals could be avoided, there may have been some information lost with regard to signal components that relax during the course of a small number of echoes. A more refined analysis appears possible, warranting future research while extending the potentials of MAS PFG NMR for this class of materials.

MAS NMR facilitated an improved resolution of IL spectra in porous CB, which allowed the identification of features so far not reported in literature. However, an exact identification of the different environments is still challenging due to the hierarchical nature of the CB structure or, more generally, of many porous electrode systems used for electrochemical experiments. Therefore, the investigated techniques provide complementary information rather than substituting established methods.

Author Contributions: Conceptualization, S.M. and J.W.; methodology, J.W., P.G. and S.M.; validation, S.M., J.W., P.G. and J.G.; formal analysis, J.W., J.G. and S.M.; investigation, S.M. and J.W.; resources, P.G.; data curation, S.M., J.W. and J.G.; writing-original draft preparation, S.M., J.W. and J.G.; writing-review and editing, S.M., J.W., P.G. and J.G.; visualization, J.W., J.G. and S.M.; supervision, S.M.; project administration, S.M.; funding acquisition, J.G. All authors have read and agreed to the published version of the manuscript.

Funding: Parts of the research were funded by BMBF [03XP0176] "FestBatt".

Data Availability Statement: The data presented in this study are available on request from the corresponding author. The data are not publicly available due to non-standard, proprietary formatting, which will necessitate explanation on sharing.

Acknowledgments: S.M. thanks Svitlana Taranenko (RWTH Aachen, Germany) for preparatory work, preliminary sample testing, and discussion and Manu Pouajen-Blakiston from the workshop (Victoria University of Wellington, NZ) for customizing the tools for MAS sample preparation. J.W. and P.G. thank the MacDiarmid Institute for contributions to research infrastructure and funding. S.M. and J.G. thank R.A. Eichel for support via "FestBatt".

Conflicts of Interest: The authors declare no conflict of interest. The funders had no role in the design of the study; in the collection, analyses, or interpretation of data; in the writing of the manuscript; or in the decision to publish the results.

Sample Availability: Samples of IL and CB compounds are available from the authors.

References

1. Li, M.; Wang, C.; Chen, Z.; Xu, K.; Lu, J. New Concepts in Electrolytes. *Chem. Rev.* **2020**, *120*, 6783–6819. [CrossRef] [PubMed]
2. Gouverneur, M.; Kopp, J.; van Wullen, L.; Schonhoff, M. Direct determination of ionic transference numbers in ionic liquids by electrophoretic NMR. *Phys. Chem. Chem. Phys.* **2015**, *17*, 30680–30686. [CrossRef] [PubMed]
3. Karger, J.; Ruthven, D.M. Diffusion in nanoporous materials: Fundamental principles, insights and challenges. *New J. Chem.* **2016**, *40*, 4027–4048. [CrossRef]
4. Hardy, E.H. *NMR Methods for the Investigation of Structure and Transport*; Springer Science & Business Media: Berlin, Germany, 2011.
5. Abragam, A. *The Principles of Nuclear Magnetism*; Oxford University Press: Clarendon, TX, USA, 1961.
6. Keeler, J. *Understanding NMR Spectroscopy*; Wiley: Chichester, UK, 2005.
7. Markley, J.L.; Opella, S.J. *Biological NMR Spectroscopy*; Oxford University Press: New York, NY, USA, 1997.
8. Günther, H. *NMR Spectroscopy: Basic Principles, Concepts and Applications in Chemistry*; Wiley: Weinheim, Germany, 2013.
9. Ernst, R.R.; Bodenhausen, G.; Wokaun, A. *Principles of Nuclear Magnetic Resonance in One and Two Dimensions*; Clerendon Press: Oxford, UK, 1987.
10. Jeener, J.; Meier, B.H.; Bachmann, P.; Ernst, R.R. Investigation of exchange processes by two-dimensional NMR spectroscopy. *J. Chem. Phys.* **1979**, *71*, 4546–4553. [CrossRef]
11. Callaghan, P.T.; Furó, I. Diffusion-diffusion correlation and exchange as a signature for local order and dynamics. *J. Chem. Phys.* **2004**, *120*, 4032–4038. [CrossRef]
12. Stallmach, F.; Galvosas, P. Spin echo NMR diffusion studies. *Annual Reports on NMR Spectroscopy*; Webb, G.A., Ed.; Academic Press: London, UK, 2007; Volume 61, pp. 51–131.
13. Stejskal, E.O.; Tanner, J.E. Spin Diffusion Measurements: Spin Echoes in the Presence of a Time-Dependent Field Gradient. *J. Chem. Phys.* **1965**, *42*, 288–292. [CrossRef]
14. Andrew, E.R.; Bradbury, A.; Eades, R.G. Removal of Dipolar Broadening of Nuclear Magnetic Resonance Spectra of Solids by Specimen Rotation. *Nature* **1959**, *183*, 1802–1803. [CrossRef]
15. Reif, B.; Ashbrook, S.E.; Emsley, L.; Hong, M. Solid-state NMR spectroscopy. *Nat. Rev. Methods Primers* **2021**, *1*, 2. [CrossRef]
16. Pampel, A.; Michel, D.; Reszka, R. Pulsed field gradient MAS-NMR studies of the mobility of carboplatin in cubic liquid-crystalline phases. *Chem. Phys. Lett.* **2002**, *357*, 131–136. [CrossRef]
17. Gratz, M.; Hertel, S.; Wehring, M.; Stallmach, F.; Galvosas, P. Mixture diffusion of adsorbed organic compounds in metal-organic frameworks as studied by magic-angle spinning pulsed-field gradient nuclear magnetic resonance. *New J. Phys.* **2011**, *13*, 45016. [CrossRef]
18. Taylor, A.J.; Granwehr, J.; Lesbats, C.; Krupa, J.L.; Six, J.S.; Pavlovskaya, G.E.; Thomas, N.R.; Auer, D.P.; Meersmann, T.; Faas, H.M. Probe-Specific Procedure to Estimate Sensitivity and Detection Limits for 19F Magnetic Resonance Imaging. *PLoS ONE* **2016**, *11*, e163704. [CrossRef]
19. Asano, A.; Hori, S.; Kitamura, M.; Nakazawa, C.T.; Kurotsu, T. Influence of magic angle spinning on T1H of SBR studied by solid state 1H NMR. *Polym. J.* **2012**, *44*, 706–712. [CrossRef]
20. Patrick, J.W. *Porosity in Carbons: Characterization and Applications*; Edward Arnold: London, UK, 1995.
21. Schröder, A.; Klüppel, M.; Schuster, R.H.; Heidberg, J. Surface energy distribution of carbon black measured by static gas adsorption. *Carbon* **2002**, *40*, 207–210. [CrossRef]
22. Sebok, E.B.; Taylor, R.L. Carbon Blacks. In *Encyclopedia of Materials: Science and Technology*; Buschow, K.H.J., Cahn, R.W., Flemings, M.C., Ilschner, B., Kramer, E.J., Mahajan, S., Veyssière, P., Eds.; Elsevier: Oxford, UK, 2001; pp. 902–906.
23. Dicks, A.L. The role of carbon in fuel cells. *J. Power Sources* **2006**, *156*, 128–141. [CrossRef]
24. Ma, Y.; Wang, H.; Ji, S.; Goh, J.; Feng, H.; Wang, R. Highly active Vulcan carbon composite for oxygen reduction reaction in alkaline medium. *Electrochim. Acta* **2014**, *133*, 391–398. [CrossRef]
25. Pan, Y.; Xu, K.; Wu, C. Recent progress in supercapacitors based on the advanced carbon electrodes. *Nanotechnol. Rev.* **2019**, *8*, 299–314. [CrossRef]
26. Liu, R.; Wang, Y.; Li, B.; Liu, B.; Ma, H.; Li, D.; Dong, L.; Li, F.; Chen, X.; Yin, X. VXC-72R/ZrO2/GCE-Based Electrochemical Sensor for the High-Sensitivity Detection of Methyl Parathion. *Materials* **2019**, *12*, 3637. [CrossRef]
27. El Khatib, K.; Hameed, R. Development of Cu_2O/Carbon Vulcan XC-72 as non-enzymatic sensor for glucose determination. *Biosens. Bioelectron.* **2011**, *26*, 3542–3548. [CrossRef]
28. Khodabakhshi, S.; Fulvio, P.F.; Andreoli, E. Carbon black reborn: Structure and chemistry for renewable energy harnessing. *Carbon* **2020**, *162*, 604–649. [CrossRef]
29. Wissler, M. Graphite and carbon powders for electrochemical applications. *J. Power Sources* **2006**, *156*, 142–150. [CrossRef]
30. Manickam, M.; Takata, M. Electrochemical and X-ray photoelectron spectroscopy studies of carbon black as an additive in Li batteries. *J. Power Sources* **2002**, *112*, 116–120. [CrossRef]

31. Fransson, L.; Eriksson, T.; Edström, K.; Gustafsson, T.; Thomas, J.O. Influence of carbon black and binder on Li-ion batteries. *J. Power Sources* **2001**, *101*, 1–9. [CrossRef]
32. Kaisheva, A.; Iliev, I. Application of Carbon-Based Materials in Metal-Air Batteries: Research, Development, Commercialization. In *New Carbon Based Materials for Electrochemical Energy Storage Systems: Batteries, Supercapacitors and Fuel Cells*; Springer: Dordrecht, Netherland, 2006; pp. 117–136.
33. Hayes, R.; Warr, G.G.; Atkin, R. Structure and Nanostructure in Ionic Liquids. *Chem. Rev.* **2015**, *115*, 6357–6426. [CrossRef]
34. Tiago, G.; Matias, I.; Ribeiro, A.; Martins, L. Application of Ionic Liquids in Electrochemistry-Recent Advances. *Molecules* **2020**, *25*, 5812. [CrossRef]
35. Torriero, A. *Electrochemistry in Ionic Liquids: Volume 1: Fundamentals*; Springer: Cham, Switzerland, 2015; pp. 1–351.
36. Torriero, A. *Electrochemistry in Ionic Liquids: Volume 2: Applications*; Springer: Cham, Switzerland, 2015; pp. 1–623.
37. MacFarlane, D.; Kar, M.; Pringle, J.M. *Fundamentals of Ionic Liquids: From Chemistry to Applications*; Wiley: Weinheim, Germany, 2017.
38. Smith, A.; Lovelock, K.; Gosvami, N.; Licence, P.; Dolan, A.; Welton, T.; Perkin, S. Monolayer to Bilayer Structural Transition in Confined Pyrrolidinium-Based Ionic Liquids. *J. Phys. Chem. Lett.* **2013**, *4*, 378–382. [CrossRef] [PubMed]
39. Zhang, S.; Zhang, J.; Zhang, Y.; Deng, Y. Nanoconfined Ionic Liquids. *Chem. Rev.* **2017**, *117*, 6755–6833. [CrossRef] [PubMed]
40. Liu, J.; Ma, L.; Zhao, Y.; Pan, H.; Tang, Y.; Zhang, H. Porous structural effect of carbon electrode formed through one-pot strategy on performance of ionic liquid-based supercapacitors. *Chem. Eng. J.* **2021**, *411*, 128573. [CrossRef]
41. Sharma, S.; Dhattarwal, H.; Kashyap, H. Molecular dynamics investigation of electrostatic properties of pyrrolidinium cation based ionic liquids near electrified carbon electrodes. *J. Mol. Liq.* **2019**, *291*, 111269. [CrossRef]
42. Sharma, S.; Kashyap, H. Interfacial Structure of Pyrrolidinium Cation Based Ionic Liquids at Charged Carbon Electrodes: The Role of Linear and Nonlinear Alkyl Tails. *J. Phys. Chem. C* **2017**, *121*, 13202–13210. [CrossRef]
43. Begic, S.; Jonsson, E.; Chen, F.; Forsyth, M. Molecular dynamics simulations of pyrrolidinium and imidazolium ionic liquids at graphene interfaces. *Phys. Chem. Chem. Phys.* **2017**, *19*, 30010–30020. [CrossRef]
44. Di Lecce, S.; Kornyshev, A.; Urbakh, M.; Bresme, F. Lateral Ordering in Nanoscale Ionic Liquid Films between Charged Surfaces Enhances Lubricity. *ACS Nano* **2020**, *14*, 13256–13267. [CrossRef]
45. Borghi, F.; Podestà, A. Ionic liquids under nanoscale confinement. *Adv. Phys. X* **2020**, *5*, 1736949. [CrossRef]
46. Anaredy, R.S.; Shaw, S.K. Long-Range Ordering of Ionic Liquid Fluid Films. *Langmuir* **2016**, *32*, 5147–5154. [CrossRef] [PubMed]
47. Zhang, Y.; Rutland, M.W.; Luo, J.; Atkin, R.; Li, H. Potential-Dependent Superlubricity of Ionic Liquids on a Graphite Surface. *J. Phys. Chem. C* **2021**, *125*, 3940–3947. [CrossRef]
48. Borghi, F.; Piazzoni, C.; Ghidelli, M.; Milani, P.; Podestà, A. Nanoconfinement of Ionic Liquid into Porous Carbon Electrodes. *J. Phys. Chem. C* **2021**, *125*, 1292–1303. [CrossRef]
49. Brutti, S. Pyr1,xTFSI Ionic Liquids (x = 1–8): A Computational Chemistry Study. *Appl. Sci.* **2020**, *10*, 8552. [CrossRef]
50. Cao, X.; He, X.; Wang, J.; Liu, H.; Roser, S.; Rad, B.; Evertz, M.; Streipert, B.; Li, J.; Wagner, R.; et al. High Voltage LiNi0.5Mn1.5O4/Li4Ti5O12 Lithium Ion Cells at Elevated Temperatures: Carbonate- versus Ionic Liquid-Based Electrolytes. *ACS Appl. Mater. Interfaces* **2016**, *8*, 25971–25978. [CrossRef]
51. Banerjee, A.; Shah, J.K. Elucidating the effect of the ionic liquid type and alkyl chain length on the stability of ionic liquid–iron porphyrin complexes. *J. Chem. Phys.* **2020**, *153*, 034306. [CrossRef]
52. Sanchora, P.; Pandey, D.K.; Kagdada, H.L.; Materny, A.; Singh, D.K. Impact of alkyl chain length and water on the structure and properties of 1-alkyl-3-methylimidazolium chloride ionic liquids. *Phys. Chem. Chem. Phys.* **2020**, *22*, 17687–17704. [CrossRef]
53. Garaga, M.N.; Nayeri, M.; Martinelli, A. Effect of the alkyl chain length in 1-alkyl-3-methylimidazolium ionic liquids on intermolecular interactions and rotational dynamics: A combined vibrational and NMR spectroscopic study. *J. Mol. Liq.* **2015**, *210*, 169–177. [CrossRef]
54. Canongia Lopes, J.N.A.; Pádua, A.A.H. Nanostructural Organization in Ionic Liquids. *J. Phys. Chem. B* **2006**, *110*, 3330–3335. [CrossRef]
55. Holstein, P.; Bender, M.; Galvosas, P.; Geschke, D.; Kärger, J. Anisotropic Diffusion in a Nematic Liquid Crystal—An Electric Field PFG NMR Approach. *J. Magn. Reson.* **2000**, *143*, 427–430. [CrossRef]
56. Umecky, T.; Saito, Y.; Matsumoto, H. Direct Measurements of Ionic Mobility of Ionic Liquids Using the Electric Field Applying Pulsed Gradient Spin−Echo NMR. *J. Phys. Chem. B* **2009**, *113*, 8466–8468. [CrossRef] [PubMed]
57. Wang, Y. Disordering and Reordering of Ionic Liquids under an External Electric Field. *J. Phys. Chem. B* **2009**, *113*, 11058–11060. [CrossRef] [PubMed]
58. Saito, Y.; Hirai, K.; Murata, S.; Kishii, Y.; Kii, K.; Yoshio, M.; Kato, T. Ionic Diffusion and Salt Dissociation Conditions of Lithium Liquid Crystal Electrolytes. *J. Phys. Chem. B* **2005**, *109*, 11563–11571. [CrossRef] [PubMed]
59. Merz, S.; Jakes, P.; Taranenko, S.; Eichel, R.-A.; Granwehr, J. Dynamics of [Pyr13][Tf2N] ionic liquid confined to carbon black. *Phys. Chem. Chem. Phys.* **2019**, *21*, 17018–17028. [CrossRef]
60. Zhu, H.; O'Dell, L.A. Nuclear magnetic resonance characterisation of ionic liquids and organic ionic plastic crystals: Common approaches and recent advances. *Chem. Commun.* **2021**, *57*, 5609–5625. [CrossRef]
61. Forse, A.C.; Griffin, J.M.; Merlet, C.; Bayley, P.M.; Wang, H.; Simon, P.; Grey, C.P. NMR Study of Ion Dynamics and Charge Storage in Ionic Liquid Supercapacitors. *J. Am. Chem. Soc.* **2015**, *137*, 7231–7242. [CrossRef] [PubMed]

62. Sánchez-González, J.; Macías-García, A.; Alexandre-Franco, M.; Mez-Serrano, V. Electrical conductivity of carbon blacks under compression. *Carbon* **2005**, *43*, 741. [CrossRef]
63. Falcon, E.; Castaing, B. Electrical conductivity in granular media and Branly's coherer: A simple experiment. *Am. J. Phys.* **2005**, *73*, 302–307. [CrossRef]
64. Creyssels, M.; Falcon, E.; Castaing, B. Experiment and Theory of the Electrical Conductivity of a Compressed Granular Metal. *AIP Conf. Proc.* **2009**, *1145*, 123–126. [CrossRef]
65. Watson, A.T.; Chang, C.T.P. Characterizing porous media with NMR methods. *Prog. Nucl. Magn. Reson. Spectrosc.* **1997**, *31*, 343–386. [CrossRef]
66. Haigh, C.W.; Mallion, R.B. Ring current theories in nuclear magnetic resonance. *Prog. Nucl. Magn. Reson. Spectrosc.* **1979**, *13*, 303–344. [CrossRef]
67. Forse, A.C.; Griffin, J.M.; Presser, V.; Gogotsi, Y.; Grey, C.P. Ring Current Effects: Factors Affecting the NMR Chemical Shift of Molecules Adsorbed on Porous Carbons. *J. Phys. Chem. C* **2014**, *118*, 7508–7514. [CrossRef]
68. Kunze, M.; Jeong, S.; Paillard, E.; Schönhoff, M.; Winter, M.; Passerini, S. New Insights to Self-Aggregation in Ionic Liquid Electrolytes for High-Energy Electrochemical Devices. *Adv. Energy Mater.* **2011**, *1*, 274–281. [CrossRef]
69. Ravula, S.; Larm, N.E.; Mottaleb, M.A.; Heitz, M.P.; Baker, G.A. Vapor Pressure Mapping of Ionic Liquids and Low-Volatility Fluids Using Graded Isothermal Thermogravimetric Analysis. *ChemEngineering* **2019**, *3*, 42. [CrossRef]
70. Veroutis, E.; Merz, S.; Eichel, R.A.; Granwehr, J. Intra- and inter-molecular interactions in choline-based ionic liquids studied by 1D and 2D NMR. *J. Mol. Liq.* **2020**, *322*, 114934. [CrossRef]
71. Palumbo, O.; Trequattrini, F.; Appetecchi, G.B.; Paolone, A. Influence of Alkyl Chain Length on Microscopic Configurations of the Anion in the Crystalline Phases of PYR1A-TFSI. *J. Phys. Chem. C* **2017**, *121*, 11129–11135. [CrossRef]
72. Bazant, M.Z.; Storey, B.D.; Kornyshev, A.A. Double Layer in Ionic Liquids: Overscreening versus Crowding. *Phys. Rev. Lett.* **2011**, *106*, 046102. [CrossRef]
73. de Souza, J.P.; Goodwin, Z.A.H.; McEldrew, M.; Kornyshev, A.A.; Bazant, M.Z. Interfacial Layering in the Electric Double Layer of Ionic Liquids. *Phys. Rev. Lett.* **2020**, *125*, 116001. [CrossRef]
74. Bytchenkoff, D.; Rodts, S. Structure of the two-dimensional relaxation spectra seen within the eigenmode perturbation theory and the two-site exchange model. *J. Magn. Reson.* **2011**, *208*, 4–19. [CrossRef]
75. Granwehr, J. Multiplicative or t1 Noise in NMR Spectroscopy. *Appl. Magn. Reson.* **2007**, *32*, 113–156. [CrossRef]
76. Ostroff, E.D.; Waugh, J.S. Multiple Spin Echoes and Spin Locking in Solids. *Phys. Rev. Lett.* **1966**, *16*, 1097–1098. [CrossRef]
77. Borgia, G.C.; Brown, R.J.S.; Fantazzini, P. Uniform-Penalty Inversion of Multiexponential Decay Data. *J. Magn. Reson.* **1998**, *132*, 65–77. [CrossRef] [PubMed]
78. Granwehr, J.; Roberts, P.J. Inverse Laplace Transform of Multidimensional Relaxation Data Without Non-Negativity Constraint. *J. Chem. Theory Comput.* **2012**, *8*, 3473–3482. [CrossRef]
79. Washburn, K.E.; Eccles, C.D.; Callaghan, P.T. The dependence on magnetic field strength of correlated internal gradient relaxation time distributions in heterogeneous materials. *J. Magn. Reson.* **2008**, *194*, 33–40. [CrossRef]
80. Carvalho, P.J.; Neves, C.M.S.S.; Coutinho, J.A.P. Surface Tensions of Bis(trifluoromethylsulfonyl)imide Anion-Based Ionic Liquids. *J. Chem. Eng. Data* **2010**, *55*, 3807–3812. [CrossRef]
81. Kolbeck, C.; Lehmann, J.; Lovelock, K.R.J.; Cremer, T.; Paape, N.; Wasserscheid, P.; Fröba, A.P.; Maier, F.; Steinrück, H.P. Density and Surface Tension of Ionic Liquids. *J. Phys. Chem. B* **2010**, *114*, 17025–17036. [CrossRef] [PubMed]
82. Stefan, C.S.; Lemordant, D.; Claude-Montigny, B.; Violleau, D. Are ionic liquids based on pyrrolidinium imide able to wet separators and electrodes used for Li-ion batteries? *J. Power Sources* **2009**, *189*, 1174–1178. [CrossRef]
83. Liu, J.; Song, P.; Ruan, M.; Xu, W. Catalytic properties of graphitic and pyridinic nitrogen doped on carbon black for oxygen reduction reaction. *Chin. J. Catal.* **2016**, *37*, 1119–1126. [CrossRef]
84. Xie, J.; More, K.L.; Zawodzinski, T.A.; Smith, W.H. Porosimetry of MEAs Made by "Thin Film Decal" Method and Its Effect on Performance of PEFCs. *J. Electrochem. Soc.* **2004**, *151*, 1841. [CrossRef]
85. Malinowski, M.; Iwan, A.; Hreniak, A.; Tazbir, I. An anode catalyst support for polymer electrolyte membrane fuel cells. Application of organically modified titanium and silicon dioxide. *RSC Adv.* **2019**, *9*, 24428–24439. [CrossRef]
86. Sepp, S.; Härk, E.; Valk, P.; Vaarmets, K.; Nerut, J.; Jäger, R.; Lust, E. Impact of the Pt catalyst on the oxygen electroreduction reaction kinetics on various carbon supports. *J. Solid State Electrochem.* **2014**, *18*, 1223–1229. [CrossRef]
87. Antxustegi, M.M.; Pierna, A.R.; Ruiz, N. Chemical activation of Vulcan® XC72R to be used as support for NiNbPtRu catalysts in PEM fuel cells. *Int. J. Hydrogen Energy* **2014**, *39*, 3978–3983. [CrossRef]
88. Takasu, Y.; Kawaguchi, T.; Sugimoto, W.; Murakami, Y. Effects of the surface area of carbon support on the characteristics of highly-dispersed Pt-Ru particles as catalysts for methanol oxidation. *Electrochim. Acta* **2003**, *48*, 3861–3868. [CrossRef]
89. Ehrburger-Dolle, F.; Lahaye, J.; Misono, S. Percolation in carbon black powders. *Carbon* **1994**, *32*, 1363–1368. [CrossRef]
90. Pantea, D.; Darmstadt, H.; Kaliaguine, S.; Sümmchen, L.; Roy, C. Electrical conductivity of thermal carbon blacks: Influence of surface chemistry. *Carbon* **2001**, *39*, 1147–1158. [CrossRef]
91. Aubert, G.; Jacquinot, J.F.; Sakellariou, D. Eddy current effects in plain and hollow cylinders spinning inside homogeneous magnetic fields: Application to magnetic resonance. *J. Chem. Phys.* **2012**, *137*, 154201. [CrossRef]
92. Yesinowski, J.P.; Ladouceur, H.D.; Purdy, A.P.; Miller, J.B. Electrical and ionic conductivity effects on magic-angle spinning nuclear magnetic resonance parameters of CuI. *J. Chem. Phys.* **2010**, *133*, 234509. [CrossRef]

93. Alam, T. HR-MAS NMR Spectroscopy in Material Science. *Adv. Aspects Spectrosc.* **2012**, *10*, 279–306.
94. Hayamizu, K.; Tsuzuki, S.; Seki, S.; Fujii, K.; Suenaga, M.; Umebayashi, Y. Studies on the translational and rotational motions of ionic liquids composed of N-methyl-N-propyl-pyrrolidinium (P13) cation and bis(trifluoromethanesulfonyl)amide and bis(fluorosulfonyl)amide anions and their binary systems including lithium salts. *J. Chem. Phys.* **2010**, *133*, 194505. [CrossRef] [PubMed]
95. Senthil Kumar, S.M.; Soler Herrero, J.; Irusta, S.; Scott, K. The effect of pretreatment of Vulcan XC-72R carbon on morphology and electrochemical oxygen reduction kinetics of supported Pd nano-particle in acidic electrolyte. *J. Electroanal. Chem.* **2010**, *647*, 211–221. [CrossRef]
96. Shukla, S.; Bhattacharjee, S.; Weber, A.Z.; Secanell, M. Experimental and Theoretical Analysis of Ink Dispersion Stability for Polymer Electrolyte Fuel Cell Applications. *J. Electrochem. Soc.* **2017**, *164*, 600–609. [CrossRef]
97. Meiboom, S.; Gill, D. Modified Spin-Echo Method for Measuring Nuclear Relaxation Times. *Rev. Sci. Instrum.* **1958**, *29*, 688–691. [CrossRef]
98. Galvosas, P.; Stallmach, F.; Kärger, J. Background gradient suppression in stimulated echo NMR diffusion studies using magic pulsed field gradient ratios. *J. Magn. Reson.* **2004**, *166*, 164–173. [CrossRef]
99. Cotts, R.M.; Hoch, M.J.R.; Sun, T.; Markert, J.T. Pulsed field gradient stimulated echo methods for improved NMR diffusion measurements in heterogeneous systems. *J. Magn. Reson.* **1989**, *83*, 252–266. [CrossRef]
100. Nicotera, I.; Oliviero, C.; Henderson, W.A.; Appetecchi, G.B.; Passerini, S. NMR Investigation of Ionic Liquid−LiX Mixtures: Pyrrolidinium Cations and TFSI- Anions. *J. Phys. Chem. B* **2005**, *109*, 22814–22819. [CrossRef]
101. Callaghan, P.T. *Translational Dynamics and Magnetic Resonance: Principles of Pulsed Gradient Spin Echo NMR*; Oxford University Press: New York, NY, USA, 2011.
102. Sørland, G. *Dynamic Pulsed-Field-Gradient NMR*; Springer: Berlin, Germany, 2014.
103. Grebenkov, D.S. NMR survey of reflected Brownian motion. *Rev. Mod. Phys.* **2007**, *79*, 1077–1137. [CrossRef]
104. Graf, M.F.; Tempel, H.; Kocher, S.S.; Schierholz, R.; Scheurer, C.; Kungl, H.; Eichel, R.-A.; Granwehr, J. Observing different modes of mobility in lithium titanate spinel by nuclear magnetic resonance. *RSC Adv.* **2017**, *7*, 25276–25284. [CrossRef]

Communication

Complete Assignment of the ^{1}H and ^{13}C NMR Spectra of Carthamin Potassium Salt Isolated from *Carthamus tinctorius* L.

Maiko Sasaki and Keiko Takahashi *

Faculty of Engineering, Tokyo Polytechnic University, 1583, Iiyma Atsugi, Kanagawa 243-027, Japan; hachikara.bisage@gmail.com
* Correspondence: takahasi@t-kougei.ac.jp

Abstract: Carthamin potassium salt isolated from *Carthamus tinctorius* L. was purified by an improved traditional Japanese method, without using column chromatography. The ^{1}H and ^{13}C nuclear magnetic resonance (NMR) signals of the pure product were fully assigned using one- and two-dimensional NMR spectroscopy, while the high purity of the potassium salt and deprotonation at the 3′ position of carthamin were confirmed by atomic adsorption spectroscopy and nano-electrospray ionization mass spectrometry.

Keywords: carthamin-3′ potassium salt; green metallic luster; fermented safflower petal tablet

1. Introduction

Carthamin, a traditional red pigment obtained from the dried petals of safflower (*Carthamus tinctorius* L.), has long been used in food colorants, dyes, and medicines worldwide. The mostly yellow appearance of the lively safflower petals reflects the prevalence of water-soluble yellow ingredients. Originally native to Asia Minor, safflower spread to central Europe (via Egypt) and Japan (via China) [1]. In Japan, a safflower-derived red pigment known as "beni" [2] was commonly traded and used in cosmetics despite being expensive and rare (Figure 1).

Figure 1. Beni, a typical Japanese cosmetic, in wet (left; **a**) and dry (right; **b**) states (top). Dried stable beni isolated and purified in this work (bottom; **c**).

In particular, the green luster of beni was viewed as evidence of its high quality, and the corresponding pigment was called "sasairo-beni" (bamboo-colored red). Today, we know that safflower petals contain yellow and red pigments [3], with the red pigment (represented by a single compound) accounting for <1% of the total pigment content. This rare red pigment, viz. carthamin, was first reported in 1846 [4]. Since then, carthamin purification methods and structure have been extensively investigated [5–9], and the correct molecular structure (C-glycoside with two glucose residues) was determined in 1979 [10,11]. Subsequently, the total synthesis of carthamin was achieved [12,13] and a hybrid bio-/organic synthesis incorporating enzymatic reactions was then proposed [14]. These syntheses confirmed the skeletal and full molecular structures of carthamin (Figure 2). To date, complete assignments of the ^{1}H/^{13}C nuclear magnetic resonance (NMR) and mass spectra of carthamin have not been reported [13,15–18], which can be ascribed to the light- and temperature-sensitive nature of this red pigment and the problems associated with its isolation. Assignment of the 15 proton signals (on 3′ 4, 4′, 5, 5′, 13, 13′, G2, G2′, G3, G3′, G4, G4′, G6 and G6′) of the hydroxyl group, which is important for considering the molecular structure, has been ignored.

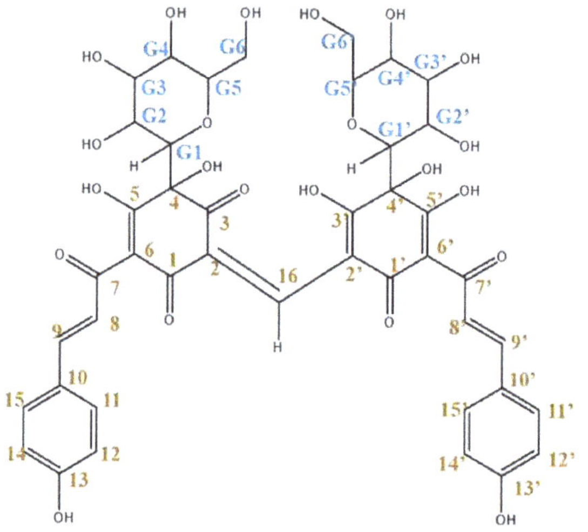

Figure 2. Molecular structure of carthamin.

Typically, the isolation of carthamin from safflower petals and its purification are performed as follows. Dry safflower petals are suspended in cold water, allowed to stand for some days to remove the yellow pigments, and repeatedly washed with running water until the disappearance of the yellow color. The petals in the filter bag are then transferred to a new vessel filled with fresh cold water containing sodium bicarbonate, and the filter bag is kneaded to release the red pigment into the solution. However, the purity of the thereby obtained product is insufficient for structure elucidation, which has inspired numerous attempts to (i) form pure crystals using various derivatizations and treatments, as well as (ii) achieve separation using column chromatography. Previously, potassium and pyridinium salts of carthamin have been reported through its purification and determination. Previous research on carthamin has been to purify or synthesize the molecular form of carthamin. There has been attention paid to the green color of carthamin (sasairo-beni) in dry conditions (Figure 1b). Japanese traditional cosmetics, sasairo-beni purified based on traditional methods, is relatively stable after the treatment of vacuum drying.

Our research initially focused on the mechanism of sasairo-beni, the bamboo color development, as the origin of carthamin's green color which is actually not a structural color but a metallic luster and has not been investigated [19–21].

Therefore, a complete assignment of all ^{1}H and ^{13}C signals in the NMR and mass spectra of carthamin was required to bring its structural study at the molecular level to the next stage. To realize full NMR and mass-spectral analyses, we used the 3′-potassium salt of carthamin, which has a green metallic luster (Figure 1c).

In this communication, we wish to indicate complete assignment of the ^{1}H and ^{13}C NMR spectra of carthamin 3′-potassium salt which is isolated from *Carthamus tinctorius* L. including 15 proton signals (on 3′ 4, 4′, 5, 5′, 13, 13′, G2, G2′, G3, G3′, G4, G4′, G6 and G6′) of the hydroxyl group.

2. Results and Discussion

2.1. Isolation of Carthamin Potassium Salt

Carthamin potassium salt was obtained from the fermented safflower petal tablet: Benimochi in Japanese, Yamagata Prefecture, Red Flower Production Association, using a modified traditional Japanese purification method [19,20]. After more than a dozen processes, using natural traditional acidic and alkaline solutions and ramie fibers, 258 mg of red pigment with a green metallic luster was yielded. Only one red spot with R_f = 0.42 was observed by TLC (eluent = 1-butanol: acetic acid: water, 4:1:5, $v/v/v$). Nano-electrospray ionization mass spectrometry (NanoESI-MS), m/z, found 987.1343, Calcd. 987.9698 for $C_{43}H_{41}O_{22}K_2$. Potassium and sodium contents that were determined using atomic absorption spectrophotometry were K; 10.1 wt% and Na; 0.038 wt%, respectively. The NanoESI-MS spectrum showed no peaks assignable to free carthamin. The above results suggested that carthamin isolated using a Japanese traditional method includes at least one potassium element as potassium salt. Moreover, the carthamin potassium salt was relatively stable when dried in vacuo (Figure 1c), and the NMR solution samples in dimethyl sulfoxide (DMSO-d_6) or pyridine-d_5 did not change after several years at r.t.

2.2. Assignment of the ^{1}H and ^{13}C NMR Spectra of Carthamin Potassium Salt

The assigned NMR spectra were almost identical to the unassigned spectra reported previously [11,14,17]. In previous studies, the NMR spectra of carthamin were recorded in pyridine-d_5 or a mixture of pyridine-d_5 and methanol-d_4 [18]. The very broad OH signals observed in pyridine [13] collapsed into a single peak upon the addition of methanol, which did not allow one to extract any information pertaining to the OH groups. Figures 3 and 4 present the assigned ^{1}H and ^{13}C NMR spectra of carthamin potassium salt in DMSO-d_6, respectively. Diffusion-ordered spectroscopy (DOSY) spectra revealed that signals below 2.8 ppm were observed on different diffusion lines and did not originate from carthamin. Therefore, these signals were ascribed to a trace impurity not detectable by TLC or NanoESI-MS.

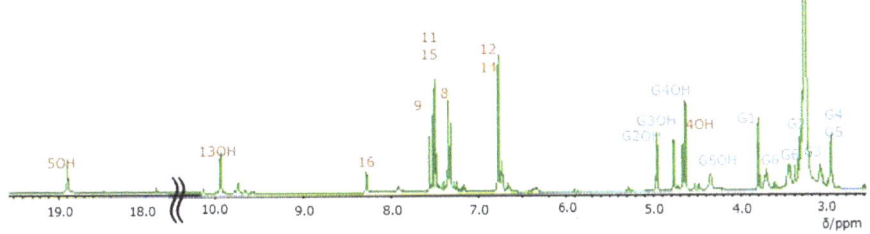

Figure 3. Assigned ^{1}H NMR spectrum of carthamin potassium salt recorded in DMSO-d_6 at 30 °C.

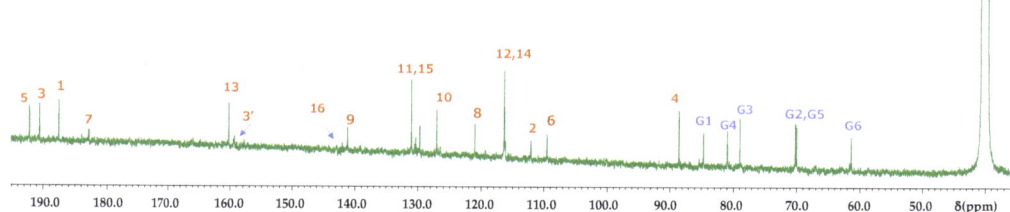

Figure 4. Assigned ^{13}C NMR spectrum of carthamin potassium salt recorded in DMSO-d_6 at 30 °C.

The carthamin protons resonated at 2.97, 3.09, 3.33, 3.46, 3.72, 3.81, 4.36, 4.64, 4.67, 4.77, 4.97, 6.87, 7.34, 7.51, 7.55, 8.28, 9.90, and 18.89 ppm. The signals at 4.36, 4.64, 4.67, 4.77, 4.97, 9.90, and 18.89 ppm disappeared after the addition of D$_2$O and were therefore assigned to OH groups. The signal at 18.89 ppm has previously been ascribed to the enolic proton of the yellow pigment safflomin A, ((4S)-4,6-di-D-glucopyranosyl-4,5-dihydroxy-2-[E-1-hydroxy-3-(4-hydroxyphenyl)prop-2-enylidene]-cyclohex-5-ene-1,3-dion) from *Carthamus tinctorius* L. [22,23], with the remarkable low-field shift attributed to hydrogen bonding between OH and C=O. The hydrogen bonding between both 5′OH and 7 and 7′ C=O has not been suggested in previous reports, because they have not paid attention to OH protons. In our case, the signal at 18.89 ppm was assigned to the 5 and 5′ enolic protons of carthamin. As the signal of the phenolic OH groups of safflomin A at 9.79 ppm was broad, the broad signal at 9.90 ppm was assigned to the 13 and 13′ phenolic OH groups in carthamin. The integrated signal intensity ratio was 1 (8.28 ppm):2 (3.09, 3.33, 3.46, 3.72, 3.81, 4.36, 4.64, 4.67, 4.77, 4.97, 7.34, 7.55, 9.90, and 18.89 ppm):4 (2.97, 6.87, and 7.51 ppm). The signal at 8.28 ppm was assigned to 16H.

2.3. COSY, HMQC, NOESY and HMBC for Assignment of the ^{1}H and ^{13}C NMR Signals

Correlation spectroscopy (COSY) revealed that cross-peaks were present not only between the signals of protons bonded to adjacent carbons, but also between the signals of protons bonded to adjacent carbon and oxygen elements [24], which allowed us to assign numerous couples (Figure 5). Between 5, 5′OH and 7, 7′C=O, hydrogen bonding exists, which regulates the structure of the carthamin molecule.

The signals of 11 and 11′ overlapped with those of 15 and 15′, while the signals of 12 and 12′ overlapped with those of 14 and 14′. In order to assign the overlapped signals of 11, 11′, 15, 15′, 12, 12′, 14, 14′, 8, 8′, and 9, 9′, nuclear Overhauser effect spectroscopy (NOESY) was used. NOESY revealed the presence of cross-peaks between signals at 7.34 and 7.51 ppm, which allowed the signals at 7.34, 7.55, 6.87, and 7.51 ppm to be ascribed to 8H, 9H, 12H overlapped with 14H, and 11H overlapped with 15H, respectively (Figure 6). According to the molecular structure model, the cross-peak was between 8H and 11, 11′H or 15, 15′H. Thus, only the signal of the 3′OH proton remained unassigned. Among the 43 carbons constituting carthamin (and affording 23 signals), six glucose-derived carbons were in almost identical environments and therefore featured the same shift, as did another group of 14 carbons (1, 2, 4–15). The cross-peaks revealed by heteronuclear multiple quantum correlation spectroscopy (HMQC) (Figure 7) allowed us to assign proton-bearing carbons on the glucose ring (G1, G1′, G2, G2′, G3, G3′, G4, G4′, G5, G5′, and G6 and G6′) as well as carbons 8, 8′, 9, 9′, 11, 11′, 12, 12′, 14, 14′, 15, 15′, 16 and 16′, whereas the long-range correlation data provided by heteronuclear multiple-bond correlation spectroscopy (HMBC) allowed us to assign carbons 1, 1′, 2, 2′, 3, 3′, 4, 4′, 5, 5′, 6, 6′, 7, 7′, 10, 10′ 13 and 13′ (Figure 8). A clear cross-peak was observed between 8.28 (^{1}H) and 142 ppm (^{13}C). The unclear noisy signal at 142 ppm in Figure 4 was not noise but real signal.

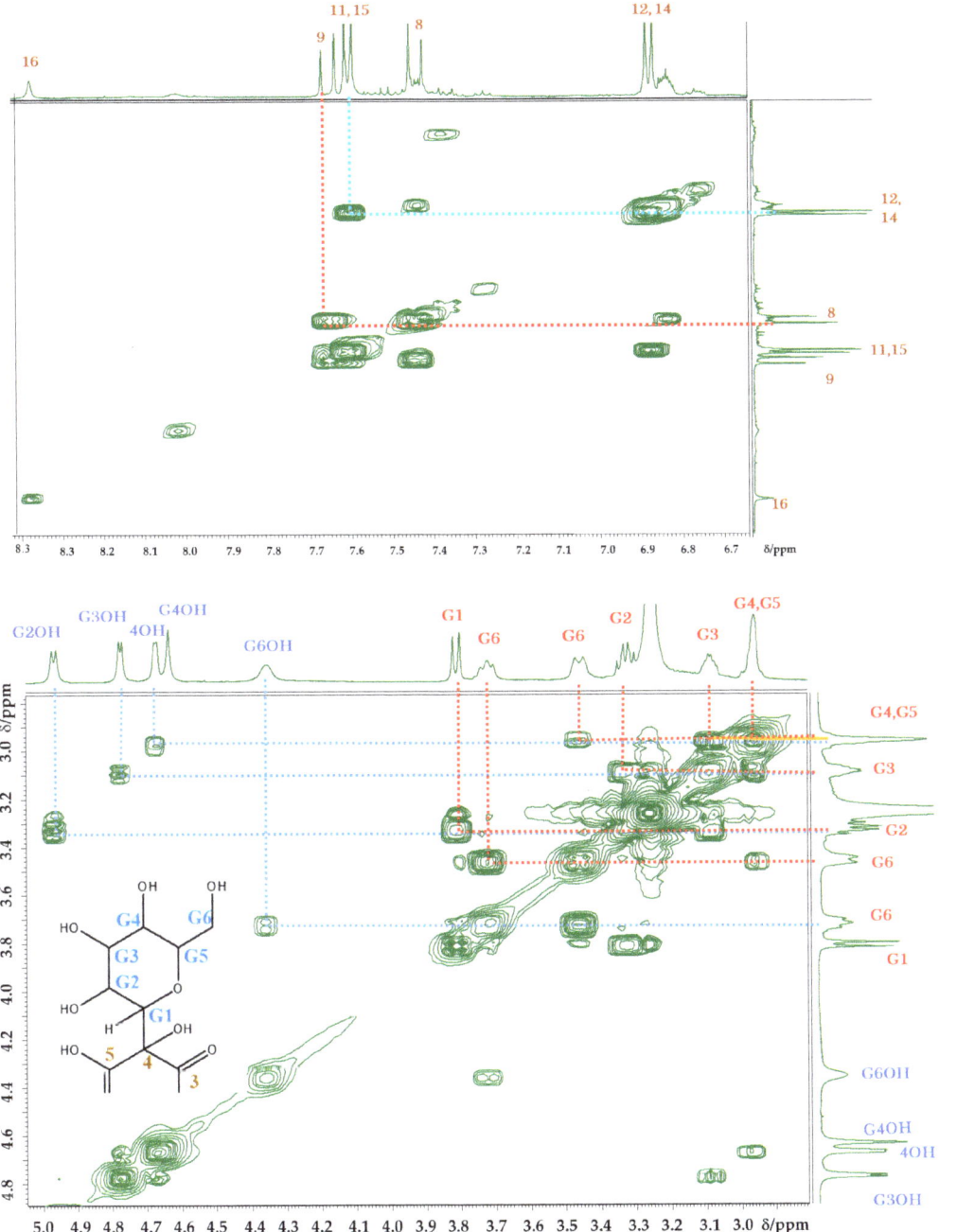

Figure 5. COSY spectra of carthamin potassium salt recorded in DMSO-d_6 at 30 °C; cross-peaks between carbon atoms (**top**) and between carbon and oxygen atoms (**bottom**).

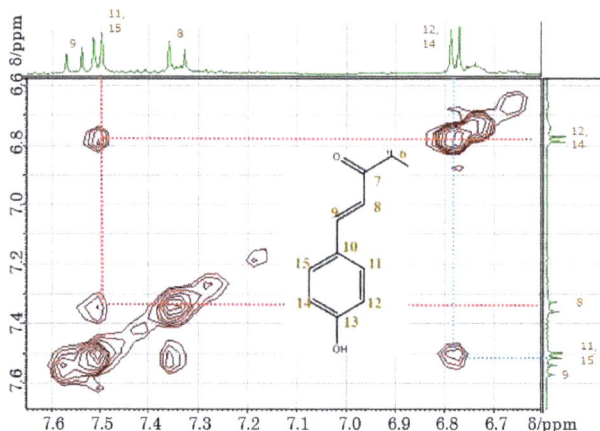

Figure 6. NOESY spectrum of carthamin potassium salt recorded in DMSO-d_6 at 30 °C.

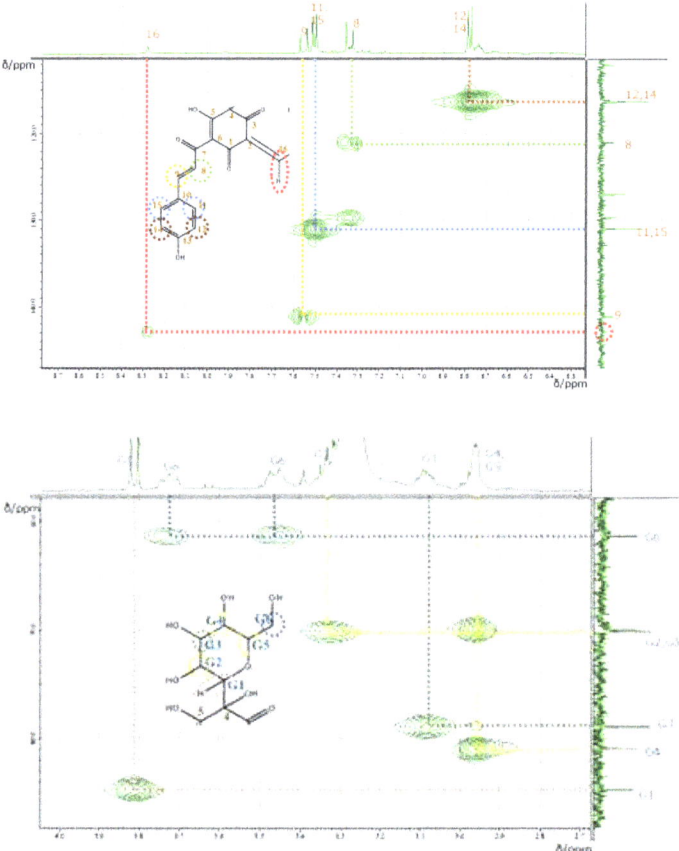

Figure 7. HMQC spectra of carthamin 3′potassium salt recorded in DMSO-d_6 at 30 °C under low (**top**) and high (**bottom**) magnetic fields.

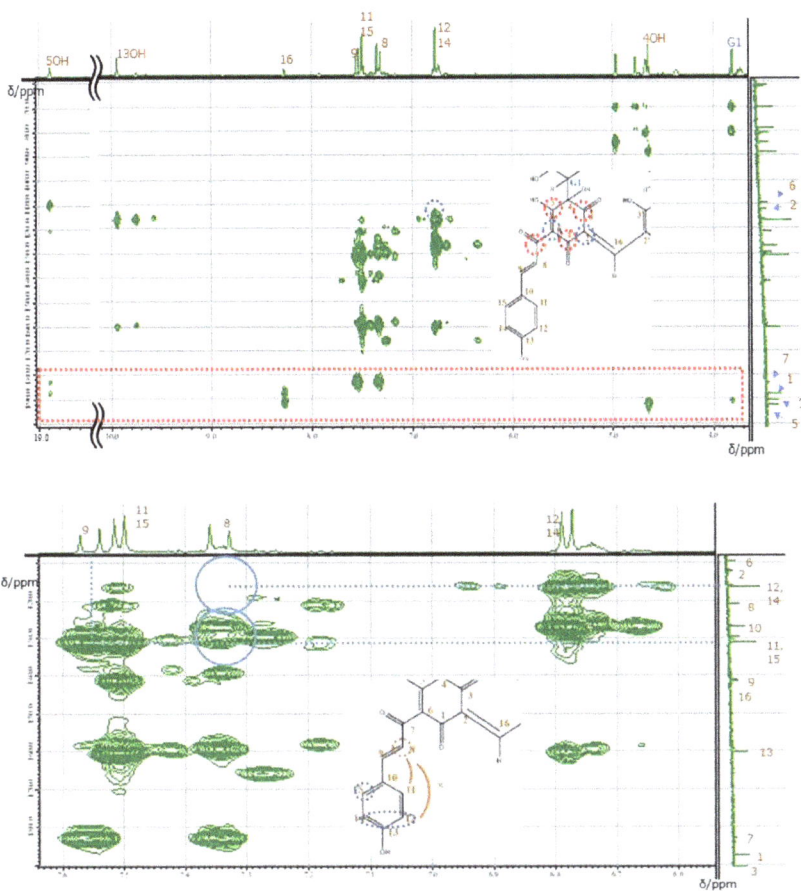

Figure 8. Full-range (**top**) and expanded (**bottom**) HMBC spectra of carthamin potassium salt recorded in DMSO-d6 at 30 °C.

The carbonyl (3) and enolic (3′) carbons were observed as two separate signals, whereas carbon 16 yielded one signal. Except for the signal derived from 3′C at 159.3 ppm, the above data agree with those reported previously in works not attempting to perform spectral assignments.

3. Materials and Methods

3.1. Isolation and Purification of Carthamin Potassium Salt

The carthamin potassium salt was extracted by a traditional method with some modifications [19,20]. A total of 100 g of the fermented dried safflower petal tablets in a cotton cloth bag was soaked in cold water at 10 °C for 48 h, and repeatedly washed with running water to remove pollen and the yellow pigments until the disappearance of the yellow color. The petals in the cotton bag were wrung to remove water, and we added the pH 12.0 alkaline solution which was prepared from plant ash taken by a similar method and plant as the Japanese traditional method, and then was kneaded to release the red pigment into the solution (pH 10.4). Natural ramie fibers were dipped in this liquid and then allowed to stand for a while. The ramie fibers dyed in red were washed with a slightly acidic aqueous solution, squeezed out of water, and then air-dried. The alkaline aqueous solution was gradually added dropwise to the dying ramie fibers to extract the red solution. The extract was filtered through a filter and then slowly neutralized with an acidic solution. At pH 6.3, a fine precipitate was formed. After being centrifuged, a red muddy pigment was yielded. The precipitate was spread on glass to block light, air-dried at room temperature, and vacuum-dried for five days to reproducibly obtain a red pigment with a green metallic luster (258 mg) [19,20]. Only one red spot with R_f = 0.42 was observed by TLC (eluent = 1-butanol:acetic acid:water, 4:1:5, $v/v/v$). Nano-electrospray ionization mass spectrometry (NanoESI-MS): m/z found 987.1343, Calcd. 987.9698 for $C_{43}H_{41}O_{22}K_2$; K; 10.1 wt%, Na; 0.038 wt%. The NanoESI-MS spectrum showed no peaks assignable to free carthamin. The red muddy substance was applied to a quartz plate, scraped off, air-dried at room temperature under shading, and then vacuum-dried to obtain 258 mg of a red pigment with a green metallic luster. Only one red spot with R_f = 0.42 was observed by TLC (eluent = 1-butanol:acetic acid: water, 4: 1: 5, $v/v/v$). Nano-electrospray ionization mass spectrometry (NanoESI-MS): m/z found 987.1343, Calcd. 987.9698 for $C_{43}H_{41}O_{22}K_2$; the NanoESI-MS spectrum showed no peaks assignable to free carthamin. Elemental analysis taken by atomic absorption spectrophotometry was K; 10.1 wt%, Na; 0.038 wt%.

3.2. Instruments

The NMR spectra were recorded on a JEOL JNM-ECZ500R (500 MHz JEOL Ltd. Tokyo, Japan) spectrometer at 30 °C. Pulse programs used for the NMR spectrometer were standard sequences taken from the JEOL Delta 5.3.1 pulse sequence library. Detailed conditions: scan times and relaxation delay times of ^{1}H, ^{13}C, COSY, HMQC, DOSY, NOESY and HMBC were 64, 5 s, 50,000, 8 s, 4, 1.5 s, 32, 4 s, 16, 7 s, 16, 4 s, 8, 16, 7 s, 256, and 4 s, respectively. DMSO-d_6, pyridine-d_5, and D$_2$O were purchased from Kanto Kagaku Co. NanoESI-MS analysis was performed on a QExactive Plus (Thermo Fisher Scientific, Waltham, MA, USA) instrument, and potassium content was determined using atomic absorption spectrophotometry (Z-2300, Hitachi High-Tech Science Co., Tokyo, Japan). Mass spectral and elemental analyses were performed at Toray Research Centre, Inc., Tokyo, Japan.

4. Conclusions

To summarize, we indicate complete assignment of the ^{1}H and ^{13}C NMR spectra of carthamin potassium salt which was isolated from *Carthamus tinctorius* L. prepared with a modified Japanese traditional method without any chromatography. Complete signal assignment was carried out by various 2D NMR spectroscopies: COSY, HMQC, NOESY, DOSY and HMBC.

Fifteen proton signals (on 3' 4, 4', 5, 5', 13, 13', G2, G2', G3, G3', G4, G4', G6 and G6') of the hydroxyl group were also assigned, which suggests the traditional Japanese cosmetic Beni is carthamin-3'-potassium salt and the existence of hydrogen bonding between both 5, 5' OH and 7C=O.

To obtain information on the content of inorganic elements and completely assign the ^{1}H and ^{13}C NMR signals of carthamin, we scrutinized the traditional methods and carried out atomic adsorption and NanoESI-MS analyses. Molecular structure analyses, including solid-state structure analysis, are currently underway. The presented information shows that carthamin is no longer "difficult to analyze" and contributes to safflower petal research and quality control.

Author Contributions: Investigation of NMR data curation was carried out by M.S. and K.T. Original draft preparation was carried out by K.T., review and editing were carried out by both M.S. and K.T., Project administration was taken by K.T. Authorship must be limited to those who have contributed substantially to the work reported. Both authors have read and agreed to the published version of the manuscript.

Funding: This research was funded by a Grant-in-Aid for Scientific Research (20K01108) and by the International Center for Science and Arts of Color, Tokyo Polytechnic University in 2020.

Institutional Review Board Statement: Not applicable.

Informed Consent Statement: Not applicable.

Data Availability Statement: Data are contained within the article.

Conflicts of Interest: The authors declare no conflict of interest.

Sample Availability: Samples of the compounds are not available from the authors.

References

1. Knowles, P.F. Safflower—Production, Processing and Utilization. *Econ. Bot.* **1955**, *9*, 273–299. [CrossRef]
2. Yoshioka, S. *Japanese Color Dictionary*; Shikousha: Tokyo, Japan, 2000.
3. Kazuma, K.; Takahashi, T.; Sato, K.; Takeuchi, H.; Matsumoto, T.; Okuno, T. Quinochalcones and flavonoids from fresh florets in different cultivars of *Carthamus tinctorius* L. *Biosci. Biotechnol. Biochem.* **2000**, *64*, 1588–1599. [CrossRef]
4. Schlieper, A. Ueber das rothe und gelbe Pigment des Saflors. *Justus Liebigs Annalen der Chemie* **1846**, *58*, 357–374. [CrossRef]
5. Kametaka, T. Study of Japanese traditional cosmetics. *Tokyo Kagakukai Zasshi* **1906**, *27*, 1202–1217.
6. Kametaka, T.; Perkin, A.G. Carthamine. Part I. *J. Chem. Soc. Trans.* **1910**, *97*, 1415–1427. [CrossRef]
7. Kuroda, C. The constitution of carthamin. *J. Chem. Soc. Jpn.* **1930**, *51*, 752–765. [CrossRef]
8. Seshadri, T.R.; Thakur, R.S. The colouring matter of the flowers of *Carthamus tinctorius*. *Curr. Sci.* **1960**, *29*, 54–55.
9. Obara, H.; Onodera, J.; Abe, S. The syntheses and properties of two analogs of Carthamin. *Chem. Lett.* **1974**, *3*, 335–338. [CrossRef]
10. Obara, H.; Onodera, J. Structure of Carthamin. *Chem. Lett.* **1979**, *8*, 201–204. [CrossRef]
11. Takahashi, Y.; Miyasaka, N.; Tasaka, S.; Miura, I.; Urano, S.; Ikura, M.; Hikichi, K.; Matsumoto, T.; Wada, M. Constitution of two coloring matters in the flower petals of *Carthamus tinctorius* L. *Tetrahedron Lett.* **1982**, *23*, 5163–5166. [CrossRef]
12. Hayashi, T.; Ohmori, K.; Suzuki, K. Synthetic study on Carthamin. 2. Stereoselective approach to C-glycosyl quinochalcone via desymmetrization. *Org. Lett.* **2017**, *19*, 866–869. [CrossRef] [PubMed]
13. Azami, K.; Hayashi, T.; Kusumi, T.; Ohmori, K.; Suzuki, K. Total synthesis of Carthamin, a traditional natural red pigment. *Angew. Chem. Int. Ed. Engl.* **2019**, *58*, 5321–5326. [CrossRef]
14. Abe, Y.; Sohtome, T.; Sato, S. Biomimetic synthesis of Carthamin, a red pigment in safflower petals, via oxidative decarboxylation. *J. Heterocycl. Chem.* **2020**, *1*, 3685–3690. [CrossRef]
15. Sicker, D.; Zeller, K.-P.; Siehl, H.-U.; Berger, S. *Natural Products*; Isolation, Structure Elucidation, History; Wiley-VCH Press: Weinheim, Germany, 2018; p. 131.
16. Kim, J.-B.; Cho, M.-H.; Hahn, T.-R.; Paik, Y.-S. Efficient purification and chemical structure identification of Carthamin from *Carthamus tinctorius*. *Agric. Chem. Biotechnol.* **1996**, *39*, 501–505.
17. Cho, M.H.; Hahn, T.R. Purification and characterization of precarthamin decarboxylase from the yellow petals of *Carthamus tinctorius* L. *Arch. Biochem. Biophys.* **2000**, *382*, 238–244. [CrossRef]
18. Yoshida, T.; Terasaka, K.; Kato, S.; Bai, F.; Sugimoto, N.; Akiyama, H.; Yamazaki, T.; Mizukami, H. Quantitative determination of Carthamin in Carthamus red by ^{1}H-nmr spectroscopy. *Chem. Pharm. Bull.* **2013**, *61*, 1264–1268. [CrossRef] [PubMed]
19. Yajima, H.; Sasaki, M.; Takahashi, K.; Hiraoka, K.; Oshima, M.; Yamada, K. Green metallic luster on the film of safflower red pigment extracted by a traditional method-approach with optical measurement. *J. Soc. Photogr. Jpn.* **2018**, *81*, 65–69.
20. Yajima, H.; Sasaki, M.; Takahashi, K.; Oshima, M.; Hiraoka, K.; Yashiro, M.; Yamada, K. Influence of photo-illumination on greenish metallic luster of safflower red pigment film. *Bull. Soc. Photogr. Imag. Jpn.* **2018**, *28*, 18–22.
21. Kaki, T.; Morii, T.; Sasaki, M.; Oshima, M.; Takahashi, K.; Hiraoka, K.; Yatsushiro, Y.; Yajima, H.; Yamada, K. Optical characteristics of metallic luster generated from non-metallic materials. *Acad. Rep. Fac. Eng. Tokyo Polytech. Univ.* **2020**, *43*, 6–9.
22. Feng, Z.M.; He, J.; Jiang, J.S.; Chen, Z.; Yang, Y.N.; Zhang, P.C. NMR solution structure study of the representative component hydroxysafflor yellow and other quinochalcone C-glycosides from *Carthamus tinctorius*. *J. Nat. Prod.* **2013**, *76*, 270–274. [CrossRef]
23. Yue, S.J.; Qu, C.; Zhang, P.X.; Tang, Y.P.; Jin, Y.; Jiang, J.S.; Yang, Y.N.; Zhang, P.C.; Duan, J.A. Carthorquinosides A and B, quinochalcone C-glycosides with diverse dimeric skeletons from *Carthamus tinctorius*. *J. Nat. Prod.* **2016**, *79*, 2644–2651. [CrossRef] [PubMed]
24. Takahashi, K.; Hamamura, K.; Sei, Y. Unique Nuclear magnetic resonance behaviour of γ-cyclodextrin in organic solvents. *J. Incl. Phenom. Macrocycl. Chem.* **2019**, *93*, 97–106. [CrossRef]

Article

Development and Validation of 2-Azaspiro [4,5] Decan-3-One (Impurity A) in Gabapentin Determination Method Using qNMR Spectroscopy

Nataliya E. Kuz'mina [1,*], Sergey V. Moiseev [1], Mikhail D. Khorolskiy [1,2] and Anna I. Lutceva [1]

1. "Scientific Centre for Expert Evaluation of Medicinal Products" of the Ministry of Health of the Russian Federation, Federal State Budgetary Institution, 8/2 Petrovsky Blvd, 127051 Moscow, Russia; MoiseevSV@expmed.ru (S.V.M.); mkhorolski@gmail.com (M.D.K.); Lutceva@expmed.ru (A.I.L.)
2. Department of Pharmaceutical and Toxicological Chemistry Named by A.P. Arzamastsev, I.M. Sechenov First Moscow State Medical University (Sechenov University), 8, Bldg. 2 St. Trubetskaya, 119991 Moscow, Russia
* Correspondence: kuzminaN@expmed.ru

Abstract: The authors developed a ^{1}H qNMR test procedure for identification and quantification of impurity A present in gabapentin active pharmaceutical ingredient (API) and gabapentin products. The validation studies helped to determine the limit of quantitation and assess linearity, accuracy, repeatability, intermediate precision, specificity, and robustness of the procedure. Spike-and-recovery assays were used to calculate standard deviations, coefficients of variation, confidence intervals, bias, Fisher's F test, and Student's t-test for assay results. The obtained statistical values satisfy the acceptance criteria for the validation parameters. The authors compared the results of impurity A quantification in gabapentin APIs and capsules by using the ^{1}H qNMR and HPLC test methods.

Keywords: gabapentin; impurity A; validation; limit of the quantitation; linearity; accuracy; repeatability; precision; specificity; robustness; qNMR; HPLC

1. Introduction

Gabapentin (2-[1-(aminomethyl) cyclohexyl] acetic acid) is a synthetic and non-benzodiazepine analogue of γ-aminobutyric acid. Gabapentin (GP) is usually used for epilepsy, symptoms of peripheral neuropathic pain, postherpetic neuralgia, diabetic peripheral neuropathy, acute alcohol withdrawal syndrome, and multiple sclerosis treatment [1–3]. Through intramolecular cyclization in solution, GP can form impurity A (ImpA)-2-azaspiro [4,5] decan-3-one, which is a European and American pharmacopoeias classification [4]. ImpA formation from crystalline GP rate depends on its polymorphic modification, temperature, moisture, shredding rate, and presence of some excipients [5–7]. Temperature and pH of medium can influence the GP cyclization. Since ImpA demonstrate some toxicity rate (LD_{50} = 300 mg/kg, white mice [5]), its content should be measured in GP drugs and substances. Identification and quantification of ImpA in GP during the pharmacopoeial analysis are carried out using the HPLC method [8]. HPLC is a highly sensitive and selective method, but the results of HPLC measurements are relative and indirect by nature. HPLC determination of ImpA requires generation of a calibration curve using a pharmacopoeial reference standard for ImpA (which accounts for the relative nature of measurements). The measurement by the HPLC method has a combined uncertainty (which accounts for the indirect nature of measurements). Sources of the total standard uncertainty are the peak area measurement in the chromatogram, the test and standard samples weighing, and solvent volumes measurement. Therefore, it would be practical to use an absolute and direct method, for example, qNMR for ImpA quantification. Absolute methods of quantitative analysis are based on known functional relationships and do not require the generation of a calibration curve using a reference standard. qNMR is considered as an absolute method for measuring the molar ratio of the analytes in a test sample, as well as the weight content of one component relative to another component, because the functional relationships between the analytes and the measurands (integrated intensities) are well-known: the molar ratio of the components in a mixture is equal to the ratio of the normalized integrated intensities of the signals of these components. qNMR quantification of an impurity relative to the main component is considered a direct method because of the direct measurement of the ratio of integrated intensities of the main component and impurity signals. Uncertainty of the test result relies only on the uncertainty of the integral intensities ratio measurement [9,10]. The aim of this article is to develop and validate an identification and quantification method of ImpA determination in GP drugs and APIs.

2. Results and Discussion

2.1. Specificity

GP and ImpA have a similar structure (Figure 1).

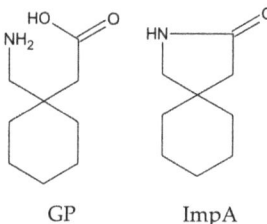

Figure 1. Chemical structures of Gabapentin (GP) and impurity A (impA).

Although the structures of GP and ImpA are similar, signal overlap on the ^{1}H spectrum can only observed be in the cyclohexane fragment range (1.25–1.70 ppm). Methylene group signals are differentiated: CH_2-N δ = 3.02 ppm (GP) and 3.24 ppm (ImpA); CH_2-C=O δ = 2.45 ppm (GP) and 2.28 ppm (ImpA). It should be noted that signals of the ImpA do not overlap with ^{13}C satellites of GP signals (Figure 2).

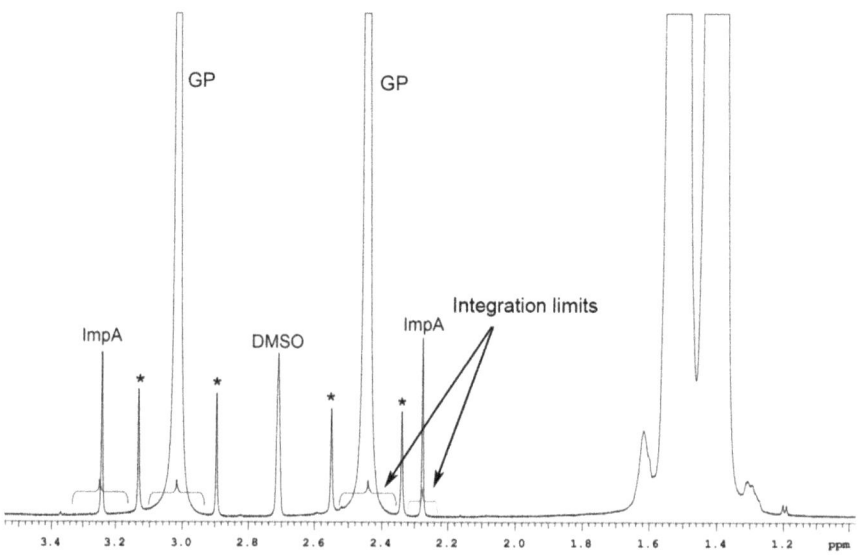

Figure 2. ^{1}H spectrum of GP and ImpA mixture (GP 40.09 mg/mL, ImpA 0.20 mg/mL). * ^{13}C satellites of GP signals.

GP drugs of different manufacturers have nonequal set of excipitents in their content. For example, capsules I (300 mg GP dose) have calcium hydrogen phosphate dihydrate, potato starch, magnesium stearate, and PEG in their content. The content of capsules II with the same active substance dose includes lactose monohydrate, corn starch, talc, and magnesium stearate. It should be noted that signals of water-soluble excipients lie outside of the methylene GP and ImpA protons range. They do not prevent ImpA identification and quantification, as can be seen in Figures 3 and 4.

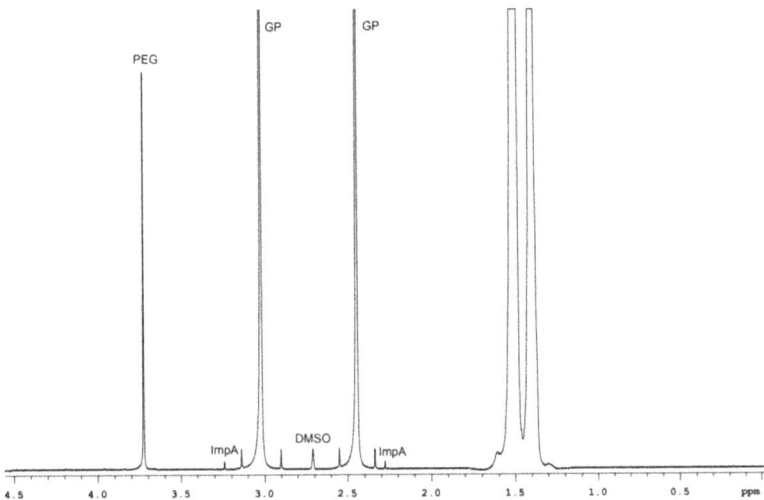

Figure 3. ^{1}H spectrum fragment of GP drug (capsules I).

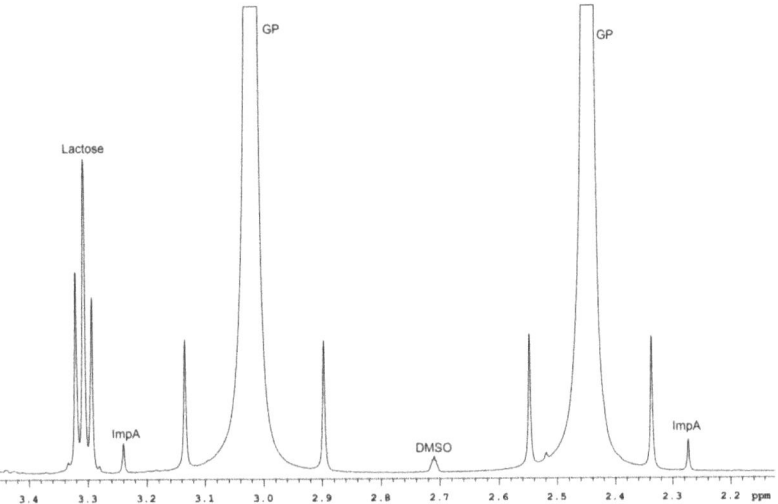

Figure 4. ^{1}H spectrum fragment of GP drug (capsules II).

In this way, signal 2.28 and 3.24 ppm are characteristic signals of ImpA. These signals help to identify the impurity presence in GP drugs and substances.

In the qNMR spectroscopy, the mass of the analyte A can be determined from a known mass of the analyte B [9]:

$$m_A = \frac{I_A}{I_B} \times \frac{N_B}{N_A} \times \frac{M_A}{M_B} \times m_B \times P \qquad (1)$$

where m_A and m_B are the mass of the analytes A and B;
M_A and M_B are the molar mass of the analytes A and B;
N_A and N_B are the number of nuclei generating the corresponding signal;
P—purity of the analyte B.

Therefore, the ImpA content in the test sample (m_{ImpA}) and its weight % (w %) relative to GP can be determined using the following formulas:

$$m_{ImpA} = 0.901(I_{ImpA}/I_{GP})\, m_{GP};\ w\ \% = 0.901(I_{ImpA}/I_{GP})100 \qquad (2)$$

where 0.901 is the relation of ImpA and GP molar masses;

I_{ImpA} is the integral intensity of any characteristic ImpA signals (2.28; 3.24 ppm) or their mean;
I_{GP} is the integral intensity of any characteristic GP signals (2.45; 3.02 ppm) or their mean;
m_{GP} is the GP content in the test sample.

2.2. Limit of Quantification

In the experimental conditions of qNMR, the limit of quantitation (LOQ) of ImpA is 10 µg/mL (0.025 weight % relative to GP content). At this concentration of ImpA, the signal-to-noise ratio is 10.3.

2.3. Linearity and the Analytical Range

The analytical range of the development method is 10–253 µg/mL or 0.025–0.63 weight % relative to GP. It corresponds to 0.25% from nominal content of ImpA in GP APIs (0.1 w %) and 158% from nominal content of ImpA in the GP drug (0.4 w %). In this range were identified validation characteristics of method. The integral intensity of signals δ 3.24 ppm (ImpA) and 3.02 ppm (GP) is used in quantification measurements. Results of linearity evaluation are shown in Table 1.

Table 1. Results of the linearity evaluating of the validated method.

Content of ImpA, µg/mL (w % Relative to GP)	I_{ImpA}	Mean Value I_{ImpA}	Content of ImpA, µg/mL (w % Relative to GP)	I_{ImpA}	Mean Value I_{ImpA}
0.0 (0.0)	0.00	0.00	50.65 (0.126)	1.40 1.41 1.41	1.41
10.13 (0.025)	0.28 0.29 0.27	0.28	101.30 (0.253)	2.74 2.69 2.71	2.71
20.26 (0.051)	0.57 0.57 0.58	0.57	151.95 (0.379)	4.33 4.35 4.31	4.33
30.39 (0.076)	0.84 0.86 0.84	0.85	202.60 (0.505)	5.68 5.70 5.74	5.71
40.52 (0.101)	1.14 1.13 1.14	1.14	253.25 (0.632)	7.03 7.05 7.01	7.03

Based on data shown in Table 1, we built a dependency graph of I_{ImpA} from ImpA content in model mixtures (Figure 5). Statistical characteristics of the established linear regression are shown in Table 2.

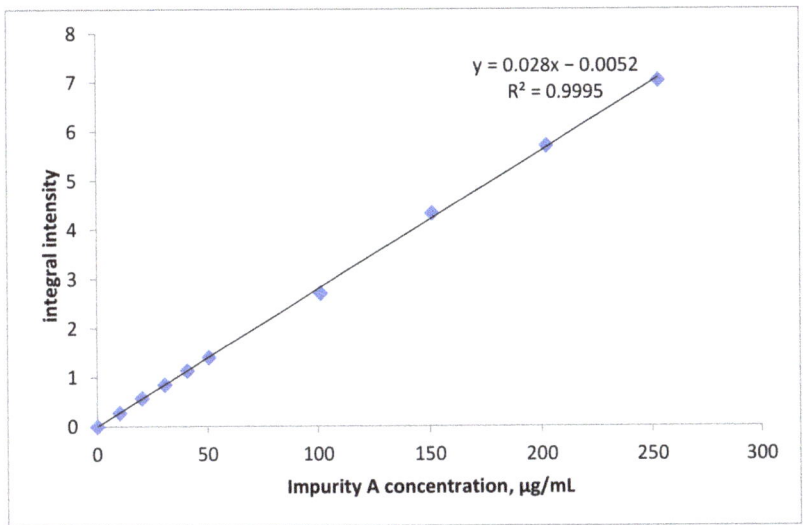

Figure 5. Dependence graph of ImpA measured integral signal intensity from its content in the sample.

Table 2. Statistical Characteristics of Linear Regression.

Statistical Characteristic	Result
Slope (b)	0.028
Segment on ordinate (a)	−0.0052
Significance interval (p = 95%)	−0.06 ÷ 0.05
Correlation coefficient (r)	0.9997

As follows from the data in Table 2, the procedure meets all linearity requirements: the correlation coefficient $r \geq 0.990$, and the value a does not exceed its confidence interval.

2.4. Accuracy

The results of accuracy evaluation are given in Table 3.

Table 3. Results of the Accuracy Studies of the method.

ImpA Added, µg/mL	ImpA Found, µg/mL	Z_i, %	ImpA Added, µg/mL	ImpA Found, µg/mL	Z_i, %
	10.11	99.80		98.94	97.67
10.13	10.47	103.36	101.30	97.13	95.88
	9.75	96.25		97.85	96.59
	20.58	101.58		156.35	102.90
20.26	20.58	101.58	151.95	157.07	103.37
	20.94	103.36		155.63	102.42
	30.33	99.80		205.09	101.23
30.39	31.05	102.17	202.60	205.82	101.59
	30.33	99.80		207.26	102.30
	41.16	101.58		253.84	100.23
40.52	40.80	100.69	253.25	254.56	100.52
	41.16	101.58		253.12	99.95
	50.55	99.80			
50.65	50.91	100.51			
	50.91	100.51			
Mean ($\bar{Z}$), %				100.63	
Systematic error (δ), %				0.63	
Standard deviation (s), %				2.067	
Coefficient of variation (R.S.D.), %				1.86	
Significant interval (Δ), %				±0.82	

The following acceptance criteria are used in method validation:
1. The systemic fault must not exceed its confidence interval (criteria of statistical insignificance);
2. The confidence interval must include 100% of the extraction coefficient value.

It follows from the data in Table 3 that the procedure being validated has acceptable accuracy, as 100% is included in the confidence interval, and the bias is statistically indistinguishable from zero ($0.63 \leq 0.82$).

2.5. Repeatability and Intra-laboratory Precision

Results of relation spiked-and-recovery (Z_i) measurements obtained in repeatability and intra-laboratory precision conditions and their statistical processing are presented in Table 4.

Table 4. Results of convergence and intralaboratory precision studies of the method being validated.

ImpA Added, µg/mL	Operator 1		Operator 2	
	Found, µg/mL	Z_i, %	Found, µg/mL	Z_i, %
10.13	10.11	99.80	10.11	99.80
	10.47	103.36	10.11	99.80
	9.75	96.25	9.39	92.69
50.65	50.55	99.80	51.27	101.22
	50.91	100.51	50.19	99.09
	50.91	100.51	50.91	100.51
253.25	253.84	100.23	253.48	100.09
	254.56	100.52	253.12	99.95
	253.12	99.95	253.84	100.23
Mean ($\overline{Z}_i$), %		100,103		99.264
Systematic error (δ), %		0.103		0.736
Standard deviation (s), %		1.809		2.532
Coefficient of variation (R.S.D.), %		1.807		2.551
Significant interval, % (Δ, p = 95%)		±1.391		±1.946
Combined mean ($\overline{Z}$), %				99.684
Combined standard deviation, %				2.20
Combined coefficient of variation, %				2.207
Combined significant interval, %				±1.555
Fisher's F test (F_{tab} = 3.44)				F_{fact} = 0.51
Student's t test (t_{tab} = 2.12)				t_{fact} = 0.81

Acceptance of intra-laboratory precision can be evaluated by the Fisher (F) and Student (t) statistical criteria by counting and comparing their actual t and F values with table values-maximal values of criteria with influence of random factors, current degrees of freedom, and given levels of significance. The data presented in Table 4, show a statistical insignificance of difference between means and standard deviations of two operators measures at a significance level of 95%, so the table F and t values substantially exceed their actual values.

2.6. Robustness

The study provided experimental evidence of the insensitivity of the procedure being validated to minor changes in the qNMR test conditions. Varying of the pulse angle, relaxation time, sample temperature, and addition of excipients do not change the position of the chemical shifts of GP and ImpA signals. The integral intensity ratio of GP and ImpA remains unchanged.

2.7. Comparative Analysis of the ImpA Content Determining by qNMR and HPLC

Results of the ImpA content in GP APIs and capsules determination, using qNMR and HPLC methods, are shown in Table 5.

Table 5. Results of the ImpA content in GP APIs and capsules determination.

Sample	Content of ImpA, w %	
	NMR	HPLC
API A	Not found	Not found
API B	BQL	BQL
Capsule I	0.10 (RSD 5.6%)	0.13 (RSD 4.9%)
Capsule II	0.08 (RSD 7.5%)	0.07 (RSD 7.2%)

It should be noted that content of ImpA in API test samples is lower than the LOQ of qNMR methods (0.025 w %) and HPLC methods (0.5 w %). The content level of lactam in capsules is more than the LOQ. ImpA content values in GP drugs, obtained by qNMR and HPLC methods, are close to each other, which is additional evidence of the accuracy of the validation method.

3. Materials and Methods

3.1. Materials

The following materials were used in the qNMR procedure development and validation: certified reference standards for GP and ImpA, manufactured by the European Pharmacopoeia (the assigned values of reference standards are 100%, the uncertainty of the assigned values is not stated), gabapentin APIs by Divis Laboratories Limited Hyderabad, India (A), gabapentin APIs by PIQ-PHARMA, Belgorod, Russia (B), gabapentin capsules by Canonpharma production PJSC, Moscow, Russia (I), and gabapentin capsules by Pharmstandard-Leksredstva JSC, Volginskiy, Russia (II). Deuterated dimethylsulfoxide (DMSO-D6, 99.90% D) and water (99.93% D) by Cambridge Isotope Laboratories, Inc. (St. Louis, MO, USA) were used in the NMR experiments.

HPLC measurements were carried out using ammonium dihydrogen phosphate, phosphoric acid, ACS grade perchloric acid, and sodium perchlorate (Sigma-Aldrich, St. Louis, MO, USA). ACS grade potassium dihydrogen phosphate was purchased from JT Baker (Philipsburg, NJ, USA). Methanol and HPLC grade acetonitrile were purchased from Fisher Scientific (Fairlawn, NJ, USA). HPLC ready 18 MΩ water was obtained, in-house, from a Milli-Q Integral 3 water purification system, Merck Millipore Corp. (Burlington, MA, USA). Syringe filters were used with PTFE membranes of 0.45 μm from Thermo Scientific Nalgene (Rochester, NY, USA).

3.2. NMR Spectroscopy Method

3.2.1. Model Solutions

GP stock solution I of 100 mg/mL was prepared by placing 501.1 mg of GP reference standard in a 5-mL flask and diluting with D_2O to volume. ImpA stock solution II (c = 2.026 mg/mL) was prepared by placing 10.13 mg of ImpA reference standard in a 5-mL flask and diluting with D_2O to volume. Solution III (c = 506.5 μg/mL) was obtained by fourfold dilution of Solution II with D_2O. Model solutions of GP and ImpA mixtures were prepared by combining different volumes of Solutions I and III and different volumes of solvents (Table 6). Trace amounts of DMSO-D6 were added as internal standards for the chemical shift scale calibration.

3.2.2. Sample Preparation

API: 20 mg of substance (an accurate amount is optional) was placed into an NMR flask, followed by adding 0.5 mL of D_2O and 10 μL of DMSO-D6, and shaking intensively to obtain a fully diluted sample.

Capsules: 1.5 mL of D_2O was added to 1/2 of capsule content (200 mg, accurate amount is optional) and shaken intensively within 10 min. We obtained suspension filtered using a membrane filter, put 0.5 mL of filtrate in an NMR flask, and added 10 μL of DMSO-D6.

Table 6. Preparation of model solutions of GP and ImpA mixtures.

№	V I mL	V III µL	V DMSO-D6 µL	V D$_2$O µL	C GP mg/mL	C ImpA µg/mL	w % ImpA Relative to GP
1	0.4	0	10	590	40.09	0	0
2	0.4	20	10	570	40.09	10.13	0.025
3	0.4	40	10	550	40.09	20.26	0.051
4	0.4	60	10	530	40.09	30.39	0.076
5	0.4	80	10	510	40.09	40.52	0.101
6	0.4	100	10	490	40.09	50.65	0.126
7	0.4	200	10	390	40.09	101.30	0.253
8	0.4	300	10	290	40.09	151.95	0.379
9	0.4	400	10	190	40.09	202.60	0.505
10	0.4	500	10	90	40.09	253.25	0.632

3.2.3. Instrumentation and Experiments Conditions

^{1}H spectra were collected on the Agilent DD2 NMR System 600 NMR spectrometer equipped with a 5 mm broadband probe and a gradient coil (VNMRJ 4.2 software). Parameters of the experiments: temperature 27 °C, spectral width 6009.6 Hz, observe pulse 90°, acquisition time 5.325 s, relaxation delay 10 s, number of scans 256, the number of analog-to-digital conversion points 64 K, exponential multiplication 0.3 Hz, zero filling 64 K, automatic linear correction of the baseline of the spectrum, manual phase adjustment, calibration of the δ scale under DMSO in D$_2$O (δ = 2.71 ppm) [11]. The manual mode was also used for the signal integration. The general rule for choosing the integration limit (64 time the half-with of a Lorentzian shape NMR signal) was not followed due to the GP ^{13}C satellites interfering effect. We took as the integration limit for ImpA the doubled distance between the center of its signal and GP ^{13}C satellites. The integration limit for GP signal was equal to the distance between the ^{13}C satellites signals (without the ^{13}C satellites). The relaxation delay value was estimated by an inversion-recovery experiment: T1 are equal 1.54 s (ImpA) and 0.89 s (GP). It was found that the experiment conditions did not affect the stability of GP: additional signals of ImpA were not detected in the spectrum of GP stock solution I.

3.3. Method Validation

Three independent experiments were run for each model solution and three values were obtained for the integral intensity of the signal. For validation, the mean value was used. Validation characteristics (specificity, linearity, accuracy, precision, limit of quantitation, range, and robustness) and validation criteria carried out according to methodological documents of GMP and US Pharmacopoeia guidance about validation of analytical procedure by qNMR [12,13]. Statistical parameters (mean value, standard deviation, coefficient of variation, significance interval, coefficient of determination, and actual and tabulated values for Fisher's F test and Student's t test) were determined at a significance level of $p = 0.05$ using MS Excel 2007.

3.3.1. Specificity

Specificity was confirmed by demonstrating absence of overlap of individual GP and ImpA signals in the ^{1}H spectrum.

3.3.2. Limit of Quantitation

The limit of quantitation (LOQ) of the procedure was determined from the signal-to-noise ratio (S/N = 10) using VNMRJ software, version 4.2.

3.3.3. Analytical Range

Range of the application method was determined by experimental value LOQ and the US Pharmacopoeia recommendation to nominal content of ImpA in GP APIs (0.1 w %) and in its marketed products (0.4 *w/w* %) [4,14,15].

3.3.4. Linearity

Graphic dependence of the integral intensity of signal ImpA versus its concentration was treated by linear least square regression analysis with 10 model solutions over a concentration range of 0–253 µg/mL for ImpA.

3.3.5. The Accuracy

The accuracy of the method was evaluated using the experimental data obtained from the linearity studies. Extraction coefficients were determined for all model solutions, i.e., the spiked-recovery ratio (Z_i), for which the systematic error (δ), standard deviation (s), coefficient of variation (RSD), and significance interval (Δ) were determined.

3.3.6. Precision

Precision was evaluated at the level of convergence and intralaboratory precision (different operators, different days). Convergence and intra-laboratory precision of the method under validation was evaluated using three model solutions with low (10 µg/mL), intermediate (100 µg/mL), and high (250 µg/mL) ImpA contents.

3.3.7. Robustness

The reliability of an analytical measurement was evaluated by analyzing of the result stability after varying observe pulse (45 and 90°), relaxation delay (±10%), probe temperature (±2 °C), possible interfering species—water soluble excipients from marketed products (polyethylenglycol 6000).

3.4. Reference Measurement with HPLC Method

3.4.1. Preparation of Solution

Diluent, a system suitability test solution, buffer solution, teste solution of samples A–D, reference solutions, and mobile phase were prepared according to USP methods [4,14].

3.4.2. Instrumentation and Chromatographic Conditions

The HPLC system consists of an Agilent Infinity 1260 series (Agilent Technologies, Wilmington, DE, USA). Data collection and analysis were performed using ChemStation software. Chromatographic conditions: column Zorbax RX-C-18 250 mm × 4.6 mm × 5 µm (Agilent Technologies, Santa-Clara, CA, USA); column temperature 40 °C; elution mode isocratic; flow rate 1 mL/min; detector UV 215 nm; injection volume 20 µL; Run time no less than 50 min.

4. Conclusions

The developed ^{1}H qNMR spectroscopy method of ImpA identification and quantification in GP APIs and GP drugs were validated by using the main parameters. It was established that the developed method is specific and has an acceptable linearity, repeatability, accuracy, precision, and robustness. Also, a limit of quantitation of developed method was established. This method can be used for carrying out GP APIs and drugs analysis.

Author Contributions: Conceptualization, N.E.K. and S.V.M.; Data curation, N.E.K. and S.V.M.; Formal analysis, N.E.K., S.V.M., and M.D.K.; Investigation, N.E.K. and S.V.M.; Methodology, N.E.K. and S.V.M.; Project administration, A.I.L.; Supervision, N.E.K. and A.I.L.; Visualization, S.V.M. and M.D.K.; Writing—original draft, N.E.K.; Writing—review & editing, S.V.M., M.D.K., and A.I.L. All authors have read and agreed to the published version of the manuscript.

Funding: This research was funded by Ministry of Health of Russia [research project No. 056-00005-21-00, R&D public accounting No. 121021800098-4]. The APC was funded by the Scientific Centre for Expert Evaluation of Medicinal Products.

Data Availability Statement: The data that support the findings of this study are available from the corresponding author upon reasonable request.

Conflicts of Interest: The authors declare no conflict of interest. The funders had no role in the design of the study; in the collection, analyses, or interpretation of data; in the writing of the manuscript, or in the decision to publish the results.

Sample Availability: Samples of gabapentin APIs, gabapentin capsules, reference standards for gabapentin and Impurity A are available from the authors.

References

1. Celikyurt, I.K.; Mutlu, O. Gabapentin, A GABA analogue, enhances cognitive performance in mice. *Neurosci. Lett.* **2011**, *492*, 124–128. [CrossRef] [PubMed]
2. Hiom, S.; Patel, G.K. Severe postherpetic neuralgia and other neuropathic pain syndromes alleviated by topical gabapentin. *Br. J. Dermatol.* **2015**, *173*, 300–302. [CrossRef] [PubMed]
3. Burns, M.L.; Kinge, E. Pharmacokinetic variability, clinical use and therapeutic drug monitoring of antiepileptic drugs in special patient groups. *Acta Neurol. Scand.* **2019**, *139*, 446–454.
4. USP43-NF38. Gabapentin. 2057. Available online: https://online.uspnf.com/ (accessed on 8 November 2020).
5. Zong, Z.; Qiu, J. Kinetic model for solid-state degradation of gabapentin. *J. Pharm. Sci.* **2012**, *101*, 2123–2133. [CrossRef] [PubMed]
6. Ranjous, Y.; Hsian, J. Improvement in the Physical and Chemical Stability of Gabapentin by Using Different Excipients. *Int. J. Pharm. Sci. Rev. Res.* **2013**, *23*, 81–86.
7. Braga, D.; Grepioni, F. Polymorphic gabapentin: Thermal behaviour, reactivity and interconversion of forms in solution and solid-state. *New J. Chem.* **2008**, *32*, 1788–1795. [CrossRef]
8. Ciavarella, A.B.; Gupta, A. Development and application of a validated HPLC method for the determination of gabapentin and its major degradation impurity in drug products. *J. Pharm. Biomed.* **2007**, *43*, 1647–1653. [CrossRef] [PubMed]
9. Malz, F.; Jancke, H. Validation of quantitative nuclear magnetic resonance. *J. Pharm. Biomed.* **2005**, *38*, 813–823. [CrossRef] [PubMed]
10. Kuz'mina, N.E.; Moiseev, S.V. Possibilities of NMR spectrometry in determining trace components of mixtures. *Russ. J. Anal. Chem.* **2014**, *69*, 1052–1060. [CrossRef]
11. Gottlieb, H.E.; Kotlyar, V. NMR Chemical Shifts of Common Laboratory Solvents as Trace Impurities. *J. Org. Chem.* **1997**, *62*, 7512–7515. [CrossRef] [PubMed]
12. WHO. *A WHO Guide to Good Manufacturing Practice (GMP) Requirements, Part 2*; Validation, WHO/VSQ/97.02; WHO: Geneva, Switzerland, 1999; Available online: https://apps.who.int/iris/bitstream/handle/10665/64465/WHO_VSQ_97.02.pdf;sequence=2 (accessed on 8 November 2020).
13. USP43-NF38. <761> Nuclear Magnetic Resonance. 6984. Available online: https://online.uspnf.com/ (accessed on 8 November 2020).
14. USP43-NF38. Gabapentin Capsules. 2058. Available online: https://online.uspnf.com/ (accessed on 8 November 2020).
15. USP43-NF38. Gabapentin Tablets. 2060. Available online: https://online.uspnf.com/ (accessed on 8 November 2020).

Article
Imaging Sequences for Hyperpolarized Solids

Xudong Lv [1], Jeffrey Walton [2], Emanuel Druga [1], Raffi Nazaryan [1], Haiyan Mao [3], Alexander Pines [1], Ashok Ajoy [1] and Jeffrey Reimer [3,4,*]

1. Department of Chemistry, University of California, Berkeley, CA 94720, USA; david.lv@berkeley.edu (X.L.); epieon@berkeley.edu (E.D.); rnazaryan@berkeley.edu (R.N.); pines@berkeley.edu (A.P.); ashokaj@berkeley.edu (A.A.)
2. Nuclear Magnetic Resonance Facility, University of California Davis, Davis, CA 95616, USA; jhwalton@ucdavis.edu
3. Department of Chemical and Biomolecular Engineering, University of California, Berkeley, CA 94720, USA; maohaiyan@berkeley.edu
4. Lawrence Berkeley National Laboratory, Materials Science Division, University of California, Berkeley, CA 94720, USA
* Correspondence: reimer@berkeley.edu

Abstract: Hyperpolarization is one of the approaches to enhance Nuclear Magnetic Resonance (NMR) and Magnetic Resonance Imaging (MRI) signal by increasing the population difference between the nuclear spin states. Imaging hyperpolarized solids opens up extensive possibilities, yet is challenging to perform. The highly populated state is normally not replenishable to the initial polarization level by spin-lattice relaxation, which regular MRI sequences rely on. This makes it necessary to carefully "budget" the polarization to optimize the image quality. In this paper, we present a theoretical framework to address such challenge under the assumption of either variable flip angles or a constant flip angle. In addition, we analyze the gradient arrangement to perform fast imaging to overcome intrinsic short decoherence in solids. Hyperpolarized diamonds imaging is demonstrated as a prototypical platform to test the theory.

Keywords: hyperpolarization; magnetic resonance imaging; flip angle

1. Introduction

NMR is central to many chemical, biological and material analysis due to the rich chemical information it can provide [1,2]. MRI, as the imaging counter part of NMR, is a powerful tool in medicine and biology [3,4]. However, the sensitivity of both techniques relies on nuclear spin polarization, which is intrinsically low at thermal equilibrium. One compelling approach to tackle this insensitivity is hyperpolarization. This approach brings the nuclear spin polarization level beyond thermal equilibrium to produce many orders of magnitude higher signal. Routes to hyperpolarization includes dynamic nuclear polarization (DNP) [5], parahydrogen induced hyperpolarization (PHIP) [6], as well as chemically-induced DNP (CIDNP) [7]. While the methods of hyperpolarization can be applied in both liquids and solids, hyperpolarized solids are particularly attractive as an imaging agent in nano-medicine [8], or as a polarization hub to deliver hyperpolarization for general chemicals [9]. However, challenges remain on how to image hyperpolarized solids given the none-replenishable nature of the polarization and short coherence times of solids.

In the work, we use diamond particles (Figure 1A) as a prototypical platform to test the imaging sequences (Figure 1C,D) and to provide some theoretical understanding of the results as well as some insight into sequence design for imaging similar hyperpolarized materials. The hyperpolarization in diamond is enabled by one type of special atom-like defect — the Nitrogen Vacancy (NV) center [10] and a recently developed protocol [9,11]. The electronic spins of NV centers are optically polarizable to ≈99% at room temperature [10], and their long coherence time ensures its efficiency at polarizing surrounding ^{13}C nuclear spins via chirped MW (Figure 1B,C). ^{13}C imaging of natural abundance diamond powders (Figure 1F) is only possible with such highly polarized signal (Figure 1E). The ability to image micron/nanodiamonds through MRI can open up possibilities in directions including physics, chemical and biological analysis. For instance, hyperpolarized diamond particles that "light up" in MRI mode can potentially be applied as a targeting and tracking agent given their bio-compatibility and surface modifiability [11–14]. Additional advantage of high surface-to-volume ratio can also enable polarization transfer to external nuclei when brought into close contact with other chemicals for high-SNR and high-resolution NMR [15].

The analysis of the imaging sequence for a diamond prototypical system relies on a theoretical framework we develop herein for imaging hyperpolarized solids in general. The theoretical framework considers two major components of an MRI sequence — flip angle and gradients (Figure 1C), which determine the quality of an MR image.

In an MRI sequence, a radio frequency pulse is normally applied at the beginning of each repetition, in order to rotate the magnetization from z direction to the xy plane, so that the nuclear Larmor precession can be detected. The angle of such rotation is referred as flip angle. In conventional MRI without hyperpolarization, the z magnetization can be recovered after each repetition by the T_1 relaxation process. In contrast, for the cases of hyperpolarization, the initial polarization is much higher beyond the equilibrium state; thus, relaxation tends to reduce it towards a much lower level. As a results, some sequence design principles in conventional MRI no long hold in such cases, and it requires careful engineering of flip angles to be suited for imaging hyperpolarized objects. The high level of the magnetization, if effectively distributed, can enhance the image SNR and resolution by orders of magnitude.

Not only does flip angle have to be designed uniquely for hyperpolarized solid state imaging, better arrangement of the gradient and pulses are critical as well. As a result of the nature of solids, static coupling between nuclei leads to short coherence times. This suggests that one has to either perform imaging rapidly or apply pulse sequences to protect coherence. We present strategies that either facilitate fast imaging or refocus signals by decoupling sequences with a focus of ^{13}C MRI in diamonds.

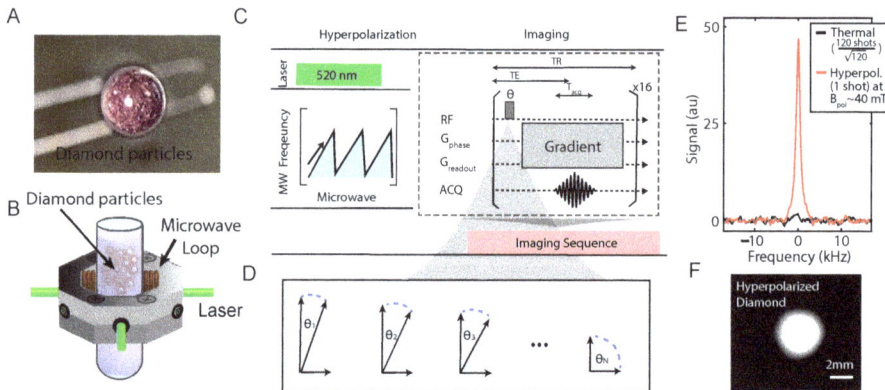

Figure 1. Experiment schematic. (**A**) A picture of diamond particles (∼200 μm in size) contained in an NMR tube as an imaging phantom (taken from the bottom of a NMR tube). (**B**) Green laser excitation and MW irradiation is applied on the sample in order to transferred polarization to lattice ^{13}C nuclei from optically polarized NV- electrons in the microscopic scale. (**C**) Experimental protocol of hyperpolarizing and imaging diamonds. ^{13}C hyperpolarization occurs at 38 mT under MW sweeps across the NV-ESR spectrum, and then transferred to a MRI machine for imaging. Flip angles and gradient arrangement determine the quality of the MRI. (**D**) Illustration of flip angles for the nth repetitions. (**E**) Typical signal enhancement by hyperpolarization, showing signal gain against signal at 7 T. For a fair comparison, the noise in both is normalized to be 1 (dash line). (**F**) A typical MR image of diamond phantom in (**A**).

2. Results

2.1. Image Equation

We analyze the dynamic of the magnetization change under certain flip angle pulses and theoretically present optimal solutions. In this section, we consider two major scenarios, i.e. dynamically changeable flip angles and a constant flip angle over different repetitions. We also consider two metrics for our optimization – total magnetization, which corresponds to total signal of the image, and the uniformity of the signal across repetitions.

More explicitly, we write down the signal equation of an MR image in terms of the xy plane magnetization M_x [16]:

$$S(k_x, k_y) = \iint M_x(x, y, k_x, k_y) e^{-i2\pi(k_x x + k_y y)} e^{-t(k_x,k_y)/T_2^*} dx dy \tag{1}$$

where $k_x = \gamma/2\pi \int_0^t G_x(t) dt$, $k_y = \gamma/2\pi \int_0^t G_y(t) dt$ (γ is the gyromagnetic ratio, and G_x, G_y are gradients along x and y axis). Note that this signal equation takes into consideration that the transverse magnetization M_x as a function of k_x and k_y can be different for each repetition. This dependence can be expressed as:

$$M_x(x, y, k_x, k_y) = K(k_x, k_y) \cdot \rho(x, y) \tag{2}$$

where $K(k_x, k_y)$ (we refer as *magnetization factor*) is the factor representing non-uniform excitation in each repetitions (for instance a progression of small tip angle pulses), and $\rho(x, y)$ is the nuclear spin density at location (x, y). Performing a Fourier transform of $S(k_x, k_y)$, we obtain the image represented in the real space:

$$\mathcal{I}(x, y) = \mathcal{F}(K) * \mathcal{F}(e^{-t(k_x,k_y)/T_2^*}) * \rho(x, y) \tag{3}$$

where $\mathcal{F}$ represents Fourier transformation and $*$ represents convolution. Note that by taking the limit of $t \ll T_2^*$, and assuming uniform excitation cross different repetition, the Equation (3) reduces to $\mathcal{I}(x, y) = \rho(x, y)$.

The image equation (Equation (3)) is different from a typical image equation as the first term represents the effect of flip angles, which is special to the case of hyperpolarization. In repetition n, we denote this effect to be K_n. In the case of Cartesian sampling, we can write $n = k_x$ without losing generality.

2.2. Flip Angle Consideration

How does K_n depend on the flip angle θ? We address this question by considering two cases: dynamically changing flip angles and a constant flip angle. In practice, whether one has the ability to program the flip angle for each repetition on the MRI machine determines which case will be utilized.

Variable flip angle — First we consider the most general scenario where one has control on the flip angle of each repetition. This stems from an intuitive demand that magnetization remains same in each repetition, similar to the magnetization in saturation recovery sequences. More specifically, if we implement an imaging sequence with a repetition time TR to a nuclear spin system with relaxation time T_1 and an equilibrium magnetization M_0, we can write the dynamic equation as following [17]:

$$\begin{cases} M_n = (M_{n-1} \cos \theta_{n-1} - M_0) e^{-\frac{TR}{T_1}} + M_0 \\ M_{x,n} = M_n \sin \theta_n \end{cases} \tag{4}$$

where we denote in the nth repetition, the flip angle to be θ_n, the longitudinal and transverse magnetization to be M_n and $M_{x,n}$ respectively. Given that magnetization can be written as multiplication of the magnetization factor and the spin density: $M_n = K_n * \rho(x, y)$, $M_{x,n} = K_{x,n} * \rho(x, y)$, and $M_0 = K_0 * \rho(x, y)$, we can eliminate the location information in $\rho(x, y)$, and simplify the dynamic equation in terms of magnetization factor K.

$$\begin{cases} K_n = (K_{n-1} \cos\theta_{n-1} - K_0)e^{-\frac{TR}{T_1}} + K_0 \\ K_{x,n} = K_n \sin\theta_n \end{cases} \quad (5)$$

The initial magnetization factor in the hyperpolarization case is K_{hp}, in contrast to the thermal polarization case K_0. With such initial condition, we solve the recurrent dynamic equation (Equation (5)) and obtain:

$$K_{x,n} = K_{hp} \left[\prod_{k=1}^{n-1}(\Gamma \cos\theta_k) + \frac{1}{K_{hp}/K_0}(1-\Gamma) \times \left\{ \sum_{i=2}^{n} \prod_{k=i}^{n-1}(\Gamma \cos\theta_k) \right\} \right] \sin\theta_n \quad (6)$$

where $\Gamma = e^{-\frac{TR}{T_1}}$. We assume hyperpolarization enhances signal much higher than thermal signal, suggesting $K_{hp} \gg K_0$. With such approximation, we have the leading order $K_{x,n} = K_{hp} \left[\prod_{k=1}^{n-1}(\Gamma \cos\theta_k) \right] \sin\theta_n$.

One of the advantages of having ability to dynamically varying the flip angle is that the transverse magnetization M_x in each repetition can be constant by carefully design the flip angles. This allows one to avoid image distortion along the phase encoding direction (further detailed in the Discussion section). Applying the condition of $K_{x,n} = $ constant, we can obtain (see Appendix A.1):

$$\tan^2\theta_n = (1-\Gamma^3) \cdot \frac{\Gamma^{2N-2n-1}}{1-\Gamma^{2N-2n-1}} \quad (7)$$

where N is the total number of repetitions.

As shown in Figure 2A, flip angles have to increase with the number of repetitions in order to maintain same transverse magnetization, and all three curves with different TR/T1 converge to 90° to saturate all the magnetization. As a result, the relative transverse magnetization stays flat throughout the scan confirmed by simulation (see Figure 2B). Such uniform magnetization factor allows $K(k_x, k_y)$ to be constant, resulting in $\mathcal{F}(K)$ to be a delta function, and the reconstructed image $\mathcal{I}(x,y)$ in Equation (3) to be: $\mathcal{I}(x,y) \propto \mathcal{F}(e^{-t(k_x,k_y)/T_2^*}) * \rho(x,y)$, immune from image blur cased by excitation.

In addition to constant magnetization, one desires to gain as large cumulative signal as possible, which leads to a different optimization problem.

$$\arg\max_{\theta_n} \left\{ S_{\text{cumulative}} = \sum_{n=1}^{N} M_{x,n} \right\} \quad (8)$$

If θ_N is optimal, it should satisfy: $\frac{\partial S_{\text{cumulative}}}{\partial \theta_N} = \Gamma^{N-1} \cos\theta_1 \cdots \cos\theta_{N-1} \cos\theta_N = 0$. Similarly, we can get:

$$\begin{aligned} \frac{\partial S_{\text{cumulative}}}{\partial \theta_{N-1}} &= \Gamma^{N-1}\cos\theta_1 \cdots (-\sin\theta_{N-1})\sin\theta_N + \Gamma^{N-2}\cos\theta_1 \cdots \cos\theta_{N-1} = 0 \\ &\Rightarrow \Gamma \sin\theta_{N-1}\sin\theta_N = \cos\theta_{N-1} \end{aligned} \quad (9)$$

In general, the relationship between two consecutive flip angles is: $\sin\theta_{n+1} = \Gamma \tan\theta_n$. Iteratively solving this sequence from the end where $\sin\theta_N = 1$ (see Appendix A.2), we have:

$$\theta_n = \tan^{-1}\sqrt{\frac{1}{\Gamma^2} \cdot \frac{1-\Gamma^2}{1-\Gamma^{2(N-n)}}} \quad (10)$$

And such results of $N = 16$ and 32 are presented in Figure 2C,D.

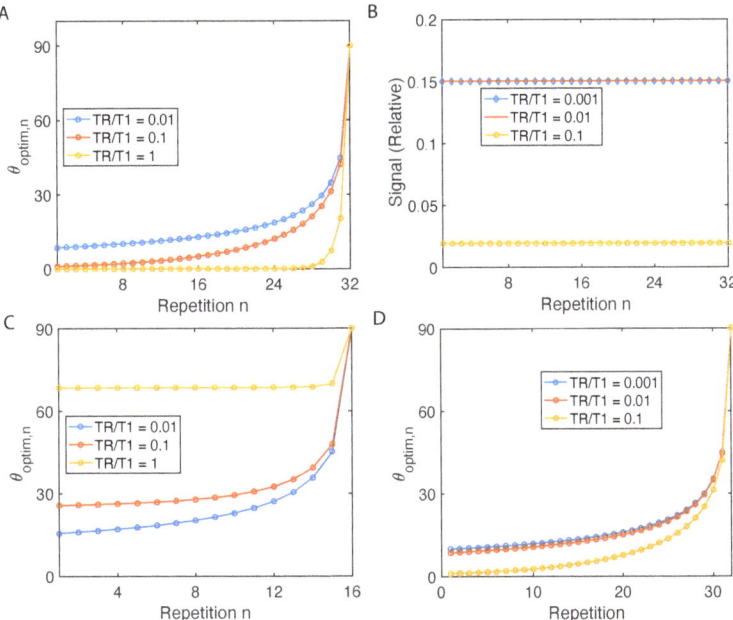

Figure 2. Variable flip angles for constant signal and maximum signal. (**A**) The flip angles to enable constant magnetization in a 32 repetition imaging sequence are determined based on Equation (7). (**B**) Implementing flip angles in (**A**), the relative transverse magnetization signal is simulated taking $K_{hp} = 1$ and $K_0 = 10^{-3}$. (**C**,**D**) The flip angles to maximize the cumulative signal under different TR/T1.

So far, we have derived the design principle of variable flip angle pulses to achieve either constant magnetization or maximum total magnetization. We here briefly comment on images we may acquire in these two cases. In the case where there is a fixed transverse magnetization in each repetition to start with, the image may display less SNR than the total signal optimized case. However, the constant signal guarantees high fidelity due to eliminated distortion in the phase encoding dimension. In contrast, in the case of maximum total magnetization, image distortion cannot be avoided but the image SNR is optimal.

Constant flip angle — In spite of the stable magnetization and high cumulative signal that is brought by variable flip angles, it posts technical challenges on MRI facilities to implement different flip angles in each repetition. A more widely used case is the constant flip angle, where the excitation pulses remain the same for all of the repetitions. We consider such case in this section and optimize the cumulative signal under such scenario.

The recurrent dynamic equation is similar despite the fact that θ is constant:

$$\begin{cases} K_n = (K_{n-1}\cos\theta - K_0)e^{-\frac{TR}{T_1}} + K_0 \\ K_{x,n} = K_n \sin\theta \end{cases} \quad (11)$$

Solving the recurrent dynamic equation:

$$K_{x,n} = \pm K_{hp}(\Gamma\cos\theta)^{n-1}\sin\theta + K_0(1-\Gamma)\sum_{k=1}^{n-1}(\Gamma\cos\theta)^{k-1}\sin\theta \quad (12)$$

Simulating this process, we observe the change of the magnetization with respect to n given a certain θ and TR/T1 in Figure A1. Note that in this case, we do not ignore the first term in Equation (12) because constant flip angle can lead to comparable magnitude of the first term with the second term.

Similarly, we calculate cumulative signal: $S_{\text{cumulative}} = \sum_{n=1}^{N} M_{x,n}$ in Figure 3. Not surprisingly, there is an optimal flip angle given certain TR/T1 and total number of scans N. Under such optimal angle, the case of $N = 32$ displays a more than 4 times higher cumulative signal than 90° pulse could (Figure 3A). When TR/T1 is less, increasing scan counts may become very effective for signal enhancement (Figure 3B). We compare this optimal flip angle with Ernst angle which is the flip angle for excitation of a particular spin that gives the maximal signal intensity in the least amount of time in the thermal polarization cases. We find that the optimal flip angle deviates from Ernst angle, however, approaching it when N increases.

It is difficult to optimize $S_{\text{cumulative}}$ analytically, and we use a gradient descent method to numerically solve the problem, and the result is shown in Figure 3C,D.

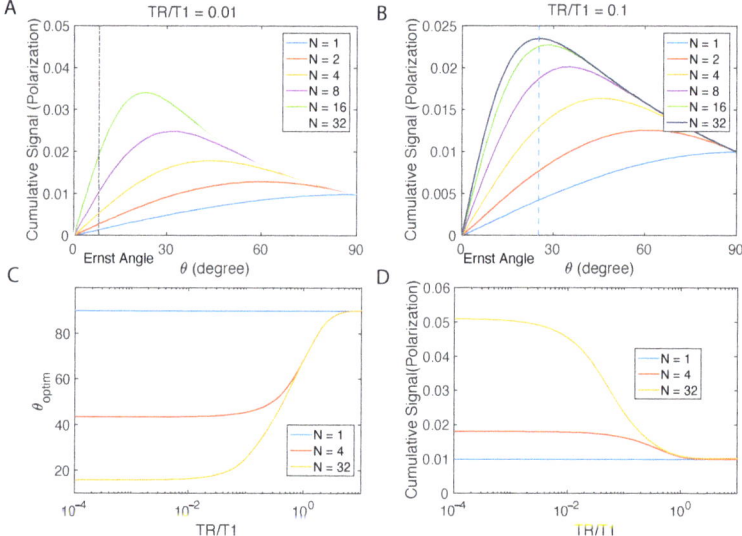

Figure 3. Constant flip angle. In the simulation, $K_{\text{hp}} = 10^{-3}$ and $K_0 = 10^{-5}$, which is on the same orders of magnitude of magnetization with our diamond imaging case at 9.4T. (**A**) Cumulative signal with different total repetitions N is displayed when the ratio of TR/T1 is fixed. The black dash line is the Ernst angle, optimal for initial magnetization to be M_0. (**B**) Fixing the ratio TR/T1, we simulate the cumulative signal with different N. (**C**) Optimal flip angles and (**D**) resultant cumulative signals when such angle is restricted to a constant are shown as a function of TR/T1.

2.3. Gradient Consideration

Gradient arrangement is another critical component in hyperpolarized solid state imaging. This determines timing for signal acquisition and k-space sampling trajectory and ultimately dictates image SNR, fidelity as well as resolution. Here we consider three categories of gradient arrangement, i.e. spin echo, gradient echo, and more exotic sequences. By analyzing different types of sequences, we provide insight into the gradient arrangement and sequence parameter determination for a given sample.

A typical spin echo sequence with small flip angle is shown in Figure 4. For a certain voxel (x, y, z), we consider the signal at the peak of the echo $S(TE)$, which is a good indication of the image SNR. This signal within one voxel is subject to decoherence posterior to the flip angle pulse, and the decay factor is e^{-TE_{se}/T_2} (see Figure 4A), where TE_{se} is the echo time of a spin echo sequence. Similarly, in a gradient echo sequence, this factor becomes e^{-TE_{ge}/T_2^*} (see Figure 4B). When assuming that both sequences use the same flip angle strategy, the decay factors imply that if $TE_{se}/T_2 < TE_{ge}/T_2^*$, the spin echo sequence is favorable for higher signal; otherwise, one should choose gradient echo given that the signal at the peak of the echo is higher in such cases.

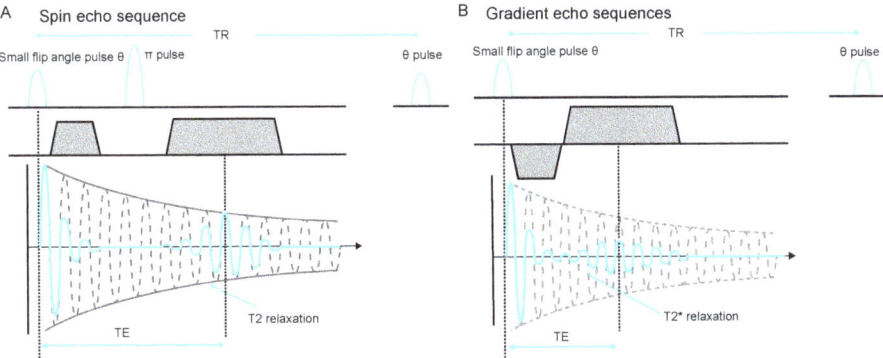

Figure 4. Spin echo and gradient echo sequence with small flip angle. (**A**) The π pulse refocuses dephasing caused by field inhomogeneity, chemical shift, and gradients. (**B**) The reversed gradient refocuses the effect of gradients. The phase encoding dimension implements same gradient arrangement for both of the two echo sequences, and is omitted here.

The RARE (Rapid Acquisition with Refocused Echoes) sequence, also known as TSE (turbo spin echo) is a sequence which takes advantage of multiple spin echo train followed by a single $\pi/2$ pulse. This sequence is originally designed for saturation recovery, can however be implemented with small flip angle excitation pulses, which may be enhance SNR in hyperpolarized solid state imaging. In this case, there can be T echo trains following a small flip angle excitation in each repetition. The cumulative signal depends on

$$\sum_{j=1}^{N}\sum_{k=1}^{T} e^{-k \cdot TE_{se}/T_2} \tag{13}$$

where two summations of j and k represent repetitions and echo trains respectively. Carefully selection of N and T can possibly enhance the cumulative signal further than conventional spin echo or gradient echo sequences.

The sequences that decouple nuclear spins in solids, which we refer to here as "exotic sequences", include magic echo sequence [18], as well as quadratic echo sequence [19]. However, those sequences are challenging to calibrate and implement due to the precise requirement of the spacing between pulses and the phase of the pulses.

Apart from forming spin echo or gradient echo, one can design the gradient arrangement so that signal acquisition can start immediately after the excitation pulse. For instance, steady gradient on both phase encoding and frequency encoding dimensions can be applied while the acquisition channel opens right after RF excitation, which corresponds to a radial trajectory in k-space. Such sequences are normally called Ultrashort TE, or UTE sequences [20]. Such methods can eliminate the decoherence happening before echo formation, although may have disadvantages in motion and gradient imperfection robustness [21].

2.4. Hyperpolarized Diamond Imaging Results

We test the above simulations using our hyperpolarized diamond imaging system [11]. A 5mm NMR tube is filled with diamond particles (average particle size $\sim$200 µm) and the particles are tightly held at the bottom of the tube. The MRI images of such phantoms are shown in Figure 5 with different flip angles. We acquired images with flip angles ranging from 13–333° by varying pulse length from 5 µs to 80 µs in Figure 5A, and we zoom in in the range of 4 μs to 19 μs to identify the optimal flip angle in Figure 5B. It turns out that the 6 μs presents the highest image fidelity and contrast. This shows agreement with Figure 3C, in which diamond particle imaging residents at low TR/T1 limit. Our diamond particles have a measured T_1 of 15s and a repetition time TR of 6ms for imaging, leading to TR/T1 $\sim 10^{-3}$, and corresponding θ_{optim} of 16°. Such flip angle can be translated as a predicted 5.5 μs pulse length. Note that according to our nutation calibration, the pulse duration $t_{tip} = \frac{\theta}{360°} \times 84.58$ μs $+ 1.73$ μs, indicating a 1.73μs delay of the pulse application by the MRI machine.

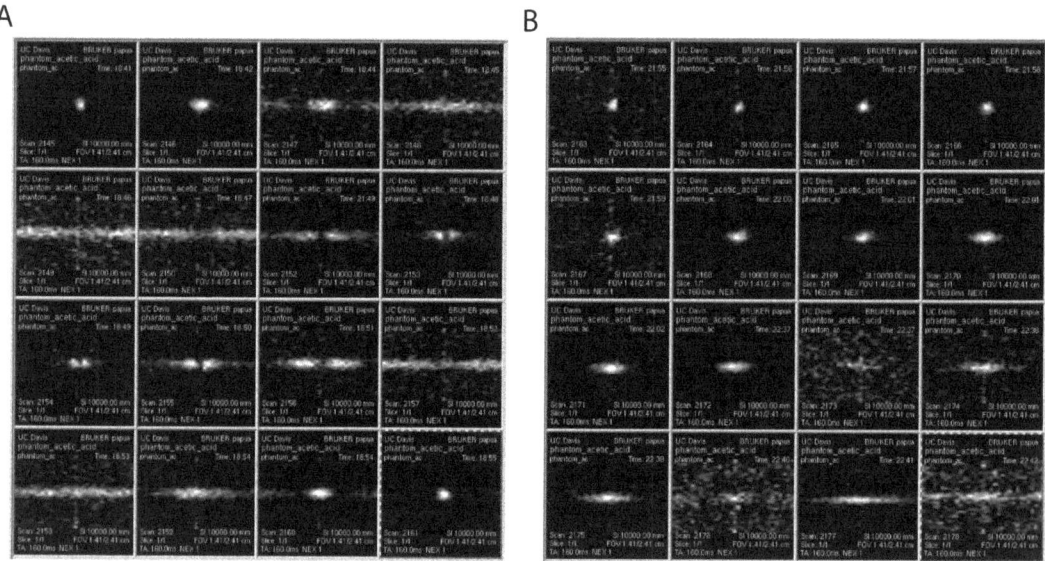

Figure 5. Diamond MRI with different flip angles. (**A**) The pulse durations are 5, 10, 15, ..., 80 μs respectively for each image. We can determine that optimal pulse duration should be within 20 μs. (**B**) The pulse durations are 4, 5, 6, ..., 19 μs respectively for each image. The text on each image is the frame number, time when the images are taken, and FOV (1.41 × 2.41 cm).

We can study the total signal in k-space and real-space by taking the integral of intensities across all the pixels, shown in Figure 6. The k-space signal maximizes at the optimal flip angle in Figure 6A,B. Note that, a 90° pulse can maximize the intensity of the center of the k-space, which is equivalent to the integral of real-space intensities (see Figure 6C,D). However such image has no high frequency information, which will be a constant in real-space along x direction. Such effect is precisely illustrated in Figure 5A first image in the second row.

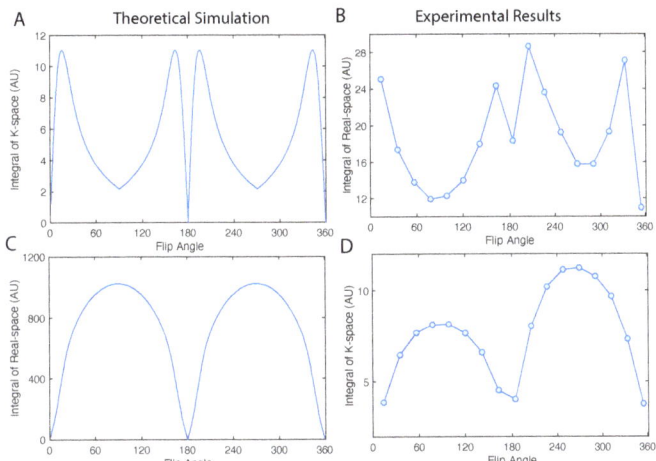

Figure 6. Total signal in k- and real- space as a function of flip angle. (**A**,**B**) The integration of absolute value in k-space is simulated and measured using the diamond particle phantom. (**C**,**D**) Display the integration of absolute value in real-space. The simulations are conducted assuming a uniform profile in real-space, i.e., $\rho(x,y)$ =constant, in which case only the effect of magnetization factor K is emphasized.

3. Discussion

In the image equation (Equation (3)), the spin density function $\rho(x,y)$ convolutes with $\mathcal{F}(e^{-t(k_y)/T_2^*})$ and $\mathcal{F}(K_x(k_x))$. The two terms correspond to two types of blur of the image. The term $\mathcal{F}(e^{-t(k_y)/T_2^*})$ caused by T_2 is similar to the linewidth in NMR spectroscopy. The Lorentzian profile leads to a resolution limit of $\propto \frac{1}{\gamma G T_2}$ in real-space, where G is imaging gradient. The second term is a Fourier transform of the profile of magnetization as a function of repetition (see Appendix B Figure A1), originating from the uniformity of the magnetization distribution over repetitions. The term will reduce to 1 when flip angles in Figure 2A is applied. In our experiment, the two types of blur happen on x and y direction respectively. Our phase encoding is on x direction, therefore, the stripe line in Figure 5 originates from the Fourier transform of the magnetization factor profile K along k_x direction. We write down $K_x(k_x) = \pm K_{hp}(\Gamma \cos\theta)^{k_x-1}\sin\theta + K_0(1-\Gamma)\sum_{j=1}^{k_x-1}(\Gamma \cos\theta)^{j-1}\sin\theta$. If we take the 5th frame in Figure 5A as an example, the flip angle of that is close to 90°. The magnetization of such pulse sequence distributes mainly on the first repetition (green line in Appendix B Figure A1). A nearly constant $\mathcal{F}(K_x(k_x))$ indicates the extreme case of blur — constant intensity along x direction when convoluting with $\rho(x,y)$.

We would also like to discuss the total signal gained by the small tip angle RARE sequence. From Equation (13), we can tell that increasing number of echo trains will increase the signal, however extends the total acquisition times at the same time. Here we try to determine the optimal sequence design to maximize the total signal given a finite total time T_{total}. We take the case where one is allowed to vary the flip angle, and the signal is constant in each repetition (as described in Equation (7)). We assume that in each repetition $TR = TE_{se} \times T$, where T is total number of echos within this repetition. We can rewrite Equation (13) to estimate total signal as a function of total repetition number $S(N)$:

$$S(N) = \sum_{j=1}^{N}\sum_{k=1}^{T(N)} e^{-k \cdot TE_{se}/T_2}$$
$$= N \cdot M_x(N) \cdot \chi \frac{1-\chi^T}{1-\chi} \qquad (14)$$

where $\chi = e^{TE_{se}/T_2}$ is a constant when minimized TE is set by instrumentation limit and T_2 is the intrinsic property of certain sample. The above derivation used the sum of a geometric sequence. In this equation $M_x = M_0 \times \sin(\theta_1)$ where θ_1 defined in Equation (7) is a function of TR, and $TR = \frac{T_{total}}{N}$. T can also be written as a function of N: $T(N) = \frac{T_{total}}{N \cdot TE_{se}}$. We plot $S(N)$ in Figure 7. We note that N values that can maximize $S(N)$ for $T_{total} = 0.1$ s and 0.2 s are ∼70 and ∼120 respectively. And when T_{total} is long enough (0.5 s), $S(N)$ is not yet saturated at $N = 256$.

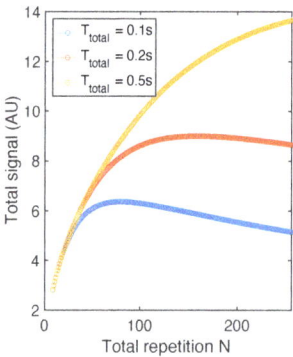

Figure 7. Total signal with RARE sequence. The simulation is conducted based on parameters close to diamonds (T_1 =50 s, T_2 = 1 ms, TE = 0.5 ms). The total signal curve $S(N)$ maximize at different N when t_{total} is set to different values.

4. Materials and Methods

4.1. Simulation and Optimization

The simulations and optimizations are conducted in Matlab, where the "fminunc" function is used to numerically optimize the flip angles in Figure 3C,D. In the simulation of Figure 6, the $K(k_x)e^{-t(k_x,k_y)/T_2^*}$ component is simulated with $K(k_x)$ in Equation (12), substituting n with k_x.

4.2. Hyperpolarization and Imaging

The diamond powder utilized in experiments in Figure 1 has ∼40 mg mass with natural abundance ^{13}C. The particles are purchased from Element6. They are enriched with ∼1 ppm NV centers and fabricated by a high pressure high temperature (HPHT) protocol. The particle size is measured in SEM (scanning electron microscopy) images. The face to face distances are 200 µm to 250 µm and diagonal edge to edge distances are approximately 400 µm.

The entire experimental setup consists of three parts: a pneumatic field-cycling device, a wide-bore 9.4T superconducting magnet, an a miniaturized hyperpolarizer [22]. The pneumatic field-cycling device [11] is uses air flow to rapidly transfer a 5mm NMR tube from low field (40 mT) to the 9.4 T detection field, within which a 10 mm ^{1}H/^{13}C volume coil is installed. The air driven by a pump flows in a quartz channel and moves the NMR tube in the channel. Diamond samples are contained in the NMR tube. A concave-shaped stopper is located at the bottom end of the channel and a rubber stopper is placed at the high field. The transport time of the sample to high field is under 1s, much shorter compared to the ^{13}C T_1 times (normally on the order of minitues). MR imaging was conducted with a Bruker DRX system equipped with microgradients running ParaVision 4 software with a modified FLASH pulse sequence. The miniaturized hyperpolarizer is a self-contained unit, which encapsulates devices for laser excitation, MW irradiation as well as an electromagnet for field fine-tuning. A 1W 520nm diode laser (Lasertack PD-01289) is employed and the beam passes through an aspheric lens and a set of anamorphic prisms to form a 4 mm diameter beam. The beam was guided by two mirrors and illuminates the sample from the bottom. MW irradiation that drives polarization transfer is generated by three voltage controlled oscillator (VCO) sources (Minicircuits ZX95-3800A+). For frequency sweeps, the VCOs are driven by phase shifted triangle waves from a home-built PIC microprocessor (PIC30F2020) driven quad ramp generator.

Please find more details of experimental methods in Ref. [11].

5. Conclusions

In this paper, we studied two major components — small flip angles and gradient arrangement in a MRI sequence in the quest for optimal sequences for hyperpolarized solids. Both variable and constant flip angles are analyzed, and strategies to achieve maximum cumulative signal or flat signal profile are provided. Beyond designing flip angle progressions to take advantage of the significant initial magnetization produced by hyperpolarization, we propose to combine these excitation pulse progressions with traditional gradient arrangements in spin echo and gradient echo sequences in order to accommodate short decoherence times in solids. Experimental results of hyperpolarized diamond MRI show agreement with theoretical analysis. Beyond diamond particles, this study can provide guidance in hyperpolarized solids MRI in systems such as such as silicon [23] and silicon carbide [24] particles.

Author Contributions: Conceptualization, X.L., A.P., A.A., J.R.; Data curation, X.L.; Formal analysis, X.L.; Investigation, X.L.; Writing-original draft, X.L.; Writing review & editing, X.L., J.W., A.A., J.R.; Methodology, X.L., J.W.; Software, X.L.; Visualization, X.L., H.M.; Resource, E.D., R.N.; Funding acquision, A.P., A.A., J.R.; Supervision, A.P., A.A., J.R.; Project administration, J.W. All authors have read and agreed to the published version of the manuscript.

Funding: We acknowledge funding by NSF 1903803. J.W. acknowledges NIH 1S10RR013871-01A1 for funding the 400 MRI instrumentation.

Data Availability Statement: The data presented in this study are available on request from the corresponding author.

Acknowledgments: We acknowledge B. Wu, K. Aryasomayajula, A. Lin, J. Chen for useful discussions.

Conflicts of Interest: The funders had no role in the design of the study; in the collection, analyses, or interpretation of data; in the writing of the manuscript, or in the decision to publish the results.

Appendix A. Derivation

Appendix A.1. Variable Flip Angle for Constant Magnetization

$$K_{x,n} = K_{\text{hp}} \left[\prod_{k=1}^{n-1} (\Gamma \cos \theta_k) \right] \sin \theta_n = \text{constant} \tag{A1}$$

Since we want to saturate the magnetization at the last pulse, we have $\sin \theta_N = 1$. Using such equation, we can first write down the $K_{x,N} = K_{x,N-1}$ as:

$$K_{\text{hp}} \left[\prod_{k=1}^{N-1} (\Gamma \cos \theta_k) \right] \sin \theta_N = K_{\text{hp}} \left[\prod_{k=1}^{N-2} (\Gamma \cos \theta_k) \right] \sin \theta_{N-1} \tag{A2}$$

This implies $\Gamma \cos \theta_{N-1} \sin \theta_N = \sin \theta_{N-1}$, and we can get: $\tan \theta_{N-1} = \sin \theta_N / \Gamma = 1/\Gamma$. Similarly, if we take the equality between $K_{x,j}$ and $K_{x,j-1}$, the recursion formula is:

$$\tan \theta_{j-1} = \Gamma \sin \theta_j \tag{A3}$$

Then, we need to solve θ_n based on the equation above. We define $a_n = \tan^2 \theta_n$, and we will have:

$$a_n = \frac{a_{n+1}}{1 + a_{n+1}} \cdot \Gamma^2 \tag{A4}$$

This is equivalent to:

$$\frac{1}{a_n} = \left(\frac{1}{a_{n+1}} - \Gamma^2\right)\Gamma^2 \tag{A5}$$

Solving the series, we can get:

$$a_n = (1 - \Gamma^3) \cdot \frac{\Gamma^{2N-2n-1}}{1 - \Gamma^{2N-2n-1}} \tag{A6}$$

which leads to:

$$\tan^2 \theta_n = (1 - \Gamma^3) \cdot \frac{\Gamma^{2N-2n-1}}{1 - \Gamma^{2N-2n-1}} \tag{A7}$$

Appendix A.2. Variable Flip Angle for Maximum Cumulative Magnetization

With

$$\Gamma \tan \theta_{j-1} = \sin \theta_j \tag{A8}$$

We define $a_n = \tan^2 \theta_n$, and we will have:

$$a_n = \frac{a_{n+1}}{1 + a_{n+1}} \cdot \frac{1}{\Gamma^2} \tag{A9}$$

This is equivalent to:

$$\frac{1}{a_n} = \left(\frac{1}{a_{n+1}} - \Gamma^2\right)/\Gamma^2 \tag{A10}$$

Solving the series, we can get:

$$a_n = \frac{1}{\Gamma^2} \cdot \frac{1 - \Gamma^2}{1 - \Gamma^{2(N-n)}} \tag{A11}$$

which leads to:

$$\tan^2 \theta_n = \frac{1}{\Gamma^2} \cdot \frac{1 - \Gamma^2}{1 - \Gamma^{2(N-n)}} \tag{A12}$$

Appendix B. Magnetization Simulation

The magnetization of each repetition when applying constant flip angle is presented.

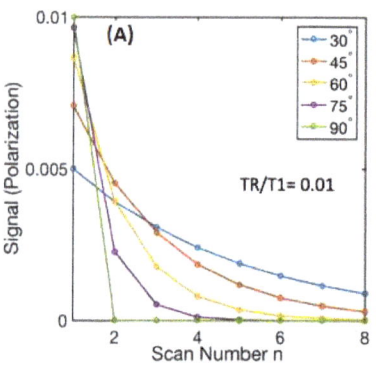
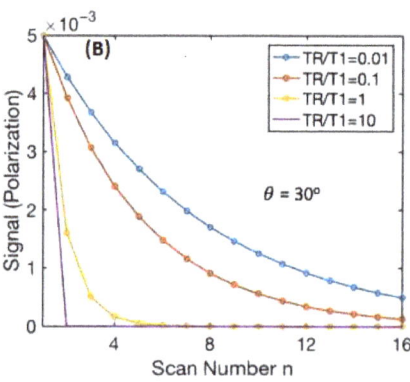

Figure A1. Simulation of signal of each individual repetition as a function of scan number of given TR/T1 in (**A**) and given θ in (**B**). In the simulation, we assume $K_{hp} = 10^{-3}$ and $K_0 = 10^{-5}$, which approximates the magnetization with our diamond imaging case at 9.4T.

References

1. Ernst, R.; Bodenhausen, G.; Wokaun, A. *Principles of Nuclear Magnetic Resonance in One and Two Dimensions*; Clarendon Press: Oxford, UK, 1987.
2. Wüthrich, K. NMR studies of structure and function of biological macromolecules (Nobel Lecture). *Angew. Chem. Int. Ed.* **2003**, *42*, 3340–3363. [CrossRef] [PubMed]
3. Morris, P.G. *Nuclear Magnetic Resonance Imaging in Medicine and Biology*; Clarendon Press: Oxford, UK, 1986.
4. Huettel, S.A.; Song, A.W.; McCarthy, G. *Functional Magnetic Resonance Imaging*; Sinauer Associates: Sunderland, MA, USA, 2004; Volume 1.
5. Abragam, A.; Goldman, M. Principles of Dynamic Nuclear Polarization. *Rep. Prog. Phys.* **1978**, *41*, 395–467. [CrossRef]
6. Theis, T.; Ganssle, P.; Kervern, G.; Knappe, S.; Kitching, J.; Ledbetter, M.; Budker, D.; Pines, A. Parahydrogen-enhanced zero-field nuclear magnetic resonance. *Nat. Phys.* **2011**, *7*, 571–575. [CrossRef]
7. Salikhov, K.M.; Molin, Y.N.; Sagdeev, R.; Buchachenko, A. *Spin Polarization and Magnetic Effects in Radical Reactions*; Elsevier Science Ltd: Amsterdam, The Netherlands, 1984.
8. Guthi, J.S.; Yang, S.G.; Huang, G.; Li, S.; Khemtong, C.; Kessinger, C.W.; Peyton, M.; Minna, J.D.; Brown, K.C.; Gao, J. MRI-visible micellar nanomedicine for targeted drug delivery to lung cancer cells. *Mol. Pharm.* **2010**, *7*, 32–40. [CrossRef] [PubMed]
9. Ajoy, A.; Liu, K.; Nazaryan, R.; Lv, X.; Zangara, P.R.; Safvati, B.; Wang, G.; Arnold, D.; Li, G.; Lin, A.; et al. Orientation-independent room temperature optical 13C hyperpolarization in powdered diamond. *Sci. Adv.* **2018**, *4*, eaar5492. [CrossRef] [PubMed]
10. Jelezko, F.; Wrachtrup, J. Single defect centres in diamond: A review. *Phys. Status Solidi (A)* **2006**, *203*, 3207–3225. [CrossRef]
11. Lv, X.; Walton, J.; Druga, E.; Wang, F.; Aguilar, A.; McKnelly, T.; Nazaryan, R.; Wu, L.; Shenderova, O.; Vigneron, D.; et al. High contrast dual-mode optical and 13C magnetic resonance imaging in diamond particles. *arXiv* **2019**, arXiv:1909.08064.
12. Chao, J.I.; Perevedentseva, E.; Chung, P.H.; Liu, K.K.; Cheng, C.Y.; Chang, C.C.; Cheng, C.L. Nanometer-sized diamond particle as a probe for biolabeling. *Biophys. J.* **2007**, *93*, 2199–2208. [CrossRef] [PubMed]
13. Miller, B.S.; Bezinge, L.; Gliddon, H.D.; Huang, D.; Dold, G.; Gray, E.R.; Heaney, J.; Dobson, P.J.; Nastouli, E.; Morton, J.J.; et al. Spin-enhanced nanodiamond biosensing for ultrasensitive diagnostics. *Nature* **2020**, *587*, 588–593. [CrossRef] [PubMed]
14. Choi, J.; Zhou, H.; Landig, R.; Wu, H.Y.; Yu, X.; Von Stetina, S.E.; Kucsko, G.; Mango, S.E.; Needleman, D.J.; Samuel, A.D.T.; et al. Probing and manipulating embryogenesis via nanoscale thermometry and temperature control. *Proc. Natl. Acad. Sci. USA* **2020**, *117*, 14636–14641. [CrossRef] [PubMed]
15. Shagieva, F.; Zaiser, S.; Neumann, P.; Dasari, D.; Stohr, R.; Denisenko, A.; Reuter, R.; Meriles, C.; Wrachtrup, J. Microwave-assisted cross-polarization of nuclear spin ensembles from optically pumped nitrogen-vacancy centers in diamond. *Nano Lett.* **2018**, *18*, 3731–3737. [CrossRef] [PubMed]
16. Nishimura, D.G. *Principles of Magnetic Resonance Imaging*; Standford Univ.: Stanford, CA, USA, 2010.
17. Abragam, A. *Principles of Nuclear Magnetism*; Oxford Univ. Press: Oxford, UK, 1961.
18. Takegoshi, K.; McDowell, C. A "magic echo" pulse sequence for the high-resolution NMR spectra of abundant spins in solids. *Chem. Phys. Lett.* **1985**, *116*, 100–104. [CrossRef]
19. Frey, M.A.; Michaud, M.; VanHouten, J.N.; Insogna, K.L.; Madri, J.A.; Barrett, S.E. Phosphorus-31 MRI of hard and soft solids using quadratic echo line-narrowing. *Proc. Natl. Acad. Sci. USA* **2012**, *109*, 5190–5195. [CrossRef] [PubMed]
20. Chang, E.Y.; Du, J.; Chung, C.B. UTE imaging in the musculoskeletal system. *J. Magn. Reson. Imaging* **2015**, *41*, 870–883. [CrossRef] [PubMed]
21. Block, K.T. *Advanced Methods for Radial Data Sampling in Magnetic Resonance Imaging*; SUB University of Goettingen: Göttingen, Germany, 2008.
22. Ajoy, A.; Nazaryan, R.; Druga, E.; Liu, K.; Aguilar, A.; Han, B.; Gierth, M.; Oon, J.T.; Safvati, B.; Tsang, R.; et al. Room temperature "optical nanodiamond hyperpolarizer": Physics, design, and operation. *Rev. Sci. Instrum.* **2020**, *91*, 023106. [CrossRef] [PubMed]
23. Dementyev, A.; Cory, D.; Ramanathan, C. Dynamic nuclear polarization in silicon microparticles. *Phys. Rev. Lett.* **2008**, *100*, 127601. [CrossRef] [PubMed]
24. Falk, A.L.; Klimov, P.V.; Ivády, V.; Szász, K.; Christle, D.J.; Koehl, W.F.; Gali, Á.; Awschalom, D.D. Optical polarization of nuclear spins in silicon carbide. *Phys. Rev. Lett.* **2015**, *114*, 247603. [CrossRef] [PubMed]

Article

Gas and Liquid Phase Imaging of Foam Flow Using Pure Phase Encode Magnetic Resonance Imaging

Alexander Adair *, Sebastian Richard and Benedict Newling

MRI Centre, University of New Brunswick, 8 Bailey Dr., Fredericton, NB E3B5A3, Canada; srichar6@unb.ca (S.R.); bnewling@unb.ca (B.N.)
* Correspondence: adair.alex@unb.ca

Abstract: Magnetic resonance imaging (MRI) is a non-invasive and non-optical measurement technique, which makes it a promising method for studying delicate and opaque samples, such as foam. Another key benefit of MRI is its sensitivity to different nuclei in a sample. The research presented in this article focuses on the use of MRI to measure density and velocity of foam as it passes through a pipe constriction. The foam was created by bubbling fluorinated gas through an aqueous solution. This allowed for the liquid and gas phases to be measured separately by probing the ^{1}H and ^{19}F behavior of the same foam. Density images and velocity maps of the gas and liquid phases of foam flowing through a pipe constriction are presented. In addition, results of computational fluid dynamics simulations of foam flow in the pipe constriction are compared with experimental results.

Keywords: foam flow; magnetic resonance imaging; velocity mapping; pipe flow; two-phase flow

Citation: Adair, A.; Richard, S.; Newling, B. Gas and Liquid Phase Imaging of Foam Flow Using Pure Phase Encode Magnetic Resonance Imaging. *Molecules* **2021**, *26*, 28. https://dx.doi.org/10.3390/molecules26010028

Academic Editors: Igor Serša and José A. González-Pérez
Received: 1 November 2020
Accepted: 19 December 2020
Published: 23 December 2020

Publisher's Note: MDPI stays neutral with regard to jurisdictional claims in published maps and institutional affiliations.

Copyright: © 2020 by the authors. Licensee MDPI, Basel, Switzerland. This article is an open access article distributed under the terms and conditions of the Creative Commons Attribution (CC BY) license (https://creativecommons.org/licenses/by/4.0/).

1. Introduction

Foams are integral to many industrial and consumer applications, including petroleum (e.g., enhanced oil recovery), cosmetics (e.g., shaving cream), and food science (e.g., beer foam). Foams have been studied extensively, with investigations commonly focusing on liquid holdup and drainage [1–6], and bubble size distribution [7,8]. Foam measurements have been performed using a variety of methods, both invasive and non-invasive, such as optical measurements with insertion probes [8], gamma ray absorption [9], and tracer particles [10]. Foam has also been studied using magnetic resonance imaging (MRI) with particular focus on foam drainage [1–4] and bubble size distribution [7]. At its core, MRI reports the position of nuclei in a sample based on their position within a magnetic field gradient. This information on the density and position of nuclei is used to create images of the sample. Data from an MRI experiment can also be sensitized to provide additional information on sample behavior. Of particular interest for the experiments described in this paper is sensitizing MRI to the motion of nuclei in a sample, allowing for velocity mapping.

MRI has several benefits for the study of foams when contrasted with other flow measurement techniques. MRI is non-optical; therefore, the internal behavior of a foam can be studied even in visually opaque foams. MRI is also non-invasive; therefore, it does not impede movement or otherwise interfere with the structure of the foam. This is especially important in the case of delicate foams which might break down under the scrutiny of an invasive measurement technique. MRI is also capable of imaging in one to three spatial dimensions. In addition, MRI measurements can be sensitized to different nuclei. This is of particular importance because separate measurements of the liquid and gas phases of a foam can be performed by choosing the liquid and gas components of the foam judiciously.

When considering measurements of foam, an important consideration is the effect that magnetic susceptibility differences between the gas and liquid phases have on results. These magnetic susceptibility differences manifest as image artifacts in typical frequency-encoding techniques. On the other hand, a pure phase encoding technique provides images free of artifacts due to magnetic susceptibility differences at the gas–liquid interfaces in foams. Of particular interest for the study of foams are MRI measurement techniques that are well-suited for measuring signals with short lifetimes. This area of study is well developed with both frequency encoding techniques (such as UTE (ultra-short echo time imaging) and ZTE (zero echo time imaging)) and phase encoding techniques (such as SPI (single point imaging) and SPRITE (Single Point Ramped Imaging with T_1-Enhancement)) [11]. While phase encoding measurement techniques like SPRITE have the benefit of reduced image artifacts due to magnetic susceptibility differences, they are typically much slower measurements compared to frequency encoding techniques like UTE and ZTE.

The SPRITE [12] measurement technique is a pure phase encoding technique and is well-suited for creating density images of foam. To study the movement of foam, data acquired during a SPRITE measurement can be sensitized to motion with the addition of pulsed field gradients (PFGs) [13]. Motion sensitization was incorporated into the measurements used in these experiments by preceding the typical SPRITE imaging with a PFG preparation phase. In particular for the study of foam flow, the alternating pulsed gradient stimulated echo (APGSTE) preparation [14] was used because it reduces the effects of the background magnetic field gradients on the data. The benefits of the APGSTE preparation are applicable to the foam flow studied for this paper, but they would be increasingly important in applied foam flows with extreme magnetic susceptibility-induced magnetic field gradients (such as enhanced oil recovery). Motion-sensitized measurements using the APGSTE preparation and SPRITE imaging have previously been used to study other two-phase flow systems and flow in porous media [15,16].

This paper reports the first use of APGSTE-SPRITE to study foam flow. In addition, this paper also reports the first use of APGSTE-SPRITE to create velocity maps of both phases of a two-phase flow. This was achieved by bubbling a fluorinated gas through a water solution to create the foam. Information on the liquid phase was acquired by measuring signals from ^1H nuclei, and information on the gas phase was acquired by measuring signals from ^{19}F nuclei. The experimental apparatus was designed such that gas and liquid phase measurements were taken sequentially, without interrupting the foam flow. This paper demonstrates the use of motion-sensitized SPRITE MRI as a measurement technique for the study of foam flow and the comparison between gas and liquid phase velocity maps. Further, this paper presents the use of the Herschel–Bulkley viscosity model as a means of modelling the foam flow through the pipe, rather than more computationally intensive methods [17,18]. Computational fluid dynamics (CFD) simulation results created using this model are presented and found to be only superficially similar to experimental results, suggesting the future use of MRI measurements to guide CFD simulations of foam behavior.

2. Results

The density and velocity maps of the flowing foam for hydrogen and fluorine signal are shown together in Figure 1. As mentioned previously, hydrogen signal reports on the liquid phase of the foam, while fluorine signal reports on the gas phase. Figure 1A,B are 2D images of the foam for hydrogen and fluorine signal, respectively. Since the images are time-averaged and two-dimensional projections, the images do not show individual bubbles in the foam, but rather, signal intensity is proportional to the averaged density (of hydrogen or fluorine) in each image. The effect of buoyancy is evident in the hydrogen density image—there is a gradation of density in the Y-direction (i.e., vertically in the lab frame) with higher density shown at the bottom of the pipe in the hydrogen image, indicating a separation of the two foam phases. Buoyancy is less obvious in the fluorine density image, likely due to a poorer signal-to-noise ratio than in the hydrogen image.

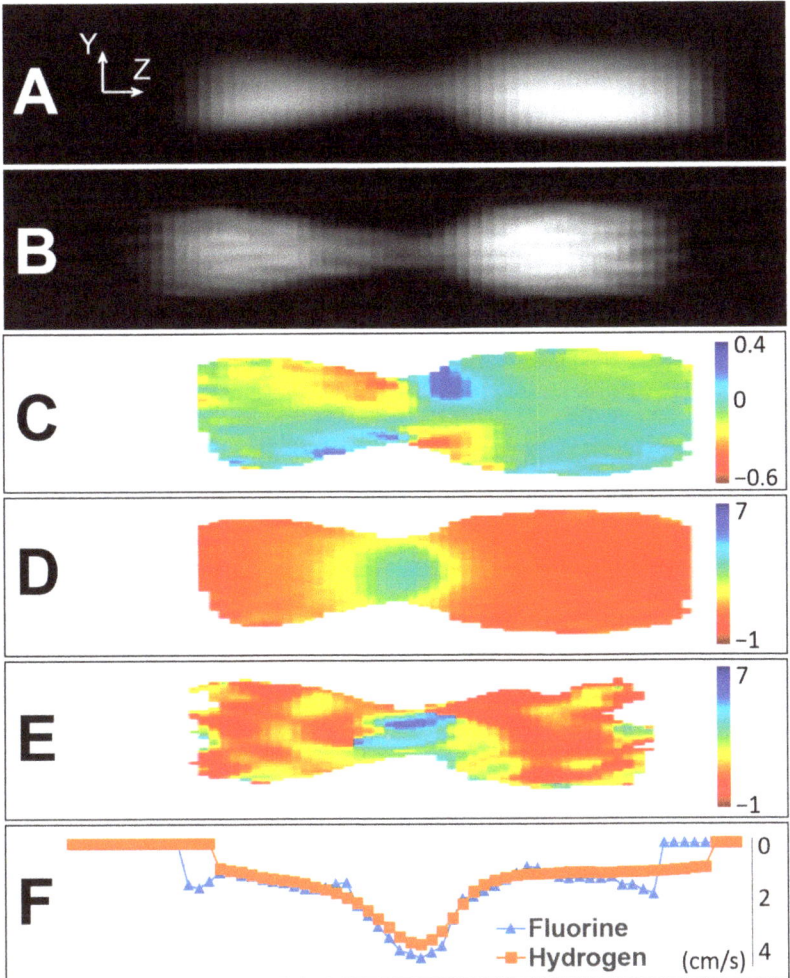

Figure 1. Density and velocity maps of the foam flow for the liquid phase (hydrogen signal) and the gas phase (fluorine signal). (**A**) Density image of the liquid phase. (**B**) Density image of the gas phase. (**C**) Y-component of velocity map of the liquid phase. (**D**) Z-component velocity map of the liquid phase. (**E**) Z-component velocity map of the gas phase. (**F**) Z-component of velocity for both phases, averaged across the width of the pipe. All speeds are in units of cm/s, and bulk flow is from left to right. Uncertainty in speed values was estimated from variations in speed in a region downstream from the constriction throat where speed should be uniform. Uncertainty in speed for (**D**) is ± 0.2 cm/s and for (**E**) is ± 1.7 cm/s.

Figure 1C is the map of the Y-component of velocity for the hydrogen signal, and it shows expected results. The foam moved in towards the pipe axis as it flowed into the constriction and the pipe diameter decreased. The foam moved away from the pipe axis as the pipe diameter increased past the narrowest point. Figure 1D is the map of the Z-component of velocity (i.e., along the pipe axis) for the hydrogen signal. The results were as expected. The flow speed along the pipe axis increased as the diameter of the pipe decreased in the constriction. The flow speed decreased as the diameter increased past the narrowest point, and a jet was present. Figure 1E is the map of the Z-component of velocity for the fluorine signal. The expected behavior of higher speeds being present at the narrowest point in the pipe was apparent, although the results were less clear than the hydrogen velocity map due to the poor signal-to-noise ratio (SNR) of the fluorine signal. Figure 1F shows velocity profiles of the Z-component for hydrogen and fluorine signal taken along the pipe axis, averaged across the width of the pipe. Although the uncertainty in velocity values for the fluorine signal was high due to the low SNR, the comparison of centerline profiles indicated that the velocities along the pipe axis were comparable with hydrogen.

Figure 2 shows the CFD simulation results of the foam flow through the pipe constriction, for which representative parameter values were taken from the literature (see Materials and Methods). The simulation results superficially resembled the experimental results but were not identical, which encourages future use of the MRI results to guide the development of more accurate simulations. The figure at the top shows the Y-component of velocity. The general behavior was similar to the results shown in Figure 1C (the Y-component velocity map of the liquid phase from experimental data). The fluid moved towards the pipe axis as the pipe diameter decreased, and it moved away from the axis as the diameter increased past the narrowest point. The figure at the bottom of Figure 2 shows the Z-component of velocity. The general behavior was similar to that of Figure 1D (the Z-component velocity map of the liquid phase, from experimental data). The flow speed attained its highest value at the narrowest point in the constriction, and the jet is present in both images.

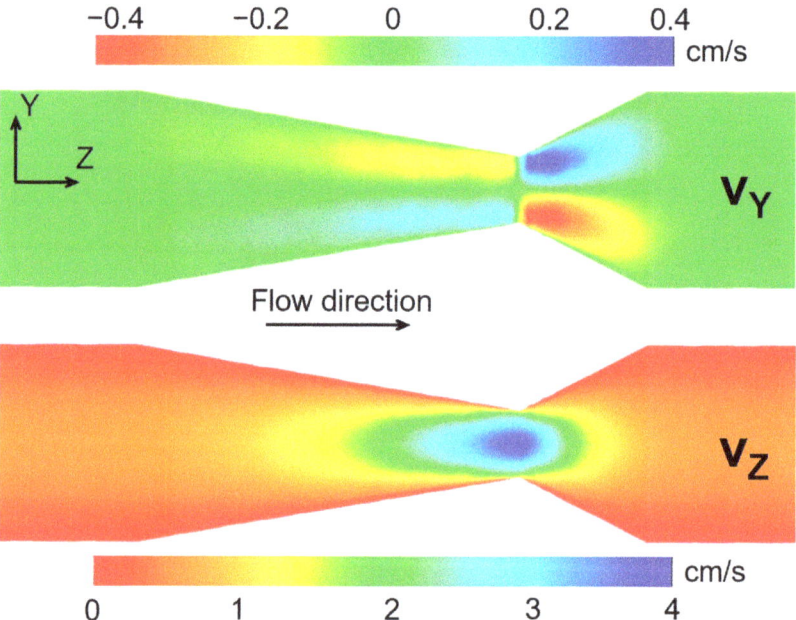

Figure 2. Simulation results of foam flow through the pipe constriction using a Herschel–Bulkley fluid viscosity model, a foam quality of 0.75, and a flow rate of 0.0009 L/s. The bulk flow direction is from left to right. The figure on top shows the results for the Y-component of velocity, and the figure on bottom shows the results for the Z-component of velocity.

3. Discussion

Magnetic resonance imaging has several advantages over other flow measurement techniques that make it well-suited for studying foam flow. It is able to map directly sample density to position (to create density images as shown in Figure 1) and is able to create density images for separate nuclei present in the sample. The hydrogen and fluorine density maps created of the foam flow demonstrate the effectiveness of MRI at studying the two phases (gas and liquid) of the foam. Buoyancy is clearly evidenced by the gradation in density, demonstrating the effect of gravity on the foam for both phases.

The APGSTE-SPRITE preparation-readout technique incorporated motion information into the images, and velocity maps of the foam flow were created. The velocity maps do not provide a snapshot image for the flow but are instead time-averaged over several hours as a consequence of signal averaging to improve SNR. The velocity maps show expected flow behavior and demonstrate that APGSTE-SPRITE is well-suited for imaging velocity. Primarily, when comparing the Z-component of velocity for the hydrogen and fluorine signals (i.e., liquid and gas phases), the velocity values are comparable, within uncertainty. In addition, CFD simulation results show only similar behavior to the foam flow when compared with experimental results. This gives incentive to continue the development of the Herschel–Bulkley viscosity model for simulating foam flow in a pipe constriction. Typical values were taken from the literature to create the initial simulations, but future work will continue improving the simulations of the foam flow by analyzing a wider range of values guided by the MRI results.

While the APGSTE-SPRITE technique has been demonstrated in this paper for the study of foam flow in a pipe constriction, it can easily be adapted to studying flow systems where the effects of magnetic susceptibility effects are more extreme, such as in enhanced oil recovery or froth flotation. Additionally, a principal focus of this work was the adaptation of the measurement technique to multi-nuclear study of flow. In future work, a multi-nuclear measurement approach could be used to study flow in rock cores [16] and bubbly flow in pipes [19]. The APGSTE-SPRITE measurement could also be used for studies in a variety of other flow systems, such as rock fractures [20] and polymer flooding [21].

4. Materials and Methods

Measurements were performed using a 2.4 T horizontal bore superconducting magnet (Nalorac, CA, USA) with homebuilt magnetic field gradient hardware capable of delivering maximum gradient amplitudes of 0.26 T/m in the Z-direction (oriented along the magnet bore) and 0.28 T/m in the Y-direction (oriented vertically). Radio frequency excitation and signal detection was accomplished using two homebuilt birdcage coils (RF probes) driven by a 2 kW Tomco amplifier (Tomco, Australia). The RF probes were designed and tuned such that one measured the signal from hydrogen (^{1}H) in the sample and the other measured the signal from fluorine (^{19}F).

A plexiglass pipe (I.D. = 1.9 cm) was placed inside the superconducting magnet with the pipe axis oriented in the Z-direction. A constriction in the pipe with a minimum inner diameter of 0.6 cm (see Figure 3) was positioned inside the imaging region of the apparatus. Foam was generated with an aqueous solution of distilled water, sodium dodecyl sulfate (SDS) (1.5 g/L), and glycerol (30 mL/L). The aqueous solution was kept in a reservoir, and a peristaltic pump (Masterflex, Cole-Parmer Montreal, Canada) moved the solution from the main reservoir to a smaller reservoir used to create the foam. The foam was created by flowing octafluorocyclobutane (C_4F_8) gas through a sparger (50 μm pore size) immersed in the aqueous solution. The flow rate of the peristaltic pump was set to maintain a constant liquid level in the smaller reservoir. The foam flowed through the plexiglass pipe inside the magnet before returning to the reservoir containing the foaming solution.

The flow rate of the C_4F_8 gas was set at the beginning of the experiment and maintained at that flow rate to ensure a consistent foam for all measurements. The flow rate was chosen to fit with experimental constraints imposed by the imaging field of view (15 cm in the Z-direction) and achievable magnetic field gradient amplitudes. A conventional RF probe design would have required stopping the foam flow and disassembling the flow network in order to switch between hydrogen and fluorine measurements. The RF probes were modified to allow them to be moved in and out of the region of interest without interrupting the foam flow, thus allowing for separate gas and liquid phase measurements of a continuously flowing foam (see Figure 3). The probe switching was implemented in the following manner: the flow apparatus was set up with a flowing foam and the

^{19}F RF probe in the imaging position. All measurements were acquired of the ^{19}F signal. Without interrupting the foam flow, the ^{19}F RF probe was manually removed from the magnet by sliding it along the pipe. The ^{1}H RF probe was then manually inserted from the opposite side of the magnet by sliding it along the pipe and into the imaging region. All measurements were then acquired of the ^{1}H signal. It took several minutes to switch the RF probes and verify their position. The process could be automated, but it was not in this proof-of-principle measurement. The foam solution was doped with gadolinium chloride (GdCl$_3$) to attain a spin-lattice relaxation time constant from the hydrogen signal of T_1 = 327 ms. The T_1-relaxation time constant of the C$_4$F$_8$ gas was T_1 = 56 ms. Both values were measured in a stationary foam using standard inversion recovery methods. The T_2^* relaxation time constant of the foam from the hydrogen signal was T_2^* = 0.4 ms, which is indicative of considerable susceptibility-induced line-broadening (typical T_2^* in homogeneous liquids being 3–5 ms in the same magnet).

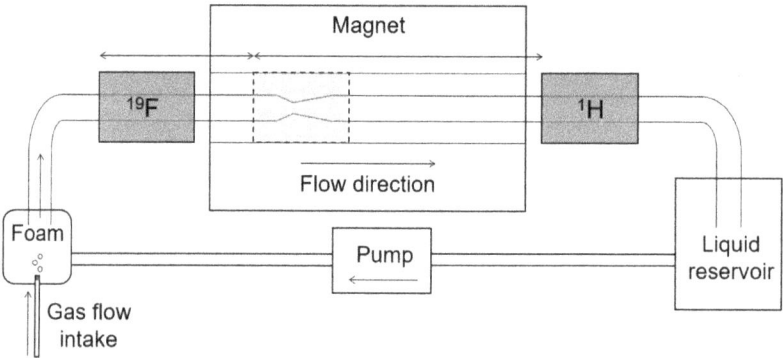

Figure 3. Schematic of the flow apparatus. Foam was created by bubbling C$_4$F$_8$ gas through a liquid solution and flowed through a pipe placed in the magnet. A constriction in the pipe was the region of interest. Fluorine signal (gas phase) was measured by inserting the ^{19}F RF probe and removing the ^{1}H RF probe. Hydrogen signal (liquid phase) was measured by inserting the ^{1}H RF probe and removing the ^{19}F probe. The shuttling of probes was accomplished without interrupting the foam flow.

Hydrogen and fluorine density images of the foam flow were acquired using the 2D Spiral SPRITE MRI measurement technique [22]. Interleaved k-space measurement trajectories were used for this experiment such that each individual trajectory had a short duration. The duration of the measurement trajectories was set to be less than the T_1-relaxation time constants (see above) in order to ensure that signal sensitization from the preparation phase persisted through all acquired imaging data points. The ^{1}H RF probe was used to create density images of the liquid phase of the foam. The hydrogen density image was created from 64 scans, with a total imaging time of 16 min. The ^{19}F RF probe was used to create density images of the gas phase of the foam. The fluorine density image was created from 1024 scans, with a total imaging time of 65 min. All images created from these measurements are time-averaged and do not show an instantaneous image of the foam. The field of view for both density images was 15 cm in the Z-direction (oriented along the pipe axis) and 3 cm in the Y-direction (oriented vertically), with a nominal resolution of approximately 2 mm/pixel in the Z-direction and 0.5 mm/pixel in the Y-direction. The phase-encoding time (the time interval between sample excitation and signal detection) was t_p = 150 µs. Flow-induced smearing effects caused by sample movement during the phase-encoding interval were not present due to the short phase-encoding time and slow flow speeds.

Velocity maps of the flowing foam were created using a preparation readout approach. Velocity information was introduced into the measured signal during the preparation stage with an alternating pulsed gradient stimulated echo (APGSTE) pulse sequence, which encoded signal phase based on the distance traveled by the sample during a set time interval. A Spiral SPRITE readout followed the preparation, which imposed spatial information onto the prepared sample magnetization. A Fourier transformation of the acquired data was used to construct an image of the sample. To ensure that motion sensitization information was contained in all acquired data, each individual, short-duration measurement trajectory of the Spiral SPRITE imaging readout was preceded by an APGSTE preparation, and sample magnetization was allowed to fully recover after the previous imaging readout and before the next preparation was applied. A schematic of the APGSTE-SPRITE measurement pulse sequence is shown in Figure 4.

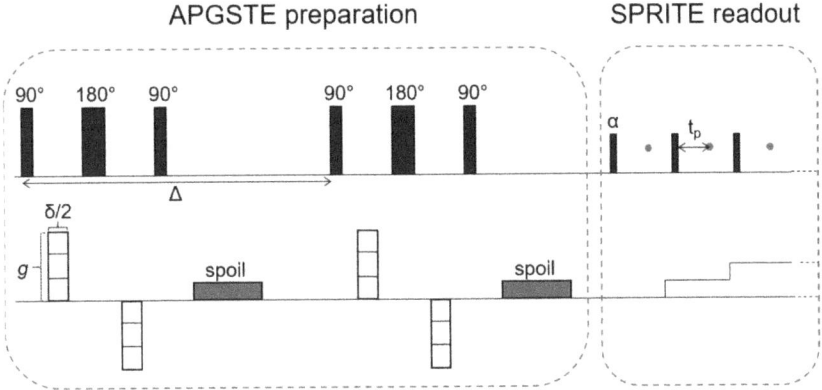

Figure 4. A schematic of the APGSTE-SPRITE pulse sequence [15]. Each individual SPRITE k-space measurement trajectory is preceded by an APGSTE preparation to ensure all measured data are motion sensitized. The parameters shown are as follows: g is the amplitude of the motion-sensitizing PFG, $\delta/2$ is the duration of the PFG, Δ is the flow evolution time, α is the flip angle of the imaging RF pulses in the SPRITE readout, and t_p is the phase encoding time (the time between sample excitation and signal detection).

The APGSTE-SPRITE measurement was repeated three times with varying amplitude of the motion-sensitizing gradient pulses (g in Figure 4). Acquiring measurements with different values of g builds up information in a reciprocal displacement space (or q-space) where q is defined as $q = \gamma \delta g (2\pi)^{-1}$, where γ is the gyromagnetic ratio of the nucleus. A mean velocity value v_{avg} was extracted for each pixel in the image by linear fitting the signal phase against q across the three measurements according to Equation (1):

$$v_{avg} = \frac{1}{2\pi} \frac{d}{dq} \left(\tan^{-1} \frac{Im(S)}{Re(S)} \right) \quad (1)$$

where $Im(S)$ and $Re(S)$ are the imaginary and real components of the measured signal, respectively [16].

Velocity maps were created in the Y- and Z-directions for the liquid phase of the foam (hydrogen signal) and in the Z-direction for the gas phase of the foam (fluorine signal). A velocity map of the gas phase in the Y-direction was not acquired due to the large-amplitude, motion-sensitizing gradients that would have been required in that case. The duration of the motion-sensitizing pulses and flow evolution time are $\delta = 0.6$ ms and $\Delta = 6.15$ ms, respectively. For the hydrogen measurements, the maximum amplitude of the motion-sensitizing gradients was 0.037 T/m in the Z-direction and 0.079 T/m in the Y-direction. For the fluorine measurements, the maximum amplitude of the motion-sensitizing gradients was 0.039 T/m in the Z-direction.

Computational fluid dynamics simulations of the foam flow were performed using the SimScale computer-aided engineering software (www.simscale.com) [23] with the OpenFOAM CFD modelling module (www.openfoam.com) [24,25]. The foam was modelled as a Herschel–Bulkley fluid, which is a non-Newtonian, shear thinning fluid. This model has previously been used to describe the flow of aqueous foams. A foam quality value (defined as the ratio of the volume of the gas in the foam to the total liquid and gas volume of the foam) of 0.75 was used for the simulations, based on parameters provided in [17]. A foam quality factor below 0.52 represents a foam that consists of spherical bubbles that are not in contact with each other. Between 0.52 and 0.96, the bubbles are in contact with each other with a corresponding increase in viscosity. Above 0.96, the foam becomes a mist with a corresponding decrease in viscosity. A range of foam quality values of 0.6 to 0.85 was also simulated, and 0.75 was chosen as representative of this range because it proved possible to mimic the experimental results with this simulation.

Author Contributions: Conceptualization, A.A. and B.N.; methodology, A.A., B.N. and S.R.; validation, A.A., B.N. and S.R.; formal analysis, A.A. and S.R.; investigation, A.A. and S.R.; resources, B.N.; data curation, A.A. and S.R.; writing—original draft preparation, A.A.; writing—review and editing, A.A. and B.N.; visualization, A.A. and S.R.; supervision, B.N.; project administration, B.N.; funding acquisition, B.N. All authors have read and agreed to the published version of the manuscript.

Funding: This research was funded by the Natural Sciences and Engineering Research Council of Canada Discovery Grant program [RGPIN-2017-04564].

Data Availability Statement: The data presented in this study are available on request from the corresponding author. The data are not publicly available due to non-standard, proprietary formatting, which will necessitate explanation on sharing.

Conflicts of Interest: The authors declare no conflict of interest. The funders had no role in the design of the study; in the collection, analyses, or interpretation of data; in the writing of the manuscript, or in the decision to publish the results.

References

1. Assink, R.A.; Caprihan, A.; Fukushima, E. Density profiles of a draining foam by nuclear magnetic resonance imaging. *AIChE J.* **1988**, *34*, 2077–2079. [CrossRef]
2. German, J.B.; McCarthy, M.J. Stability of aqueous foams: Analysis using magnetic resonance imaging. *J. Agric. Food Chem.* **1989**, *37*, 1321–1324. [CrossRef]
3. Gonatas, C.P.; Leigh, J.S.; Yodh, A.D. Magnetic resonance images of coarsening inside a foam. *Phys. Rev. Lett.* **1995**, *75*, 573–576. [CrossRef]
4. Stevenson, P.; Mantle, M.D.; Sederman, A.J.; Gladden, L.F. Quantitative measurements of liquid holdup and drainage in foam using NMRI. *AIChE J.* **2006**, *53*, 290–296. [CrossRef]
5. Bhakta, A.; Ruckenstein, E. Decay of standing foams: Drainage, coalescence and collapse. *Adv. Colloid Interfac.* **1997**, *70*, 1–124. [CrossRef]
6. Koehler, S.A.; Hilgenfeldt, S.; Stone, H.A. A generalized view of foam drainage: experiment and theory. *Langmuir* **2000**, *16*, 6327–6341. [CrossRef]
7. Stevenson, P.; Sederman, A.J.; Mantle, M.D.; Li, X.; Gladden, L.F. Measurement of bubble size distribution in a gas–liquid foam using pulsed-field gradient nuclear magnetic resonance. *J. Colloid. Interf. Sci.* **2010**, *352*, 114–120. [CrossRef]
8. Bisperink, C.G.J.; Ronteltap, A.D.; Prins, A. Bubble-size distributions in foams. *Adv. Colloid Interfac.* **1992**, *38*, 13–32. [CrossRef]
9. Deshpande, N.S.; Barigou, M. Foam flow phenomena in sudden expansions and contractions. *Int. J. Multiphas. Flow* **2001**, *27*, 1463–1477. [CrossRef]
10. Tsai, Y.; Chou, F.; Cheng, S. Using tracer technique to study the flow behavior of surfactant foam. *J. Hazard. Mater.* **2009**, *166*, 1232–1237. [CrossRef]
11. Weiger, M.; Pruessmann, K.P. Short-T_2 MRI: Principles and recent advances. *Prog. Nucl. Mag. Res. Spectrosc.* **2019**, *114*, 237–270. [CrossRef] [PubMed]
12. Balcom, B.J.; MacGregor, R.P.; Beyea, S.D.; Green, D.P.; Armstrong, R.L.; Bremner, T.W. Single-point ramped imaging with T_1 enhancement (SPRITE). *J. Magn. Reson. Ser. A* **1996**, *123*, 131–134. [CrossRef] [PubMed]
13. Newling, B.; Poirier, C.C.; Zhi, Y.; Rioux, J.A.; Coristine, A.J.; Roach, D.; Balcom, B.J. Velocity imaging of highly turbulent gas flow. *Phys. Rev. Lett.* **2004**, *93*, 154503. [CrossRef] [PubMed]
14. Cotts, R.M.; Hoch, M.J.R.; Sun, T.; Markert, J.T. Pulsed field gradient stimulated echo methods for improved NMR diffusion measurements in heterogeneous systems. *J. Magn. Reson.* **1989**, *83*, 252–266. [CrossRef]
15. Li, L.; Chen, Q.; Marble, A.E.; Romero-Zeron, L.; Newling, B.; Balcom, B.J. Flow imaging of fluids in porous media by magnetization prepared centric-scan SPRITE. *J. Magn. Reson.* **2009**, *197*, 1–8. [CrossRef]
16. Romanenko, K.; Xiao, D.; Balcom, B.J. Velocity field measurements in sedimentary rock cores by magnetization prepared 3D SPRITE. *J. Magn. Reson.* **2012**, *223*, 120–128. [CrossRef]

17. Dallagi, H.; Gheith, R.; Al Saabi, A.; Faille, C.; Augustin, W.; Benezech, T.; Aloui, F. CFD characterization of a wet foam flow rheological behavior. In Proceedings of the 5th Joint US-European Fluids Engineering Summer Conference, 2018, FEDSM18, Montréal, QC, Canada, 18 July 2018.
18. Herschel, W.H.; Bulkley, R. Konsistenz-messungen von Gummibenzöllösungen. *Kolloid Z.* **1926**, *39*, 291–300. [CrossRef]
19. Sankey, M.; Yang, Z.; Gladden, L.; Johns, M.L.; Lister, D.; Newling, B. SPRITE MRI of bubbly flow in a horizontal pipe. *J. Magn. Reson.* **2009**, *199*, 126–135. [CrossRef] [PubMed]
20. Fheed, A.; Klodowski, K.; Krzyzak, A. Fracture orientation and fluid flow direction recognition in carbonates using diffusion-weighted nuclear magnetic resonance imaging: An example from Permian. *J. Appl. Geophys.* **2020**, *174*, 103964. [CrossRef]
21. Liu, H.; Ding, Y.; Wang, W.; Ma, Y.; Zhu, T.; Ma, D. Dynamic monitoring of polymer flooding using magnetic resonance imaging technology. *Appl. Magn. Reson.* **2020**, 1–17. [CrossRef]
22. Szomolanyi, P.; Goodyear, D.; Balcom, B.J.; Matheson, D. Spiral-SPRITE: A rapid single point MRI technique for application to porous media. *Magn. Reson. Imaging* **2001**, *19*, 423–428. [CrossRef]
23. SimScale. Available online: www.simscale.com (accessed on 28 October 2020).
24. OpenFOAM. Available online: www.openfoam.com (accessed on 28 October 2020).
25. Weller, H.G.; Tabor, G.; Jasak, H.; Fureby, C. A tensorial approach to computational continuum mechanics using object-oriented techniques. *Comput. Phys.* **1998**, *12*, 620–631. [CrossRef]

Article

Magnetic Resonance Imaging of Water Content and Flow Processes in Natural Soils by Pulse Sequences with Ultrashort Detection

Sabina Haber-Pohlmeier [1], David Caterina [2,3], Bernhard Blümich [4] and Andreas Pohlmeier [2,*]

1. Institute for Modelling Hydraulic and Environmental Systems, University of Stuttgart, Pfaffenwaldring 61, D-70569 Stuttgart, Germany; sabina.haber-pohlmeier@iws.uni-stuttgart.de
2. Institute for Bio- and Geosciences Agrosphere (IBG-3), Research Center Jülich, D-52425 Jülich, Germany; David.Caterina@uliege.be
3. Département UEE, Faculté des Sciences Appliquées, Université de Liège, B-4000 Liège, Belgium
4. Institute for Technical and Macromolecular Chemistry, RWTH Aachen University, Worringer Weg 2, D-52074 Aachen, Germany; bluemich@itmc.rwth-aachen.de
* Correspondence: a.pohlmeier@fz-juelich.de

Abstract: Magnetic resonance imaging is a valuable tool for three-dimensional mapping of soil water processes due to its sensitivity to the substance of interest: water. Since conventional gradient- or spin-echo based pulse sequences do not detect rapidly relaxing fractions of water in natural porous media with transverse relaxation times in the millisecond range, pulse sequences with ultrafast detection open a way out. In this work, we compare a spin-echo multislice pulse sequence with ultrashort (UTE) and zero-TE (ZTE) sequences for their suitability to map water content and its changes in 3D in natural soil materials. Longitudinal and transverse relaxation times were found in the ranges around 80 ms and 1 to 50 ms, respectively, so that the spin echo sequence misses larger fractions of water. In contrast, ZTE and UTE could detect all water, if the excitation and detection bandwidths were set sufficiently broad. More precisely, with ZTE we could map water contents down to 0.1 cm^3/cm^3. Finally, we employed ZTE to monitor the development of film flow in a natural soil core with high temporal resolution. This opens the route for further quantitative imaging of soil water processes.

Keywords: magnetic resonance imaging; natural soil material; fast relaxation times; water content; water flow

1. Introduction

Water content and flow in soils belong to the most important processes controlling plant growth and crop yield. They take place on different scales ranging from the field scale down to the pore scale. On a coarse-grained scale the pore system is a continuum, which one may classify in five categories, of which the three most important ones are made up from macropores with voids >75 μm, such as wormholes or dead root holes, mesopores with voids between 75 and 30 μm, and micropores, smaller than 30 μm [1]. The water content of the meso- and micropores in the soil matrix is controlled by capillary suction, whereas macropores are mostly empty with thin water films on the walls. Water flows predominantly through the pore system in the soil matrix, i.e., in the micro- and mesopores. Under certain conditions, such as re-wetting after severe desiccation combined with the formation of cracks or high irrigation rates, preferential flow through the macropore system can significantly contribute to the total water flow. This may take place on each scale between the macropores (10^{-3} m), the core scale (10^{-1} m) up to the pedon and field [2]. While the larger scales >10^0 m are conveniently investigated by, e.g., TDR probe arrays, surface NMR [3] or geophysical methods [4], there is need for high resolution, non-invasive imaging addressing the core and soil aggregate scales.

Three-dimensional, non-invasive imaging techniques are promising tools to improve our understanding of the interplay between soil structure, water content and flow processes (Figure 1). These methods include X-ray CT, an excellent tool for studying the microstructure of soils, i.e., the spatial distribution of minerals and pore networks. Sample and detector sizes control the resolution, where the current limit is about one micrometer for 1 mm wide samples [5,6]. While the contrast in XCT images relies on the density difference between soil particles and voids, magnetic resonance imaging (MRI) and its special subdiscipline magnetic resonance microscopy (MRM) directly probe the local dynamics of the molecule of interest in the pore void: water [7–10]. This makes MRI especially convenient for the investigation of stationary and mobile water in the soil pore system (Figure 1).

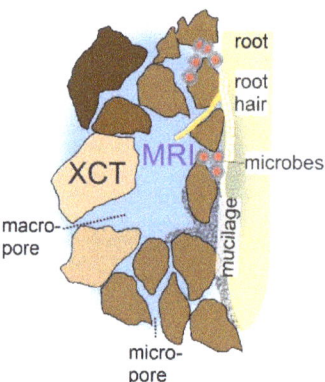

Figure 1. Sketch of the pore system and the domains addressed by different non-invasive imaging methods. X-ray computed microtomography (XCT) is sensitive to the solid soil structure and magnetic resonance imaging (MRI) is sensitive to the fluid in the pore system and its dynamics.

The NMR principle is that certain atomic nuclei possess a quantum-mechanical property called spin, which is linked to a nuclear magnetic moment and interacts with the external magnetic field B_0. Many spins form an ensemble with a macroscopic magnetization aligned with the direction of B_0 in equilibrium. This equilibrium can be disturbed by the application of a radio frequency pulse if its frequency matches the Larmor-frequency of the spin system, $\nu_0 = \gamma B_0 / 2\pi$, where γ is the gyromagnetic ratio of the nucleus under consideration. It is, in most cases, the hydrogen atom ^{1}H, which is abundant in water or hydrocarbons. After excitation, the nuclear magnetization induces a current in the surrounding *rf* coil, which produces the NMR-signal. Further excitation pulses, frequently combined with magnetic field gradient pulses can modify the signal—for instance, to create one or more echoes. In parallel, the equilibrium is re-established by characteristic relaxation processes, denoted as T_1 and T_2 relaxation, sensitive to the physical and chemical environments. The observable signal relies on the interplay between magnetic and dynamic (rotational and translational diffusion, flow) properties of the actual system and method-specific, adjustable parameters. The spatial coordinates of an image are encoded in MRI by switching additional magnetic field gradients before or during signal acquisition. The sequential application of all pulses is termed the MRI pulse sequence. Eventually, Fourier transformation of a time series of NMR signals obtained with systematic variation of the pulsed magnetic field gradients yields the 2D or 3D image. The versatility of the information of MRI is due to the fact that the signal intensity of the individual image pixels, i.e., the contrast, is controlled by factors such as the volumetric water content, the NMR relaxation times T_1 and T_2, diffusion coefficients, and flow velocity, which allows tuning the MRI pulse sequence to be sensitive to certain of these parameters. For instance, the so-called spin-echo multislice sequence (SEMS) is sensitive to the water content if the shortest T_2 relaxation time in the sample is significantly longer than the echo time t_E. On the other hand, if T_2 becomes equal to or shorter than t_E, the pixel intensity is weighted by the local relaxation properties. This has consequences for MRI of natural soil material since soils are natural porous media with a broad distribution of pore sizes and significant fractions of organic matter, clay minerals and paramagnetic ions such as iron or manganese, which greatly impact the soil relaxation times. Thus, the pixel intensity of a SEMS sequence can significantly decrease for such systems with short T_2 since t_E cannot be minimized beyond certain limits, and water content is not reliably mapped [11,12].

The limitation caused by fast T_2 relaxation times has driven the development of a family of different pulse sequences with ultrashort detection times. They have in common that they avoid the time-consuming creation of a spin-echo so that the image intensity is given by the free induction decay (FID), which decays with the relaxation time $T_2{}^*$, and the longitudinal relaxation time T_1. One may differentiate between two classes. Single point imaging methods (SPI, SPRITE) are purely phase encoded and thus do not suffer from susceptibility artefacts [13,14], but require relative long measurement times, especially for 3D imaging. On the other hand, sequences such as ultrashort echo time imaging (UTE) [15], sweep imaging with Fourier transformation (SWIFT) [16], and zero echo time (ZTE) [17,18] allow very short measurement times of some minutes by using frequency encoding of the image dimensions in two or three directions. Their disadvantage is their sensitivity for artefacts caused by internal magnetic field gradients occurring at interfaces between structures with significantly different magnetic susceptibilities. Strategies to address this are the use of extremely short and broadband rf-pulses and short acquisition times. Besides medical applications such as the imaging of bones, tendons, or certain parts of the brain [19,20], these methods are of high convenience in the geo- and material sciences. Examples are fluid content imaging using SPRITE in rock cores [21–23] or mortar [24]. SPRITE is also especially convenient for rapid moisture profiling combined with relaxation time analysis in soil cores [25]. The relative long measuring times motivated other groups to use 3D ZTE instead for imaging fractures in rocks [26] or moisture ingress in rock cores [27,28] where the interpretation is strengthened by correlation with X-ray CT images of the solid matrix. Since ZTE detects also signal from plastic cuvette materials, difference images of the sample and the empty plastic sample holder can be computed so that the scattering of intensity from the plastic holder into the sample was minimized [27].

To overcome the issue caused by the rapid transverse relaxation of water in natural soil materials in the millisecond range [11,12], the objective of this study was to explore the usefulness and applicability of the ultrashort pulse sequences UTE and ZTE for mapping water contents in natural soils. First, we started with the investigation of the relaxation behavior of a selected soil material followed by the determination of optimal pulse sequence parameters for ZTE and UTE with special focus on the interplay between the excitation and detection of pulse bandwidths, acquisition time and flip angles. We continued with the question, if water content can be mapped quantitatively with these methods for different degrees of unsaturation. The final example is the application of ZTE in a case study of water ingress into a natural soil core to answer the question if it is suitable to monitor rapid transient changes of water content occurring in macropore flow.

2. Results

2.1. Relaxometric Imaging, Sample S1

Soil materials frequently possess short transverse and longitudinal relaxation times due to their considerable content of paramagnetic ions and clay minerals [11,12]. Therefore, the relaxation properties of the sample need to be determined before acquiring images for an optimal setup of the imaging pulse sequence parameters. As an example, Figure 2a,b show the FID of the saturated sandy loam sample S1 (Table 1) and the corresponding spectrum. The FID is short with an average $T_2{}^*$ of about 0.15 ms, and the corresponding spectrum has a width at half height of 2.09 kHz whereby the line is not a single Lorentzian.

This makes imaging pulse sequences using the FID such as ZTE or UTE prone for T_2^* blurring so that the acquisition time should be kept as short as possible. Furthermore, gradient echo sequences, frequently used in biomedical imaging, will deliver inferior results, since most of the signal will have decayed at the time the echo is created. Another family of imaging pulse sequences relies on the Hahn echo, whose intensity is controlled by the transverse relaxation time T_2. The transverse relaxation spectrum in Figure 2c shows a very fast component at 3 ms associated with clay-bound and micropore water and a slower component at 20 ms caused by immobile water in the capillary pores. Larger fractions of free water in the macropores were not visible since they were drained under the given conditions. The high fraction of fast relaxing water indicates a considerable loss of image intensity also in spin-echo sequences since the echo time cannot be reduced below a technical limit. The average longitudinal relaxation time T_1 was 50 ms so that the Ernst angle is 28° for a repetition time of 6 ms. Therefore, if the adjusted flip angle is significantly smaller, no interference of saturation on the signal is expected and the signal intensity will be proportional to the volumetric water content.

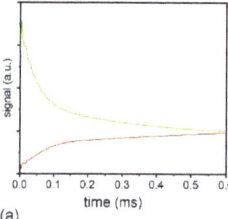
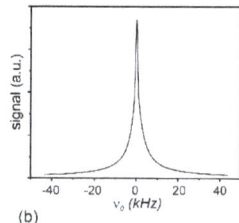
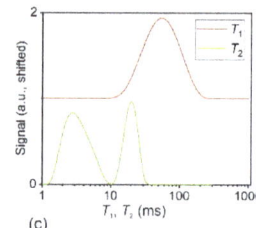

Figure 2. Saturated sandy loam sample S1. (**a**) FID with real and imaginary part in green and red. (**b**) NMR magnitude spectrum. The linewidth is 2.09 kHz corresponding to T_2^* = 0.15 ms. (**c**) Relaxation spectra from a selected region of interest (ROI) centered in one cuvette. The T_2 spectrum was obtained by inverse Laplace transformation of the average intensity of a ROI mapped by a multislice multiecho imaging pulse sequence using an echo time of 1.6 ms and 32 echoes. The T_1 spectrum was obtained from a series of images measured by a multislice single echo imaging pulse sequence with IR preparation with inversion times between 4.2 ms and 300 ms. The average T_1 is 50 ms, T_2 exhibits a fast mode of 3 ms and a slower mode at 20 ms.

Table 1. Description of the samples.

Sample	Type	Texture (Weight-%)			Porosity (cm³/cm³)
		Sand	Silt	Clay	
S1	Kaldenkirchen sandy loam	73	23	4	0.4
S2	Selhausen silt loam	13	70	17	0.45
	Kaldenkirchen sandy loam	73	23	4	0.4
	FH31, sand	100	0	0	0.36
S3	Kaldenkirchen sandy loam	73	23	4	0.4
S4	Selhausen soil core, A-horizon	13	70	17	0.55
R1	0.1 M $CuSO_4$ in 75%/25% D_2O/H_2O	-	-	-	-

2.2. Comparison of MSME and ZTE, Sample S1

Following the initial relaxation analysis, the effects of relaxation on images acquired with different methods are checked. The MSME images of sample S1 yielded an image with sufficient intensity only for the first echo at t_E = 1.6 ms, whereas the intensities of the subsequent echoes decreased significantly (Figure 3a–c). For elucidation, Figure 3e depicts the intensity profiles of some echo images along the yellow line. Even the intensity of the first echo image from the soil was not proportional to the volumetric water content due to the loss of immobile water associated with clay and micropores as revealed by comparing the image intensities with the intensities of water in the marker tubes. The latter remained approximately constant and reflected the volumetric water content of 0.33 cm^3/cm^3. In contrast, the ZTE image (Figure 3d) represented the correct water content, as can be seen by the green profile in Figure 3e. One also notes a slight decay of the ZTE intensity with increasing distance from the center of the field of view. It was caused by the spatial heterogeneity of the *rf*-irradiation intensity, which had to be compensated for by normalizing the ZTE images to an image of a homogeneous reference sample (cf. next section).

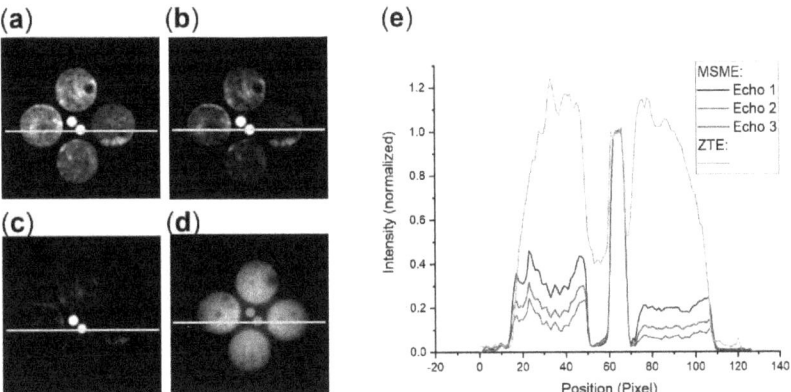

Figure 3. Comparison of MSME and ZTE images of sample S1. The marker tubes in the middle contained 33% water. (**a**–**c**) MSME images of a central axial slice through S1 of echoes 1, 2, and 3. Echo time t_E = 1.6 ms, slice thickness = 2 mm, matrix size 128^2, FOV: 7 × 7 cm^2. (**d**) ZTE image of an axial slice, FOV 7 × 7 × 9 cm^3, matrix size = 128^3 resulting in an in-plane resolution of 0.55 mm and a slice thickness of 0.7 mm. (**e**) Intensity profiles along the horizontal yellow lines, normalized to the intensity of the reference tube containing 0.33 vol-% water.

2.3. Comparison ZTE and UTE Sequences, Sample S2

The rapid T_2^* decay causes not only blurring but makes the ultrashort pulse sequence prone for artefacts if the acquisition and excitation bandwidths are too small. This becomes clear when investigating the composite sample S2 (Table 1) with soil materials of different texture, iron content and T_2^* (Figure 4a). In the top-left image of Figure 4b (i) the fine-textured silt-loam soil is practically invisible, as well as the marker tube in the top compartment. The marker tube becomes more visible in the central compartment filled with the medium textured sandy loam material, although the intensity is distributed laterally due to the susceptibility artefact. The susceptibility difference between soil and marker tube creates local magnetic field gradients, which shift the local resonance frequency and lead to shifted intensities in frequency-encoded images which UTE and ZTE are.

The marker tube is represented correctly only in the sand compartment. The situation improves when the acquisition bandwidths are increased so that the space-encoding gradients become stronger and exceed the internal gradients (Figure 4b (ii) to (iv)). The same is true for the ZTE sequence (Figure 4c). Using a small bandwidth of 100 kHz results in strong artefacts, which are considerably reduced by increasing the bandwidth to 300 kHz. As an interim conclusion both sequences are convenient for soil systems; however, we performed the following experiments in this paper with ZTE.

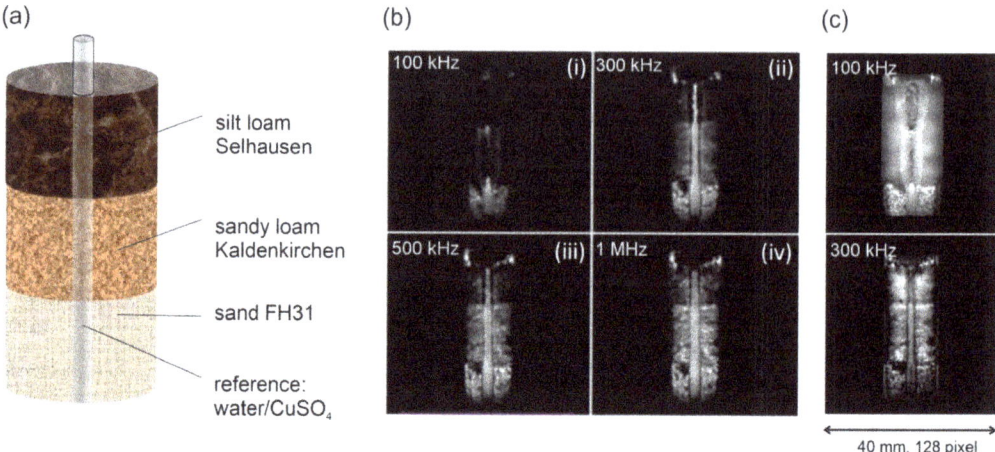

Figure 4. Imaging a composite soil sample S2 with UTE and ZTE. (**a**) Sketch of the phantom. (**b**) UTE images of coronal cross-sections showing the three soil materials and the marker tube in the center. Matrix size and resolution are 128^3 and isotropic 0.31 mm, respectively. Increasing the receiver bandwidth (reduction of acquisition time) from 0.1, 0.3, 0.5, to 1.0 MHz decreased T_2^* blurring artefacts especially for the silt loam soil material. (**c**) ZTE Images: Increasing the receiver bandwidth from 100 to 300 kHz reduced the T_2^* blurring and improved the image quality.

2.4. Check for Volumetric Water Content, Samples S3 and R1

To check the expected linearity between the MRI-ZTE signal with volumetric water content, we scanned in the next step a bundle of four soil-containing cuvettes with different water contents (Sample S3, Table 1). It turned out that the homogeneity of the rf-field was too low so that the intensities near the inner wall of the resonator and more than 2 cm below and above its center decreased, and the image intensities had to be corrected accordingly. For that purpose, we additionally scanned the homogeneity phantom sample R1 with identical pulse-sequence parameters. The 3D ZTE image of the bundle was normalized to this 3D reference image (Figure 5a,b). The slight blurring which is visible in the axial cross sections in Figure 5b is due to the rapid T_2^* decay during the acquisition time of 0.213 ms. This means that the signal of the data points acquired in the outer *k*-space, which determine the resolution, is attenuated with respect to the signal in the center. In contrast to the soil cuvettes, the marker tubes yielded sharp images since T_2^* of the $CuSO_4$ solutions is considerably longer (T_2^* = 1 ms). Finally, we divided the normalized image by the intensity of one of the marker tubes in the center with the known water/D_2O ratio yielding the volumetric water content map of the bundle. The results for a selected central slice are plotted in Figure 5c, confirming the expected linearity between the given and the MRI-ZTE water content.

2.5. Infiltration into a Natural Soil Core, Sample S4

After the optimized parameters for ZTE had been determined, we applied the method on an infiltration experiment in a natural soil core (Table 1). The core (Figure 6a) was initially saturated from the bottom and scanned. The central large wormhole and a neighboring macropore, probably caused be a degraded root, are clearly visible in Figure 6b (green arrows). The soil surface is uneven, and declined to the left side (dotted orange line in Figure 6b).

After some time of equilibration, the first irrigation period of one hour started, monitored by frequent MRI-ZTE scans. During this low irrigation rate of 9 mm/h the water content remains almost constant since the difference images did not change even after 30 min (Figure 6c). The situation changed when doubling of the irrigation rate (Figure 6d). After 6 min ponding water accumulated on the top left corner of the soil column with a minor degree of penetration into the topmost soil area. After 13 min the ponding increased giving rise to film flow in the large wormhole. This effect was even more pronounced after 30 min—see the black arrow in Figure 6d, right panel. When the irrigation rate was further increased to 36 mm/h, the ponding was more pronounced, as well as the film flow inside the wormhole (Figure 6e). After 13 min further ponding at the lower end of the wormhole was noticed, which further intensified after 30 min. Summarizing the observations from all images, two points become clear: (1) Steady states were established after a few minutes—this means that the water content patterns did not change significantly after 13 min. Not shown, but worth mentioning is that the reverse process, the desiccation after stopping the irrigation took place as quickly as the irrigation, i.e., a steady state was established in the same period as after starting the irrigation; (2) The onset of film flow inside the wormhole at an irrigation rate $I > 18$ mm/s proves that at this point the irrigation rate exceeded the hydraulic conductivity of the soil matrix. The hydraulic conductivity was sufficiently high below this critical rate to transport all incoming water through the soil matrix so that no preferential flow was visible. Only if the irrigation rate exceeded this value did the macropore flow contribute to a significant fraction of the overall water flux.

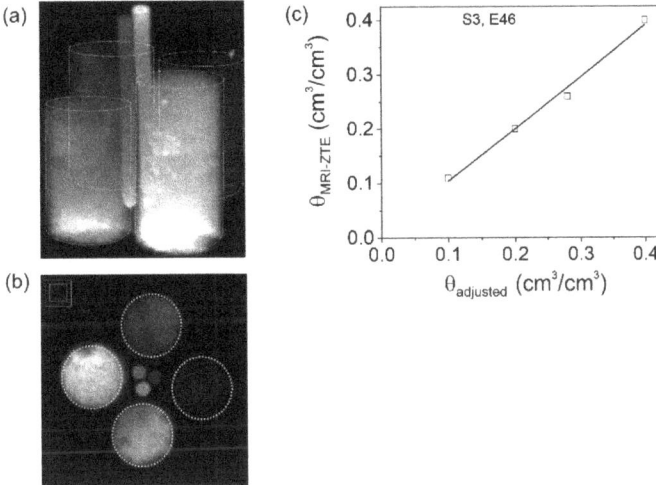

Figure 5. Quantification of water content in sandy loam by means of MRI-ZTE, sample S3. (**a**) The maximum intensity projection shows the four cuvettes with volumetric water contents of 0.10, 0.19, 0.28, and 0.40 cm^3/cm^3. The raw images were normalized firstly on the homogeneity phantom R1 to compensate for radial signal-intensity inhomogeneities and secondly on the signal intensity from the reference cuvettes visible in the center with known volumetric water contents. The yellow wireframes indicate the positions of the cuvettes. (**b**) Axial cross section through the ZTE image. (**c**) Water content obtained from MRI-ZTE versus volumetric water content from sample weights.

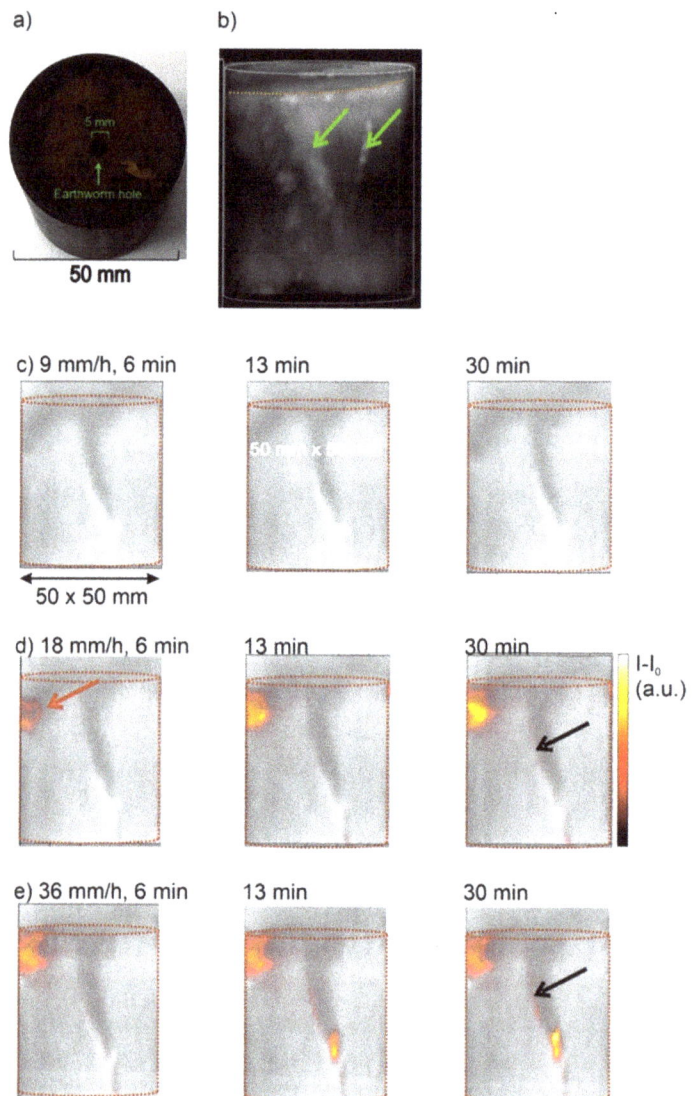

Figure 6. Imaging flow in a natural soil core during irrigation. (**a**) Setup. (**b**) ZTE image of the natural core (maximum intensity projection) with no irrigation. Matrix size 128^3, FOV: $70 \times 70 \times 90$ mm. The dotted orange line indicates the soil surface, the green arrows point to the large wormhole and a smaller macropore, probably caused by a degraded root. (**c**–**e**) Central vertical slice with different irrigation rates (top to bottom) and time-points after start (left to right). Shown are difference images at the given time-points minus t = 0. The red arrow indicates ponding water near the left edge. The black arrows indicate film flow at the walls of the large wormhole.

3. Discussion

Natural soil cores exhibit fast NMR relaxation processes characterized by the longitudinal relaxation time T_1 and the transverse relaxation time T_2. In terms of NMR imaging the transverse relaxation time is the most critical parameter since values on the order of some ms significantly reduce the signal intensity of conventional, echo-based MRI images of the water in the soil matrix. The relaxation time spectrum of the examined sandy loam exhibited short T_2 components in the range of 3 ms and an average value of $T_2^* = 0.15$ ms, which fit well in the range reported for many other soil materials by Hall et al. [11]. One should further keep in mind that T_2 depends strongly on the echo time at higher field strengths due to a significant contribution of diffusional attenuation of the spin-echo by the Bloch–Torrey term. Therefore, a further reduction in T_2 is expected with increasing echo-time. Summarizing, larger water fractions might become undetectable, and spin-echo based pulse sequences are only convenient for the detection of preferential pathways in macropores or fractures [29,30]. A way out for quantitative water content mapping are pulse sequences with ultrashort detection time, which probe the FID directly a few microseconds after the excitation pulse. Although several methods, such as SPI, SPRITE, or UTE are suitable, we focused on ZTE due to its robust implementation and short total scan time.

Pulse sequences with ultrashort detection time are inherently T_1 weighted since they employ short repetition times to allow multiple data acquisition periods for improving the S/N ratio. Therefore, the flip angle should be significantly smaller than the Ernst angle for the given ratio of repetition time t_R and average longitudinal relaxation time T_1 to minimize T_1-weighting. For example, the Ernst angle of the sandy loam soil samples S1, S2, and S3 was 28° for a repetition time of 6 ms and a mean T_1 of 50 ms. Thus we employed flip angles in the range of 3° in the UTE and ZTE protocols to assure the proportionality between pixel signal and volumetric water content.

The sample holder and probe head are frequently made of hydrogen atom-containing plastic materials. These contribute to very rapidly decaying FIDs and thus remain visible with ultrashort detection pulse sequences. In fact, such materials lead to diffuse, heterogeneous image backgrounds that should be corrected. A first correction is a sufficiently large field of view that covers at least a part of the probehead, since the rf-coil always excites these probehead components [18]. Secondly, one can normalize the image of interest on a homogeneous phantom, for instance a water/D_2O mixture in a PTFE cuvette, which has been recorded with identical parameter settings. This strategy was used to validate the quantitative water content imaging in Figure 5. The linearity in Figure 5c between adjusted and MRI measured water content confirms the suitability of ZTE for soil material. Alternatively, if changes in the water content are of interest, one can also calculate difference images of the state at a given point in time and a reference state, which has been successfully applied for water infiltration and desiccation in rock cores [27,28]. We used this procedure in our work to monitor film flow formation as a function of irrigation rate in a core of natural soil (Figure 6). This example shows that it is possible to follow such transient processes with high temporal resolution.

ZTE and UTE are pure frequency encoding sequences, which makes them prone for artefacts by adjacent pixels with a high contrast in T_2^*. An example is shown in Figure 4 where the vertical water-filled NMR tube became essentially invisible when the acquisition bandwidth was set too low. A way out are stronger read-out gradients accompanied by shorter dwell time and acquisition time, which reduces T_2^* blurring. Likewise noteworthy is that the excitation in the ZTE sequence takes place in the presence of strong gradients by short rf-pulses with high power (hard or block pulses). Therefore, one should use a wide excitation bandwidth (here 1.28 MHz) so that the approximate linear range of the central lobe covers the entire field of view.

An alternative to ZTE or UTE is the use of pure phase encoding sequences, i.e., the family of single point imaging pulse sequences. They were originally developed for imaging solids, but also proved very useful for imaging rapidly relaxing fluids in natural porous media [13,20,31,32]. They do not suffer from the artefacts addressed above but demand a long overall measuring time. For instance, using a repetition time of $t_R = 6$ ms and a matrix size of 64^3 pixels the total time for a single scan is about 1/2 h. While this is acceptable for samples in stationary state, for one-dimensional profiling, and 2D imaging [25,33], it might be too long for monitoring rapid transient processes such as film flows. However, SPI in combination with compressed sensing could be also a useful tool [34].

4. Materials and Methods

4.1. MRI Methods and Image Processing

Most samples (with the exception of sample S2) were scanned using a Bruker super-widebore scanner (SWB) at B_0 = 4.7 T, operated by an Avance III console and controlled by Paravision 6 software (Bruker Microimaging, Rheinstetten, Germany). The gradient system had a maximum strength of 0.6 T/m, and we used a ^{1}H probehead with 66 mm internal diameter. We used the following pulse sequences provided by the manufacturer: multislice multiecho (MSME), zero-TE (ZTE) and ultrashort TE (UTE) with the parameters specified below. Figure 7 depicts schematically the pulse sequence diagrams. Additionally, sample S2 was scanned with UTE and ZTE in the applications lab of Bruker in Rheinstetten using a WB scanner at B_0 = 7T. The ZTE and UTE image were reconstructed by re-gridding to Cartesian coordinates of 128^3 points without further filtering or zero-filling by Paravision, and finally displayed by Fiji [31].

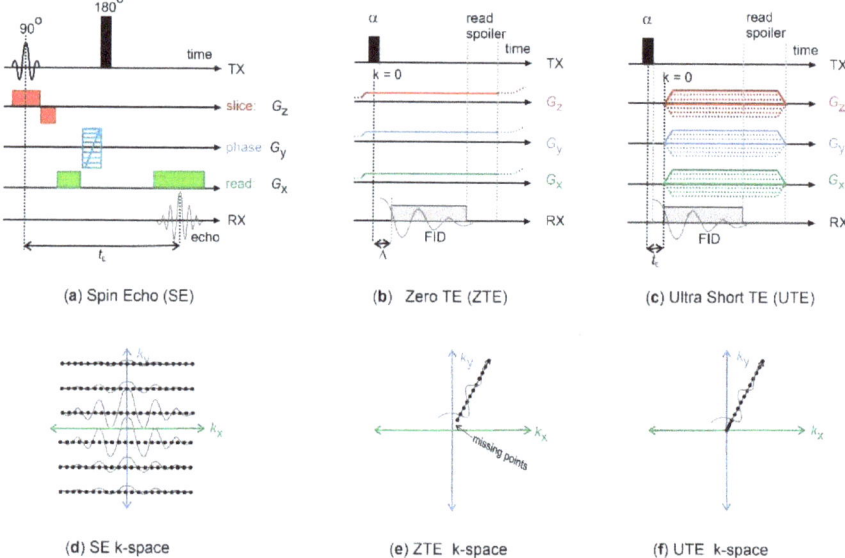

Figure 7. Simplified MRI pulse sequence diagrams. (**a**) Spin-echo (SE) sequence, also termed multislice multiecho sequence (MSME), if more than one echo more recorded per excitation and phase-encoding step. A combination of the 90° rf pulse with the slice gradient G_z excites an individual slice, the 180° rf pulse refocuses the dephased magnetization in the xy-plane and creates an echo after echo time t_E, which is read-out in presence of the read gradient G_x and stored. The phase gradient G_y encodes the 3rd dimension. Thus, the corresponding k-space (**d**) is filled in Cartesian coordinates. (**b**) In the ZTE sequence a non-slice selective hard rf-pulse excites the spin system with a flip angle α after the three spatially encoding gradients, G_x, G_y, and G_z have been switched on. Instead of an echo, the FID is monitored by n_{acq} complex data points after the dead time Δ. Different combinations of the gradients separated by the spoiling time achieve spatial encoding. (**e**) The k-space is filled radially, shown is one exemplary projection line. (**c**) In the UTE sequence a non-slice selective hard rf-pulse excites the spin system with a flip angle α before the three spatially encoding gradients, G_x, G_y, and G_z are switched on. Like in ZTE, the FID is monitored and stored, and space is encoded by different combinations of the gradients. (**f**) The k-space is filled radially.

4.2. Sample S1

We assembled a bundle of four quartzglass cuvettes with 20 mm inner diameter, filled with repacked soil material from Kaldenkirchen. Its textural composition is summarized in Table 1. Iron content was 0.25% [32]. The bulk density was 1.55 g/cm^3 corresponding to a porosity of 0.41 cm^3/cm^3. The cuvettes were initially saturated from the bottom with tap water so that the volumetric water content was θ = 0.36 cm^3/cm^3 and closed with Parafilm. First, the FID was measured using a single pulse scan. We inserted two 5 mm NMR tubes as markers into this sample, which were filled with 0.04 and 0.1 M CuSO$_4$ in a 33% water/67% D$_2$O mixture. Next, the bundle was scanned with a multislice multiecho pulse sequence (MSME) to determine T_2 using following parameters: echo time t_E = 1.6 ms, representing the shortest possible value of the multislice imaging sequence for these systems, number of echoes n_E = 32, 20 slices with a thickness of 2 mm, and a gap of 0.1 mm. The receiver bandwidth was 300 kHz and 16 scans were averaged using a repetition time of 1.0 s. Carr–Purcell–Meiboom–Gill (CPMG) curves were obtained by plotting the average intensities in a ROI inside one column using Fiji.

To determine T_1, we used a single echo multislice sequence with the same settings as described above with the only exception that only one echo was monitored. We set 18 different inversion times in the range between 4.2 ms and 300 ms, plus one reference scan without inversion–recovery (IR) preparation. The T_1 relaxation curve was constructed by normalization of the IR prepared scans on the reference scan in a selected ROI. Finally, the relaxation spectra were obtained by inverse Laplace transformation of the time domain data using Prospa (Magritek, Wellington, New Zealand).

Finally, sample S1 was scanned with ZTE to compare the ZTE images to the MSME images using following parameters: FoV 70 × 70 × 90 mm^3, excitation bandwidth 1.2 MHz at a block pulse length of 1 µs for a flip angle of 2.5°. The receiver bandwidth of 300 kHz and the number of 64 points per spoke in radial k-space resulted in an acquisition time of 0.213 ms. The number of projections was 51897. Repetition time was t_R = 6 ms, and 8 scans were averaged.

4.3. Sample S2

Sample S2 served as test phantom for the comparison of different pulse sequences at the application lab of Bruker. We filled medium sand FH31, sandy loam, and silt loam into a 20 mm wide cuvette plus a 5 mm NMR marker tube filled with 25% water/D$_2$O mixture, see Table 1. The sample was scanned in saturated state by UTE and ZTE pulse sequences with different ratios of excitation and receiver bandwidths to demonstrate the different extend of T_2* blurring artefacts.

4.4. Sample S3

To determine the MRI images of different water contents, we assembled a new sample analogous to sample S1 but with three additional 5 mm NMR reference tubes filled with 10%, 20%, and 30% water/D$_2$O mixtures and 0.2 M CuSO$_4$ (Table 1, Figure 8a). Water was sucked out of three cuvettes to adjust volumetric water contents of θ = 0.10, 0.19, 0.28, and 0.40 cm^3/cm^3. These data were obtained by normalization of the mass differences between dry and wet soil cuvettes on the bulk volume of the packed soil. This sample was scanned with ZTE using following parameters: FoV 90 × 90 × 90 mm^3, excitation bandwidth 1.2 MHz at a block pulse length of 1 µs for a flip angle of 3°. The receiver bandwidth of 300 kHz and the number of 64 points per spoke in radial k-space resulted in an acquisition time of 0.213 ms. The number of projections was 51897. The repetition time was t_R = 6 ms and 4 scans were averaged so that the total measuring time for one scan was 20 min 45 s.

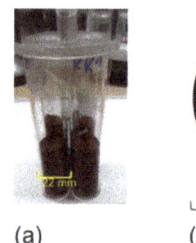
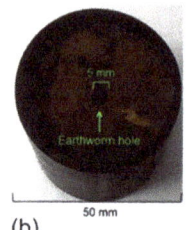

(a) (b)

Figure 8. Test samples and soil core. (**a**) Sample S3 consisting of four 22 mm × 100 mm cuvettes filled with sandy loam soil and three reference standard 5 mm NMR tubes. (**b**) Sample S4: Soil core from Selhausen test site in a 100 cm^3 PVC cutting cylinder (silt-loam).

4.5. Sample S4

A natural soil core, sample S4, was taken from the topsoil at 5 cm depth in the lower part of the test site Niederzier-Selhausen, Germany (Figure 8b). The texture is silt loam with few inclusions of gravel and stones (Table 1). The cutting cylinder was produced from PVC and had an inner diameter of 50 mm and a total volume of 100 cm^3 (Figure 3b). It was closed at the bottom by a filter plate so that the core could be saturated gently with water from bottom without disturbing the macropore structure. After placing it into the scanner, it was irrigated from top at different rates and the percolating water was collected below outside the magnet.

Scan parameters of the ZTE sequence were a field of view of 70 × 70 × 90 mm^3. The excitation bandwidth was 1.2 MHz at a block pulse length of 1 μs for a flip angle of 3°, receiver bandwidth of 300 kHz resulting in an acquisition time of 0.213 ms. The number of projections (spokes of 64 points length in radial k-space) was 51897. Repetition time was t_R = 2 ms for 1 scan so that the total measuring time for one scan was 1 min 43 s We recorded 35 individual scans during irrigation periods of 1 h to monitor the infiltration and desiccation processes with high temporal resolution. The data were reconstructed with Paravision and further data processing and displayed with Matlab (The Mathworks Inc., Natick, MA, USA) and Fiji [33].

4.6. Sample R1: Homogeneity Phantom

To compensate for radial inhomogeneities of the ZTE and UTE sequences resulting from inhomogeneous *rf* irradiation and the reconstruction process we set up a phantom consisting of 55 mm internal diameter Teflon cylinder, filled with 0.1 M CuSO$_4$ in 75%/25% D$_2$O/H$_2$O mixture to a height of 53 mm. This was scanned with identical parameters as the soil samples. The raw images obtained with ZTE and UTE were normalized on this homogeneity phantom.

5. Conclusions

We have shown that quantitative 3D imaging of water content in natural soil samples is possible by MRI-ZTE even for unsaturated soil materials. The obtained resolution in the range of 0.5 mm was controlled by the size of the FOV divided by the number of points, typically 128. This results in a minimum total measurement time of 7 min for a repetition time of 8 ms, sufficient for monitoring rapid infiltration processes. The length of the FID decay may limit the effective resolution so that one should keep the acquisition time as short as possible by adjusting large acquisition bandwidths. This in turn requires parallel high excitation bandwidths on the order of 1 MHz to excite the entire sample homogeneously in presence of the space-encoding gradients, which are already switched on at the time of the excitation pulse. For the image processing normalization to the image of a homogeneity phantom or the creation of difference-images are advantageous to compensate for inherent inhomogeneities of the rf-field and intensity scattering over the entire FOV by solidlike ^{1}H containing material. With these requirements taken into account, the method is a valuable instrument for monitoring changes in water content patterns in natural soil cores by infiltration, desiccation or root-soil processes.

Author Contributions: Conceptualization, S.H.-P., B.B. and A.P.; methodology, S.H.-P.; validation, S.H.-P., A.P. and D.C.; formal analysis, all coauthors.; investigation, S.H.-P., A.P. and D.C.; resources, S.H.-P., A.P. and D.C.; data curation, S.H.-P., A.P., D.C.; writing—original draft preparation, all coauthors; writing—review and editing, all coauthors; visualization, A.P. and D.C.; supervision, S.H.-P. and A.P.; project administration, S.H.-P., A.P.; funding acquisition, S.H.-P., A.P. and B.B. All authors have read and agreed to the published version of the manuscript.

Funding: This research was funded by The German Research Fund, DFG within the frame of the Transregional Collaborative Research Center DFG TR32, partial project T1.

Data Availability Statement: Not applicable.

Acknowledgments: The authors thank Dieter Gross, Volker Lehmann, Thomas Oerther, and Klaus Zick, Bruker AG, for valuable assistance with the setup of pulse sequences as well as for the opportunity to conduct test measurements in their application lab in Rheinstetten, Germany within the frame of TR32-T1. See also Figure 4.

Conflicts of Interest: The authors declare no conflict of interest.

Sample Availability: Samples of the compounds described in Table 1 are available from the authors.

References

1. Kirkham, M.B. *Principles of Soil and Plant Water Relations*; Elsevier: New Jersey, NJ, USA, 2005.
2. Clothier, B.E.; Green, S.R.; Deurer, M. Preferential flow and transport in soil: Progress and prognosis. *Eur. J. Soil Sci.* **2008**, *59*, 2–13. [CrossRef]
3. Yaramanci, U.; Legchenko, A.; Roy, J. Magnetic Resonance Sounding Special Issue of Journal of Applied Geophysics, 2008. *J. Appl. Geophys.* **2008**, *66*, 71–72. [CrossRef]
4. Vanderborght, J.; Huisman, J.A.; van der Kruk, J.; Vereecken, H. Geophysical Methods for Field-Scale Imaging of Root Zone Properties and Processes. In *Soil-Water-Root Processes: Advances in Tomography and Imaging*; Anderson, S.H., Hopmans, J.W., Eds.; Soil Science Society of America, Inc.: Wisconsin, WI, USA, 2013; Volume 61, pp. 247–282.
5. Keyes, S.D.; Daly, K.R.; Gostling, N.J.; Jones, D.L.; Talboys, P.; Pinzer, B.R.; Boardman, R.; Sinclair, I.; Marchant, A.; Roose, T. High resolution synchrotron imaging of wheat root hairs growing in soil and image based modelling of phosphate uptake. *New Phytol.* **2013**, *198*, 1023–1029. [CrossRef]
6. Peth, S. Applications of Microtomography in Soils and Sediments. In *Developments in Soil Science*; Singh, B., Gräfe, M., Eds.; Elsevier B.V.: Amsterdam, The Netherlands, 2010; Volume 34, pp. 73–191.
7. Callaghan, P.T. *Principles of Nuclear Magnetic Resonance Microscopy*; Oxford University Press: Oxford, UK, 1991.
8. Blümich, B. *NMR Imaging of Materials*; Clarendon Press: Oxford, OU, UK, 2000.
9. Blümich, B.; Haber-Pohlmeier, S.; Zia, W. *Compact NMR*; De Gruyter: Berlin, Germany, 2014.
10. Haber-Pohlmeier, S.; Blümich, B.; Ciobanu, L. (Eds.) *Magnetic Resonance Microscopy: Instrumentation and Application in Engineering, Life Science and Energy Research*; Wiley-VCH: Weinheim, Germany, 2021.
11. Hall, L.D.; Amin, M.H.G.; Dougherty, E.; Sanda, M.; Votrubova, J.; Richards, K.S.; Chorley, R.J.; Cislerova, M. MR properties of water in saturated soils and resulting loss of MRI signal in water content detection at 2 tesla. *Geoderma* **1997**, *80*, 431–448. [CrossRef]
12. Haber-Pohlmeier, S.; Stapf, S.; Van Dusschoten, D.; Pohlmeier, A. Relaxation in a Natural soil: Comparison of Relaxometric Imaging, T1–T2 Correlation and Fast-Field Cycling NMR. *Open Magn. Reson. J.* **2010**, *3*, 57–62. [CrossRef]
13. Balcom, B.J.; MacGregor, R.P.; Beyea, S.D.; Green, D.P.; Armstrong, R.L.; Bremner, T.W. Single-point ramped imaging with T-1 enhancement (SPRITE). *J. Magn. Reson. Ser. A* **1996**, *123*, 131–134. [CrossRef]
14. Muir, C.E.; Balcom, B.J. A comparison of magnetic resonance imaging methods for fluid content imaging in porous media. *Magn. Reson. Chem.* **2013**, *51*, 321–327. [CrossRef]
15. Robson, M.D.; Gatehouse, P.D.; Bydder, M.; Bydder, G.M. Magnetic Resonance: An Introduction to Ultrashort TE (UTE) Imaging. *J. Comput. Assist. Tomogr.* **2003**, *27*, 825–846. [CrossRef] [PubMed]
16. Garwood, M. MRI of fast relaxing spins. *J. Magn. Res.* **2013**, *229*, 49–54. [CrossRef]
17. Weiger, M.; Pruessmann, K.P.; Hennel, F. MRI with Zero Echo Time: Hard versus Sweep Pulse Excitation. *Magn. Res. Med.* **2011**, *66*, 379–389. [CrossRef]
18. Weiger, M.; Pruessmann, K.P. MRI with Zero Echo Time. *eMagRes* **2012**, *1*, 311–322. [CrossRef]
19. Weiger, M.; Brunner, D.O.; Dietrich, B.E.; Müller, C.F.; Pruessmann, K.P. ZTE Imaging in Humans. *Manetic Reson. Med.* **2013**, *70*, 328–332. [CrossRef] [PubMed]
20. Mastikhin, I.V.; Balcom, B.J. Centric SPRITE MRI of Biomaterials with Short T2*. *eMagRes* **2007**, *1*, 783–788. [CrossRef]

21. Marica, F.; Goora, F.G.; Balcom, B.J. FID-SPI pulse sequence for quantitative MRI of fluids in porous media. *J. Magn. Res.* **2014**, *240*, 61–66. [CrossRef]
22. Li, L.Q.; Chen, Q.; Marble, A.E.; Romero-Zerón, L.; Newling, B.; Balcom, B. Flow imaging if fluids in porous media by magnetization prepared centric-scan SPRITE. *J. Magn. Res.* **2009**, *197*, 1–8. [CrossRef]
23. Romanenko, K.V.; Balcom, B.J. Permeability mapping in porous media by magnetization prepared centric-scan SPRITE. *Exp. Fluids* **2011**, *50*, 301–312. [CrossRef]
24. Enjilela, R.; Cano-Barrita, P.F.J.; Komar, A.; Boyd, A.J.; Balcom, B.J. Wet front penetration with unsteady state wicking in mortar studied by Magnetic Resonance Imaging (MRI). *Mater. Struct.* **2018**, *51*, 1–16. [CrossRef]
25. Merz, S.; Pohlmeier, A.; Balcom, B.J.; EnJilela, R.; Vereecken, H. Drying of a Natural Soil Under Evaporative Conditions: A Comparison of Different Magnetic Resonance Methods. *Appl. Magn. Res.* **2016**, *47*, 121–138. [CrossRef]
26. Covington, K.L.; Goroncy, A.K.; Lehmann, T.E.; Kou, Z.H.; Wang, H.; Alvarado, V. Analysis of ZTE MRI application to sandstone and carbonate. *AIChE J.* **2021**, *67*. [CrossRef]
27. Weglarz, W.P.; Krzyzak, A.; Machowski, G.; Stefaniuk, M. ZTE MRI in high magnetic field as a time effective 3D imaging technique for monitoring water ingress in porous rocks at sub-millimetre resolution. *Magn. Reson. Imaging* **2018**, *47*, 54–59. [CrossRef]
28. Weglarz, W.P.; Krzyzak, A.; Stefaniuk, M. ZTE imaging of tight sandstone rocks at 9.4 T-Comparison with standard NMR analysis at 0.05 T. *Magn. Reson. Imaging* **2016**, *34*, 492–495. [CrossRef]
29. Jelinkova, V.; Snehota, M.; Pohlmeier, A.; van Dusschoten, D.; Cislerova, M. Effects of entrapped residual air bubbles on tracer transport in heterogeneous soil: Magnetic resonance imaging study. *Org. Geochem.* **2011**, *42*, 991–998. [CrossRef]
30. Haber-Pohlmeier, S.; Bechtold, M.; Stapf, S.; Pohlmeier, A. Water Flow Monitored by Tracer Transport in Natural Porous Media Using MRI. *Vadose Zone J.* **2010**, *9*, 835–845. [CrossRef]
31. Schindelin, J.; Arganda-Carreras, I.; Frise, E.; Kaynig, V.; Longair, M.; Pietzsch, T.; Preibisch, S.; Rueden, C.; Saalfeld, S.; Schmid, B.; et al. Fiji: An open-source platform for biological-image analysis. *Nat. Methods* **2012**, *9*, 676–682. [CrossRef]
32. Pohlmeier, A.; Haber-Pohlmeier, S.; Stapf, S. A Fast Field Cycling Nuclear Magnetic Resonance Relaxometry Study of Natural Soils. *Vadose Zone J.* **2009**, *8*, 735–742. [CrossRef]
33. Rosin-Paumier, S.; Leclerc, S.; Abdallah, A.; Stemmelen, D. Determination of the unsaturated hydrraulic conductivity of clayey soil columns using Magnetic Resonance Imaging (MRI). In *Unsaturated Soils: Research and Applications*; Khalil, N., Russell, A., Khoshgalb, A., Eds.; Taylor and Francis: London, UK, 2014; pp. 1155–1162.
34. Rioux, J.A.; Beyea, S.D.; Bowen, C.V. 3D single point imaging with compressed sensing provides high temporal resolution R_2^* mapping for in vivo preclinical applications. *Magn. Reson. Mater. Phys. Biol. Med.* **2017**, *30*, 41–55. [CrossRef]

Article

Comparison of Five Conductivity Tensor Models and Image Reconstruction Methods Using MRI

Nitish Katoch [1], Bup-Kyung Choi [1], Ji-Ae Park [2,*], In-Ok Ko [2] and Hyung-Joong Kim [1,*]

1. Department of Biomedical Engineering, Kyung Hee University, Seoul 02447, Korea; nitish@khu.ac.kr (N.K.); josh_bk@naver.com (B.-K.C.)
2. Division of Applied RI, Korea Institute of Radiological and Medical Science, Seoul 01812, Korea; inogi99@kirams.re.kr
* Correspondence: jpark@kirams.re.kr (J.-A.P.); bmekim@khu.ac.kr (H.-J.K.)

Abstract: Imaging of the electrical conductivity distribution inside the human body has been investigated for numerous clinical applications. The conductivity tensors of biological tissue have been obtained from water diffusion tensors by applying several models, which may not cover the entire phenomenon. Recently, a new conductivity tensor imaging (CTI) method was developed through a combination of B1 mapping, and multi-b diffusion weighted imaging. In this study, we compared the most recent CTI method with the four existing models of conductivity tensors reconstruction. Two conductivity phantoms were designed to evaluate the accuracy of the models. Applied to five human brains, the conductivity tensors using the four existing models and CTI were imaged and compared with the values from the literature. The conductivity image of the phantoms by the CTI method showed relative errors between 1.10% and 5.26%. The images by the four models using DTI could not measure the effects of different ion concentrations subsequently due to *prior* information of the mean conductivity values. The conductivity tensor images obtained from five human brains through the CTI method were comparable to previously reported literature values. The images by the four methods using DTI were highly correlated with the diffusion tensor images, showing a coefficient of determination (R^2) value of 0.65 to 1.00. However, the images by the CTI method were less correlated with the diffusion tensor images and exhibited an averaged R^2 value of 0.51. The CTI method could handle the effects of different ion concentrations as well as mobilities and extracellular volume fractions by collecting and processing additional B1 map data. It is necessary to select an application-specific model taking into account the pros and cons of each model. Future studies are essential to confirm the usefulness of these conductivity tensor imaging methods in clinical applications, such as tumor characterization, EEG source imaging, and treatment planning for electrical stimulation.

Keywords: electrical conductivity; anisotropy; magnetic resonance imaging (MRI); diffusion tensor imaging (DTI); conductivity tensor imaging (CTI)

Citation: Katoch, N.; Choi, B.-K.; Park, J.-A.; Ko, I.-O.; Kim, H.-J. Comparison of Five Conductivity Tensor Models and Image Reconstruction Methods Using MRI. *Molecules* **2021**, *26*, 5499. https://doi.org/10.3390/molecules26185499

Academic Editor: Igor Serša

Received: 28 June 2021
Accepted: 7 September 2021
Published: 10 September 2021

Publisher's Note: MDPI stays neutral with regard to jurisdictional claims in published maps and institutional affiliations.

Copyright: © 2021 by the authors. Licensee MDPI, Basel, Switzerland. This article is an open access article distributed under the terms and conditions of the Creative Commons Attribution (CC BY) license (https://creativecommons.org/licenses/by/4.0/).

1. Introduction

Electrical conductivity of biological tissues is determined by the cell density, extracellular volume fraction, composition and amount of extracellular matrix materials, and membrane characteristics as well as concentrations and mobility of ions in the extracellular and intracellular fluids [1]. The apparent macroscopic conductivity of such a composite material has been studied since the early 1900s and can be expressed as a weighted sum of conductivity values of its components based on the volume fractions and other factors [2,3]. The extracellular and intracellular fluids are conductors through which conductivity values are determined by concentrations and mobilities of ions and other mobile charge carriers. Cells with thin membranes behave like an insulator and lossy dielectric at low and high frequencies, respectively. Extracellular matrix materials are lossy dielectrics. Therefore, the macroscopic tissue conductivity exhibits frequency dependency [4,5].

When elongated cells are aligned towards a certain direction, movements of ions in the extracellular fluid are consequently hindered. Under a low-frequency electric field, the ions in the extracellular space are forced to move along the longitudinal direction, thereby making their mobilities direction-dependent. Therefore, in the white matter and muscle, the conductivity exhibits anisotropic properties at low frequencies. However, at high frequencies, the insulating cell membranes behave like a capacitor, and the anisotropic properties disappear above 1 MHz, for example [1,6]. In this paper, we approximately express the low-frequency conductivity of biological tissue as a tensor that is a symmetric positive definite 3×3 matrix [7,8]. At a high-frequency above 1 MHz, the conductivity is expressed as a scalar quantity [9].

Electrical conductivity is a passive material property whose measurement requires a probing current to generate a signal affected by the conductivity. In impedance imaging area, there are two different approaches in conductivity imaging using MRI. Magnetic resonance electrical impedance tomography (MREIT) reconstructs an image of low-frequency isotropic conductivity (σ_L) distribution by injecting low-frequency currents into a subject and measuring the induced magnetic flux density distributions using an MRI scanner [10–12]. Magnetic resonance electrical properties tomography (MREPT) produces high-frequency isotropic conductivity (σ_H) and permittivity (ϵ_H) images by generating a radio-frequency (RF) eddy current that is affected by σ_H and ϵ_H and measuring an induced RF magnetic field using a B1 mapping method [13–15].

In the diffusion tensor imaging (DTI) area, conductivity tensor image reconstructions have been investigated based on a physical relationship that conductivity and water diffusion tensors, denoted as **C** and **D**, respectively, have the same eigenvectors [16]. Based on this, Tuch et al. derived the first linear model between **C** and **D**, which enabled transformation of a diffusion tensor image into a conductivity tensor image [17,18]. Based on the idea of cross-property relation [17], three more conductivity tensor models and image reconstruction methods using DTI were developed [19–22]. These conductivity tensor models are based on a linear relation of $\mathbf{C} = \eta \mathbf{D}$ and the corresponding image reconstruction methods determine the scale factor η.

Although MREIT and MREPT reconstruct images of σ_L and σ_H, respectively, through measurement of the effects of both ion concentrations and mobilities, neither of these two methods can produce an image of **C**. On the other hand, the four methods using DTI produce an image of **C**, but the effects of ion concentrations are not adequately demonstrated. To overcome these limitations, Sajib et al. developed a novel low-frequency conductivity tensor imaging (CTI) method [23]. In the first step of the CTI method, MREPT is used to reconstruct an image of σ_H including the effects of both ion concentrations and mobilities at the Larmor frequency. In the second step, they apply a multi-b diffusion weighted imaging method to obtain an image of **D** and pixel-by-pixel information about extracellular and intracellular spaces. Combining all of these, the scale factor η between **C** and **D** is determined for every pixel using the following: (1) a physical relationship between the water diffusivity and ion mobility and (2) a model-based relation between σ_H and σ_L [6,24].

Recently, Wu et al. reviewed the DTI-based reconstruction models [25]. They reported that the conversion coefficient of water diffusion tensor to electrical conductivity tensor should require the information on ion concentration and extracellular volume fraction [25]. In addition, knowledge of accurate anisotropic conductivity can achieve reliable volume conduction models of electrical brain stimulation and EEG dipole reconstructions [26,27]. Over the years, considerable variation in the conductivity reconstruction models used for such approaches has been observed [27–29]. Consequently, an in-depth investigation of conductivity tensor models was required.

In this paper, we compared the accuracy of the five methods in reconstructed conductivity images of two phantoms using a 9.4 T research MRI scanner. The data acquired from five human subjects using a clinical 3 T MRI scanner were used to reconstruct conductivity tensor images of the brains using the five methods. The reconstructed conductivity tensor images of the human brains were analyzed and compared with each other and also with previous literature values.

2. Five Conductivity Tensor Models

2.1. Linear Eigenvalue Model (LEM)

Adopting the physical analysis that **C** and **D** have the same eigenvectors [16], Tuch et al. assumed that the conductivity of the intracellular space is negligible, that is, $\sigma_i \approx 0$ at low frequencies, and as a result derived the following relation between the eigenvalues of **C** and **D** [17]:

$$c_m \approx \frac{\sigma_e}{d_e}\left[d_m\left(\frac{d_i}{3d_e}+1\right)+\frac{d_m^2 d_i}{3d_e^2}-\frac{2}{3}d_i\right] \tag{1}$$

where c_m and d_m for $m = 1, 2, 3$ are the eigenvalues of **C** and **D**, respectively; σ_e is the extracellular conductivity; and d_i and d_e are the intracellular and extracellular water diffusion coefficients, respectively. Assuming that $d_i \approx 0$, the following linear relation between **C** and **D** was derived:

$$\mathbf{C} = \frac{\sigma_e}{d_e}\mathbf{D} = \eta\mathbf{D}. \tag{2}$$

Without measuring σ_e and d_e in (2), the empirically-calculated scale factor $\eta = 0.844\,\text{S}\cdot\text{s}/\text{mm}^3$ was applied to all pixels [17].

Since this empirical scale factor does not consider any intra-subject and inter-subject variabilities, a modified approach was proposed where the volume of the conductivity tensor ellipsoid was matched with the cubed value of the isotropic conductivity using the least square method [18]. For the human brain, the scale factor η was determined as

$$\eta = \frac{d_{WM}\sigma_{WM} + d_{GM}\sigma_{GM}}{d_{WM}^2 + d_{GM}^2} \tag{3}$$

where d_{WM} and d_{GM} are the measured water diffusion coefficient of the white matter (WM) and gray matter (GM), respectively. In this paper, we used (3) with the literature values of $\sigma_{WM,GM} = 0.14$ and 0.27 S/m, respectively, for human brain imaging experiments [30–32]. For a conductivity phantom, σ_{WM} and σ_{GM} were replaced by measured conductivity values using an impedance analyzer.

2.2. Force Equilibrium Model (FEM)

The relation between the diffusion coefficient d and viscosity ν is given by the Stokes–Einstein relation as follows:

$$d = \frac{k_B T}{6\pi\nu r} \tag{4}$$

where k_B is Boltzmann's constant, T is the absolute temperature, and r is the radius of a spherical particle [19]. From the equilibrium condition between the electric and viscous forces, the conductivity σ can be expressed as

$$\sigma = \frac{J}{E} = \frac{q^2 N}{6\pi r\nu} \tag{5}$$

where J and E are the magnitude of the current density and electric field, respectively, $q = 1.6 \times 10^{-19}$ C and $N = 2 \times 10^{25}$ m^{-3}. Applying (4) and (5) to the extracellular space, the following relation can be derived:

$$\sigma_e = \frac{0.76q^2 N}{k_B T}d_e \tag{6}$$

where $k_B T = 4.1 \times 10^{-21}$ J. For an anisotropic case, Sekino et al. assumed that the following relation can be inferred from (6) [19]:

$$\mathbf{C} = \frac{0.76q^2 N}{k_B T}\mathbf{D}_e = \eta \mathbf{D}_e. \tag{7}$$

To measure $\mathbf{D}_e$ in (7) for each pixel, Sekino et al. [20] adopted the following bi-exponential model for a diffusion-weighted MRI signal S_b with a given b value [33,34]:

$$\frac{S_i(b)}{S_i(0)} = v_{f,i} e^{-bd_{f,i}} + v_{s,i} e^{-bd_{s,i}} \tag{8}$$

where $S_i(b)$ denote the MR signal with a diffusion gradient ($b \neq 0$) along the ith direction, $S_i(0)$ is the MRI signal without applying a diffusion gradient ($b = 0$), $v_{f,i}$ and $d_{f,i}$ are the volume fraction and diffusion coefficient of a fast diffusion component along the ith direction, respectively, and $v_{s,i}$ and $d_{s,i}$ are the volume fraction and diffusion coefficient of a slow diffusion component along the ith direction, respectively. Using a curve fitting method, the fast diffusion tensor $\mathbf{D}_f$ is extracted from the measured data of $S_i(b)$ and $S_i(0)$ in three orthogonal directions. In (7), $\mathbf{D}_e$ is replaced by $\mathbf{D}_f$ assuming that the fast diffusion corresponds to the extracellular diffusion.

2.3. Volume Constraint Model (VCM)

Miranda et al. [21] suggested a method to determine the scaling factor η at each pixel using the measured water diffusion tensor $\mathbf{D}$ and the values of equivalent isotropic conductivity σ_{iso} for different brain tissues from the given literature [30–32,35] as follows:

$$\mathbf{C} = \frac{3\sigma_{iso}}{\text{trace}(\mathbf{D})}\mathbf{D} = \eta \mathbf{D} \tag{9}$$

where $\text{trace}(\mathbf{D})$ is the sum of the three eigenvalues of $\mathbf{D}$. In this paper, the conductivity phantom, σ_{iso} was replaced by a measured conductivity value using an impedance analyzer. For the human brain, we used the literature values of σ_{iso} = 0.14, 0.27, and 1.79 S/m for the WM, GM, and cerebrospinal fluid (CSF), respectively, [30–32,35].

2.4. Volume Fraction Model (VFM)

For the brain tissue, Wang et al. assumed that movements of water molecules and ions are constrained by its multi-compartment environment including axons, glial cells, and CSF [22]. The measured diffusion tensor $\mathbf{D}$ was expressed as

$$\mathbf{D} = \mathbf{S}\,\text{diag}(d_l, d_{t1}, d_{t2})\,\mathbf{S}^T \tag{10}$$

where d_l, d_{t1}, and d_{t2} are the eigenvalues of $\mathbf{D}$ along the longitudinal and two transversal directions, respectively. Although the WM may include myelinated axons with various directions, the researchers assumed that three groups of the WM exist with their longitudinal directions in parallel to the longitudinal and two transversal directions of the measured diffusion tensor $\mathbf{D}$. In addition, they assumed that the volume fractions of the three WM groups and the remaining isotropic tissues are $\alpha_l, \alpha_{t1}, \alpha_{t2}$, and α_{iso}. Therefore, for this multi-compartment model, the following equations can be obtained:

$$\begin{cases} \alpha_l e^{-bd_l^W} + \alpha_{t1} e^{-bd_{t1}^W} + \alpha_{t2} e^{-bd_{t2}^W} + (1 - \alpha_l - \alpha_{t1} - \alpha_{t2}) e^{-bd_{iso}} = e^{-bd_l} \\ \alpha_l e^{-bd_l^W} + \alpha_{t1} e^{-bd_{t1}^W} + \alpha_{t2} e^{-bd_{t2}^W} + (1 - \alpha_l - \alpha_{t1} - \alpha_{t2}) e^{-bd_{iso}} = e^{-bd_{t1}} \\ \alpha_l e^{-bd_{t2}^W} + \alpha_{t1} e^{-bd_{t1}^W} + \alpha_{t2} e^{-bd_l^W} + (1 - \alpha_l - \alpha_{t1} - \alpha_{t2}) e^{-bd_{iso}} = e^{-bd_{t2}} \end{cases} \tag{11}$$

where d_{iso} is the diffusion coefficient of the isotropic tissues; d_l^W, d_{t1}^W, and d_{t2}^W are the diffusion coefficients of the WM in the longitudinal and two transversal directions, respectively; and $\alpha_l + \alpha_{t1} + \alpha_{t2} + \alpha_{iso} = 1$.

The eigenvalues of the conductivity tensor **C** are expressed as a weighted sum of the conductivity values of the four compartments as:

$$\begin{cases} \sigma_l = \alpha_l \sigma_l^W + \alpha_{t1} \sigma_{t1}^W + \alpha_{t2} \sigma_{t2}^W + \alpha_{iso} \sigma_{iso} \\ \sigma_{t1} = \alpha_l \sigma_{t1}^W + \alpha_{t1} \sigma_l^W + \alpha_{t2} \sigma_{t2}^W + \alpha_{iso} \sigma_{iso} \\ \sigma_{t2} = \alpha_l \sigma_{t2}^W + \alpha_{t1} \sigma_{t1}^W + \alpha_{t2} \sigma_l^W + \alpha_{iso} \sigma_{iso} \end{cases} \quad (12)$$

where σ_l^W, σ_{t1}^W, and σ_{t2}^W are the conductivity values of the WM in the longitudinal and two transversal directions, respectively. The values of σ_l^W, σ_{t1}^W, σ_{t2}^W, and σ_{iso} are not measured; instead, the literature values are adopted [32]. Note that the VFM method does not use the scale factor η and can be used only for the brain tissue including the WM. For the GM and CSF, isotropic conductivity values from existing literature are used. The constraint of $\sigma_{t1} = \sigma_{t2}$ is applied in the implementation of the VFM method.

2.5. Conductivity Tensor Imaging (CTI) Model

The CTI method derives a low-frequency conductivity tensor **C** by using a high-frequency isotropic conductivity σ_H obtained using the MREPT technique and the information about water diffusion obtained by the multi-b diffusion weighted imaging method. Since the details of its basic theory and algorithm are available in [6,23,24], we simply introduce the following CTI formula in this paper:

Here, the low-frequency conductivity tensor is expressed as

$$\mathbf{C} = \alpha \bar{c}_e \mathbf{D}_e \quad (13)$$

Using the reference value of $\beta = 0.41$ [23,24], the apparent extracellular ion concentration ($\bar{c}_e$) of (13) can be estimated as

$$\bar{c}_e = \frac{\sigma_H}{\alpha d_e^w + (1-\alpha) d_i^w \beta} \quad (14)$$

Using $\bar{c}_e$ from (14) in (13), low-frequency conductivity tensor can be expressed as

$$\mathbf{C} = \alpha \bar{c}_e \mathbf{D}_e = \frac{\alpha \sigma_H}{\alpha d_e^w + (1-\alpha) d_i^w \beta} \mathbf{D}_e = \eta \mathbf{D}_e \quad (15)$$

where σ_H is the high-frequency conductivity at the Larmor frequency, α is the extracellular volume fraction, $\bar{c}_e$ is apparent extracellular ion concentration, β is the ion concentration ratio of intracellular and extracellular spaces, d_e^w and d_i^w are the extracellular and intracellular water diffusion coefficients, respectively, and $\mathbf{D}_e$ is the extracellular water diffusion tensor.

3. Imaging Experiments and Data Processing

3.1. Phantom Imaging

To compare the accuracy of the reconstructed conductivity images based on five methods described in Section 2, we used two conductivity phantoms, each with three compartments of known conductivity values. The compartments were filled with electrolytes or giant vesicle suspensions. The giant vesicles were cell-like materials with thin insulating membranes [36].

Phantom #1 comprised of two compartments of electrolytes (EL$_1$ and EL$_2$) and one compartment of a giant vesicle suspension (GVS$_1$) where giant vesicles were suspended in the electrolyte EL$_1$. Phantom #2 comprised of two compartments of different electrolytes (EL$_3$ and EL$_4$) and one compartment of a different giant vesicle suspension (GVS$_2$). Table 1 shows the concentrations of NaCl and CuSO$_4$, the extracellular volume fraction, mobility, and low-frequency conductivity values, which are measured by an impedance analyzer (SI1260A, AMETEK, West Sussex, UK) at 10 Hz. For the electrolyte EL$_3$, we increased its viscosity by adding 2 g/L of hyaluronic acid and 10 g/L of polyethylene glycol (PEG, average Mv 8000) solution, thereby decreased the ion mobility and also its conductivity value. In the giant vesicle suspension GVS$_2$, we used electrolyte EL$_3$.

Table 1. Compositions of two conductivity phantoms. The electrolyte and giant vesicle suspension are denoted as EL and GVS, respectively. EL$_1$, EL$_2$, and GVS$_1$ were used in phantom #1. EL$_3$, EL$_4$, and GVS$_2$ were used in phantom #2.

Compartment	EL$_1$	EL$_2$	GVS$_1$	EL$_3$	EL$_4$	GVS$_2$
NaCl (g/L)	7.5	3.5	7.5	3	3	3
CuSO$_4$ (g/L)	0	1	0	0	0	0
Extracellular volume fraction (%)	100	100	10	100	100	50
Mobility	high	high	high	low	high	low
σ at 10 Hz (S/m)	1.56	0.83	0.29	0.55	0.70	0.45

A 9.4 T research MRI scanner (Agilent Technologies, Santa Clara, CA, USA) equipped with a single-channel birdcage coil (Model: V-HQS-094-00638-029, RAPID Biomedical GmbH, Rimpar, Germany) with a cubic voxel having 0.5 mm edge length was used for phantom imaging. For multi-b diffusion weighted imaging, we used the single-shot spin-echo echo-planar imaging (SS-SE-EPI) pulse sequence. The imaging parameters were as follows: repetition time (TR)/echo time (TE) = 2000/70 ms, number of signal acquisitions = 2, field-of-view (FOV) = 65 × 65 mm^2, slice thickness = 0.5 mm, flip angle = 90°, and image matrix size = 128 × 128. The number of diffusion-weighting gradient directions was 30 with b-values of 50, 150, 300, 500, 700, 1000, 1400, 1800, 2200, 2600, 3000, 3600, 4000, 4500, and 5000 s/mm^2. For high-frequency conductivity image reconstructions in the CTI method, B1 phase maps were acquired using the multi-slice multi-echo spin-echo (MS-ME-SE) pulse sequence. The imaging parameters were as follows: TR/TE = 2200/22 ms, number of signal acquisitions = 5, FOV = 65 × 65 mm^2, slice thickness = 0.5 mm, flip angle = 90°, and image matrix size = 128 × 128. More details of the phantom preparation and data acquisition are described in [6,24].

3.2. In Vivo Human Imaging

Five healthy volunteers were recruited based on the protocol approved by the Institutional Review Board (IRB) at Kyung Hee University (KHSIRB-18-073). Informed consent forms were obtained from the volunteers before conducting the studies. In vivo human brain imaging experiments were performed using a clinical 3 T MRI scanner (Magnetom Trio A Tim, Siemens Medical Solution, Erlangen, Germany) with body coil in transmitting mode and an 8-channel head coil in receiving mode (3D head matrix, A Tim Coil, Siemens Medical Solution, Erlangen, Germany).

For multi-b diffusion weighted imaging, we used the SS-SE-EPI pulse sequence. The number of diffusion-weighting gradient directions was 15 with similar b-values used in the phantom experiments. The imaging parameters were as follows: TR/TE = 2000/70 ms, slice thickness = 4 mm, flip angle = 90°, number of averaging = 2, number of slices = 5, and acquisition matrix = 64 × 64. The matrix size of 64 × 64 was extended to 128 × 128 for subsequent data processing steps. The imaging time was 23 min for the multi-b diffusion data acquisition. For B1 mapping, the MS-ME-SE pulse sequence was used. The imaging parameters were as follows: TR/TE = 1500/15 ms, number of echoes = 6, number of averaging = 5, slice thickness = 4 mm, number of slices = 5, acquisition matrix = 128 × 128, and FOV = 240 × 240 mm^2 using a scan duration of 16 min. For anatomical reference, a conventional T$_2$-weighted scan was obtained. The total imaging time for each human subject was about 41 min. More details of the in vivo scans are described in [24].

3.3. Data Processing

The acquired multi-b diffusion data were preprocessed using the MRtrix3 (www.mrtrix.org, accessed on 10 March 2021) [37] and FMRIB software library (FSL, www.fmrib.ox.ac.uk/fsl, accessed on 10 March 2021) [38]. The preprocessing steps included MP-PCA denoising [39], Gibbs-ringing correction [40], and eddy current distortion correction [38]. The averaged images at $b = 0$ were linearly coregistered to the T_2-weighted images using the FLIRT method, and the affine transformation matrix was used to nonlinearly coregister the diffusion weighted images at $b \neq 0$ to the T_2-weighted images using the FNIRT method [38]. The b-value of 700 s/mm^2 was used for diffusion tensor image reconstructions in the LEM, VCM, and VFM methods. In the FEM and CTI methods, we calculated $\mathbf{D}_f$ and $\mathbf{D}_s$ using a bi-exponential model [24,34].

For high-frequency conductivity image reconstructions in CTI method, the acquired phase map data were first corrected for Gibbs-ringing artifacts [40], and then multi-channel multiple echoes were combined to achieve the best signal-to-noise ratio (SNR) [41,42]. The high-frequency conductivity images were reconstructed using the method proposed by Gurler et al. [14] in order to suppress boundary artifacts.

For human brain images, T_2-weighted images were segmented into the WM, GM, and CSF regions using the MICO [43]. The summary of numbers of pixels in the regions of interest (ROIs) are given in Table 2. In each ROI, we excluded the outermost layer of two-pixel width to reduce partial volume effects. All the data processing steps were implemented using the MATLAB software (Mathworks, Natick, MA, USA). For the image reconstructions using the CTI method, we used the MRCI toolbox (https://iirc.khu.ac.kr/toolbox.html, accessed on 10 March 2021) [44].

Table 2. Numbers of pixels in the WM, GM, and CSF ROI of the five human brains. The reconstructed conductivity values in these ROIs were compared based on the five image reconstruction methods and also with existing literature values.

ROI	Subject				
	#1	#2	#3	#4	#5
WM	1093	1026	1257	1070	1047
GM	1057	918	785	927	1023
CSF	209	135	180	237	234

4. Results

4.1. Two Phantoms

Figure 1 shows the reconstructed conductivity tensor and diffusion tensor images of phantoms #1 and #2. For both tensors, we plotted their longitudinal and two transversal components. The images of the scale factor η between two tensors were also plotted. Note that the VFM method could not be used for the phantoms since it is only applicable to an anisotropic object. The ROIs were defined as shown in Figure 2a,b corresponding to the three different compartments of phantoms #1 and #2, respectively. The mean and standard deviation (SD) of the reconstructed conductivity values for each ROI were calculated and compared with those that were independently measured using the impedance analyzer at 10 Hz as in Figure 2.

In the case of phantom #1 shown in Figures 1a and 2a, the electrolytes EL_1 and EL_2 had different NaCl concentrations as shown in Table 1. The LEM and FEM methods failed to distinguish the difference in concentration, and the VCM and CTI methods were able to differentiate the difference. In phantom #2 shown in Figures 1b and 2b, the conductivity of EL_3 was reduced compared to that of EL_4 due to the reduced mobility in EL_3 as shown in Table 1. Since the diffusion coefficient was altered by this change in mobility, all four methods could distinguish the difference between EL_3 and EL_4 with the same NaCl concentration in phantom #2. Only the CTI method was able to measure the changes in both concentration and mobility without using *prior* information of the conductivity values measured by the impedance analyzer.

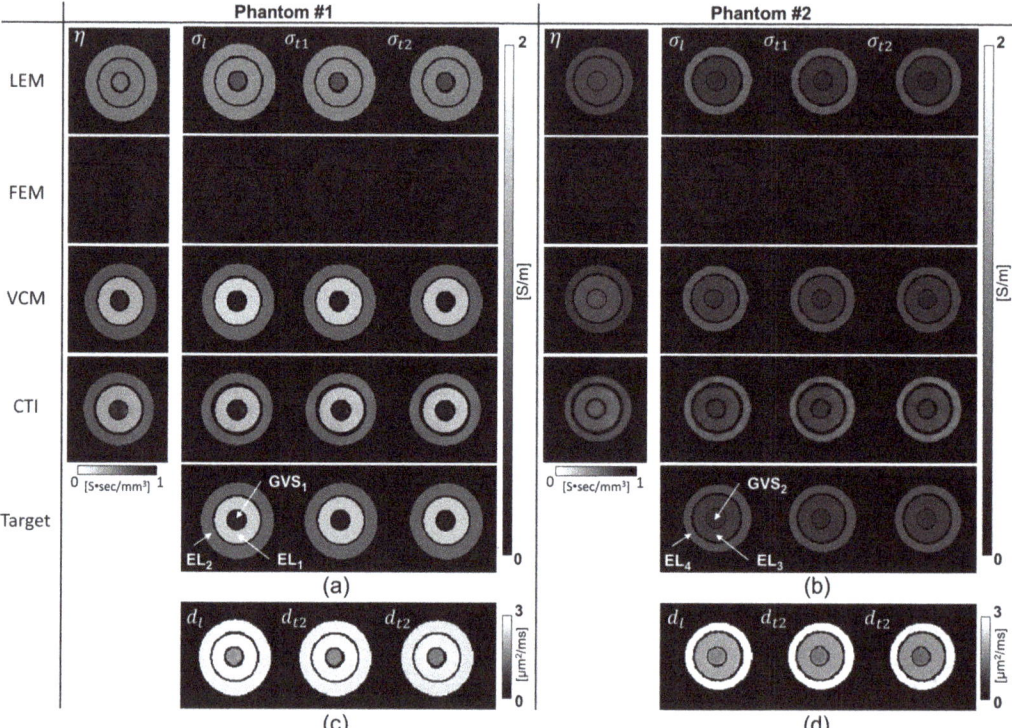

Figure 1. Reconstructed conductivity tensor images of phantoms #1 (**a**) and #2 (**b**). Target images are the true conductivity tensor images generated using the conductivity values measured by the impedance analyzer. (**c**,**d**) are the diffusion tensor images of phantoms #1 and #2, respectively. The VFM method was not applicable to the phantoms since it was specifically designed for the brain tissues.

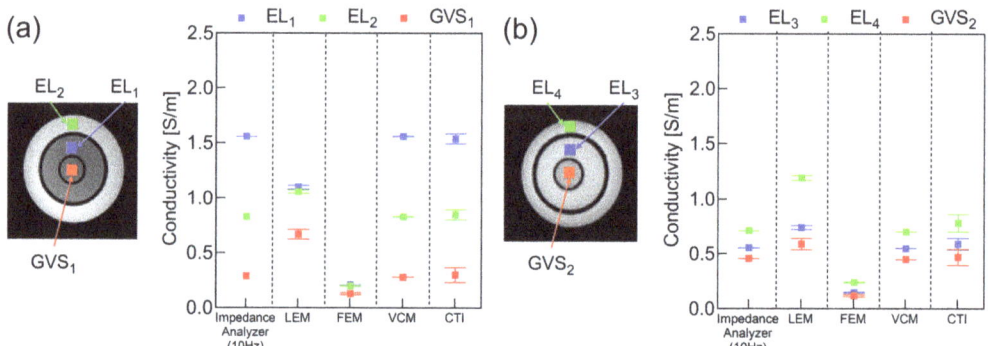

Figure 2. Values of $\sigma_L = \frac{\sigma_l + \sigma_{t1} + \sigma_{t2}}{3}$ in the reconstructed conductivity tensor images of phantoms #1 (**a**) and #2 (**b**). The values were compared with the conductivity values measured at 10 Hz using the impedance analyzer. The square symbol indicates the mean value and the bar represents the SD.

Table 3 shows the errors in the reconstructed conductivity values with respect to the reference values measured by the impedance analyzer (SI1260A, AMETEK Inc., Berwyn,

PA, USA). Since the measured conductivity values using the impedance analyzer were used in place of $\sigma_{WM,GM,CSF}$ in (3), the values of η in the LEM method were different from each other for two phantoms. However, the LEM method could not account for the effects of the concentration difference in phantom #1 despite being able to detect the mobility difference in phantom #2. In the VCM method, the errors in the four electrolyte regions were zero since the measured conductivity values were used in place of σ_{iso} in (9). The CTI method could recover the conductivity values with an error of 1.10% to 5.26% in all the regions of the electrolytes and giant vesicle suspensions.

Table 3. Errors of the reconstructed conductivity values $\sigma_L = \frac{\sigma_l + \sigma_{t1} + \sigma_{t2}}{3}$ with respect to the reference values measured by the impedance analyzer at 10 Hz. Note that the reference values were themselves used in the VCM method for the compartments of EL_1, EL_2, EL_3, and EL_4.

ROI	LEM (%)	FEM (%)	VCM (%)	CTI (%)
EL_1	29.60	86.24	0	1.10
EL_2	28.14	75.57	0	4.42
GVS_1	131.17	54.83	3.45	1.74
EL_3	32.97	73.27	0	3.39
EL_4	67.82	66.27	0	5.26
GVS_2	28.16	74.22	2.02	2.13

4.2. Five Human Brains

Figure 3 shows the images of the longitudinal (σ_l) and transversal (σ_{t1} and σ_{t2}) components of the reconstructed conductivity tensor images of the brains of the first and second subject using the five methods. Figure 4 shows the mean and SD values of the reconstructed conductivity tensors in the WM, GM, and CSF regions, which are summarized in Table 2 for all five subjects. For the WM region, the mean and SD values of σ_l, σ_{t1}, and σ_{t2} were plotted for the five different methods. For the GM and CSF regions, we plotted the mean and SD values of $\sigma_L = \frac{\sigma_l + \sigma_{t1} + \sigma_{t2}}{3}$.

In Figure 4, the LEM and FEM methods underestimated the conductivity values of the CSF region compared with the literature values [35]. For the WM and GM regions, the VFM method overestimated conductivity values compared to the literature values [30–32,45]. The conductivity values observed in the VCM method, especially for GM and CSF regions, were comparable to the literature [31,32,45]. Nevertheless, in the LEM and VCM method, we used the literature values of σ_{iso} for the WM, GM, and CSF regions for all five subjects, and this resulted in a small amount of inter-subject variability. The CTI method produced conductivity values that were comparable to the existing literature values without using *prior* information of the literature values. In the WM region, the value of σ_l obtained by the CTI method was between 0.19 and 0.32 S/m and the values of σ_{t1} and σ_{t2} were between 0.07 and 0.19 S/m, respectively. In the GM region, the value of σ_L was between 0.23 and 0.30 S/m. The value of σ_L in the CSF region was between 1.59 and 1.82 S/m with the mean value of 1.72 S/m for all the five subjects, consistent with those found in the existing literature [35,45].

We performed a correlation analyses to visualize the simultaneous influence of water diffusion tensor in reconstructed conductivity tensor. Figure 5 shows the plots of the linear regression analyses between σ_l and d_l, the longitudinal component of both tensors and averaged transversal components $\sigma_t = (\sigma_{t1} + \sigma_{t2})/2$ and $d_t = (d_{t1} + d_{t2})/2$ for the WM, GM, and CSF regions in all five human subjects (Appendix A). For the LEM and FEM methods, the coefficient of determination (R^2) was 0.99 and 1.00, respectively. Therefore, these two methods, may not provide additional information that is not available in the water diffusion tensor. For the VCM and VFM methods, R^2 ranges from 0.65 to 0.79, respectively, whereas the CTI method showed lesser correlation between $\sigma_{l,t}$ and $d_{l,t}$ with R^2 ranges from 0.46 and 0.56. In the CTI method, the magnitudes of **C** and **D** provided moderately dependent information although they have the same directional property, i.e., the same eigenvectors. This is also because the conductivity tensor in the CTI method has a predominant effect of apparent ion concentrations.

To analyze the directional property, we also computed the anisotropy ratio (AR) in the WM region expressed as $AR_C = \frac{2\sigma_l}{\sigma_{t1}+\sigma_{t2}}$ and $AR_D = \frac{2d_l}{d_{t1}+d_{t2}}$ for the conductivity tensor and diffusion tensor, respectively [46]. The mean and SD value of AR_D were 2.52 ± 0.76, whereas the mean and SD values of AR_C were 2.53 ± 0.77, 2.52 ± 0.78, 2.50 ± 0.78, 1.73 ± 0.47, and 2.52 ± 0.84 for the LEM, FEM, VCM, VFM, and CTI methods, respectively, for the five human subjects. The lower value of AR_C in the VFM model can be attributed to the assumption of $\sigma_{t1} = \sigma_{t2}$. For the other four methods, the values of AR_C and AR_D were similar since the structural property primarily determined the anisotropy.

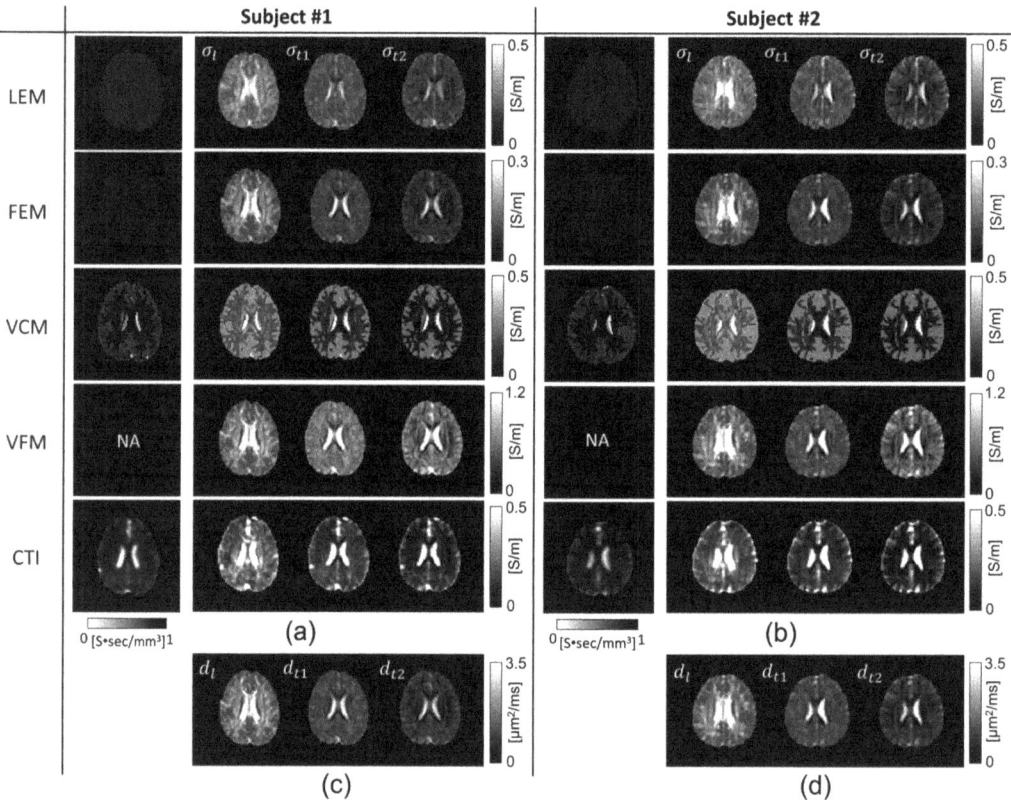

Figure 3. (**a**,**b**) are the longitudinal (σ_l) and transversal (σ_{t1} and σ_{t2}) components of the reconstructed conductivity tensor images of the human brains from subjects #1 and #2, respectively, using the five methods of LEM, FEM, VCM, VFM, and CTI. (**c**,**d**) show the images of the water diffusion tensors.

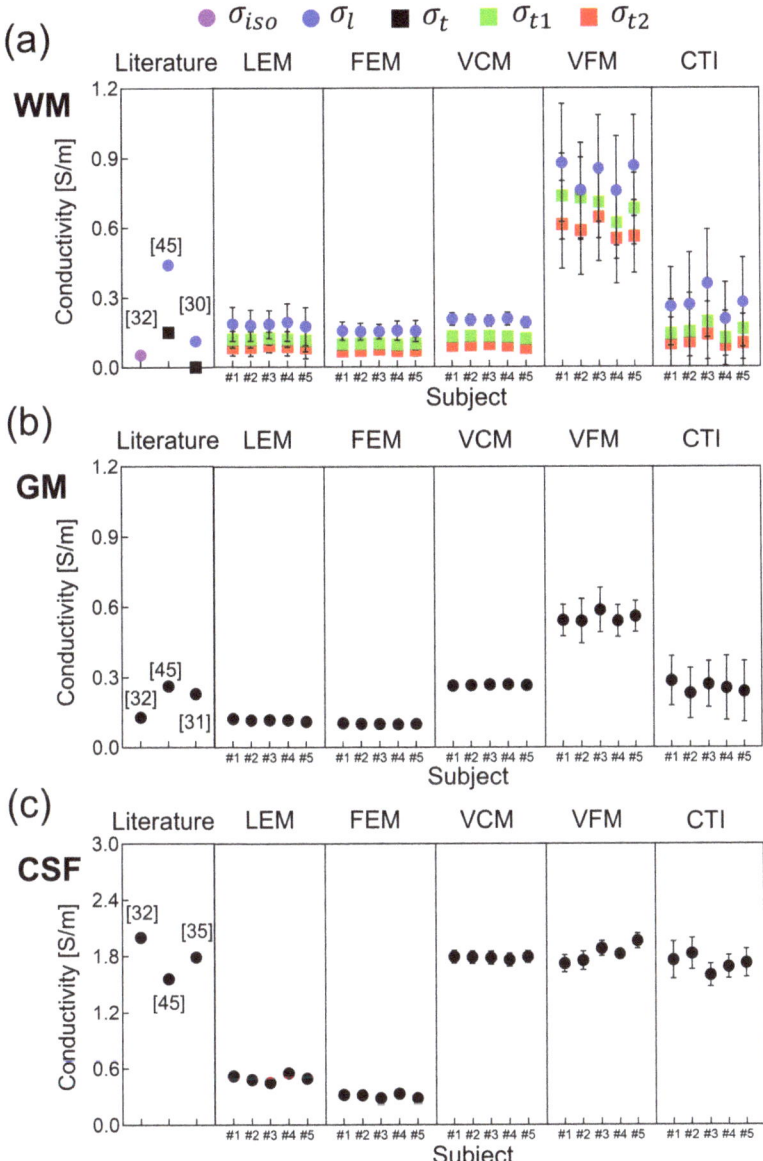

Figure 4. Mean and SD values of the reconstructed conductivity tensors using the five methods for all five human subjects are compared with the existing literature values. (**a**) are conductivity values from WM region, (**b**,**c**) are from GM and CSF, respectively. For the WM region, the mean and SD values of σ_l, σ_{t1}, and σ_{t2} are plotted. For the GM and CSF regions, the mean and SD values of $\sigma_L = \frac{\sigma_l + \sigma_{t1} + \sigma_{t2}}{3}$ are plotted. The vertical bar represents the SD.

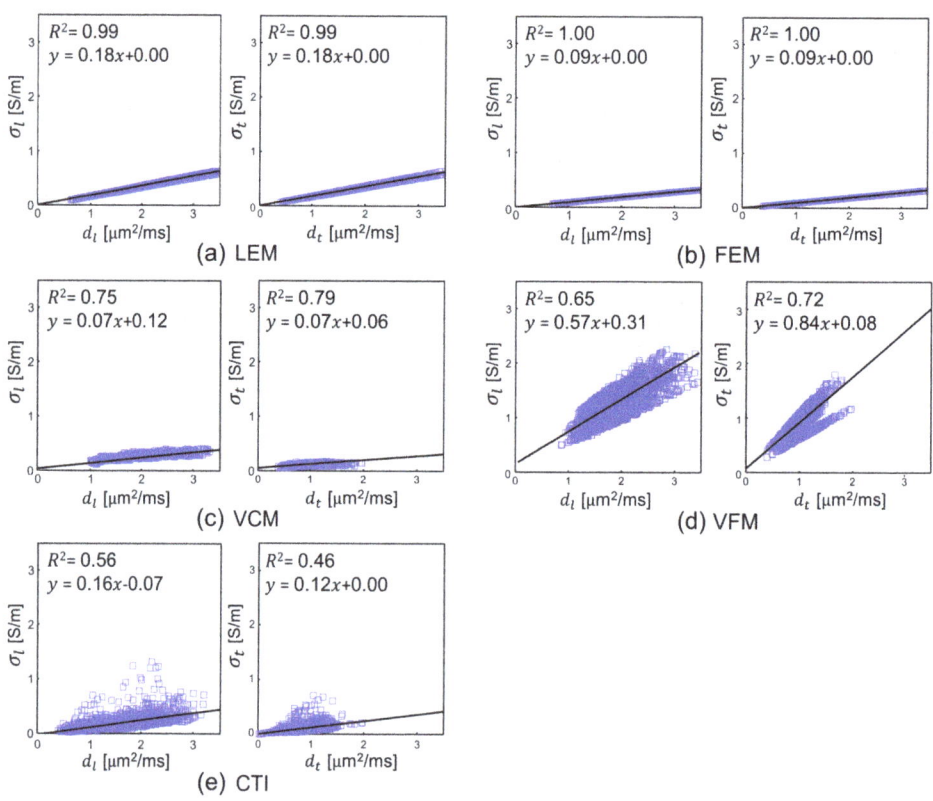

Figure 5. Results of the linear regression analyses between s_l and d_l, the longitudinal component of both tensors and averaged transversal components $\sigma_t = \frac{\sigma_{t1}+\sigma_{t2}}{2}$ and $d_t = \frac{d_{t1}+d_{t2}}{2}$ in the reconstructed conductivity and water diffusion tensor images of all five human subjects pooled together.

Using the CTI method as a reference, we plotted the images of the relative difference (rd_{DIR}), which are defined as

$$rd_{DIR} = \frac{\sigma_{DIR,CTI} - \sigma_{DIR,MTH}}{\sigma_{DIR,CTI}} \times 100\ (\%) \tag{16}$$

where DIR is l, $t1$, or $t2$ and $\sigma_{DIR,CTI}$ and $\sigma_{DIR,MTH}$ are the reconstructed σ_l, σ_{t1}, or σ_{t2} using the CTI method and one of the LEM, FEM, VCM, and VFM methods, respectively (Appendix A). Figure 6 shows the images of rd_{DIR} for all five subjects pooled together. Table 4 summarizes the absolute mean values of $|rd_{DIR}|$ for the WM, GM, and CSF regions for each subject. The bar plots at top of Figure 6 show absolute mean values of $|rd_{DIR}|$ for all five subjects for WM, GM, and CSF regions, and images at the bottom of Figure 6 show the relative difference with sign of deviation to CTI method. The mean and SD of the relative difference for the entire brain slice of the five subjects were 69.63 ± 31.11, 104.94 ± 35.41, 53.35 ± 25.15, and 68.83 ± 14.38% for the LEM, FEM, VCM, and VFM methods, respectively.

Table 4. Mean values of the absolute relative differences ($|rd_{DIR}|$) in the WM, GM, and CSF regions for all five human subjects. The relative differences of the LEM, FEM, VCM, and VFM methods were computed with respect to the CTI method.

Subject	LEM and CTI (%)			FEM and CTI (%)			VCM and CTI (%)			VFM and CTI (%)		
	WM	GM	CSF	WM	GM	CSF	WM	GM	CSF	WM	GM	CSF
#1	44.25	50.99	130.24	63.10	87.90	171.39	50.27	68.60	30.58	91.86	89.73	31.86
#2	49.90	48.29	112.21	58.92	95.20	151.74	53.93	73.25	39.03	91.60	89.08	40.66
#3	40.33	48.87	121.07	63.35	99.29	165.46	51.63	66.47	35.23	92.71	90.61	37.09
#4	43.30	44.00	106.83	49.36	63.41	130.46	49.91	65.17	35.30	92.26	90.56	50.81
#5	46.98	47.66	158.90	56.63	101.28	207.59	62.19	73.10	40.66	89.04	80.15	39.43

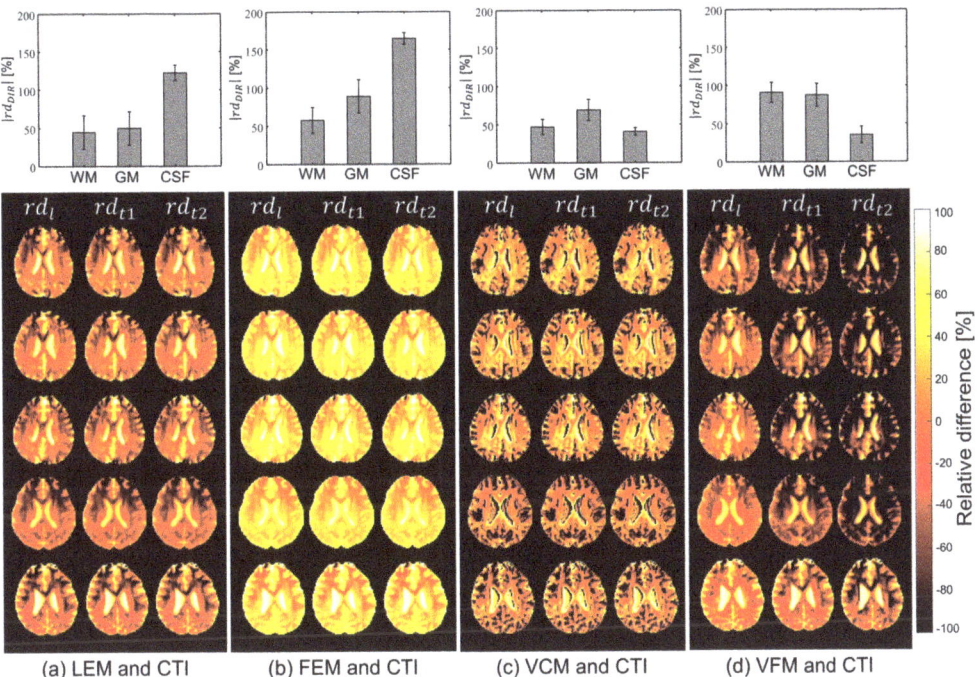

(a) LEM and CTI (b) FEM and CTI (c) VCM and CTI (d) VFM and CTI

Figure 6. Images of the relative differences in the longitudinal (e_l) and transversal (e_{t1} and e_{t2}) components between one of the LEM, FEM, VCM, and VFM methods and the CTI method. The graphs at the top are the absolute mean and SD values of the relative differences ($|rd_{DIR}|$) in the WM, GM, and CSF regions. The bar represents the SD.

5. Discussion

In this paper, we compared five conductivity tensor models and image reconstruction methods using two phantoms and five human subjects. The data from the phantom experiments were used to evaluate the accuracy of the reconstructed conductivity images against the measured conductivity values obtained using the impedance analyzer. In vivo human experiment data were used to compare the values of the reconstructed conductivity tensors with existing literature values and also with each other using the CTI method as a reference.

In the phantom experiments, the LEM and FEM methods could not distinguish the electrolytes with different NaCl concentrations. The errors in the reconstructed conductivity images using the LEM and FEM were relatively large. The errors of the VCM method were the smallest since we used the conductivity values measured using the impedance analyzer in place of σ_{iso} in (9). In most clinical applications where the value of σ_{iso} is unknown, the VCM method may fail to recover the inhomogeneous conductivity values. In addition, living tissues are heterogeneous, and their conductivity varies with the pathophysiology condition. Thus, a single global value of conductivity to the entire tissue type may not sufficient. The VFM method could not be applied to the phantoms since it was explicitly designed for anisotropic brain tissues. Without relying on *prior* information of mean isotropic conductivity values, the CTI method recovered conductivity values with an error ranging from 1.10% to 5.26% for all six different compartments in the two phantoms having different ion concentrations and mobilities. Furthermore, the CTI method could properly handle the effects of different cell densities in two giant vesicle suspensions with different extracellular volume fractions.

Using the data from in vivo human brain imaging experiments, we produced conductivity tensor images of the brains as shown in Figure 3 using the five methods. Compared to the conductivity values of the WM, GM, and CSF from existing literature, the values using the LEM and FEM methods had large amounts of bias in the WM and GM regions. Since the global scale factor was used for all pixels in the FEM method, the method's ability to handle intra-subject and inter-subject variability appeared to be primarily limited. Although the scale factor η in the LEM method contains inter-subject variability, but still lacks intra-subject tissue heterogeneity. Furthermore, the global scale factor does not account for position dependence, and its deviation in the tissues only explains the water diffusion alterations. The conductivity values in CSF from both LEM and FEM methods were significantly underestimated compared to the literature values. The calculated conductivity tensor in both these methods was simply linearly scaled water diffusion tensor, and hence may explain their considerable low conductivity measurements [17,19]. The conductivity values measured in our study were matched with previous studies by Rullman et al. and Sekino et al. [18,20].

In the volume constrained method (VCM), conductivity values of GM and CSF were matched with the literature and also with the CTI method in our study. In contrast, conductivity values of WM were underestimated at longitudinal directions. This might be due to the use of similar literature values of σ_{iso} in all three fiber directions. This method still lacks the intra-voxel heterogenic property of the tissues, and segmentation inaccuracies may end up assigning incorrect σ_{iso} values. An introduction of the isotropic extracellular electrical conductivity from MREIT could be helpful to overcome this limitation [12]. In the volume fraction method (VFM), conductivity values in WM and GM were highly overestimated. This stems from the fact that the eigenvalues of conductivity tensor were computed as a weighted sum of conductivity values associated with four compartments of individual fiber bundle direction. VFM method is computationally complicated, and the ill-posed problem may lead to failed fit in noisy voxels [22]. In addition, the assumption of constant σ_L and $\sigma_{t1} = \sigma_{t2}$, which were then set to be literature conductivity value, somehow neglected the intra-voxel heterogeneity, and reduced the anisotropic ratio. Instead of such limitations, VFM method effectively handled the partial volume effect.

Although the scale factor η in the LEM, FEM, and VCM methods was computed for each pixel, their ability to handle the intra-subject and inter-subject variability was also limited as a result of fixed literature conductivity values of the WM, GM, and CSF regions for all subjects. The VFM method manages cross-subject variability but failed with intra-subject flexibility. In contrast to the four methods using DTI only, the CTI method produced conductivity tensor images with values that were comparable to existing literature values. Without using *prior* information about tissue conductivity values, the CTI method appeared to properly handle the intra-subject and inter-subject variability. However, the intra-subject and inter-subject variabilities, may have been affected in all methods by measurement noise, and systematic artifacts.

The recovered conductivity tensors using the LEM and FEM methods in Figure 5a to Figure 5b shows similar results as the water diffusion tensor in our study. In spite of the calculating the pixel-dependent scale factor in the VCM method, a higher correlation with water diffusion was observed (Figure 5c). Considering that the local concentration of a charge carrier can affect the linearity between **C** and **D**, the CTI method appeared to properly reflect the effects of position-dependent concentration differences by incorporating the actually measured high-frequency conductivity σ_H into the conductivity tensor image reconstruction. Note that these linear regression analyses should not be considered as a general validity of $C = \eta D$. In CTI method, the mutual restriction of both the ionic and the water mobility by the geometry of the brain medium builds the basis for the relationship of **C** and **D**. The voxels in conductivity tensor map are expressed as a sum of products of the carrier concentration, extracellular volume fraction, and the mobility tensor. Since all of the five methods utilized the mobility information embedded in the water diffusion tensor **D**, the anisotropy ratio of **C** was primarily determined by the anisotropy ratio of **D**.

Data acquisition and processing steps are more involved in the CTI method than in those other four methods using DTI. Given that there are various sources of error in the data acquisition and processing stages, subsequent studies are necessary to rigorously validate the performance of the CTI method in terms of errors due to measurement noise and artifacts, partial volume effects, and coregistration of different images. In particular, high-frequency conductivity images using MREPT suffer from errors at boundaries of two regions with different conductivity values and also assumes the piecewise constant conductivity [47], which does not hold in practice. Although the MREPT method we adopted in this paper could reduce such boundary artifacts [14] at the expense of smooth images, future improvements in high-frequency conductivity image reconstruction algorithms are needed to enhance the accuracy of the CTI method in human subjects. The parameter β in (15) was estimated from the literature values and assumed to be constant for all pixels. Although the sensitivity of a reconstructed conductivity tensor image to β appeared to be small [24], clinical studies are needed to find out any practical limitations imposed by this assumption. The CTI method can provide clinically useful information about pathological and physiological changes in cells and cellular structures associated with disease progression. It is one of most promising clinical applications is tumor imaging in terms for early detection, better characterization, and monitoring after a treatment [48,49]. The recovered low-frequency conductivity tensor could be also used to predict internal current pathways and electric field distributions subject to externally injected or induced therapeutic currents in electrical stimulation [28,29].

This study has several limitations that should be considered in the future works. The distribution of conductivity within the giant vesicle phantom does not reflect an adequate anisotropic environment. Future studies should include phantoms mimicking microstructural properties similar to the brain [50]. Diffusion MRI suffers from various systematic errors, such as gradient inhomogeneity [51]. Although we corrected the data using commonly used correction methods, a study of systematic error propagation or noise analysis is needed to enhance the accuracy of the conductivity images. In human brain, conventional DTI measures water diffusion assuming that displacement distribution of water molecules in a given time is a Gaussian function. However, this assumption may not be valid in complex biological tissue where water molecules often show non-Gaussian diffusion [52]. DTI also has limitations in characterizing the diffusion process in areas of low anisotropy and complex fiber structure in a voxel. Future studies with diffusion kurtosis imaging (DKI) or high angular methods (HARDI) can provide better characterization of human brain architecture [52,53]. Cell membranes are assumed to resist low-frequency currents in the CTI approach. As a result, the CTI approach may underestimate the low-frequency conductivity value of tissue including cells with leaky membranes. It would be worthwhile to investigate a more sophisticated CTI model involving such cells.

6. Conclusions

In this study, we provided an overview of the current state of the art in MRI-based conductivity tensor reconstruction. The accuracy of five conductivity tensor model was investigated using two phantoms with four electrolytes and two giant vesicles suspensions with known internal conductivity values. The findings show that methods with pixel-dependent scale factors work better than methods with global scale factors. The experimental results showed that the accuracy of the VCM and CTI methods was superior to that of the LEM and FEM methods. Contrary to the four methods using only DTI, the CTI method did not use *prior* information on mean isotropic conductivity values, and produced conductivity images with errors ranging from 1.10% to 5.26%. From in vivo human brain imaging experiments, the reconstructed conductivity values of the white and gray matter using the LEM, VCM, and CTI methods were comparable with the values available in the literature. Except for LEM and FEM, all methods yielded conductivity values of the CSF similar to those of literature. Methods using only the water diffusion tensor and prior knowledge of the isotropic mean conductivity values varied depending on the parameter value used. The CTI method was able to properly handle the effects of different ion concentrations as well as mobilities and extracellular volume fractions in our study. Although the data processing of the CTI method is more involved, it appeared to be the most accurate among the five methods in this study. Future research is required to confirm the clinical utility of these low-frequency conductivity tensor image reconstruction approaches in diagnostic imaging and bioelectromagnetic modeling.

Author Contributions: N.K. designed the MR experiment, processed and analyzed the data. and drafted the manuscript. B.-K.C. and I.-O.K. prepared the giant vesicle, performed the MR experiment, and analyzed the data. J.-A.P. and H.-J.K. analyzed the results and edited the manuscript. All authors have read and agreed to the published version of the manuscript.

Funding: This work was supported by the National Research Foundation of Korea (NRF), the Ministry of Health and Welfare of Korea, and Korea Institute of Radiological and Medical Sciences (KIRAMS) grants funded by the Korea government (No. 2019R1A2C2088573, 2020R1A2C200790611, 2021R1A2C2004299, HI18C2435, and 50461-2021).

Institutional Review Board Statement: The protocols are approved by the Institutional Review Board at Kyung Hee University(KHSIRB-18-073).

Informed Consent Statement: Informed consent was obtained from all subjects involved in the study.

Data Availability Statement: Some of the data used in this study are available on https://iirc.khu.ac.kr/toolbox.html, (accessed on 10 March 2021) or available from corresponding author upon reasonable request.

Acknowledgments: The authors thank Prof EunAh Lee for the helpful advice on giant vesicle suspension.

Conflicts of Interest: The authors declare no conflict of interest.

Abbreviations

This section highlights the abbreviation used in the manuscript:

LEM	Linear Eigenvalue Model
FEM	Force Equilibrium Model
VCM	Volume Constraint Model
VFM	Volume Fraction Model
CTI	Conductivity Tensor Imaging
WM	White Matter
GM	Gray Matter
CSF	Cerebrospinal Fluid
MREPT	Magnetic Resonance Electrical Properties Tomography

Appendix A

In this section, the main symbols used in this paper can be found with description and units.

σ_L Isotropic low-frequency conductivity (S/m)
σ_l Longitudinal component of conductivity tensor (S/m)
σ_t Transversal component of conductivity tensor (S/m)
d_l Longitudinal component of water diffusion tensor ($\mu m^2/ms$)
d_t Transversal component of water diffusion tensor ($\mu m^2/ms$)
σ_{iso} Isotropic low-frequency conductivity value from the literature (S/m)
d_e Isotropic extracellular diffusion coefficient ($\mu m^2/ms$)
$\mathbf{D}_e$ Extracellular water diffusion tensor ($\mu m^2/ms$)

References

1. Grimnes, S.; Martinsen, O.G. *Bioimpedance and Bioelectricity Basics*; Academic Press: London, UK, 2015.
2. Kerner, E.H. The electrical conductivity of composite materials. *Proc. Phys. Soc. B* **1956**, *69*, 802–807. [CrossRef]
3. Lux F. Models proposed to explain the electrical conductivity of mixtures made of conductive and insulating materials. *J. Mater. Sci.* **1993**, *28*, 285–301. [CrossRef]
4. Schwan, H.P. Electrical properties of tissue and cell suspensions. *Adv. Biol. Med. Phys.* **1957**, *5*, 147–209. [PubMed]
5. Gabreil, C.; Peyman, A.; Grant, E.H. Electrical conductivity of tissue at frequency below 1 MHz. *Phys. Med. Biol.* **2009**, *54*, 4863–4878. [CrossRef] [PubMed]
6. Choi, B.K.; Katoch, N.; Park, J.E.; Ko, I.O.; Kim, H.J.; Kwon, O.I.; Woo, E.J. Validation of conductivity tensor imaging using giant vesicle suspensions with different ion mobilities. *Biomed. Eng. OnLine* **2020**, *19*, 1–17. [CrossRef] [PubMed]
7. Basser, P.J.; Mattiello, J.; Le bihan, D. MR diffusion tensor spectroscopy and imaging. *Biophys. J.* **1994**, *66*, 259–267. [CrossRef]
8. Kwon, O.I.; Jeong, W.C.; Sajib, S.Z.K.; Kim, H.J.; Woo, E.J. Anisotropic conductivity tensor imaging in MREIT using directional diffusion rate of water molecules. *Phys. Med. Biol.* **2014**, *59*, 2955–2974. [CrossRef]
9. Liu, J.; Yang, Y.; Katscher, U.; He, B. Electrical properties tomography based on B1 maps in MRI: Principles, applications, and challenges. *IEEE Trans. Biomed. Eng.* **2017**, *64*, 2515–2530. [CrossRef]
10. Seo, J.K.; Woo, E.J. Magnetic resonance electrical impedance tomography (MREIT). *SIAM Rev.* **2011**, *53*, 40–68. [CrossRef]
11. Seo, J.K.; Kim, D.H.; Lee, J.; Kwon, O.I.; Sajib, S.Z.K.; Woo, E.J. Electrical tissue property imaging using MRI at dc and Larmor frequency. *Inv. Prob.* **2012**, *28*, 084002. [CrossRef]
12. Seo, J.K.; Woo, E.J. Electrical tissue property imaging at low frequency using MREIT. *IEEE Trans. Biomed. Eng.* **2014**, *61*, 1390–1399. [PubMed]
13. Katscher, U.; Voigt, T.; Findeklee, C.; Vernickel, P.; Nehrke, K.; Dossel, O. Determination of electrical conductivity and local SAR via B1 mapping. *IEEE Trans. Med. Imag.* **2009**, *28*, 1365–1374. [CrossRef] [PubMed]
14. Gurler, N.; Ider, Y.Z. Gradient-based electrical conductivity imaging using MR phase. *Mag. Reson. Med.* **2016**, *77*, 137–150. [CrossRef] [PubMed]
15. Leijsen, R.; Brink, W.; van den Berg, C.; Webb, A.; Remis, R. Electrical properties tomography: A methodological review. *Diagnostics* **2021**, *11*, 176. [CrossRef] [PubMed]
16. Sen, A.K.; Torquato, S. Effective conductivity of anisotropic two-phase composite media. *Phys. Rev. B* **1989**, *39*, 4504–4515. [CrossRef] [PubMed]
17. Tuch, D.S.; Wedeen, V.J.; Dale, A.M.; George, J.S.; Belliveau, J.W. Conductivity tensor mapping of the human brain using diffusion tensor MRI. *Proc. Nat. Acad. Sci. USA* **2001**, *98*, 11697–11701. [CrossRef] [PubMed]
18. Rullmann, M.; Anwander, A.; Dannhauer, M.; Warfield, S.K.; Duffy, F.H.; Wolters, C.H. EEG source analysis of epileptiform activity using a 1 mm anisotropic hexahedra finite element head model. *NeuroImage* **2009**, *44*, 399–410. [CrossRef]
19. Sekino, M.; Yamaguchi, K.; Iriguchi, N.; Ueno, S. Conductivity tensor imaging of the brain using diffusion-weighted magnetic resonance imaging. *J. App. Phys.* **2003**, *93*, 6430–6732. [CrossRef]
20. Sekino, M.; Inoue, Y.; Ueno, S. Magnetic resonance imaging of electrical conductivity in the human brain. *IEEE Trans. Mag.* **2005**, *41*, 4203–4205. [CrossRef]
21. Miranda, P.C.; Pajevic, S.; Hallett, M.; Basser, P. The distribution of currents induced in the brain by magnetic stimulation: A finite element analysis incorporating DT-MRI-derived conductivity data. *Proc. Int. Soc. Mag. Res. Med.* **2001**, *9*, 1540.
22. Wang, K.; Zhu, S.; Mueller, B.A.; Lim, K.O.; He, B. A new method to derive white matter conductivity from diffusion tensor MRI. *IEEE Trans. Biomed. Eng.* **2008**, *55*, 2481–2486. [CrossRef] [PubMed]
23. Sajib, S.Z.K.; Kwon, O.I.; Kim, H.J.; Woo, E.J. Electrodeless conductivity tensor imaging (CTI) using MRI: Basic theory and animal experiments. *Biomed. Eng. Lett.* **2018**, *8*, 273–282. [CrossRef] [PubMed]
24. Katoch, N.; Choi, B.K.; Sajib, S.Z.K.; Lee, E.; Kim, H.J.; Kwon, O.I.; Woo, E.J. Conductivity tensor imaging of *in vivo* human brain and experimental validation using giant vesicle suspension. *IEEE Trans. Biomed. Eng.* **2018**, *38*, 1569–1577. [CrossRef] [PubMed]

25. Wu, Z.; Liu, Y.; Hong, M.; Yu, X. A review of anisotropic conductivity models of brain white matter based on diffusion tensor imaging. *Med. Biol. Eng. Comput.* **2018**, *56*, 1325–1332. [CrossRef] [PubMed]
26. Wolters, C.H.; Anwander, A.; Tricoche, X.; Weinstein, D.; Koch, M.A.; Macleod, R.S. Influence of tissue conductivity anisotropy on eeg/meg field and return current computation in a realistic head model: A simulation and visualization study using high-resolution finite element modeling. *NeuroImage* **2006**, *30*, 813–826. [CrossRef]
27. Vorwerk, J.; Aydin, Ü.; Wolters, C.H.; Butson, C.R. Influence of head tissue conductivity uncertainties on EEG dipole reconstruction. *Front. Neurosci.* **2009**, *13*, 531.
28. Shahid, S.; Wen, P.; Ahfock, T. Numerical investigation of white matter anisotropic conductivity in defining current distribution under tDCS. *Comput. Meth. Prog. Biomed.* **2013**, *9*, 48–64. [CrossRef]
29. Lee, W.H.; Liu, Z.; Mueller, B.A.; Lim, K.O.; He, B. Influence of white matter anisotropic conductivity on EEG source localization: comparison to fMRI in human primary visual cortex. *Clin. Neurophysiol.* **2009**, *120*, 2071–2081. [CrossRef]
30. Nicholson, P.W. Specific impedance of cerebral white matter. *Exp. Neurol.* **1965**, *13*, 386–401. [CrossRef]
31. Geddes, L.A.; Baker, L.E. The specific resistance of biological material: A compendium of data for the biomedical engineer and physiologist. *Phys. Med. Biol.* **1967**, *44*, 271–193. [CrossRef]
32. Gabriel, C.; Gabriel, S.; Corthout, E. The dielectric properties of biological tissues: I. Literature survey. *Phys. Med. Biol.* **1996**, *41*, 2231–2249. [CrossRef]
33. Niendorf, T.; Dijkhuizen, R.M.; Norris, D.G.; Campagne, M.V.K.; Nicolay, K. Biexponential diffusion attenuation in various states of brain tissue: Implications for diffusion-weighted imaging. *Mag. Reson. Med.* **1996**, *36*, 847–857. [CrossRef]
34. Clark, C.A.; Hedehus, M.; Moseley, M.E. In vivo mapping of the fast and slow diffusion tensors in human brain *Mag. Reson. Med.* **2002**, *47*, 623–628. [CrossRef]
35. Baumann, S.B.; Wozny, D.R.; Kelly, S.K.; Meno, F.M. The electrical conductivity of human cerebrospinal fluid at body temperature. *IEEE Trans. Biomed. Eng.* **1997**, *44*, 220–223. [CrossRef]
36. Moscho, A.; Orwar, O.; Chiu, D.T.; Modi, B.P.; Zare, R.N. Rapid preparation of the giant unilamellar vesicles. *Proc. Nat. Acad. Sci. USA* **1996**, *93*, 11443–11447. [CrossRef] [PubMed]
37. Tournier, J.D.; Smith, R.; Raffelt, D.; Tabbara, R.; Dhollander, T.; Pietsch, M.; Christiaens, D.; Jeurissen, B.; Yeh, C.H.; Connelly, A. MRtrix3: A fast, flexible and open software framework for medical image processing and visualisation. *NeuroImage* **2019**, *202*, 116137. [CrossRef]
38. Smith, S.M.; Jenkinson, M.; Woolrich, M.W.; Beckmann, C.F.; Behrens, T.E.J.; Johansen-Berg, H.; Bannister, P.R.; De Luca, M.; Drobnjak, I.; Flitney, D.E.; Niazy, R.K.; Saun-ders, J.; Vickers, J.; Zhang, Y.; De Stefano, N.; Brady, J.M.; Matthews, P.M. Advances in functional and structural MR image analysis and implementation as FSL. *NeuroImage* **2004**, *23*, 208–219. [CrossRef] [PubMed]
39. Veraart, J.; Novikov, D.S.; Christiaens, D.; Ades-Aron, B.; Sijbers, J.; Fieremans, E. Denoising of diffusion MRI using random matrix theory. *NeuroImage* **2016**, *142*, 394–406. [CrossRef] [PubMed]
40. Kellner, E.; Dhital, B.; Kiselev, V.G.; Reisert, M. Gibbs-ringing artifact removal based on local subvoxel-shifts. *Magn. Reson. Med.* **2016**, *76*, 1574–1581. [CrossRef] [PubMed]
41. Walsh, D.O.; Gmitro, A.F.; Marcellin, M.W. Adaptive reconstruction of phased array MR imagery. *Magn. Reson. Med.* **2000**, *43*, 682–690. [CrossRef]
42. Kwon, O.I.; Jeong, W.C.; Sajib, S.Z.K.; Kim, H.J.; Woo, E.J.; Oh, T.I. Reconstruction of dual-frequency conductivity by optimization of phase map in MREIT and MREPT. *Biomed. Eng. Online* **2014**, *13*, 1–15. [CrossRef] [PubMed]
43. Li, C.; Liu, Z.; Gore, J.C.; Davatzikos, C. Multiplicative intrinsic component optimization (MICO) for MRI bias field estimation and tissue segmentation. *Mag. Res. Imag.* **2014**, *32*, 913–923. [CrossRef] [PubMed]
44. Sajib, S.Z.K.; Katoch, N.; Kim, H.J.; Kwon, O.I.; Woo, E.J. Software toolbox for low-frequency conductivity and current density imaging using MRI. *IEEE Trans. Biomed. Eng.* **2017**, *64*, 2505–2514. [PubMed]
45. Chauhan, M.; Indahlastari, A.; Kasinadhuni, A.K.; Schar, M.; Mareci, T.H.; Sadleir, R.J. Low-frequency conductivity tensor imaging of the human head in vivo using DT-MREIT: first study. *IEEE Trans. Med. Imag.* **2018**, *37*, 966–976. [CrossRef]
46. Pierpaoli, C.; Basser, P.J. Toward a quantitative assessment of diffusion anisotropy. *Mag. Res. Imag.* **1996**, *36*, 893–906. [CrossRef]
47. Seo, J.K.; Kim, M.O.; Lee, J.S.; Choi, N.; Woo, E.J.; Kim, H.J.; Kwon, O.I.; Kim, D.H. Error analysis of non constant admittivity for MR-based electric property imaging. *IEEE Trans. Med. Imag.* **2012**, *31*, 430–437.
48. Tha, K.K.; Katscher, U.; Yamaguchi, S.; Stehning, C.; Terasaka, S.; Fujima, N.; Kudo, K.; Kazumata, K.; Yamamoto, T.; Van Cauteren, M.; Shirato, H. Noninvasive electrical conductivity measurement by MRI: A test of its validity and the electrical conductivity characteristics of glioma. *Eur. Radiol.* **2018**, *28*, 348–355. [CrossRef]
49. Lesbats, C.; Katoch, N.; Minhas, A.S.; Taylor, A.; Kim, H.J.; Woo, E.J.; Poptani, H. High-frequency electrical properties tomography at 9.4T as a novel contrast mechanism for brain tumors. *Mag. Reson. Med.* **2021**, *86*, 382–892. [CrossRef] [PubMed]
50. Fieremans, E.; Lee, H.H. Physical and numerical phantoms for the validation of brain microstructural MRI: A cookbook. *Neuroimage* **2018**, *182*, 39–61. [CrossRef]
51. Dowell, N.; Tofts P. *Quality Assurance for Diffusion MRI*; Oxford University Press Inc.: Oxford, UK, 2011.
52. Veraart, J.; Poot, D.H.; Van Hecke, W.; Blockx, I.; Van der Linden, A.; Verhoye, M.; Sijbers, J. More accurate estimation of diffusion tensor parameters using diffusion kurtosis imaging. *Mag. Reson. Med.* **2011**, *65*, 138–145. [CrossRef]
53. Tuch D.S.; Reese T.G.; Wiegell M.R.; Makris N; Belliveau J.W; Wedeen V.J. High angular resolution diffusion imaging reveals intravoxel white matter fiber heterogeneity. *Mag. Reson. Med.* **2002**, *48*, 577–582. [CrossRef] [PubMed]

Review

The Renal Clearable Magnetic Resonance Imaging Contrast Agents: State of the Art and Recent Advances

Xiaodong Li [1], Yanhong Sun [2], Lina Ma [2], Guifeng Liu [1,*] and Zhenxin Wang [2,*]

1. Department of Radiology, China-Japan Union Hospital of Jilin University, Xiantai Street, Changchun 130033, China; xiaodong20@mails.jlu.edu.cn
2. State Key Laboratory of Electroanalytical Chemistry, Changchun Institute of Applied Chemistry, Chinese Academy of Sciences, Changchun 130022, China; yhyan@ciac.ac.cn (Y.S.); malina@ciac.ac.cn (L.M.)
* Correspondence: gfliu@jlu.edu.cn (G.L.); wangzx@ciac.ac.cn (Z.W.); Tel.: +86-431-8499-5700 (G.L.); +86-431-8526-2243 (Z.W.)

Academic Editor: Igor Serša
Received: 12 October 2020; Accepted: 26 October 2020; Published: 1 November 2020

Abstract: The advancements of magnetic resonance imaging contrast agents (MRCAs) are continuously driven by the critical needs for early detection and diagnosis of diseases, especially for cancer, because MRCAs improve diagnostic accuracy significantly. Although hydrophilic gadolinium (III) (Gd^{3+}) complex-based MRCAs have achieved great success in clinical practice, the Gd^{3+}-complexes have several inherent drawbacks including Gd^{3+} leakage and short blood circulation time, resulting in the potential long-term toxicity and narrow imaging time window, respectively. Nanotechnology offers the possibility for the development of nontoxic MRCAs with an enhanced sensitivity and advanced functionalities, such as magnetic resonance imaging (MRI)-guided synergistic therapy. Herein, we provide an overview of recent successes in the development of renal clearable MRCAs, especially nanodots (NDs, also known as ultrasmall nanoparticles (NPs)) by unique advantages such as high relaxivity, long blood circulation time, good biosafety, and multiple functionalities. It is hoped that this review can provide relatively comprehensive information on the construction of novel MRCAs with promising clinical translation.

Keywords: magnetic resonance imaging contrast agents; renal clearance; nanodots; gadolinium (III)-based composites

1. Introduction

Magnetic resonance imaging (MRI) is extensively used as a noninvasive, nonionizing, and radiation-free clinical diagnosis tool for detection and therapeutic response assessment of various diseases including cancer, because it can provide anatomical and functional information of regions-of-interest (ROI) with high spatial resolution through manipulating the resonance of magnetic nucleus (e.g., ^1H) in the body via an external radiofrequency pulse magnetic field [1–7]. Although it is possible to generate high-contrast images of soft tissues for diagnosis by manipulating pulse sequences alone, MRI is able to further highlight the anatomic and pathologic features of ROI through utilized in concert with contrast agents. Magnetic resonance imaging contrast agents (MRCAs) play an extremely important role in modern radiology because the growth of contrast-enhanced MRI has been remarkable since 1988 [3–8]. Up to date, the commercially approved MRCAs and the MRCAs in clinical trial are shown in Table 1.

Table 1. Typical magnetic resonance imaging contrast agents (MRCAs) clinically approved or in clinical trials.

Trade Name	Generic Name	Chemical Code	MRI Mode	Clinical Trial	Clinically Approved
Dotarem/Clariscan	Gadoterate meglumine	Gd-DOTA	T_1-weighted	-	Yes
ProHance	Gadoteridol	Gd-HPDO3A	T_1-weighted	-	Yes
Gadovist	Gadobutrol	Gd-DO3A-butrol	T_1-weighted	-	Yes
Magnevist	Gadopentetate dimeglumine	Gd-DTPA	T_1-weighted	-	Yes
Omniscan	Gadodiamide	Gd-DTPA-BMA	T_1-weighted	-	Yes
Optimark	Gadoversetamide	Gd-DTPA-BMEA	T_1-weighted	-	Yes
Multihance	Gadobenate dimeglumine	Gd-BOPTA	T_1-weighted	-	Yes
Combidex/Sinerem	Ferumoxtran	Dextran coated SPION	T_2-weighted	Yes	-
Resovist/Cliavist	Ferucarbotran/Ferrixan	Carboxydextran coated SPION	T_2-weighted	-	Yes
Feridex I.V./Endorem	Ferumoxide	Dextran	T_2-weighted	-	Yes
Feraheme/Rienso	Ferumoxytol	Carboxymethyl-dextran coated SPION	T_2-weighted	-	Yes
Clariscan	Feruglose	PEGylated starch coated SPION	T_2-weighted	Yes	-
Lumirem/GastroMARK	Ferumoxsil	Siloxane coated SPION	T_2-weighted	-	Yes
Abdoscan	-	Sulfonated poly (styrene-divinylbenzene) copolymer coated SPION	T_2-weighted	-	Yes

In general, there are two imaging modes of MRI, named as T_1- or T_2-weighted MRI, which have been employed to acquire the restored or residual magnetization by adjusting parameters in either the longitudinal direction or the transverse plane, respectively. The T_1-weighted MR image shows bright signal contrast (recovered magnetization), while the T_2-weighted MR image shows dark signal contrast (residual magnetization). The T_1- or T_2-weighted MRI can be performed in one machine by simply adjusting the acquisition sequences during the MRI process. Consequently, the MRCAs are divided into two categories based on their dominant functions in T_1- or T_2-weighted MRI. Paramagnetic metal nanomaterials/complexes are usually designated as T_1-weighted MRCAs, which cause bright contrast in T_1-weighted MR images [3,4,8–10]. For example, gadolinium (III) (Gd^{3+}) has seven unpaired electrons and a long electron spin relaxation time, which can efficiently promote the longitudinal ^1H relaxation. Gd^{3+}-diethylenetriamine penta-acetic acid (Gd-DTPA) was synthesized as the first T_1-weighted MRCA and used for contrast-enhanced T_1-weighted MRI of intracranial lesions in 1988 [8]. Superparamagnetic nanoparticles (e.g., superparamagnetic iron-oxide nanoparticles (SPIONs)) are normally used as T_2-weighted MRCAs, which provide dark contrast in MR images [11–19]. The advantages and disadvantages of Gd^{3+}- and SPION-based MRCAs have been discussed in the reviews which are published elsewhere [3–19]. In particular, the advent of nephrogenic systemic fibrosis (NSF) and bone/brain deposition has led to increased regulatory scrutiny of the safety of commercial Gd^{3+} chelates [9]. Figure 1 shows the typical paramagnetic cations including transition metallic cations and lanthanide cations, which are capable of enhancing contrast on MR images. The cations contain unpaired electrons in 3d electron orbitals (transition metallic cations) and/or 4f electron orbitals (lanthanide cations). In addition, several strategies have been proposed for development of T_1-/T_2-weighted dual-mode MRCAs, because T_1-/T_2-weighted dual-mode MRI can provide an accurate match of spatial and temporal imaging parameters [12]. Therefore, the accuracy and reliability of disease diagnosis can be clearly improved by synergistically enhancing both T_1-/T_2-weighted contrast effects.

Because of their unique physiochemical and magnetic properties, magnetic nanoparticles (MNPs) have attracted considerable attention in the construction of MRCAs with high performance during the last two decades (as shown in Figure 2), and they exhibit high potential for clinical applications in MRI-guided therapy. The synthesis, properties, functionalization strategies, and different application potentials of MNPs have been reviewed in detail in the literature published elsewhere [11–19]. For example, the MNPs can be used as excellent MRCAs for sensitive detection of tumors because MNPs can efficiently accumulate in tumor through the leaky vasculature of tumor (also known as the enhanced permeability and retention (EPR) effect). The contrast efficacy and in vivo fate of the MNP-based MRCAs are strongly dependent on their physical and chemical features including shape, size, surface charge, surface coating material, and chemical/colloidal stability. For example, the PEGylated

MNPs normally have relatively longer blood circulation time and higher colloidal/chemical stability than those of uncoated MNPs [11]. Very recently, the structure-relaxivity relationships of magnetic nanoparticles for MRI have been summarized in detail by Chen et al. [15]. Among all the characteristics of MNP, size plays a particularly important role in the biodistribution and blood circulation half-life of MNP [15,18,20]. The long blood circulation half-life of MNP can significantly increase the time window of imaging. However, the nonbiodegradable MNPs with large hydrodynamic size (more than 10 nm) exhibit high uptake in the reticuloendothelial system (RES) organs such as lymph nodes, spleen, liver, and lung, which causes slow elimination through hepatobiliary excretion [21]. The phenomenon increases the likelihood of toxicity in vivo [22], and severely hampers translating MNPs into clinical practices because the United States Food and Drug Administration (FDA) requires that any imaging agent (administered into the body) should be completely metabolized/excreted from the body just after their intended medical goals such as image-guided therapy [23]. The conundrum could be sorted out by the development of renal clearable MNPs since renal elimination enables rapidly clear intravenously administered nanoparticles (NPs) from circulation to be excreted from body. Due to the pore size limit of glomerular filtration in the kidney, only the MNPs with small hydrodynamic diameter (less than 10 nm) or biodegradable ability are able to balance the long blood circulation half-life time for imaging and efficient renal elimination for biosafety. In addition, although they have similar small hydrodynamic diameters, the negatively charged or neutral NPs are more difficult to be eliminated by the kidney than their positively charged counterparts.

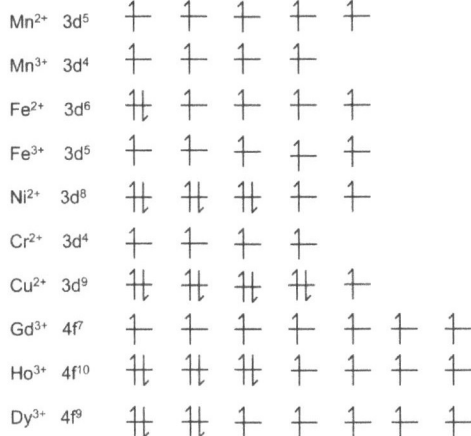

Figure 1. The electron subshell diagrams of typical paramagnetic cations. Generally, the larger number of unpaired electrons leads to stranger magnetic resonance (MR) contrast.

Herein, this review provides the state of the art of renal clearable composites/MNPs-based MRCAs with particular focus on several typical formats, namely Gd^{3+}-complex-based composites and magnetic metal nanodots (MNDs), by using illustrative examples. The MNDs mean MNPs with ultrasmall hydrodynamic size (typically less than 10 nm in diameter) such as ultrasmall Gd_2O_3 NPs, ultrasmall $NaGdF_4$ NPs, ultrasmall Fe_2O_3/Fe_3O_4 NPs and ultrasmall polymetallic oxide NPs. The gathered data clearly demonstrate that renal clearable composites/MNDs offer great advantages in MRI, which shows a great impact on the development of theranostic for various diseases, in particular for cancer diagnosis and treatment. In addition, we also discussed current challenges and gave an outlook on potential opportunities in the renal clearable composites/MNDs-based MRCAs.

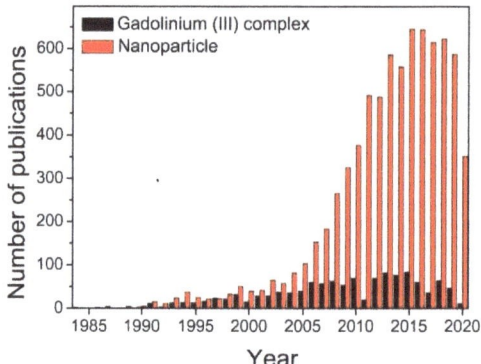

Figure 2. The number of publications searching for "magnetic resonance imaging and contrast agents" plus "gadolinium (III) complex" or "nanoparticle" in the "Web of Science". Development of nanoparticle-based MRCAs has been the hot topic of the area since 2005.

2. Gadolinium (III)-Complex-Based Composites

Up to date, all of the MRCAs commercially available in clinic are small molecule Gd^{3+}-complexes, which are used in about 40% of all MRI examinations (i.e., about 40 million administrations of Gd^{3+}-complex-based MRCA) [24]. However, the small molecule Gd^{3+}-complexes generally have short blood circulation time with a typical elimination half-life of about 1.5 h [8,9,25–27]. The rapid clearance characteristic of small molecule Gd^{3+}-complex makes it difficult to conduct the time-dependent MRI studies. Conjugation of Gd^{3+}-complexes with biodegradable materials (e.g., polymers) and/or renal clearable NDs can not only significantly prolong blood circulation of Gd^{3+}-complexes, but also improve accumulation amount of Gd^{3+}-complexes in solid tumors through an EPR effect [28–52]. Therefore, the Gd^{3+}-complex-based composites enable one to provide an imaging window of a few hours before it is cleared from the body, and additionally enhance the resulting MR signal of a tumor. In particular, biodegradability or small size that ensures the complete clearance of Gd^{3+}-complex-based composites within a relatively short time (i.e., a day) by renal elimination after diagnostic scans.

As early as 2012, Li and coauthors synthesized a poly(N-hydroxypropyl-L-glutamine)-DTPA-Gd (PHPG-DTPA-Gd) composite through conjugation of Gd-DTPA on the poly(N-hydroxypropyl-L-glutamine) (PHPG) backbone [34]. The longitudinal relaxivity (r_1) of PHPG-DTPA-Gd (15.72 $mM^{-1} \cdot S^{-1}$) is 3.7 times higher than that of DTPA-Gd (Magnevist^T). PHPG-DTPA-Gd has excellent blood pool activity. The in vivo MRI of C6 glioblastom-bearing nude mouse exhibited significant enhancement of the tumor periphery after administration of PHPG-DTPA-Gd at a dose of 0.04 mmol Gd kg^{-1} via tail vein, and the mouse brain angiography was clearly delineated up to 2 h after injection of PHPG-DTPA-Gd. Degradation of PHPG-DTPA-Gd by lysosomal enzymes and hydrolysis of side chains lead to complete clearance of Gd-DTPA moieties from the body within 24 h though the renal route. Shi and coauthors synthesized a Gd-chelated poly(propylene imine) dendrimer composite (PPI-MAL DS-DOTA(Gd)) through chelation of Gd^{3+} with tetraazacyclododecane tetraacetic acid (DOTA) modified fourth generation poly(propylene imine) (PPI) glycodendrimers (as shown in Figure 3a) [36]. The r_1 of PPI-MAL DS-DOTA(Gd) is 10.2 $mM^{-1} \cdot s^{-1}$, which is 3.0 times higher than that of DOTA(Gd) (3.4 $mM^{-1} \cdot s^{-1}$). The as-synthesized PPI-MAL DS-DOTA(Gd) can be used as an efficient MRCA for enhanced MRI of blood pool (aorta/renal artery, as shown in Figure 3b) and organs in vivo. The PPI-MAL DS-DOTA(Gd) exhibits good cytocompatibility and hemocompatibility, which can be metabolized and cleared out of the body at 48 h post-injection.

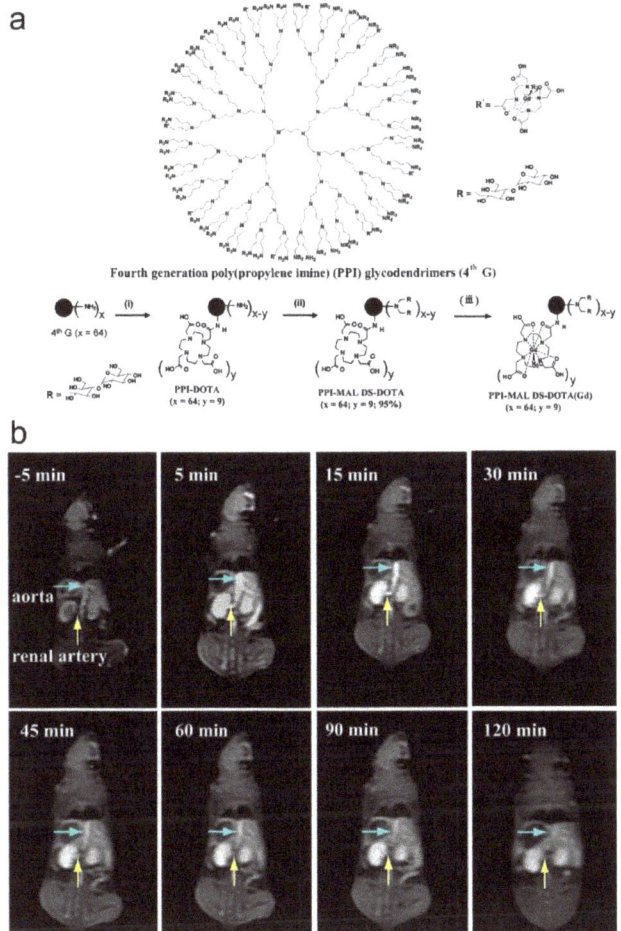

Figure 3. (a) Schematic illustration of the approach used to synthesize the PPI-MAL DS-DOTA(Gd), and (b) T_1-weighted MR images of mouse aorta and renal artery at 5 min before injection (−5 min) and at 15, 30, 45, 60, 90, and 120 min post-injection of the PPI-MAL DS-DOTA(Gd) ([Gd^{3+}] = 2 mg mL^{-1} in 0.2 mL saline through tail vein, adapted from Shi et al. 2016 [36], Copyright 2016 The Royal Society of Chemistry and reproduced with permission).

In comparison with small molecule Gd^{3+}-complexes, Gd^{3+}-complex-based nanocomposites provide significant advantages for contrast-enhanced MRI such as increased Gd^{3+} payload, prolonged blood circulation, enhanced r_1 and improved uptake of Gd^{3+} [28–30]. Generally, the contrast efficiency of Gd^{3+}-complex-based nanocomposite is affected by the structure and surface chemistry of the used nanomaterial. For example, the MRI contrast capability of Gd^{3+}-complex-metal-organic framework (MOF) nanocomposite is strongly dependent on the size and pore shape of MOF [38]. Furthermore, the nanocomposites of Gd^{3+}-complexes with NDs not only have relatively long blood circulation time, but also integrate the properties of both ND and Gd^{3+}-complexes [37–52]. The relatively easy modification property of ND offers opportunities for generation of theranostics for various biomedical applications such as targeted delivery, multimodal imaging, and imaging-guided therapy. For instance, coupling of Gd^{3+}-complexes with optical NDs (e.g., quantum dots (QDs), noble metal nanoclusters, and carbon NDs) we can generate contrast agents for MR/fluorescence dual-mode imaging [39,42,44–48]. Using bovine serum albumin (BSA) as templates, Liang and Xiao synthesized Gd-DTPA functionalized gold nanoclusters (BAG), which show intense red fluorescence emission (4.9 ± 0.8 quantum yield (%)) and high r_1 (9.7 mM^{-1}·s^{-1}) [47]. The in vivo MRI demonstrates that BAG circulate freely in the blood pool with negligible accumulation in the liver and spleen and can be removed from the body through renal clearance, when BAG were injected intravenously into a Kunming mouse at a dose of 0.008 mmol Gd kg^{-1} via the tail vein. The unique properties of BAG make it an ideal dual-mode fluorescence/MR imaging contrast agent, suggesting its potential in practical bioimaging applications in the future. Very recently, Basilion and coauthors constructed a targeted nanocomposite, Au-Gd^{3+}-prostate-specific membrane antigen (PSMA) NPs for MRI-guided radiotherapy of prostate cancer by immobilization of the Gd^{3+}-complex and prostate-specific membrane antigen (PSMA) targeting ligands on the monodispersing Au NPs (as shown in Figure 4a) [50]. Because the hydrodynamic diameter of Au-Gd^{3+}-PSMA NPs is 7.8 nm, Au-Gd^{3+}-PSMA NPs enable efficient accumulation into the tumor site through the EPR effect and ligand-antigen binding, and can be excreted through the renal route. The r_1 of Au-Gd^{3+}-PSMA NPs (20.6 mM^{-1}·s^{-1}) is much higher than that of free Gd^{3+}-complexes (5.5 mM^{-1}·s^{-1}). In addition, both of the Au and Gd^{3+} atoms can serve as sensitizers of radiotherapy. The Au-Gd^{3+}-PSMA NPs show good tumor-targeting specificity, high MR contrast, significant in vivo radiation dose amplification, and renal clearance ability, which exhibit great potential in the clinical MR-guided radiotherapy of PSMA-positive solid tumors ((as shown in Figure 4b).

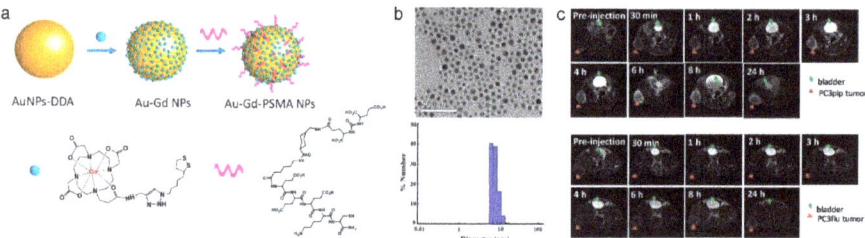

Figure 4. Au-Gd^{3+}-prostate-specific membrane antigen (PSMA) NPs for MR-guided radiation therapy. (**a**) Schematic representation of Au-Gd^{3+}-PSMA NPs. (**b**) TEM micrograph indicates that the average core size of Au-Gd^{3+}-PSMA NPs is 5 nm, and DLS shows that the hydrodynamic diameter of Au-Gd^{3+}-PSMA NPs is 7.8 nm. (**c**) In vivo tumor targeting of Au-Gd^{3+}-PSMA NPs and MR imaging of PC3pip tumor-bearing mouse (up) and PC3flu tumor-bearing mouse (bottom) obtained at 7 T. PC3pip tumor cell expresses high level of PSMA, while PC3flu tumor cell expresses low level of PSMA (the mice were injected with Au-Gd^{3+}-PSMA NPs at 60 µmol Gd^{3+}·kg^{-1} through the tail vein, adapted from Basilion et al. 2020 [50], Copyright 2020 The American Chemical Society and reproduced with permission).

As a rising star of carbon nanomaterial, carbon quantum dots (CDs) have drawn tremendous attention because of their excellent optical property, high physicochemical stability, good biocompatibility, and the ease of surface functionalization [53–57]. Recently, various Gd^{3+}-doped CDs (Gd-CD) have been synthesized for fluorescence/MR dual-mode imaging by the low temperature pyrolysis of precursors containing Gd^{3+} (such as Gd^{3+}-complexes) and carbon [58–61]. Zou and coauthors reported Gd-CD-based theranostics for MRI-guided radiotherapy of a tumor (as shown in Figure 5) [58]. The Gd-CDs were synthesized through a one-pot pyrolysis of glycine and Gd-DTPA at 180 °C, which exhibited stable photoluminescence (PL) at the visible region, relatively long circulation time (~6 h), and efficient passive tumor-targeting ability. The r_1 value of Gd-CDs was calculated to be 6.45 mM^{-1}·s^{-1}, which was higher than that of Magnevist^T (4.05 mM^{-1}·s^{-1}) under the same conditions. An in vivo experiment demonstrated that the Gd-CDs could provide better anatomical and pathophysiologic detection of tumor and precisely positioning for MRI-guided radiotherapy, when the mice were injected intravenously with the Gd-CD solution at a dose of 10 mg Gd kg^{-1}. In addition, the efficient renal clearance of Gd-CDs meant it was finally excreted from the body by urine.

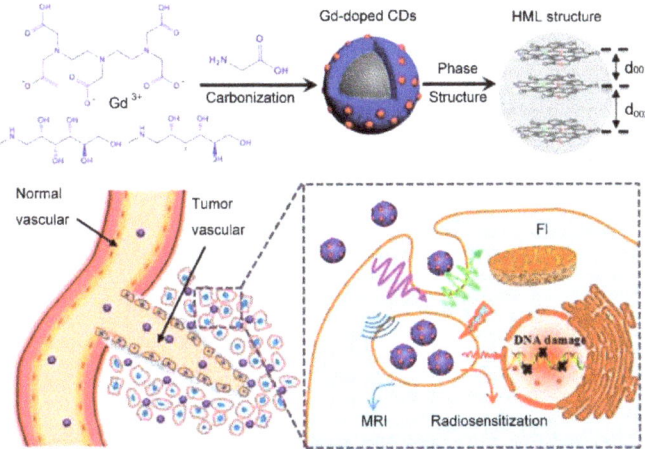

Figure 5. Schematic synthesis and application of Gd-doped CDs through one-pot pyrolysis of glycine and Gd-DTPA at 180 °C (adapted from Zou et al. 2017 [58], Copyright 2017 The Elsevier Ltd. and reproduced with permission).

3. Paramagnetic Metal Nanodots

During the last two decades, several MNDs have been synthesized and used as MRCAs because of their advanced imaging properties compared to small molecule Gd^{3+}-complexes. For example, the MNDs exhibit high surface-to-volume ratios, which allow the ^{1}H to interact with a large number of paramagnetic ions in a tiny volume, resulting in high signal-to-noise ratios at ROI. It has been demonstrated that the size of MND is directly associated with the MRI contrast capability, biodistribution, blood circulation time, and clearance rate [5,14,15,18,23,25,27]. In order to finely control over the aforementioned parameters, the as-developed MND-based MRCAs for clinical applications are highly required to be monodisperse. The hydrophilic MNDs can be directly synthesized through reactions of paramagnetic metal ionic precursors with hydrophilic coating/functionalizing agents. Although this approach is simple and direct, it is difficult to control size and uniformity of MNDs in the aqueous synthetic methods. In order to achieve narrow size distribution and low crystalline defect, most of the MNDs are synthesized by nonaqueous synthesis routes. Because of their inherent hydrophobicity, MNDs should be coated/functionalized with hydrophilic and biocompatible ligands (shells) for biomedical applications. The post-synthetic modification strategies generally involve ligand exchange with hydrophilic molecules as well as encapsulation by hydrophilic shells.

3.1. Gadolinium Nanodots

Generally, there are two protocols for the synthesis of inorganic Gd^{3+} NDs: (1) direct synthesis of hydrophilic Gd^{3+} NDs in water or polyol using a stabilizing agent which allows for crystal growth, followed by ligand exchange with a more robust stabilizing agent to improve the colloidal stability of Gd^{3+} NDs in complex matrixes [62–81], and (2) preparation of hydrophobic Gd^{3+} NDs in high boiling organic solvents by the pyrolysis methods, and subsequent transfer of the hydrophobic Gd^{3+} NDs into aqueous phase by using a hydrophilic/amphiphilic ligand to render them water-dispersible [82–90].

3.1.1. Gadolinium Oxide Nanodots

Several methods have been developed to synthesize Gd_2O_3 NDs for the application in MRCAs [62–72,74,77–80]. As early as 2006, Uvdal and coauthors synthesized Gd_2O_3 NDs (5 to 10 nm in diameter) by thermal decomposition of Gd^{3+} precursors ($Gd(NO_3)_3 \cdot 6H_2O$ or $GdCl_3 \cdot 6H_2O$) in the diethylene glycol (DEG) [62]. Both r_1 and r_2 of the DEG capped Gd_2O_3 NDs are approximately 2 times higher than those of Gd-DTPA. Lee's group developed a one-pot synthesis method with two steps for preparing a series of hydrophilic Gd_2O_3 NDs with high colloidal stability: (1) using $GdCl_3 \cdot 6H_2O$ as Gd^{3+} as precursors, Gd_2O_3 NDs were firstly synthesized in tripropylene/triethylene glycol (TPG/TEG) under alkaline condition, and aged by H_2O_2 and/or an O_2 flow; (2) the as-prepared Gd_2O_3 NDs were then stabilized by different coating materials including amino acid, polymers, and carbon [63–70]. As early as 2009, Lee and coauthors synthesized D-glucuronic acid coated Gd_2O_3 NDs in TPG by using D-glucuronic acid as stabilizer [63]. The D-glucuronic acid coated Gd_2O_3 NDs with 1.0 nm in diameter have a high r_1 (9.9 $mM^{-1} \cdot s^{-1}$), which can easily cross the brain-blood barrier (BBB), causing a strong contrast enhancement in the brain tumor within two hours. Due to the excretion of the injected D-glucuronic acid coated Gd_2O_3 NDs, the MRI contrast began to decrease at 2 h post-injection. The in vivo MR images of brain tumor, kidney, bladder, and aorta demonstrate that D-glucuronic acid coated Gd_2O_3 NDs have good tumor-targeting ability, strong blood pool effect, and high renal clearance when the mice were injected intravenously Gd_2O_3 NDs solution at a dose of 0.07 mmol Gd kg^{-1}. Subsequently, Lee and coauthors successfully synthesized hydrophilic polyacrylic acid (PAA) coated Gd_2O_3 NDs (average diameter = 2.0 nm) with high biocompatibility by electrostatically binding of -COO^- groups of PAA with Gd^{3+} on the Gd_2O_3 NDs (as shown in Figure 6) [69]. Because the magnetic dipole interaction between surface Gd^{3+} and ^1H is impeded by the PAA, both r_1 and r_2 values of Gd_2O_3 NDs are decreased with increasing the molecular weight of PAA (Mw = 1200, 5100 and 15,000 Da). Very recently, Lee and coauthors synthesized a kind of carbon-coated Gd_2O_3 NDs (Gd_2O_3@C, average diameter = 3.1 nm) by using dextrose as a carbon source in the aqueous solution (as shown in Figure 7) [70]. The Gd_2O_3@C have excellent colloidal stability, very high r_1 value (16.26 $mM^{-1} \cdot s^{-1}$, r_2/r_1 = 1.48), and strong photoluminescence (PL) in the visible region, which can be used as renal clearable MR/PL dual-mode imaging agent.

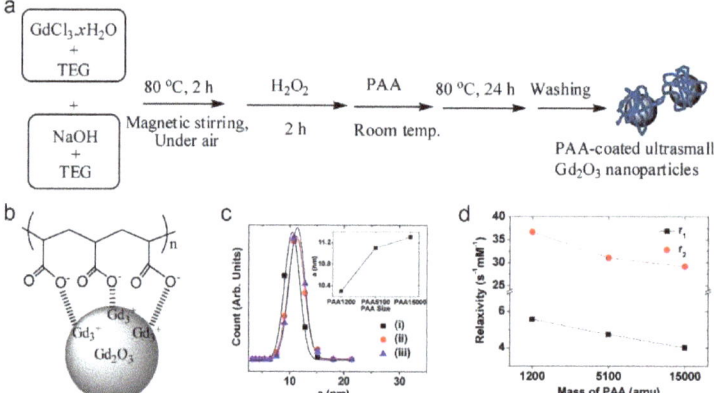

Figure 6. (a) Schematic representation of the one-pot synthesis of polyacrylic acid (PAA) coated Gd_2O_3 NDs, (b) the representative PAA surface-coating structure, (c) DLS measurement of (i) PAA 1200, (ii) PAA 5100, and (iii) PAA 15,000 coated Gd_2O_3 NDs, inset of (c) is the hydrodynamic diameter of PAA coated Gd_2O_3 NDs as a function of PAA molecular weight, and (d) r_1 and r_2 values as a function of PAA molecular weight (adapted from Lee et al. 2019 [69], Copyright 2017 the Elsevier B.V. and reproduced with permission).

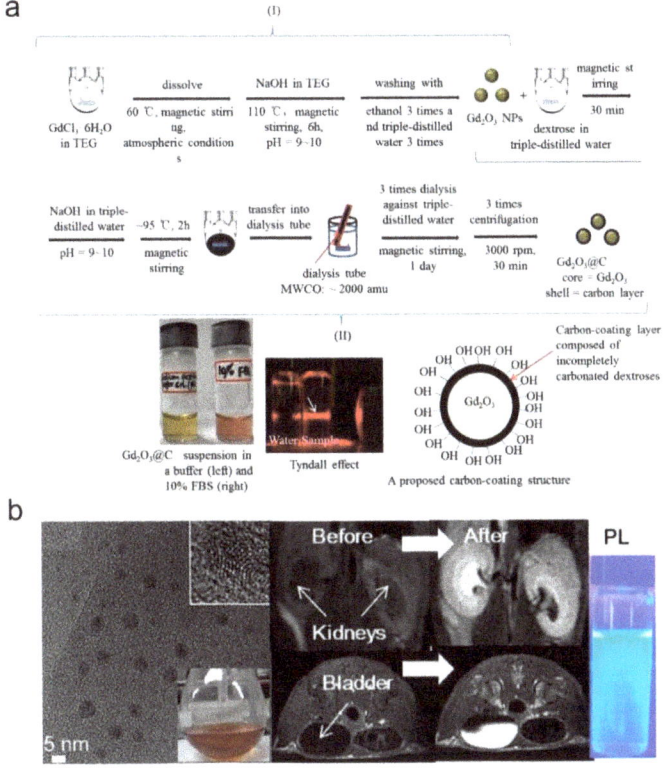

Figure 7. (**a**) Two-step synthesis of Gd_2O_3@C: (i) the synthesis of Gd_2O_3 NDs in triethylene glycol (TEG), and (ii) carbon coating on the Gd_2O_3 ND surfaces in aqueous solution. From left to right at the bottom of (**a**), photographs of the Gd_2O_3@C in sodium acetate buffer solution (pH = 7.0) and a 10% FBS in RPMI1640 medium, the Tyndall effect (or laser light scattering) as indicated with an arrow in the Gd_2O_3@C in water sample (right) with no light scattering in the reference triple-distilled water (left), and a proposed carbon-coating structure of Gd_2O_3@C. (**b**) High resolution transmission electron microscope (HRTEM) micrograph of Gd_2O_3@C, photograph of Gd_2O_3@C in water, in vivo MRI of mouse before or after intravenously administrated by Gd_2O_3@C and PL of Gd_2O_3@C at a dose of 0.1 mmol Gd kg^{-1} (adapted from Lee et al. 2020 [70], Copyright 2019 the Elsevier B.V. and reproduced with permission).

3.1.2. NaGdF$_4$ Nanodots

Hydrophobic NaGdF$_4$ NDs with narrow size distribution are readily synthesized by pyrolysis methods via a high boiling binary solvent mixture [82–89]. In the presence of oleic acid, the approach has been demonstrated to be able to produce NaGdF$_4$ NDs with high quality (e.g., low crystal defect and good monodispersity) through the thermal decomposition of Gd^{3+} precursors, NaOH, and NH$_4$F in octadecene. In order to transfer the as-prepared hydrophobic NaGdF$_4$ NDs into aqueous phase, the NaGdF$_4$ NDs surfaces should be capped with appropriate surface-coating materials such as amphiphilic polymers and biomacromolecules. As early as 2011, Veggel and coauthors reported a size-selective synthesis of paramagnetic NaGdF$_4$ NPs with four different sizes (between 2.5 and 8.0 nm in diameter) with good monodispersity by adjusting the concentration of the coordinating ligand (i.e., oleic acid), reaction time, and temperature [82]. The oleate coated NaGdF$_4$ NPs were then transferred into aqueous phase by using polyvinylpyrrolidone (PVP) as phase transfer agent. They found that the r$_1$ values of PVP coated NaGdF$_4$ NPs are decreased from 7.2 to 3.0 mM^{-1}·s^{-1} with increasing the NP size from 2.5 to 8.0 nm. In particular, the r_1 value of PVP coated 2.5 nm NaGdF$_4$ NDs is about twice as high as that of Magnevist^T under same conditions.

Shi and coauthors successfully employed the amphiphilic molecule, PEG-phospholipids, (such as DSPE-PEG$_{2000}$ (1,2-Distearoyl-sn-glycero-3-phosphoethanolamine-Poly(ethylene glycol)$_{2000}$) and/or the mixture of DSPE-PEG$_{2000}$ and DSPE-PEG$_{2000}$-NH$_2$ (amine functionalized DSPE-PEG$_{2000}$)) to convert 2 nm NaGdF$_4$ NDs from hydrophobic to hydrophilic through the van der Waals interactions of the two hydrophobic tails of phospholipid groups of PEG-phospholipids and the oleic acids on the ND surface [84,85]. The PEG-phospholipids/amine functionalized PEG-phospholipids coated NaGdF$_4$ NDs can be further functionalized by other molecules via suitable physical and chemical reactions. For example, poly-L-lysine (PLL) coated NaGdF$_4$ NDs (NaGdF$_4$@PLL NDs) were developed as dual-mode MRCA by layer-by-layer (LbL) self-assembly of positively charged PLL on the negatively charged DSPE-PEG$_{5000}$ coated NaGdF$_4$ NDs (as shown in Figure 8) [85]. The NaGdF$_4$@PLL ND exhibits high r_1 (6.42 mM$^{-1} \cdot$s^{-1}) for T$_1$-weighted MRI, while the PLL shows an excellent sensitive chemical exchange saturation transfer (CEST) effect for pH mapping (at +3.7 ppm). The results of in vivo small animal experiments demonstrate that NaGdF$_4$@PLL NDs can be used as highly efficient MRCA for precise measurement of in vivo pH value and diagnosis of kidney and brain tumor. Moreover, the NaGdF$_4$@PLL NDs could be excreted through urine with negligible toxicity to body tissues, which holds great promise for future clinical applications.

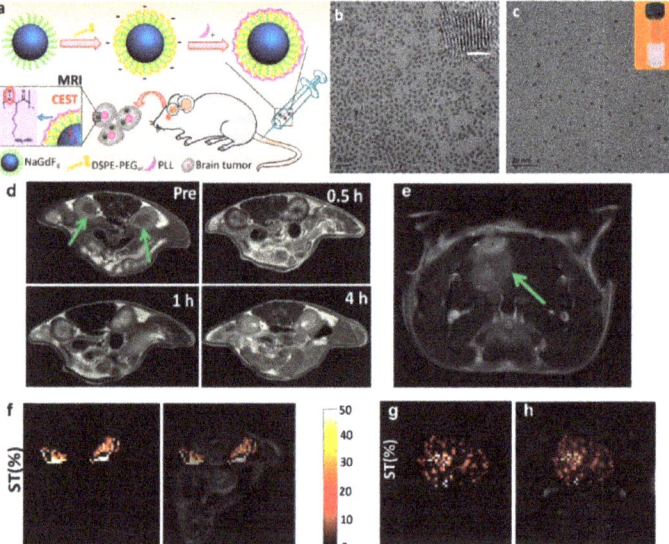

Figure 8. (a) Schematic representation of the synthesis of NaGdF$_4$@PLL NDs for T$_1$-weighted and chemical exchange saturation transfer (CEST) MR imaging. (b) TEM and HRTEM (inset, scale bar: 1 nm) micrograph oleic acid-coated NaGdF$_4$ NDs in chloroform. (c) TEM micrograph and photograph (inset) of NaGdF$_4$@PLL NDs in water. (d) In vivo T$_1$-weighted MRI of kidneys of mouse (as arrowed) before and after the intravenous administration of NaGdF$_4$@PLL NDs at a dose of 5 mg Gd kg^{-1}. (e) In vivo T$_1$-weighted MRI of brain tumor (as arrowed) after the intravenous administration of NaGdF$_4$@PLL NDs at a dose of 10 mg Gd kg^{-1}. (f) CEST contrast difference map between pre/post-injection following radio frequency (RF) irradiation at 3.0 µT. Only the kidney signal is displayed in color on the grayscale image to highlight the CEST effect. (g) CEST ST difference map between pre/post-injection at 3.0 µT. Only the brain ventricle signal is displayed in color on the grayscale image to highlight the CEST effect. (h) Merged image of (e) and (g) (adapted from Shi et al. 2016 [85], Copyright 2016 the American Chemical Society and reproduced with permission).

Taking benefit from robust Gd^{3+}-phosphate coordination bonds, our group developed a facile method for transferring hydrophobic $NaGdF_4$ NDs into aqueous phase though ligand exchange reaction between oleate and phosphopeptides in tryptone [86,87]. The tryptone-coated $NaGdF_4$ NDs (tryptone-$NaGdF_4$ NDs) have excellent colloidal stability, low toxicity, outstanding MRI enhancing performance (r_1 = 6.745 $mM^{-1} \cdot s^{-1}$), efficient renal clearance, and EPR effect-based passive tumor-targeting ability. Importantly, the tryptone-$NaGdF_4$ NDs can be easily functionalized by other molecules including organic dyes and bioaffinity ligands. As shown in Figure 9, the tryptone-$NaGdF_4$ NDs were further functionalized by hyaluronic acid (HA, a naturally occurring glycosaminoglycan) [87]. The HA functionalized tryptone-$NaGdF_4$ NDs (tryptone-$NaGdF_4$ ND@HAs) exhibit high binding affinity with CD44-positive cancer cells, good paramagnetic property (r_1 = 7.57 $mM^{-1} \cdot s^{-1}$) and reasonable biocompatibility. Using MDA-MB-231 tumor-bearing mouse as a model, the in vivo experimental results demonstrate that the tryptone-$NaGdF_4$ ND@HAs cannot only efficiently accumulate in tumor (ca. 5.3% injection dosage (ID) g^{-1} at 2 h post-injection), but also have an excellent renal clearance efficiency (ca. 75% ID at 24 h post-injection). The approach provides a useful strategy for the preparation of renal clearable MRCAs with positive tumor-targeting ability. Using a similar phase transferring principle, the peptide functionalized $NaGdF_4$ NDs (pPeptide-$NaGdF_4$ NDs) were prepared by conjugation of the hydrophobic oleate coated $NaGdF_4$ NDs (4.2 nm in diameter) with the mixture of phosphorylated peptides including a tumor targeting phosphopeptide (pD-SP5) and a cell penetrating phosphopeptide (pCLIP6) [88]. Due to its high isoelectric point, the pCLIP6 can enhanced the cellular uptake of pPeptide-$NaGdF_4$ NDs. The pD-SP5 can improve the tumor-targeting ability of pPeptide-$NaGdF_4$ NDs because it has high binding affinity to human tumor cells. The pPeptide-$NaGdF_4$ NDs show low toxicity, outstanding MRI enhancing performance (r_1 = 13.2 $mM^{-1} \cdot s^{-1}$) and positive-tumor targeting ability, which were used successfully as efficient MRCA for an in vivo imaging small drug induced orthotopic colorectal tumor (c.a., 195 mm^3), when the mice were then intravenously injected with pPeptide-$NaGdF_4$ NDs at a dose of 5 mg Gd kg^{-1}. In addition, the half-life of pPeptide-$NaGdF_4$ NDs in blood is found to be 0.5 h, and more than 70% Gd is excreted with the urine after 24 h intravenous administration.

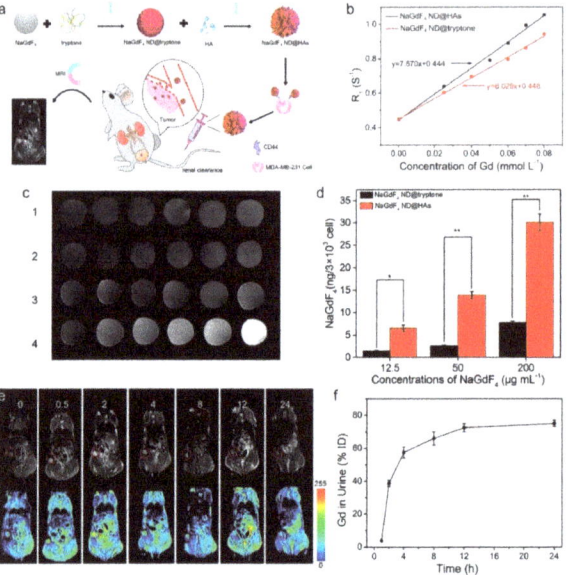

Figure 9. (**a**) Schematic representation of NaGdF$_4$ ND@HAs synthesis, and the application in MRI of tumor through recognizing the overexpressed CD44 on cancer cell membrane. (**b**) R$_1$ relaxivities of NaGdF$_4$ ND@tryptone and NaGdF$_4$ ND@HAs as a function of the molar concentration of Gd^{3+} in solution, respectively. (**c**) MR images of (1) NaGdF$_4$ ND@HA-stained MCF-7 cells, (2) free HA + MDA-MB-231 cells + NaGdF$_4$ ND@HAs, (3) NaGdF$_4$ ND@tryptone-stained MDA-MB-231 cells, (4) NaGdF$_4$ ND@HA-stained MDAMB-231 cells. (**d**) The amounts of Gd element in the NaGdF$_4$ ND-stained MDA-MB-231cells. Error bars mean standard deviations (n = 5, * p < 0.05 or ** p < 0.01 from an analysis of variance with Tukey's post-test). (**e**) In vivo MR images and corresponding pseudo color images of Balb/c mouse bearing MDA-MB-231 tumor after intravenous injection of NaGdF$_4$ ND@HAs (10 mg Gd kg^{-1}) at different timed intervals (0 (pre-injection), 0.5, 2, 4, 8, 12 and 24 h post-injection), respectively. (**f**) The total amounts of NaGdF$_4$ ND@HAs in mouse urine as a function of post-injection times. Error bars mean standard deviations (n = 5) (adapted from Yan et al. 2020 [87], Copyright 2020 The Royal Society of Chemistry and reproduced with permission).

3.2. Iron Nanodots

In 1996, the U. S. FDA approved Ferumoxides (SPIONs) as T$_2$-weighted MR CAs for the diagnosis of liver disease [91]. After that, various strategies based on physical, chemical, and biological methods have been developed for synthesizing the SPIONs including FeNDs for using as T$_2$-weighted MRCAs [91–95]. Because of the negative contrast effect and magnetic susceptibility artifacts, it is still a great challenge when the SPION-enhanced T$_2$-weighted MRI is employed to distinguish the lesion region in the tissues with low background MR signals such as bone and vasculature. Recently, several methods have developed for synthesis of FeNDs, which are showing increasing potential as alternatives to Gd^{3+}-based T$_1$-weighted MRCAs because of low magnetization by a strong size-related surface spin-canting effect [96–111]. For instance, Wu and coauthors synthesized a silica-coated Fe$_3$O$_4$ ND (4 nm in diameter), which exhibited a good r$_1$ relaxivity of 1.2 mM^{-1}·s^{-1} with a low r$_2$/r$_1$ ratio of 6.5 [101]. The result of in vivo T$_1$-weighted MR imaging of heart, liver, kidney, and bladder in a mouse demonstrated that silica-coated FeNDs exhibited strong MR enhancement capability, when the mice were then intravenously injected with silica-coated FeNDs at a dose of 2.8 mg Fe kg^{-1}. Shi and coauthors reported a zwitterion L-cysteine (Cys) coated Fe$_3$O$_4$ ND (3.2 nm in diameter, Fe$_3$O$_4$-PEG-Cys) with r$_1$ relaxivity of 1.2 mM^{-1}·s^{-1} [102]. The Fe$_3$O$_4$-PEG-Cys are able to resist macrophage cellular uptake, and display a prolonged blood circulation time with a half-life of 6.2 h. In vivo experimental results demonstrated that the Fe_3O_4-PEG-Cys were able to be used as a T_1-weighted MRCA for enhanced blood pool and tumor MR imaging, when the mice were then intravenously injected with Fe_3O_4-PEG-Cys at a dose of 0.05 mmol Fe kg^{-1}. Bawendi and coauthors prepared zwitterion-coated ultrasmall superparamagnetic Fe_2O_3 NDs (ZES-SPIONs) with hydrodynamic size of 5.5 nm for use as Gd^{3+} free T_1-weighted MRCA [103]. The ZES-SPIONs have strong T_1-weighted MRI enhancement capacity (r_1 = 5.2 $mM^{-1} \cdot s^{-1}$), and the majority of ZES-SPIONs are cleared through the renal route within 24 h intravenous administration at a dose of 0.2 mmol Fe kg^{-1}. As shown in Figure 10, Zhang and coauthors reported a large-scalable (up to 10 L) albumin-constrained strategy to synthesize monodispersed 3 nm ferrous sulfide NDs (FeS@BSA) with an ultralow magnetization by reaction of the mixture of BSA and $FeCl_2$ with Na_2S under ambient conditions (pH 11 and 37 °C) [104]. In this case, BSA plays crucial roles in the synthesis process including as a constrained microenvironment reactor for particle growth, a water-soluble ligand for colloidal stability, and a carrier for multifunctionality. FeS@BSAs exhibit high MRI enhancing performance (r_1 = 5.35 $mM^{-1} \cdot s^{-1}$), good photothermal conversion efficiency (η = 30.04%), strong tumor-targeting ability, and an efficient renal clearance characteristic. In vivo experiments show that FeS@BSAs have good performance of T_1-weighted MR/phototheranostics dual-mode imaging-guided photothermal therapy (PTT) of mouse-bearing 4T1 tumor, demonstrating FeS@BSA to be an efficient T_1-weighted MR/PA/PTT theranostic agent.

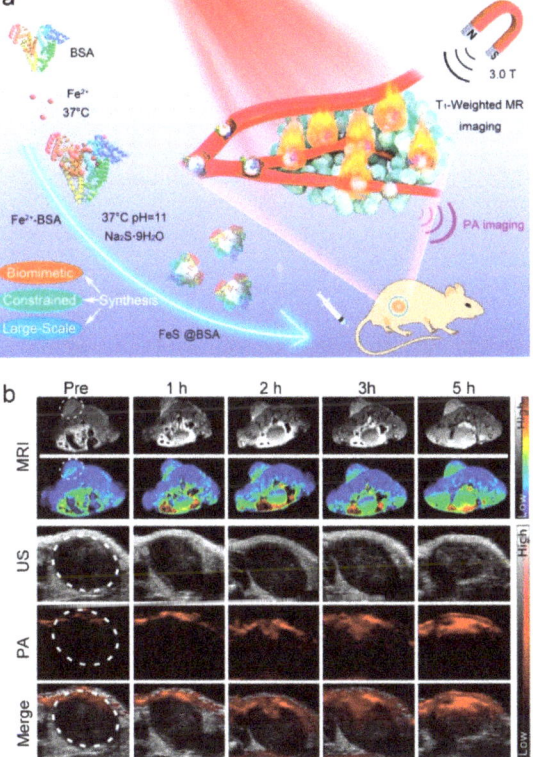

Figure 10. (**a**) Schematic representation of bovine serum albumin (BSA)-constrained biomimetic synthesis of 3 nm FeS@BSA for in vivo T_1-weighted MR/phototheranostics dual-mode imaging-guided photothermal therapy (PTT) of tumors. (**b**) In vivo T_1-weighted MR, US, PA, and merged (US and PA) images of tumor after intravenous injection of at a dose of 20 mg FeS@BSA kg^{-1} (adapted from Zhang et al. 2020 [104], Copyright 2020 Elsevier Ltd. and reproduced with permission).

Fe^{3+} coordination polymer NDs (Fe-CPNDs) were synthesized through the self-assembling the multidental Fe^{3+}-polyvinyl pyrrolidone (PVP) complexes and Fe^{3+}-gallic acid (GA) complexes and used as nanotheranostics for MRI-guided PTT [107–109]. For instance, our group has developed a simple and scalable method for synthesizing the pH-activated Fe coordination polymer NDs (Fe-CPNDs) by the coordination reactions among Fe^{3+}, GA, and PVP at ambient conditions (as shown in Figure 11) [107]. The Fe-CPNDs exhibit an ultrasmall hydrodynamic diameter (5.3 nm), nearly neutral Zeta potential (−3.76 mV), pH-activatable MRI imaging contrast (r_1 = 1.9 $mM^{-1} \cdot s^{-1}$ at pH 5.0), and outstanding photothermal performance. The Fe-CPNDs were successfully used as T_1-weighted MRCA to detect mouse-bearing tumors as small as 5 mm^3 in volume, and as a PTT agent to completely suppress tumor growth by MRI-guided PTT, demonstrating that Fe-CPNDs constitute a new class of renal clearable theranostics. In addition, the Fe-CPND-enhanced MRI was successfully employed for noninvasively monitoring the kidney dysfunction by drug (daunomycin)-induced kidney injury, which further highlights the potential in clinical applications of the Fe-CPND [108].

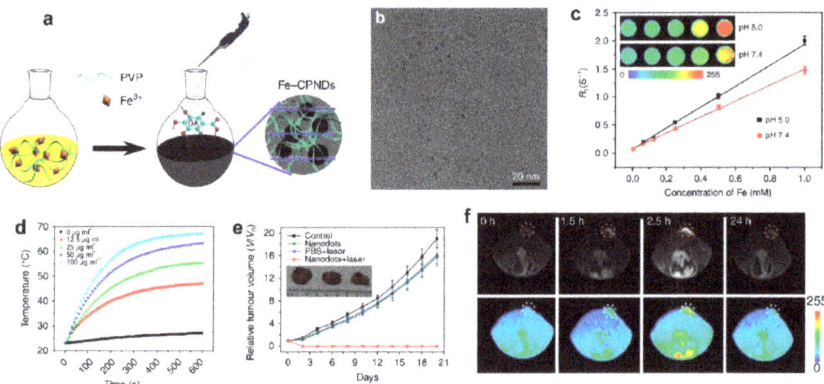

Figure 11. (**a**) Schematic representation of the synthesis of Fe-CPNDs. (**b**) TEM micrograph of Fe-CPNDs. (**c**) The effect of solution pH on the R_1 relaxivity of Fe-CPNDs. (**d**) Temperature elevation of Fe-CPNDs solutions with various concentrations under 1.3 W cm^{-2} 808 nm NIR laser irradiation for 10 min. (**e**) Tumor growth curves of different groups of mice after intravenous treatments. The inset shows the digital photographs of tumors collected from different groups of mice at day 20th post-administration. (**f**) In vivo MR images of the SW620 tumor-bearing nude mouse after intravenous injection of at a dose of 0.25 mg Fe kg^{-1} (The tumor was marked by circle, which was about 5 mm^3 in volume.) after the intravenous injection of Fe-CPNDs at different time intervals (0 h indicates pre-injection) (adapted from Liu et al. 2015 [107], Copyright 2015 Macmillan Publishers Ltd. and reproduced with permission).

Development of contrast agents for simultaneously T_1-/T_2-weighted dual-mode MRI may circumvent the drawbacks of single imaging modalities. Iron-based NDs are also able to serve as T_1-/T_2-weighted dual-mode MRCAs [110–113]. Uvdal and coauthors synthesized a series of water-dispersible PAA-coated Fe_3O_4 NDs via a modified one-step coprecipitation approach [110]. In particular, the 2.2 nm PAA coated Fe_3O_4 NDs have relatively high relaxivities (r_1 = 6.15 $mM^{-1} \cdot s^{-1}$ and r_2 = 28.62 $mM^{-1} \cdot s^{-1}$) and low r_2/r_1 ratio (4.65). Using a mouse model, the in vivo experiments indicate that the 2.2 nm PAA-coated Fe_3O_4 NDs exhibit long-term circulation, low toxicity, and great contrast enhancement (brightened on the T_1-weighted and darkened on the T_2-weighted MR images), when the mice were then intravenously injected with PAA-coated Fe_3O_4 NDs with a dose of 0.0125 mmol Fe kg^{-1}. The in vivo results demonstrate that the 2.2 nm PAA-coated Fe_3O_4 NDs have great potential as T_1-/T_2-weighted dual-mode MRCA for clinical applications including diagnosis of renal failure, myocardial infarction, atherosclerotic plaque, and tumor. Very recently, Shi and coauthors developed a strategy for preparing switchable T_1/T_2-weighted dual-mode MRCA by formation of cystamine dihydrochloride (Cys) cross-linked Fe_3O_4 NDs clusters [113]. The Fe_3O_4 NDs clusters, with a hydrodynamic size of 134.4 nm, and can be dissociated to single 3.3 nm Fe_3O_4 NDs under a reducing microenvironment (e.g., 10 mmol L^{-1} glutathione (GSH)) because of redox-responsiveness of the disulfide bond of Cys. The Fe_3O_4 NDs clusters exhibit a dominant T_2-weighted MR effect with an r_2 of 26.4 $mM^{-1} \cdot s^{-1}$, while Fe_3O_4 NDs have a strong T_1-weighted MR effect with an r_1 of 3.9 $mM^{-1} \cdot s^{-1}$. Due to the reductive tumor microenvironment, the Fe_3O_4 NCs can be utilized for dynamic precision imaging of a subcutaneous tumor model in vivo, and pass through the kidney filter, when the mice were then intravenously injected Fe_3O_4 NCs with a dose of 2.5 mmol Fe kg^{-1}.

3.3. Other Paramagnetic Metal-Based Nanomaterials

Due to their in vivo safety, Mn^{2+}-based T_1-weighted MRCAs have attracted increasing attention [114–134]. Although free Mn^{2+} has a higher r_1 than those of Mn-based nanomaterials, Mn^{2+} exhibits low accumulation and poor performance for disease contrast because of its short blood retention time in vivo [114–119]. The increase of Mn^{2+} accumulation in a tumor can be achieved through the design of pH/GSH-activated Mn-based nanomaterials because a tumor has weakly acidic microenvironment and high concentration of GSH, and nanomaterials can efficiently accumulate in a tumor by EPR effect [123–129]. Therefore, several tumor microenvironment activatable Mn-based nanotheranostic systems have been constructed for T_1-weighted MRI-guided therapy. We have synthesized the polydopamine@ultrathin manganese dioxide/methylene blue nanoflowers (PDA@ut-MnO_2/MB NFs) for the T_1-weighted MRI-guided PTT/photodynamic therapy (PDT) synergistic therapy of tumor (as shown in Figure 12) [125]. In the presence of 5 mmol·L^{-1} GSH, the r_1 of PDA@ut-MnO_2/MB NFs is significantly increased from 0.79 to 5.64 $mM^{-1} \cdot s^{-1}$ since the ultrathin MnO_2 nanosheets on PDA@ut-MnO_2/MB NFs can be reduced into Mn^{2+} ions by GSH. Due to the performance in response to the tumor microenvironment, the PDA@ut-MnO_2/MB NFs show an enhancement 3 times that of the T_1-weighted MRI signal at the tumor site at 4 h post-injection. The Mn^{2+} can be diffused from the tumor to the circulatory system and excreted from the body through renal clearance. In addition, doping of Mn^{2+} into the matrix of other metallic NDs and/or formation of Mn^{2+}-complexes/nanocomposites can also improve its T_1-weighted MR contrast ability [115,130–134]. For example, Wang and coauthors developed a Mn^{2+}-based MRCA (MNP-PEG-Mn) through chelation of Mn^{2+} with 5.6 nm water-soluble melanin NDs [134]. The r_1 (20.56 $mM^{-1} \cdot s^{-1}$) of as-prepared MNP-PEG-Mn is much higher than that of Gadodiamide (6.00 $mM^{-1} \cdot s^{-1}$). Using a 3T3 tumor-bearing mouse as model, in vivo MRI experiments demonstrated that MNP-PEG-Mn (200 µL of 8 mg mL^{-1} MNP-PEG-Mn PBS solution) showed excellent tumor-targeting specificity, and could be efficiently excreted via renal and hepatobiliary pathways with negligible toxicity.

Because Dy^{3+} and Ho^{3+} ions exhibit relatively high magnetic moments among of lanthanide (III) (Ln^{3+}) ions, Dy and Ho NDs are believed as promising candidates for T_2-weighted MRCAs with renal excretion [135–139]. As early as 2011, Lee and coauthors developed a facile one-pot method for synthesis of D-glucuronic acid-coated Ln_2O_3 NDs (Ln = Eu, Gd, Dy, Ho, and Er) [138]. They demonstrated that the D-glucuronic acid-coated 3.2 nm Dy_2O_3 NDs have high r_2 (65.04 $mM^{-1} \cdot s^{-1}$) and very low r_1 (0.008 $mM^{-1} \cdot s^{-1}$). After injection of the Dy_2O_3 NDs at a dose of 0.05 mmol Dy kg^{-1} through tail vein of mouse, the clearly negative MR contrast enhancement in both liver and kidneys of mouse are observed in in vivo T_2-weighted MRI. The Dy_2O_3 NDs are also excreted from the body through the renal route, which is prerequisite for clinical applications as a MRCA.

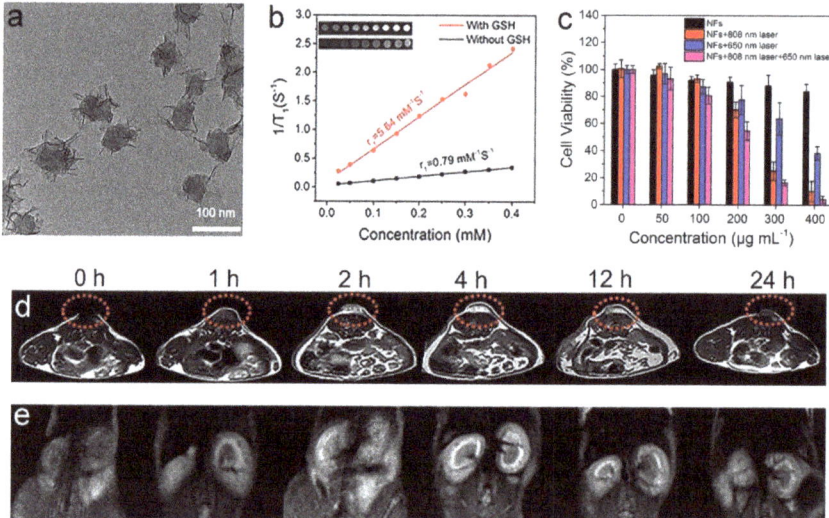

Figure 12. (**a**) TEM micrographs of PDA@ut-MnO$_2$/MB NFs. (**b**) The r$_1$ values of PDA@ut-MnO$_2$/MB NFs with and without 5 mmol L^{-1} GSH. (**c**) An assessment of PDT/PTT efficacy using PDA@ut-MnO$_2$/MB NFs via CCK-8 assays. In vivo MR images of HCT 116 tumor-bearing mouse (**d**) tumor and (**e**) kidneys different time points after the intravenous injection of PDA@ut-MnO$_2$/MB NFs (0, 1, 2, 4, 12, and 24 h; 0 h means pre-injection) at a dose of 10 mg Mn kg^{-1} (adapted from Sun et al. 2019 [125], Copyright 2019 The Royal Society of Chemistry and reproduced with permission).

4. Dual Paramagnetic Metal Nanodots

The MR contrast capabilities of nanomaterials can be further improved while two paramagnetic metallic ions are integrated into one nanoplatform [140–155]. For instance, Zhou and coauthors constructed a 4.95 nm triple-mode imaging platform by doping of Mn^{2+} and ^{68}gallium (III) (^{68}Ga^{3+}) into the matrix of copper sulfide (CuS) NDs using BSA as the synthetic template [143]. Although the Mn^{2+}/^{68}Ga^{3+}-CuS@BSA NDs have relatively low r$_1$ (0.1119 mM^{-1}·s^{-1}, the ratio of r$_2$/r$_1$ = 1.67), in vivo experimental results of SKOV-3 ovarian tumor-bearing mouse demonstrated that the as-prepared Mn^{2+}/^{68}Ga^{3+}-CuS@BSA NDs (150 μL solution at 2 OD concentration of NDs) could be used as an excellent agent for T$_1$-weihted MR/positron emission tomography (PET)/photoacoustic (PAT) triple-mode imaging-guided PTT of the tumor, and were efficiently cleared via the renal-urinary route. Very recently, our group constructed a multifunctional nanotheranostic (MnIOMCP) for active tumor-targeting T$_1$-/T$_2$-weighted dual-mode MRI-guided biological-photothermal therapy (bio-PTT) through bioconjugation of the monocyclic peptides (MCP, the CXC chemokine receptor 4 (CXCR4) antagonist) with 3.8 nm manganese-doped iron oxide NDs (MnIO NDs) [146]. The MnIOMCP displays reasonable T$_1$-/T$_2$-weighted MR contrast abilities (r$_1$ = 13.1 mM^{-1}·s^{-1}, r$_2$ = 46.6 mM^{-1}·s^{-1} and r$_2$/r$_1$ = 3.56), good photothermal conversion efficiency (η = 28.8%), strong tumor-targeting ability (~15.9% ID g^{-1} at 1 h after intravenous injection with a dose of 10 mg [Mn + Fe] kg^{-1}) and clear inhibition of CXCR4-positive tumor growth. In addition, the MnIOMCP at can be rapidly excreted from the body through renal clearance (about 75% ID of MnIOMCP found in urine at 24 h post-injection), which illuminates a new pathway for the development of efficient nanotheranostics with high biosafety. In addition, the r$_1$ and/or r$_2$ values of Ln^{3+}-based nanomaterials could be increased by mixing paramagnetic transition metal ions into them owing to unpaired 3d-electrons of transition metal ions [151–155]. Gao and coauthors reported a facile strategy to design and synthesis of zwitterionic dopamine sulfonate-coated 4.8 nm gadolinium-embedded iron oxide NDs (GdIO@ZDS), which showed a hydrodynamic diameter of about 5.2 nm in both PBS buffer and BSA solution [155]. The combination of the spincanting effects and the collection of Gd^{3+} within small-sized GdIO NDs led to a strongly enhanced T_1-weighted MR contrast effect, which exhibited a high r_1 of 7.85 $mM^{-1} \cdot s^{-1}$ and a low r_2/r_1 ratio of 5.24. Using SKOV3-bearing mouse as a model, the in vivo experimental results demonstrated that the GdIO@ZDS with a dose of 2.0 mg GdIO@ZDS kg^{-1} are suitable candidates as excellent T_1-weighted MRCAs for tumor imaging and disease diagnosis because they have relatively long circulation half-life (~50 min), passive-tumor targeting capacity, and efficient renal clearance ability.

5. Conclusions and Outlook

In summary, we have illustrated the recent advances in the development of renal clearable MRCAs including Gd^{3+}-based composites and MNDs. Compared to conventional small molecule Gd^{3+}-complex contrast agents, the composites/MNDs have demonstrated improved MR signal intensity, targeting ability, and longer circulation time both in vitro and in small animal disease models, especially for cancer diagnosis. In particular, with the help of nanotechnology, the recent research developments have progressed towards construction of renal clearable MRCAs with multifunctionality, which enables us to integrate several functions at the same time, such as simultaneous disease targeting, multimodal imaging, and therapy. For instance, several unique characteristics of MNDs enable their use for selective cancer theragnostics, which include: (a) their size, which leads to preferential accumulation of MNDs in tumors though EPR effect, wide MRI time window by prolonging circulation time, and completely excreted from the body within a reasonable period of time (i.e., within a few of days) through renal clearance; (b) high surface-to-volume ratios, which result in strong enhancement of MR signal at ROI through increasing the interaction opportunities of ^1H with paramagnetic ions; and (c) large surface area, which exhibits the possibility to load different molecular therapeutics for MRI-guided therapy and/or functionalize with cancer-homing ligands for achieving positive-tumor targeting. The performance of MND-based MRCAs could be further improved through optimizing one or more of the above-mentioned characteristics of the MNDs. The renal clearable composite-/MND-based MRCAs have, indeed, bright prospects regarding their possibilities for biomedical applications, which have already been demonstrated in the scientific literature.

Unfortunately, very few MND-based MRCAs were approved for clinical application except SPIONs (Ferumoxides). On the other hand, the need for specific molecular information is greater than ever, since we are entering into the era of precision medicine. The noninvasive diagnostic or screening methods such as MRI can efficiently help patients to avoid ineffective and/or costly treatments because many of the newly developed powerful and expensive therapies are only effective in a subset of patients. This situation strongly requires the development of high-performance contrast agents to improve the accuracy of molecular imaging. The clinical translation of MND-based MRCAs research is generally impeded by the significant heterogeneity in the construction of these agents. Up to date, the MND-based MRCAs are still in the experimental stage. There are several technical challenges regarding the MND-based MRCAs for clinical trials that need to be clearly addressed before they are evaluated in humans. For example, the safety and efficacy of the MND-based MRCAs should be comprehensively evaluated. Ongoing research should focus on evaluating the biodistribution, pharmacokinetics, and ultimate fates in vivo, improving targeting specificity while minimizing toxicity, and demonstrating the translational potential with appropriate animal models. It is critical to develop cost-effective methods for the kilogram-scale production of MND-based MRCAs since the MRCAs are normally administered in gram quantities. This matter could be solved by systematic optimization of synthesis conditions of polyol methods and thermal decomposition methods. The leakage of Gd^{3+} should be minimized while Gd^{3+} NDs were used as MRCAs. Future efforts should aim to synthesize chelating agents with a high Gd^{3+} binding constant, and/or develop coating materials with highly stable physicochemical properties and excellent biocompatibility. In addition, integration of two or more paramagnetic metallic elements into a single hybrid ND is beneficial to circumvent their individual drawbacks and potentially provide more comprehensive imaging information through T_1-/T_2-weighted dual-mode MRI. The MNDs are providing revolutionary potential as new MRCAs, which could achieve various clinical applications through close cooperation among of multidisciplinary teams of chemists, materials scientists, biologists, pharmacists, physicians, and imaging experts, and have a strongly positive impact on human health.

Author Contributions: Writing—original draft preparation, X.L., Y.S., G.L. and Z.W.; writing—review and editing, X.L., L.M., G.L. and Z.W. All authors have read and agreed to the published version of the manuscript.

Funding: This research was funded by JILIN PROVINCIAL SCIENCE AND TECHNOLOGY DEPARTMENT and NATIONAL NATURAL SCIENCE FOUNDATION OF CHINA, grant numbers 20190701052GH and 21775145.

Acknowledgments: The authors would like to thank the Jilin Provincial Science and Technology Department, and National Natural Science Foundation of China for financial support.

Conflicts of Interest: The authors declare no conflict of interest.

References

1. Weissleder, R.; Mahmood, U. Molecular imaging. *Radiology* **2001**, *219*, 316–333. [CrossRef] [PubMed]
2. Smith, S.M.; Jenkinson, M.; Woolrich, M.W.; Beckmann, C.F.; Behrens, T.E.J.; Johansen-Berg, H.; Bannister, P.R.; De Luca, M.; Drobnjak, I.; Flitney, D.E.; et al. Advances in functional and structural MR image analysis and implementation as FSL. *Neuroimage* **2004**, *23*, S208–S219. [CrossRef] [PubMed]
3. Caravan, P.; Ellison, J.J.; McMurry, T.J.; Lauffer, R.B. Gadolinium(III) chelates as MRI contrast agents: Structure, dynamics, and applications. *Chem. Rev.* **1999**, *99*, 2293–2352. [CrossRef] [PubMed]
4. Werner, E.J.; Datta, A.; Jocher, C.J.; Raymond, K.N. High-relaxivity MRI contrast agents: Where coordination chemistry meets medical imaging. *Angew. Chem. Int. Ed.* **2008**, *47*, 8568–8580. [CrossRef]
5. Frey, N.A.; Peng, S.; Cheng, K.; Sun, S. Magnetic nanoparticles: Synthesis, functionalization, and applications in bioimaging and magnetic energy storage. *Chem. Soc. Rev.* **2009**, *38*, 2532–2542. [CrossRef]
6. Louie, A.Y. Multimodality imaging probes: Design and challenges. *Chem. Rev.* **2010**, *110*, 3146–3195. [CrossRef] [PubMed]
7. Law, G.-L.; Wong, W.-T. An introduction to molecular imaging. In *The Chemistry of Molecular Imaging*; John Wiley & Sons, Inc.: Hoboken, NJ, USA, 2014; pp. 1–24.
8. Lohrke, J.; Frenzel, T.; Endrikat, J.; Alves, F.C.; Grist, T.M.; Law, M.; Lee, J.M.; Leiner, T.; Li, K.-C.; Nikolaou, K.; et al. 25 Years of contrast-enhanced MRI: Developments, current challenges and future perspectives. *Adv. Ther.* **2016**, *33*, 1–28. [CrossRef]
9. Wahsner, J.; Gale, E.M.; Rodriguez-Rodriguez, A.; Caravan, P. Chemistry of MRI contrast agents: Current challenges and new frontiers. *Chem. Rev.* **2019**, *119*, 957–1057. [CrossRef]
10. Brasch, R.C. Methods of contrast enhancement for NMR imaging and potential applications-A subject review. *Radiology* **1983**, *147*, 781–788. [CrossRef]
11. Gupta, A.K.; Gupta, M. Synthesis and surface engineering of iron oxide nanoparticles for biomedical applications. *Biomaterials* **2005**, *26*, 3995–4021. [CrossRef]
12. Xu, W.; Kattel, K.; Park, J.Y.; Chang, Y.; Kim, T.J.; Lee, G.H. Paramagnetic nanoparticle T-1 and T-2 MRI contrast agents. *Phys. Chem. Chem. Phys.* **2012**, *14*, 12687–12700. [CrossRef]
13. Knezevic, N.Z.; Gadjanski, I.; Durand, J.-O. Magnetic nanoarchitectures for cancer sensing, imaging and therapy. *J. Mater. Chem. B* **2019**, *7*, 9–23. [CrossRef]
14. Bao, Y.; Sherwood, J.A.; Sun, Z. Magnetic iron oxide nanoparticles as T-1 contrast agents for magnetic resonance imaging. *J. Mater. Chem. C* **2018**, *6*, 1280–1290. [CrossRef]
15. Zhou, Z.; Yang, L.; Gao, J.; Chen, X. Structure-relaxivity relationships of magnetic nanoparticles for magnetic resonance imaging. *Adv. Mater.* **2019**, *31*, 1804567. [CrossRef] [PubMed]
16. Han, X.; Xu, K.; Taratula, O.; Farsad, K. Applications of nanoparticles in biomedical imaging. *Nanoscale* **2019**, *11*, 799–819. [CrossRef]
17. Sun, C.; Lee, J.S.H.; Zhang, M. Magnetic nanoparticles in MR imaging and drug delivery. *Adv. Drug Deliv. Rev.* **2008**, *60*, 1252–1265. [CrossRef] [PubMed]
18. Longmire, M.; Choyke, P.L.; Kobayashi, H. Clearance properties of nano-sized particles and molecules as imaging agents: Considerations and caveats. *Nanomedicine* **2008**, *3*, 703–717. [CrossRef] [PubMed]
19. Qiao, R.; Yang, C.; Gao, M. Superparamagnetic iron oxide nanoparticles: From preparations to in vivo MRI applications. *J. Mater. Chem.* **2009**, *19*, 6274–6293. [CrossRef]

20. Jain, T.K.; Reddy, M.K.; Morales, M.A.; Leslie-Pelecky, D.L.; Labhasetwar, V. Biodistribution, clearance, and biocompatibility of iron oxide magnetic nanoparticles in rats. *Mol. Pharmaceut.* **2008**, *5*, 316–327. [CrossRef]
21. Zhang, Y.-N.; Poon, W.; Tavares, A.J.; McGilvray, I.D.; Chan, W.C.W. Nanoparticle-liver interactions: Cellular uptake and hepatobiliary elimination. *J. Control. Release* **2016**, *240*, 332–348. [CrossRef]
22. Buchman, J.T.; Hudson-Smith, N.V.; Landy, K.M.; Haynes, C.L. Understanding nanoparticle toxicity mechanisms to inform redesign strategies to reduce environmental impact. *Acc. Chem. Res.* **2019**, *52*, 1632–1642. [CrossRef] [PubMed]
23. Choi, H.S.; Frangioni, J.V. Nanoparticles for biomedical imaging: Fundamentals of clinical translation. *Mol. Imaging* **2010**, *9*, 291–310. [CrossRef]
24. Runge, V.M. Critical questions regarding gadolinium deposition in the brain and body after injections of the gadolinium-based contrast agents, safety, and clinical recommendations in consideration of the EMA's pharmacovigilance and risk assessment committee recommendation for suspension of the marketing authorizations for 4 linear agents. *Investig. Radiol.* **2017**, *52*, 317–323.
25. Na, H.B.; Song, I.C.; Hyeon, T. Inorganic nanoparticles for MRI contrast agents. *Adv. Mater.* **2009**, *21*, 2133–2148. [CrossRef]
26. Lu, Z.R.; Parker, D.L.; Goodrich, K.C.; Wang, X.H.; Dalle, J.G.; Buswell, H.R. Extracellular biodegradable macromolecular gadolinium(III) complexes for MRI. *Magn. Reson. Med.* **2004**, *51*, 27–34. [CrossRef]
27. Villaraza, A.J.L.; Bumb, A.; Brechbiel, M.W. Macromolecules, dendrimers, and nanomaterials in magnetic resonance imaging: The interplay between size, function, and pharmacokinetics. *Chem. Rev.* **2010**, *110*, 2921–2959. [CrossRef] [PubMed]
28. Kelkar, S.S.; Reineke, T.M. Theranostics: Combining imaging and therapy. *Bioconjug. Chem.* **2011**, *22*, 1879–1903. [CrossRef]
29. Detappe, A.; Kunjachan, S.; Sancey, L.; Motto-Ros, V.; Biancur, D.; Drane, P.; Guieze, R.; Makrigiorgos, G.M.; Tillement, O.; Langer, R.; et al. Advanced multimodal nanoparticles delay tumor progression with clinical radiation therapy. *J. Control. Release* **2016**, *238*, 103–113. [CrossRef]
30. Pellico, J.; Ellis, C.M.; Davis, J.J. Nanoparticle-based paramagnetic contrast agents for magnetic resonance imaging. *Contrast Media Mol. Imaging* **2019**, 1845637. [CrossRef]
31. Zhu, W.; Artemov, D. Biocompatible blood pool MRI contrast agents based on hyaluronan. *Contrast Media Mol. Imaging* **2011**, *6*, 61–68. [CrossRef]
32. Grogna, M.; Cloots, R.; Luxen, A.; Jerome, C.; Desreux, J.-F.; Detrembleur, C. Design and synthesis of novel DOTA(Gd^{3+})-polymer conjugates as potential MRI contrast agents. *J. Mater. Chem.* **2011**, *21*, 12917–12926. [CrossRef]
33. Schopf, E.; Sankaranarayanan, J.; Chan, M.; Mattrey, R.; Almutairi, A. An extracellular MRI polymeric contrast agent that degrades at physiological pH. *Mol. Pharmaceut.* **2012**, *9*, 1911–1918. [CrossRef] [PubMed]
34. Zhang, G.; Zhang, R.; Melancon, M.P.; Wong, K.; You, J.; Huang, Q.; Bankson, J.; Liang, D.; Li, C. The degradation and clearance of Poly(N-hydroxypropyl-L-glutamine)-DTPA-Gd as a blood pool MRI contrast agent. *Biomaterials* **2012**, *33*, 5376–5383. [CrossRef] [PubMed]
35. Xiao, Y.; Xue, R.; You, T.; Li, X.; Pei, F. A new biodegradable and biocompatible gadolinium (III)-polymer for liver magnetic resonance imaging contrast agent. *Magn. Reson. Imaging* **2015**, *33*, 822–828. [CrossRef] [PubMed]
36. Xiong, Z.; Wang, Y.; Zhu, J.; He, Y.; Qu, J.; Effenberg, C.; Xia, J.; Appelhans, D.; Shi, X. Gd-Chelated poly(propylene imine) dendrimers with densely organized maltose shells for enhanced MR imaging applications. *Biomater. Sci.* **2016**, *4*, 1622–1629. [CrossRef]
37. Moussaron, A.; Vibhute, S.; Bianchi, A.; Guenduez, S.; Kotb, S.; Sancey, L.; Motto-Ros, V.; Rizzitelli, S.; Cremillieux, Y.; Lux, F.; et al. Ultrasmall nanoplatforms as calcium-responsive contrast agents for magnetic resonance imaging. *Small* **2015**, *11*, 4900–4909. [CrossRef]
38. McLeod, S.M.; Robison, L.; Parigi, G.; Olszewski, A.; Drout, R.J.; Gong, X.; Islamoglu, T.; Luchinat, C.; Farha, O.K.; Meade, T.J. Maximizing magnetic resonance contrast in Gd(III) nanoconjugates: Investigation of proton relaxation in zirconium metal-organic frameworks. *ACS Appl. Mater. Inter.* **2020**, *12*, 41157–41166. [CrossRef]

39. Shi, Y.; Pan, Y.; Zhong, J.; Yang, J.; Zheng, J.; Cheng, J.; Song, R.; Yi, C. Facile synthesis of gadolinium (III) chelates functionalized carbon quantum dots for fluorescence and magnetic resonance dual-modal bioimaging. *Carbon* **2015**, *93*, 742–750. [CrossRef]
40. Zhang, M.; Liu, X.; Huang, J.; Wang, L.; Shen, H.; Luo, Y.; Li, Z.; Zhang, H.; Deng, Z.; Zhang, Z. Ultrasmall graphene oxide based T-1 MRI contrast agent for in vitro and in vivo labeling of human mesenchymal stem cells. *Nanomedicine* **2018**, *14*, 2475–2483. [CrossRef]
41. Cao, T.; Zhou, X.; Zheng, Y.; Sun, Y.; Zhang, J.; Chen, W.; Zhang, J.; Zhou, Z.; Yang, S.; Zhang, Y.; et al. Chelator-free conjugation of Tc-99m and Gd^{3+} to pegylated nanographene oxide for dual-modality SPECT/MR imaging of lymph nodes. *ACS Appl. Mater. Inter.* **2017**, *9*, 42612–42621. [CrossRef]
42. Chen, H.; Wang, G.D.; Tang, W.; Todd, T.; Zhen, Z.; Tsang, C.; Hekmatyar, K.; Cowger, T.; Hubbard, R.B.; Zhang, W.; et al. Gd-encapsulated carbonaceous dots with efficient renal clearance for magnetic resonance imaging. *Adv. Mater.* **2014**, *26*, 6761–6766. [CrossRef]
43. Liang, G.; Ye, D.; Zhang, X.; Dong, F.; Chen, H.; Zhang, S.; Li, J.; Shen, X.; Kong, J. One-pot synthesis of Gd^{3+}-functionalized gold nanoclusters for dual model (fluorescence/magnetic resonance) imaging. *J. Mater. Chem. B* **2013**, *1*, 3545–3552. [CrossRef] [PubMed]
44. Zhu, Q.; Pan, F.; Tian, Y.; Tang, W.; Yuan, Y.; Hu, A. Facile synthesis of Gd(III) metallosurfactant-functionalized carbon nanodots with high relaxivity as bimodal imaging probes. *RSC Adv.* **2016**, *6*, 29441–29447. [CrossRef]
45. Hu, D.-H.; Sheng, Z.-H.; Zhang, P.-F.; Yang, D.-Z.; Liu, S.-H.; Gong, P.; Gao, D.-Y.; Fang, S.-T.; Ma, Y.-F.; Cai, L.-T. Hybrid gold-gadolinium nanoclusters for tumor-targeted NIRF/CT/MRI triple-modal imaging in vivo. *Nanoscale* **2013**, *5*, 1624–1628. [CrossRef]
46. Zhang, J.; Hao, G.; Yao, C.; Hu, S.; Hu, C.; Zhang, B. Paramagnetic albumin decorated $CuInS_2$/ZnS QDs for CD133(+) glioma bimodal MR/fluorescence targeted imaging. *J. Mater. Chem. B* **2016**, *4*, 4110–4118. [CrossRef]
47. Liang, G.; Xiao, L. Gd^{3+}-Functionalized gold nanoclusters for fluorescence-magnetic resonance bimodal imaging. *Biomater. Sci.* **2017**, *5*, 2122–2130. [CrossRef]
48. Gao, A.; Kang, Y.-F.; Yin, X.-B. Red fluorescence-magnetic resonance dual modality imaging applications of gadolinium containing carbon quantum dots with excitation independent emission. *New J. Chem.* **2017**, *41*, 3422–3431. [CrossRef]
49. Truillet, C.; Bouziotis, P.; Tsoukalas, C.; Brugiere, J.; Martini, M.; Sancey, L.; Brichart, T.; Denat, F.; Boschetti, F.; Darbost, U.; et al. Ultrasmall particles for Gd-MRI and Ga-68-PET dual imaging. *Contrast Media Mol. Imaging* **2015**, *10*, 309–319. [CrossRef]
50. Luo, D.; Johnson, A.; Wang, X.; Li, H.; Erokwu, B.O.; Springer, S.; Lou, J.; Ramamurthy, G.; Flask, C.A.; Burda, C.; et al. Targeted radiosensitizers for MR-guided radiation therapy of prostate cancer. *Nano Lett.* **2020**. [CrossRef]
51. Vivero-Escoto, J.L.; Taylor-Pashow, K.M.L.; Huxford, R.C.; Della Rocca, J.; Okoruwa, C.; An, H.; Lin, W.; Lin, W. Multifunctional mesoporous silica nanospheres with cleavable Gd(III) chelates as MRI contrast agents: Synthesis, characterization, target-specificity, and renal clearance. *Small* **2011**, *7*, 3519–3528. [CrossRef]
52. Ma, Y.; Mou, Q.; Sun, M.; Yu, C.; Li, J.; Huang, X.; Zhu, X.; Yan, D.; Shen, J. Cancer theranostic nanoparticles self-assembled from amphiphilic small molecules with equilibrium shift-induced renal clearance. *Theranostics* **2016**, *6*, 1703–1716. [CrossRef]
53. Baker, S.N.; Baker, G.A. Luminescent carbon nanodots: Emergent nanolights. *Angew. Chem. Int. Ed.* **2010**, *49*, 6726–6744. [CrossRef]
54. Lim, S.Y.; Shen, W.; Gao, Z. Carbon quantum dots and their applications. *Chem. Soc. Rev.* **2015**, *44*, 362–381. [CrossRef]
55. Garg, B.; Bisht, T. Carbon nanodots as peroxidase nanozymes for biosensing. *Molecules* **2016**, *21*, 1653. [CrossRef]
56. Pardo, J.; Peng, Z.; Leblanc, R.M. Cancer targeting and drug delivery using carbon-based quantum dots and nanotubes. *Molecules* **2018**, *23*, 378. [CrossRef]
57. Nekoueian, K.; Amiri, M.o.; Sillanpaa, M.; Marken, F.; Boukherroub, R.; Szunerits, S. Carbon-based quantum particles: An electroanalytical and biomedical perspective. *Chem. Soc. Rev.* **2019**, *48*, 4281–4316. [CrossRef]
58. Du, F.; Zhang, L.; Zhang, L.; Zhang, M.; Gong, A.; Tan, Y.; Miao, J.; Gong, Y.; Sun, M.; Ju, H.; et al. Engineered gadolinium-doped carbon dots for magnetic resonance imaging-guided radiotherapy of tumors. *Biomaterials* **2017**, *121*, 109–120. [CrossRef]

59. Yu, C.; Xuan, T.; Chen, Y.; Zhao, Z.; Liu, X.; Lian, G.; Li, H. Gadolinium-doped carbon dots with high quantum yield as an effective fluorescence and magnetic resonance bimodal imaging probe. *J. Alloys Compd.* **2016**, *688*, 611–619. [CrossRef]
60. Zhao, Y.; Hao, X.; Lu, W.; Wang, R.; Shan, X.; Chen, Q.; Sun, G.; Liu, J. Facile preparation of double rare earth-doped carbon dots for MRI/CT/FI multimodal imaging. *ACS Appl. Nano Mater.* **2018**, *1*, 2544–2551. [CrossRef]
61. Gong, N.; Wang, H.; Li, S.; Deng, Y.; Chen, X.; Ye, L.; Gu, W. Microwave-assisted polyol synthesis of gadolinium-doped green luminescent carbon dots as a bimodal nanoprobe. *Langmuir* **2014**, *30*, 10933–10939. [CrossRef]
62. Engstrom, M.; Klasson, A.; Pedersen, H.; Vahlberg, C.; Kall, P.-O.; Uvdal, K. High proton relaxivity for gadolinium oxide nanoparticles. *MAGMA* **2006**, *19*, 180–186. [CrossRef] [PubMed]
63. Park, J.Y.; Baek, M.J.; Choi, E.S.; Woo, S.; Kim, J.H.; Kim, T.J.; Jung, J.C.; Chae, K.S.; Chang, Y.; Lee, G.H. Paramagnetic ultrasmall gadolinium oxide nanoparticles as advanced T-1 MR1 contrast agent: Account for large longitudinal relaxivity, optimal particle diameter, and in vivo T-1 MR images. *ACS Nano* **2009**, *3*, 3663–3669. [CrossRef] [PubMed]
64. Xu, W.; Park, J.Y.; Kattel, K.; Bony, B.A.; Heo, W.C.; Jin, S.; Park, J.W.; Chang, Y.; Do, J.Y.; Chae, K.S.; et al. A T-1, T-2 magnetic resonance imaging (MRI)-fluorescent imaging (FI) by using ultrasmall mixed gadolinium-europium oxide nanoparticles. *New J. Chem.* **2012**, *36*, 2361–2367. [CrossRef]
65. Kim, C.R.; Baeck, J.S.; Chang, Y.; Bae, J.E.; Chae, K.S.; Lee, G.H. Ligand-size dependent water proton relaxivities in ultrasmall gadolinium oxide nanoparticles and in vivo T-1 MR images in a 1.5 T MR field. *Phy. Chem. Chem. Phy.* **2014**, *16*, 19866–19873. [CrossRef] [PubMed]
66. Ahmad, M.W.; Kim, C.R.; Baeck, J.S.; Chang, Y.; Kim, T.J.; Bae, J.E.; Chaed, K.S.; Lee, G.H. Bovine serum albumin (BSA) and cleaved-BSA conjugated ultrasmall Gd_2O_3 nanoparticles: Synthesis, characterization, and application to MRI contrast agents. *Colloid. Surf. A* **2014**, *450*, 67–75. [CrossRef]
67. Miao, X.; Ho, S.L.; Tegafaw, T.; Cha, H.; Chang, Y.; Oh, I.T.; Yaseen, A.M.; Marasini, S.; Ghazanfari, A.; Yue, H.; et al. Stable and non-toxic ultrasmall gadolinium oxide nanoparticle colloids (coating material = polyacrylic acid) as high-performance T-1 magnetic resonance imaging contrast agents. *RSC Adv.* **2018**, *8*, 3189–3197. [CrossRef]
68. Ho, S.L.; Cha, H.; Oh, I.T.; Jung, K.-H.; Kim, M.H.; Lee, Y.J.; Miao, X.; Tegafaw, T.; Ahmad, M.Y.; Chae, K.S.; et al. Magnetic resonance imaging, gadolinium neutron capture therapy, and tumor cell detection using ultrasmall Gd_2O_3 nanoparticles coated with polyacrylic acid-rhodamine B as a multifunctional tumor theragnostic agent. *RSC Adv.* **2018**, *8*, 12653–12665. [CrossRef]
69. Miao, X.; Xu, W.; Cha, H.; Chang, Y.; Oh, I.T.; Chae, K.S.; Tegafaw, T.; Ho, S.L.; Kim, S.J.; Lee, G.H. Ultrasmall Gd_2O_3 nanoparticles surface-coated by polyacrylic acid (PAA) and their PAA-size dependent relaxometric properties. *Appl. Surf. Sci.* **2019**, *477*, 111–115. [CrossRef]
70. Yue, H.; Marasini, S.; Ahmad, M.Y.; Ho, S.L.; Cha, H.; Liu, S.; Jang, Y.J.; Tegafaw, T.; Ghazanfari, A.; Miao, X.; et al. Carbon-coated ultrasmall gadolinium oxide (Gd_2O_3@C) nanoparticles: Application to magnetic resonance imaging and fluorescence properties. *Colloid. Surf. A* **2020**, *586*, 124261. [CrossRef]
71. Ahren, M.; Selegard, L.; Klasson, A.; Soderlind, F.; Abrikossova, N.; Skoglund, C.; Bengtsson, T.; Engstrom, M.; Kall, P.-O.; Uvdal, K. Synthesis and characterization of PEGylated Gd_2O_3 nanoparticles for MRI contrast enhancement. *Langmuir* **2010**, *26*, 5753–5762. [CrossRef]
72. Le Duc, G.; Miladi, I.; Alric, C.; Mowat, P.; Braeuer-Krisch, E.; Bouchet, A.; Khalil, E.; Billotey, C.; Janier, M.; Lux, F.; et al. Toward an image-guided microbeam radiation therapy using gadolinium-based nanoparticles. *ACS Nano* **2011**, *5*, 9566–9574. [CrossRef] [PubMed]
73. Lux, F.; Mignot, A.; Mowat, P.; Louis, C.; Dufort, S.; Bernhard, C.; Denat, F.; Boschetti, F.; Brunet, C.; Antoine, R.; et al. Ultrasmall rigid particles as multimodal probes for medical applications. *Angew. Chem. Int. Ed.* **2011**, *50*, 12299–12303. [CrossRef] [PubMed]
74. Faucher, L.; Tremblay, M.; Lagueux, J.; Gossuin, Y.; Fortin, M.-A. Rapid synthesis of PEGylated ultrasmall gadolinium oxide nanoparticles for cell labeling and tracking with MRI. *ACS Appl. Mater. Inter.* **2012**, *4*, 4506–4515. [CrossRef] [PubMed]
75. Viger, M.L.; Sankaranarayanan, J.; de Gracia Lux, C.; Chan, M.; Almutairi, A. Collective activation of MRI agents via encapsulation and disease-triggered release. *J. Am. Chem. Soc.* **2013**, *135*, 7847–7850. [CrossRef] [PubMed]

76. Li, Y.; Chen, T.; Tan, W.; Talham, D.R. Size-dependent MRI relaxivity and dual imaging with $Eu_{0.2}Gd_{0.8}PO_4$ center dot H_2O nanoparticles. *Langmuir* **2014**, *30*, 5873–5879. [CrossRef]
77. Fang, J.; Chandrasekharan, P.; Liu, X.-L.; Yang, Y.; Lv, Y.-B.; Yang, C.-T.; Ding, J. Manipulating the surface coating of ultra-small Gd_2O_3 nanoparticles for improved T-1-weighted MR imaging. *Biomaterials* **2014**, *35*, 1636–1642. [CrossRef]
78. Vahdatkhah, P.; Hosseini, H.R.M.; Khodaei, A.; Montazerabadi, A.R.; Irajirad, R.; Oghabian, M.A.; Delavari, H.H. Rapid microwave-assisted synthesis of PVP-coated ultrasmall gadolinium oxide nanoparticles for magnetic resonance imaging. *Chem. Phys.* **2015**, *453*, 35–41. [CrossRef]
79. Dufort, S.; Le Duc, G.; Salome, M.; Bentivegna, V.; Sancey, L.; Brauer-Krisch, E.; Requardt, H.; Lux, F.; Coll, J.-L.; Perriat, P.; et al. The high radiosensitizing efficiency of a trace of gadolinium-based nanoparticles in tumors. *Sci. Rep.* **2016**, *6*, 29678. [CrossRef]
80. Cheng, Y.; Lu, T.; Wang, Y.; Song, Y.; Wang, S.; Lu, Q.; Yang, L.; Tan, F.; Li, J.; Li, N. Glutathione-mediated clearable nanoparticles based on ultrasmall Gd_2O_3 for MSOT/CT/MR imaging guided photothermal/radio combination cancer therapy. *Mol. Pharmaceut.* **2019**, *16*, 3489–3501. [CrossRef]
81. Bony, B.A.; Miller, H.A.; Tarudji, A.W.; Gee, C.C.; Sarella, A.; Nichols, M.G.; Kievit, F.M. Ultrasmall mixed Eu-Gd oxide nanoparticles for multimodal fluorescence and magnetic resonance imaging of passive accumulation and retention in TBI. *ACS Omega* **2020**, *5*, 16220–16227. [CrossRef]
82. Johnson, N.J.J.; Oakden, W.; Stanisz, G.J.; Prosser, R.S.; van Veggel, F.C.J.M. Size-tunable, ultrasmall $NaGdF_4$ nanoparticles: Insights into their T-1 MRI contrast enhancement. *Chem. Mater.* **2011**, *23*, 3714–3722. [CrossRef]
83. Liu, Q.; Feng, W.; Yang, T.; Yi, T.; Li, F. Upconversion luminescence imaging of cells and small animals. *Nat. Protoc.* **2013**, *8*, 2033–2044. [CrossRef]
84. Xing, H.; Zhang, S.; Bu, W.; Zheng, X.; Wang, L.; Xiao, Q.; Ni, D.; Zhang, J.; Zhou, L.; Peng, W.; et al. Ultrasmall $NaGdF_4$ nanodots for efficient MR angiography and atherosclerotic plaque imaging. *Adv. Mater.* **2014**, *26*, 3867–3872. [CrossRef] [PubMed]
85. Ni, D.; Shen, Z.; Zhang, J.; Zhang, C.; Wu, R.; Liu, J.; Yi, M.; Wang, J.; Yao, Z.; Bu, W.; et al. Integrating anatomic and functional dual mode magnetic resonance imaging: Design and applicability of a bifunctional contrast agent. *ACS Nano* **2016**, *10*, 3783–3790. [CrossRef] [PubMed]
86. Liu, F.; He, X.; Zhang, J.; Zhang, H.; Wang, Z. Employing tryptone as a general phase transfer agent to produce renal clearable nanodots for bioimaging. *Small* **2015**, *11*, 3676–3685. [CrossRef]
87. Yan, Y.; Ding, L.; Liu, L.; Abualrejal, M.M.A.; Chen, H.; Wang, Z. Renal-clearable hyaluronic acid functionalized $NaGdF_4$ nanodots with enhanced tumor accumulation. *RSC Adv.* **2020**, *10*, 13872–13878. [CrossRef]
88. Chen, H.; Li, X.; Liu, F.; Zhang, H.; Wang, Z. Renal clearable peptide functionalized $NaGdF_4$ nanodots for high-efficiency tracking orthotopic colorectal tumor in mouse. *Mol. Pharmaceut.* **2017**, *14*, 3134–3141. [CrossRef]
89. Chen, Y.; Fu, Y.; Li, X.; Chen, H.; Wang, Z.; Zhang, H. Peptide-functionalized $NaGdF_4$ nanoparticles for tumor-targeted magnetic resonance imaging and effective therapy. *RSC Adv.* **2019**, *9*, 17093–17100. [CrossRef]
90. Sun, W.; Luo, L.; Feng, Y.; Qiu, Y.; Shi, C.; Meng, S.; Chen, X.; Chen, H. Gadolinium-rose bengal coordination polymer nanodots for MR-/Fluorescence-image-guided radiation and photodynamic therapy. *Adv. Mater.* **2020**, *32*, 2000377. [CrossRef]
91. Tassa, C.; Shaw, S.Y.; Weissleder, R. Dextran-coated iron oxide nanoparticles: A versatile platform for targeted molecular imaging, molecular diagnostics, and therapy. *Acc. Chem. Res.* **2011**, *44*, 842–852. [CrossRef]
92. Wong, X.Y.; Sena-Torralba, A.; Alvarez-Diduk, R.; Muthoosamy, K.; Merkoci, A. Nanomaterials for nanotheranostics: Tuning their properties according to disease needs. *ACS Nano* **2020**, *14*, 2585–2627. [CrossRef] [PubMed]
93. Farzin, A.; Etesami, S.A.; Quint, J.; Memic, A.; Tamayol, A. Magnetic nanoparticles in cancer therapy and diagnosis. *Adv. Healthc. Mater.* **2020**, *9*, 1901058. [CrossRef] [PubMed]
94. Dardzinski, B.J.; Schmithorst, V.J.; Holland, S.K.; Boivin, G.P.; Imagawa, T.; Watanabe, S.; Lewis, J.M.; Hirsch, R. MR imaging of murine arthritis using ultrasmall superparamagnetic iron oxide particles. *Magn. Reson. Imaging* **2001**, *19*, 1209–1216. [CrossRef]
95. Wang, Y.; Xu, C.; Chang, Y.; Zhao, L.; Zhang, K.; Zhao, Y.; Gao, F.; Gao, X. Ultrasmall superparamagnetic iron oxide nanoparticle for T-2-weighted magnetic resonance imaging. *ACS Appl. Mater. Interfaces* **2017**, *9*, 28959–28966. [CrossRef]

96. Kim, B.H.; Lee, N.; Kim, H.; An, K.; Park, Y.I.; Choi, Y.; Shin, K.; Lee, Y.; Kwon, S.G.; Na, H.B.; et al. Large-scale synthesis of uniform and extremely small-sized iron oxide nanoparticles for high-resolution T-1 magnetic resonance imaging contrast agents. *J. Am. Chem. Soc.* **2011**, *133*, 12624–12631. [CrossRef]
97. Li, P.; Chevallier, P.; Ramrup, P.; Biswas, D.; Vuckovich, D.; Fortin, M.-A.; Oh, J.K. Mussel-inspired multidentate block copolymer to stabilize ultrasmall superparamagnetic Fe_3O_4 for magnetic resonance imaging contrast enhancement and excellent colloidal stability. *Chem. Mater.* **2015**, *27*, 7100–7109. [CrossRef]
98. Yoon, J.; Cho, S.H.; Seong, H. Multifunctional ultrasmall superparamagnetic iron oxide nanoparticles as a theranostic agent. *Colloid. Surf. A* **2017**, *520*, 892–902. [CrossRef]
99. Tromsdorf, U.I.; Bruns, O.T.; Salmen, S.C.; Beisiegel, U.; Weller, H. A highly effective, nontoxic T-1 MR contrast agent based on ultrasmall pegylated iron oxide nanoparticles. *Nano Lett.* **2009**, *9*, 4434–4440. [CrossRef]
100. Vangijzegem, T.; Stanicki, D.; Panepinto, A.; Socoliuc, V.; Vekas, L.; Muller, R.N.; Laurent, S. Influence of experimental parameters of a continuous flow process on the properties of very small iron oxide nanoparticles (VSION) designed for T_1-weighted magnetic resonance imaging (MRI). *Nanomaterials* **2020**, *10*, 757. [CrossRef]
101. Iqbal, M.Z.; Ma, X.; Chen, T.; Zhang, L.e.; Ren, W.; Xiang, L.; Wu, A. Silica-coated super-paramagnetic iron oxide nanoparticles (SPIONPs): A new type contrast agent of T-1 magnetic resonance imaging (MRI). *J. Mater. Chem. B* **2015**, *3*, 5172–5181. [CrossRef]
102. Ma, D.; Chen, J.; Luo, Y.; Wang, H.; Shi, X. Zwitterion-coated ultrasmall iron oxide nanoparticles for enhanced T-1-weighted magnetic resonance imaging applications. *J. Mater. Chem. B* **2017**, *5*, 7267–7273. [CrossRef]
103. Wei, H.; Bruns, O.T.; Kaul, M.G.; Hansen, E.C.; Barch, M.; Wisniowska, A.; Chen, O.; Chen, Y.; Li, N.; Okada, S.; et al. Exceedingly small iron oxide nanoparticles as positive MRI contrast agents. *Proc. Natl. Acad. Sci. USA* **2017**, *114*, 2325–2330. [CrossRef] [PubMed]
104. Yang, W.; Xiang, C.; Xu, Y.; Chen, S.; Zeng, W.; Liu, K.; Jin, X.; Zhou, X.; Zhang, B. Albumin-constrained large-scale synthesis of renal clearable ferrous sulfide quantum dots for T-1-Weighted MR imaging and phototheranostics of tumors. *Biomaterials* **2020**, *255*, 120186. [CrossRef] [PubMed]
105. Wei, R.; Cai, Z.; Ren, B.W.; Li, A.; Lin, H.; Zhang, K.; Chen, H.; Shan, H.; Ai, H.; Gao, J. Biodegradable and renal-clearable hollow porous iron oxide nanoboxes for in vivo imaging. *Chem. Mater.* **2018**, *30*, 7950–7961. [CrossRef]
106. Luo, Y.; Yang, J.; Yan, Y.; Li, J.; Shen, M.; Zhang, G.; Mignani, S.; Shi, X. RGD- functionalized ultrasmall iron oxide nanoparticles for targeted T-1-weighted MR imaging of gliomas. *Nanoscale* **2015**, *7*, 14538–14546. [CrossRef] [PubMed]
107. Liu, F.; He, X.; Chen, H.; Zhang, J.; Zhang, H.; Wang, Z. Gram-scale synthesis of coordination polymer nanodots with renal clearance properties for cancer theranostic applications. *Nat. Commun.* **2015**, *6*, 8003. [CrossRef]
108. Li, X.; Chen, H.; Liu, F.; Chen, Y.; Zhang, H.; Wang, Z. Accurate monitoring of renal injury state through in vivo magnetic resonance imaging with ferric coordination polymer nanodots. *ACS Omega* **2018**, *3*, 4918–4923. [CrossRef]
109. Chen, L.; Chen, J.; Qiu, S.; Wen, L.; Wu, Y.; Hou, Y.; Wang, Y.; Zeng, J.; Feng, Y.; Li, Z.; et al. Biodegradable nanoagents with short biological half-life for SPECT/PAI/MRI multimodality imaging and PTT therapy of tumors. *Small* **2018**, *14*, 1702700. [CrossRef]
110. Wang, G.; Zhang, X.; Skallberg, A.; Liu, Y.; Hu, Z.; Mei, X.; Uvdal, K. One-step synthesis of water-dispersible ultra-small Fe_3O_4 nanoparticles as contrast agents for T-1 and T-2 magnetic resonance imaging. *Nanoscale* **2014**, *6*, 2953–2963. [CrossRef]
111. Zhou, H.; Tang, J.; Li, J.; Li, W.; Liu, Y.; Chen, C. In vivo aggregation-induced transition between T-1 and T-2 relaxations of magnetic ultra-small iron oxide nanoparticles in tumor microenvironment. *Nanoscale* **2017**, *9*, 3040–3050. [CrossRef]
112. Li, X.; Lu, S.; Xiong, Z.; Hu, Y.; Ma, D.; Lou, W.; Peng, C.; Shen, M.; Shi, X. Light-addressable nanoclusters of ultrasmall iron oxide nanoparticles for enhanced and dynamic magnetic resonance imaging of arthritis. *Adv. Sci.* **2019**, *6*, 1901800. [CrossRef]
113. Ma, D.; Shi, M.; Li, X.; Zhang, J.; Fan, Y.; Sun, K.; Jiang, T.; Peng, C.; Shi, X. Redox-sensitive clustered ultrasmall iron oxide nanoparticles for switchable T-2/T-1-Weighted magnetic resonance imaging applications. *Bioconjugate Chem.* **2020**, *31*, 352–359. [CrossRef] [PubMed]
114. Zhen, Z.; Xie, J. Development of manganese-based nanoparticles as contrast probes for magnetic resonance imaging. *Theranostics* **2012**, *2*, 45–54. [CrossRef] [PubMed]

115. Botta, M.; Carniato, F.; Esteban-Gomez, D.; Platas-Iglesias, C.; Tei, L. Mn(II) compounds as an alternative to Gd-based MRI probes. *Future Med. Chem.* **2019**, *11*, 1461–1483. [CrossRef] [PubMed]
116. Na, H.B.; Lee, J.H.; An, K.; Park, Y.I.; Park, M.; Lee, I.S.; Nam, D.-H.; Kim, S.T.; Kim, S.-H.; Kim, S.-W.; et al. Development of a T-1 contrast agent for magnetic resonance imaging using MnO nanoparticles. *Angew. Chem. Int. Ed.* **2007**, *46*, 5397–5401. [CrossRef]
117. Shin, J.; Anisur, R.M.; Ko, M.K.; Im, G.H.; Lee, J.H.; Lee, I.S. Hollow manganese oxide nanoparticles as multifunctional agents for magnetic resonance imaging and drug delivery. *Angew. Chem. Int. Ed.* **2009**, *48*, 321–324. [CrossRef] [PubMed]
118. Letourneau, M.; Tremblay, M.; Faucher, L.; Rojas, D.; Chevallier, P.; Gossuin, Y.; Lagueux, J.; Fortin, M.-A. MnO-labeled cells: Positive contrast enhancement in MRI. *J. Phys. Chem. B* **2012**, *116*, 13228–13238. [CrossRef]
119. Lu, Y.; Zhang, L.; Li, J.; Su, Y.-D.; Liu, Y.; Xu, Y.-J.; Dong, L.; Gao, H.-L.; Lin, J.; Man, N.; et al. MnO nanocrystals: A platform for integration of MRI and genuine autophagy induction for chemotherapy. *Adv. Funct. Mater.* **2013**, *23*, 1534–1546. [CrossRef]
120. Chevallier, P.; Walter, A.; Garofalo, A.; Veksler, I.; Lagueux, J.; Begin-Colin, S.; Felder-Flesch, D.; Fortin, M.A. Tailored biological retention and efficient clearance of pegylated ultra-small MnO nanoparticles as positive MRI contrast agents for molecular imaging. *J. Mater. Chem. B* **2014**, *2*, 1779–1790. [CrossRef]
121. McDonagh, B.H.; Singh, G.; Hak, S.; Bandyopadhyay, S.; Augestad, I.L.; Peddis, D.; Sandvig, I.; Sandvig, A.; Glomm, W.R. L-DOPA-coated manganese oxide nanoparticles as dual MRI contrast agents and drug-delivery vehicles. *Small* **2016**, *12*, 301–306. [CrossRef]
122. Zhan, Y.; Zhan, W.; Li, H.; Xu, X.; Cao, X.; Zhu, S.; Liang, J.; Chen, X. In Vivo Dual-modality fluorescence and magnetic resonance imaging-guided lymph node mapping with good biocompatibility manganese oxide nanoparticles. *Molecules* **2017**, *22*, 2208. [CrossRef]
123. Cai, X.; Gao, W.; Ma, M.; Wu, M.; Zhang, L.; Zheng, Y.; Chen, H.; Shi, J. A prussian blue-based core-shell hollow-structured mesoporous nanoparticle as a smart theranostic agent with ultrahigh ph-responsive longitudinal relaxivity. *Adv. Mater.* **2015**, *27*, 6382–6389. [CrossRef]
124. Wang, D.; Lin, H.; Zhang, G.; Si, Y.; Yang, H.; Bai, G.; Yang, C.; Zhong, K.; Cai, D.; Wu, Z.; et al. Effective pH-activated theranostic platform for synchronous magnetic resonance imaging diagnosis and chemotherapy. *ACS Appl. Mater. Interfaces* **2018**, *10*, 31114–31123. [CrossRef] [PubMed]
125. Sun, Y.; Chen, H.; Liu, G.; Ma, L.; Wang, Z. The controllable growth of ultrathin MnO_2 on polydopamine nanospheres as a single nanoplatform for the MRI-guided synergistic therapy of tumors. *J. Mater. Chem. B* **2019**, *7*, 7152–7161. [CrossRef] [PubMed]
126. Gong, F.; Cheng, L.; Yang, N.; Betzer, O.; Feng, L.; Zhou, Q.; Li, Y.; Chen, R.; Popovtzer, R.; Liu, Z. Ultrasmall oxygen-deficient bimetallic oxide $MnWO_X$ nanoparticles for depletion of endogenous GSH and enhanced sonodynamic cancer therapy. *Adv. Mater.* **2019**, *31*, 1900730. [CrossRef]
127. Zheng, S.; Zhang, M.; Bai, H.; He, M.; Dong, L.; Cai, L.; Zhao, M.; Wang, Q.; Xu, K.; Li, J. Preparation of AS1411 aptamer modified Mn-MoS_2 QDs for targeted MR imaging and fluorescence labelling of renal cell carcinoma. *Int. J. Nanomed.* **2019**, *14*, 9513–9524. [CrossRef]
128. Li, J.; Wu, C.; Hou, P.; Zhang, M.; Xu, K. One-pot preparation of hydrophilic manganese oxide nanoparticles as T-1 nano-contrast agent for molecular magnetic resonance imaging of renal carcinoma in vitro and in vivo. *Biosens. Bioelectron.* **2018**, *102*, 1–8. [CrossRef] [PubMed]
129. Meng, J.; Zhao, Y.; Li, Z.; Wang, L.; Tian, Y. Phase transfer preparation of ultrasmall MnS nanocrystals with a high performance MRI contrast agent. *RSC Adv.* **2016**, *6*, 6878–6887. [CrossRef]
130. Chen, A.; Sun, J.; Liu, S.; Li, L.; Peng, X.; Ma, L.; Zhang, R. The effect of metal ions on endogenous melanin nanoparticles used as magnetic resonance imaging contrast agents. *Biomater. Sci.* **2020**, *8*, 379–390. [CrossRef]
131. Sun, J.; Xu, W.; Li, L.; Fan, B.; Peng, X.; Qu, B.; Wang, L.; Li, T.; Li, S.; Zhang, R. Ultrasmall endogenous biopolymer nanoparticles for magnetic resonance/photoacoustic dual-modal imaging-guided photothermal therapy. *Nanoscale* **2018**, *10*, 10584–10595. [CrossRef]
132. Jin, M.; Li, W.; Spillane, D.E.M.; Geraldes, C.F.G.C.; Williams, G.R.; Bligh, S.W.A. Hydroxy double salts intercalated with Mn(II) complexes as potential contrast agents. *Solid State Sci.* **2016**, *53*, 9–16. [CrossRef]
133. Wu, Y.; Xu, L.; Qian, J.; Shi, L.; Su, Y.; Wang, Y.; Li, D.; Zhu, X. Methotrexate-Mn^{2+} based nanoscale coordination polymers as a theranostic nanoplatform for MRI guided chemotherapy. *Biomater. Sci.* **2020**, *8*, 712–719. [CrossRef] [PubMed]

134. Xu, W.; Sun, J.; Li, L.; Peng, X.; Zhang, R.; Wang, B. Melanin-manganese nanoparticles with ultrahigh efficient clearance in vivo for tumor-targeting T-1 magnetic resonance imaging contrast agent. *Biomater. Sci.* **2018**, *6*, 207–215. [CrossRef] [PubMed]
135. Bottrill, M.; Nicholas, L.K.; Long, N.J. Lanthanides in magnetic resonance imaging. *Chem. Soc. Rev.* **2006**, *35*, 557–571. [CrossRef] [PubMed]
136. Kattel, K.; Park, J.Y.; Xu, W.; Kim, H.G.; Lee, E.J.; Bony, B.A.; Heo, W.C.; Jin, S.; Baeck, J.S.; Chang, Y.; et al. Paramagnetic dysprosium oxide nanoparticles and dysprosium hydroxide nanorods as T-2 MRI contrast agents. *Biomaterials* **2012**, *33*, 3254–3261. [CrossRef]
137. Yue, H.; Park, J.Y.; Chang, Y.; Lee, G.H. Ultrasmall europium, gadolinium, and dysprosium oxide nanoparticles: Polyol synthesis, properties, and biomedical imaging applications. *Mini Rev. Med. Chem.* **2020**. [CrossRef]
138. Kattel, K.; Park, J.Y.; Xu, W.; Kim, H.G.; Lee, E.J.; Bony, B.A.; Heo, W.C.; Lee, J.J.; Jin, S.; Baeck, J.S.; et al. A facile synthesis, in vitro and in vivo MR studies of D-glucuronic acid-coated ultrasmall Ln_2O_3 (Ln = Eu, Gd, Dy, Ho, and Er) nanoparticles as a new potential MRI contrast agent. *ACS Appl. Mater. Interfaces* **2011**, *3*, 3325–3334. [CrossRef]
139. Das, G.K.; Zhang, Y.; D'Silva, L.; Padmanabhan, P.; Heng, B.C.; Loo, J.S.C.; Selvan, S.T.; Bhakoo, K.K.; Tan, T.T.Y. Single-phase Dy_2O_3:Tb^{3+} nanocrystals as dual-modal contrast agent for high field magnetic resonance and optical imaging. *Chem. Mater.* **2011**, *23*, 2439–2446. [CrossRef]
140. Hu, F.; Zhao, Y.S. Inorganic nanoparticle-based T-1 and T-1/T-2 magnetic resonance contrast probes. *Nanoscale* **2012**, *4*, 6235–6243. [CrossRef]
141. Shokrollahi, H. Contrast agents for MRI. *Mat. Sci. Eng. C-Mater.* **2013**, *33*, 4485–4497. [CrossRef]
142. Zeng, L.; Ren, W.; Zheng, J.; Cui, P.; Wu, A. Ultrasmall water-soluble metal-iron oxide nanoparticles as T-1-weighted contrast agents for magnetic resonance imaging. *Phys. Chem. Chem. Phys.* **2012**, *14*, 2631–2636. [CrossRef] [PubMed]
143. Zhou, B.; Zhao, J.; Qiao, Y.; Wei, Q.; He, J.; Li, W.; Zhong, D.; Ma, F.; Li, Y.; Zhou, M. Simultaneous multimodal imaging and photothermal therapy via renal-clearable manganese-doped copper sulfide nanodots. *Appl. Mater. Today* **2018**, *13*, 285–297. [CrossRef]
144. Li, Z.; Wang, S.X.; Sun, Q.; Zhao, H.L.; Lei, H.; Lan, M.B.; Cheng, Z.X.; Wang, X.L.; Dou, S.X.; Lu, G.Q. Ultrasmall manganese ferrite nanoparticles as positive contrast agent for magnetic resonance imaging. *Adv. Healthc. Mater.* **2013**, *2*, 958–964. [CrossRef] [PubMed]
145. Zhang, H.; Li, L.; Liu, X.L.; Jiao, J.; Ng, C.-T.; Yi, J.B.; Luo, Y.E.; Bay, B.-H.; Zhao, L.Y.; Peng, M.L.; et al. Ultrasmall ferrite nanoparticles synthesized via dynamic simultaneous thermal decomposition for high-performance and multifunctional T-1 magnetic resonance imaging contrast agent. *ACS Nano* **2017**, *11*, 3614–3631. [CrossRef] [PubMed]
146. Fu, Y.; Li, X.; Chen, H.; Wang, Z.; Yang, W.; Zhang, H. CXC chemokine receptor 4 antagonist functionalized renal clearable manganese-doped iron oxide nanoparticles for active-tumor-targeting magnetic resonance imaging-guided bio-photothermal therapy. *ACS Appl. Bio Mater.* **2019**, *2*, 3613–3621. [CrossRef]
147. Xiao, S.; Yu, X.; Zhang, L.; Zhang, Y.; Fan, W.; Sun, T.; Zhou, C.; Liu, Y.; Liu, Y.; Gong, M.; et al. Synthesis of PEG-Coated, Ultrasmall, manganese-doped iron oxide nanoparticles with high relaxivity for T-1/T-2 dual-contrast magnetic resonance imaging. *Int. J. Nanomed.* **2019**, *14*, 8499–8507. [CrossRef] [PubMed]
148. Miao, Y.; Xie, Q.; Zhang, H.; Cai, J.; Liu, X.; Jiao, J.; Hu, S.; Ghosal, A.; Yang, Y.; Fan, H. Composition-tunable ultrasmall manganese ferrite nanoparticles: Insights into their in vivo T-1 contrast efficacy. *Theranostics* **2019**, *9*, 1764–1776. [CrossRef]
149. Tan, L.; Wan, J.; Guo, W.; Ou, C.; Liu, T.; Fu, C.; Zhang, Q.; Ren, X.; Liang, X.-J.; Ren, J.; et al. Renal-clearable quaternary chalcogenide nanocrystal for photoacoustic/magnetic resonance imaging guided tumor photothermal therapy. *Biomaterials* **2018**, *159*, 108–118. [CrossRef]
150. Tegafaw, T.; Xu, W.; Ahmad, M.W.; Baeck, J.S.; Chang, Y.; Bae, J.E.; Chae, K.S.; Kim, T.J.; Lee, G.H. Dual-mode T-1 and T-2 magnetic resonance imaging contrast agent based on ultrasmall mixed gadolinium-dysprosium oxide nanoparticles: Synthesis, characterization, and in vivo application. *Nanotechnology* **2015**, *26*, 365102. [CrossRef]
151. Jin, X.; Fang, F.; Liu, J.; Jiang, C.; Han, X.; Song, Z.; Chen, J.; Sun, G.; Lei, H.; Lu, L. An ultrasmall and metabolizable PEGylated $NaGdF_4$:Dy nanoprobe for high-performance T-1/T-2-weighted MR and CT multimodal imaging. *Nanoscale* **2015**, *7*, 15680–15688. [CrossRef]

152. Bony, B.A.; Baeck, J.S.; Chang, Y.; Bae, J.E.; Chae, K.S.; Lee, G.H. Water-soluble D-glucuronic acid coated ultrasmall mixed Ln/Mn (Ln = Gd and Dy) oxide nanoparticles and their application to magnetic resonance imaging. *Biomater. Sci.* **2014**, *2*, 1287–1295. [CrossRef] [PubMed]
153. Wang, X.; Hu, H.; Zhang, H.; Li, C.; An, B.; Dai, J. Single ultrasmall Mn^{2+}-doped $NaNdF_4$ nanocrystals as multimodal nanoprobes for magnetic resonance and second near-infrared fluorescence imaging. *Nano Res.* **2018**, *11*, 1069–1081. [CrossRef]
154. Yang, M.; Liu, Y.; Wang, M.; Yang, C.; Sun, S.; Zhang, Q.; Guo, J.; Wang, X.; Sun, G.; Peng, Y. Biomineralized Gd/Dy composite nanoparticles for enhanced tumor photoablation with precise T-1/T-2-MR/CT/thermal imaging guidance. *Chem. Eng. J.* **2020**, *391*, 123562. [CrossRef]
155. Zhou, Z.; Wang, L.; Chi, X.; Bao, J.; Yang, L.; Zhao, W.; Chen, Z.; Wang, X.; Chen, X.; Gao, J. Engineered iron-oxide-based nanoparticles as enhanced T-1 contrast agents for efficient tumor imaging. *ACS Nano* **2013**, *7*, 3287–3296. [CrossRef]

Publisher's Note: MDPI stays neutral with regard to jurisdictional claims in published maps and institutional affiliations.

© 2020 by the authors. Licensee MDPI, Basel, Switzerland. This article is an open access article distributed under the terms and conditions of the Creative Commons Attribution (CC BY) license (http://creativecommons.org/licenses/by/4.0/).

MDPI
St. Alban-Anlage 66
4052 Basel
Switzerland
Tel. +41 61 683 77 34
Fax +41 61 302 89 18
www.mdpi.com

Molecules Editorial Office
E-mail: molecules@mdpi.com
www.mdpi.com/journal/molecules

www.ingramcontent.com/pod-product-compliance
Lightning Source LLC
LaVergne TN
LVHW070245100526
838202LV00015B/2179